Construmat 2018

IOP Conference Series: Materials Science and Engineering
Volume 385

Herlany, Slovakia
13-15 June 2018

Editors:

Adriana Estokova
Jozef Junak
Alena Sicakova

ISBN: 978-1-5108-6754-3
ISSN: 1757-8981

Printed from e-media with permission by:

Curran Associates, Inc.
57 Morehouse Lane
Red Hook, NY 12571

Printed by Curran Associates, Inc. (2018)

For permission requests, please contact the Institute of Physics
at the address below.

Institute of Physics
Dirac House, Temple Back
Bristol BS1 6BE UK

Phone: 44 1 17 929 7481
Fax: 44 1 17 920 0979

techtracking@iop.org

Additional copies of this publication are available from:

Curran Associates, Inc.
57 Morehouse Lane
Red Hook, NY 12571 USA
Phone: 845-758-0400
Fax: 845-758-2633
Email: curran@proceedings.com
Web: www.proceedings.com

TABLE OF CONTENTS

PREFACE

The 24th conference Construmat 2018, was held in Herlany, Slovakia on 13th – 15th of June 2018. The traditional meeting of academic departments working in the field of building material engineering, technology, and testing, was organised, this year, by the Department of Material Engineering, Institute of Environmental Engineering at the Faculty of Civil Engineering of the Technical University of Košice, Slovakia.

The Construmat 2018 offered an international platform for the dissemination of original research results, new ideas and discovered advances in the field of material engineering and related interdisciplinary topics. In addition, the valuable professional programme was extended by a social program including the interesting technical excursions.

The Conference assembled more than 60 delegates, from both the Slovak and the Czech republics, including PhD students and representatives from praxis. The scientific contributions involved a wide range of research fields: development and testing the advanced building materials, durability of building materials, trends in technology of modern building materials, progressive methods of diagnostics of the constructions, use of alternative materials in building materials production as well as evaluation of the negative aspects of building industry and building materials to the environment. This issue provides an integrated view of the trends in the material engineering and solving the problems associated with the achievement of sustainability.

70 contributions have been selected to be published in the volume of Journal IOP Conference Series: Materials Science and Engineering. The organizing committee appreciates the enormous work of both the editorial board and reviewers in the preparation of this volume. Many thanks also to all the authors for their flexibility and meeting deadlines.

Adriana Eštoková
Alena Sičáková
Jozef Junák

SPONSORS

PREFA invest, a.s. Sučany

Považská cementáreň, a.s. Ladce

EDITORS

prof. RNDr. Adriana EŠTOKOVÁ, PhD.	Technical University of Košice
assoc. prof. Ing. Alena SIČÁKOVÁ, Ph.D.	Technical University of Košice
Ing. Jozef JUNÁK, Ph.D.	Technical University of Košice

Technical editors

Ing. Michaela SMOLÁKOVÁ	Technical University of Košice
Ing. Martina WOLFOVÁ	Technical University of Košice
Ing. Marek KOVÁČ	Technical University of Košice

SCIENTIFIC COMMITTEE

Chairperson

prof. RNDr. Adriana EŠTOKOVÁ, Ph.D.	Technical University of Košice, Faculty of Civil Engineering – Department of Material Engineering

Co-chairpersons

assoc. prof. Ing. Alena SIČÁKOVÁ, Ph.D.	Technical University of Košice, Faculty of Civil Engineering – Department of Material Engineering
prof. RNDr. Nadežda ŠTEVULOVÁ, Ph.D.	Technical University of Košice, Faculty of Civil Engineering – Department of Material Engineering

Members

assoc. prof. Ing. Vlastimil BÍLEK, Ph.D.	Technical University of Ostrava, Faculty of civil engineering – Department of Building Materials and Diagnostics of Structures
prof. Ing. Ján BUJŇÁK, CSc.	University of Žilina, Faculty of civil engineering – Department of Structures and Bridges
prof. Ing. Robert ČERNÝ, DrSc.	Czech Technical University in Prague, Faculty of civil engineering – Department of Materials Engineering and Chemistry
prof. Ing. Rostislav DROCHYTKA, CSc.	Brno University of Technology, Faculty of civil engineering – Institute of Technology of Building Materials and Components
assoc. prof. Ing. Jiří KOLÍSKO, Ph.D.	Czech Technical University in Prague, Klokner Institute
prof. Ing. Petr KONVALINKA, CSc.	Czech Technical University in Prague, Faculty of Civil Engineering, Experimental centre
assoc. prof. RNDr. Jitka MALÁ, Ph.D.	Brno University of Technology, Faculty of Civil Engineering, Institute of Chemistry
prof. RNDr. Pavla ROVNANÍKOVÁ, Ph.D.	Brno University of Technology, Faculty of Civil Engineering, Institute of Chemistry
assoc. prof. Ing. Pavel SCHMID, Ph.D.	Brno University of Technology, Faculty of Civil Engineering, Institute of Building testing

IOP Conf. Series: Materials Science and Engineering **385** (2018) 011001 doi:10.1088/1757-899X/385/1/011001

LIST OF REVIEWERS

prof. Ing. Jiří Adámek, CSc.
prof. RNDr. Magdaléna Bálintová, Ph.D.
prof. Ing. Pavol Ďurica CSc.
prof. Ing. Pavel Fiala, Ph.D.
prof. RNDr. Adriana Eštoková, Ph.D.
prof. Ing. Ružica Nikolić, Ph.D.
prof. Ing. Martin-Tchingnabé Palou, Ph.D.
prof. RNDr. Pavla Rovnaníková, CSc.
prof. RNDr. Nadežda Števulová, Ph.D.
assoc. prof. Ing. Vlastimil Bílek, Ph.D.
assoc. prof. Ing. Lenka Bodnarova, Ph.D.
assoc. prof. Ing. Jiří Dohnálek CSc.
assoc. prof. Ing. Daniel Dobiáš, Ph.D.
assoc. prof. Ing. Petr Drexler, Ph.D.
assoc. prof. Ing. Karel Dvořák, Ph.D.
assoc. prof. Ing. Martin Keppert, Ph. D.
assoc. prof. Ing. Tomáš Klečka, CSc.
assoc. prof. Ing. Karel Kolář, CSc.
assoc. prof. Ing. Jiří Kolísko, Ph.D.
assoc. prof. Ing. Dr. techn. Azra Korjenic
assoc. prof. Ing. Boris Lakatoš, Ph.D.
assoc. prof. Ing. Peter Paulík, Ph.D.
assoc. prof. Ing. Alena Pauliková, Ph.D.
assoc. prof. Ing. Alena Sičáková, Ph.D.
assoc. prof. Ing. Július Strigáč, Ph.D.
assoc. prof. Ing. Martin Šimkovič, Ph.D.
assoc. prof. Ing. Igor Štubňa, CSc.
assoc. prof. Ing. Girum Solomon Urgessa
assoc. prof. Ing. Vojtěch Václavík, Ph.D.
assoc. prof. Ing. Jan Vanerek, Ph.D.
assoc. prof. Ing. Eva Vejmelkova, Ph.D.
assoc. prof. Ing. Silvia Vilčeková, Ph.D.
assoc. prof. Ing. Jan Vodička, CSc.
assoc. prof. Ing. Jaroslav Výborný, Ph.D.
Dr. Manuel Torres Carrasco
Dr. inż. Teresa Zych
Ing. Milada Balkova, Ph.D.
Ing. Johannes Bester, Ph.D.
Ing. Andrej Bisták
Ing. Zdenek Brachař
Ing. Miroslav Brodňan, Ph.D.
Ing. Iva Broukalová, CSc.
Ing. Peter Demeter, Ph.D.
Ing. Pavel Dohnálek, Ph.D.
Ing. Tomáš Dvorský, Ph.D.
Ing. Silvia Ďubek, Ph.D.
Ing. Róbert Figmig
Ing. Ján Fleischhacker
Ing. Valéria Gregorová, Ph.D.
Ing. Hana Hanzlová, Ph.D.
Ing. Věra Heřmánková, Ph.D.
Ing. Marián Hudák, Ph.D.
Ing. Petr Huňka, Ph.D.
Ing. Kateřina Jiroušková
Ing. Jozef Junák, Ph.D.
Ing. Natália Junáková, Ph.D.
Ing. Lucie Keršnerová, Ph.D.
Ing. Filip Khestl, Ph.D.
Ing. Michal Kloiber, Ph.D.
Ing. Václav Kočí, Ph.D.
Ing. Michaela Kostelecká, Ph.D.
Ing. Marek Kováč
Ing. Karel Lang, CSc.
Ing. Miriam Ledererová, Ph.D.
Ing. Martin Luňáček, Ph.D.
Ing. Jozef Mitterpach, Ph.D.
Ing. Vlasta Ondrejka Harbuľáková, Ph.D.
Ing. Marcela Ondová, Ph.D.
Ing. Eva Panulinova, Ph.D.
Ing. Vít Petránek, Ph.D.
Ing. Katarína Prokopčákova, Ph.D.
Ing. Stanislav Řeháček
Ing. Pavel Reiterman, Ph.D.
Ing. Konstantinos Sotiriadis, Ph.D.
Ing. Ing. Alena Struhárová, Ph.D
Ing. Zuzana Struková, Ph.D.
Ing. Jiří Šafrata, Ph.D.
Ing. Pavel Šiler, Ph.D.
Ing. Matej Špak, Ph.D.
Ing. Lukáš Tvrdík, Ph.D.
Ing. Daniel Urbán, Ph.D.
Ing. Lenka Vavrušková, Ph.D.
Ing. Tomáš Vlach
Ing. František Vlašic, Ph.D.
Ing. Michala Užáková, Ph.D.
Mgr. Dalibor Všianský, Ph.D.
Mgr. Ján Ondruška, Ph.D.
Mgr. Martin Boháč, Ph.D.
Mgr. Petr Holub
RNDr. Eva Terpáková, PhD.
RNDr. Jakub Štainbruch, Ph.D.
RNDr. Jana Fridrichová Ph.D.

IOP Conf. Series: Materials Science and Engineering **385** (2018) 011002 doi:10.1088/1757-899X/385/1/011002

Delegate list

Construmat 2018, June 13 – 15, 2018, Herlany

1. Ing. Alena Struhárová, PhD.
2. assoc. prof. Ing. Vladimír Pavlík, PhD.
3. Ing. Miriam Ledererová, PhD.
4. Ing. Zuzana Štefunková
5. Ing. Valéria Gregorová, PhD.
6. Ing. Patrik Mihala
7. assoc. prof. Ing. Peter Koteš, PhD.
8. Ing. Miroslav Brodňan PhD.
9. Ing. Petra Bujňáková, PhD.
10. Ing. Patrik Kotula, PhD.
11. Ing. Ondřej Anton, Ph.D.
12. assoc. prof.. Ing. Petr Cikrle, Ph.D.
13. Ing. Věra Heřmánková, Ph.D.
14. assoc. prof. Ing. Michal Stehlík, Ph.D.
15. prof. RNDr. Pavla Rovnaníková, Ph.D.
16. Mgr. Martin Vyšvařil, Ph.D.
17. Ing. Tomáš Žižlavský
18. Ing. Olesia Mikhailova
19. assoc. prof. Ing. Vlastimil Bílek, Ph.D.
20. Ing. Markéta Bambuchová
21. Ing. Filip Khestl, Ph.D.
22. Ing. David Pytlík
23. Ing. Libor Žídek
24. Ing. Josef Koňařík
25. assoc. prof. Ing. Jaroslav Výborný, Ph.D.
26. assoc. prof. Ing. Eva Vejmelková, Ph.D.
27. Ing. Lukáš Fiala, Ph.D.
28. Ing. Jan Kočí, Ph.D.
29. Ing. Michaela Petříková
30. Ing. Lenka Scheinherrova
31. doc. Ing. Tomáš Klečka, CSc.
32. Ing. Stanislav Řeháček
33. Ing. David Čítek
34. prof. Ing. Petr Konvalinka, CSc.
35. assoc. prof.. Ing. Karel Kolář, CSc.
36. Ing. Pavel Reiterman, Ph.D.
37. assoc. prof. Ing. JiříLitoš, Ph.D.
38. assoc. prof. Ing.Karel Dvořák, PhD.
39. Ing. Vít Černý, PhD.
40. Ing. Dominik Gazdič, PhD.
41. Ing. Žlebek
42. Ing. Jančaříková
43. Ing. Martin Nguyen
44. Ing. Ravaszová
45. Ing. Sklenářová
46. prof. Ing. Jaromír Havlica, DrSc.

47. Ing. Tomáš Opravil, PhD.
48. assoc. prof. Ing. Vojtěch Václavík, Ph.D.
49. Ing. Tomáš Dvorský, Ph.D.
50. assoc. prof. Ing. Július Strigáč, PhD.
51. prof. RNDr. Adriana Eštoková, PhD.
52. prof. RNDr. Nadežda Števulová, PhD.
53. assoc. prof. Ing. Alena Sičáková, PhD.
54. RNDr. Eva Terpáková, PhD.
55. Ing. Jozef Junák, PhD.
56. Ing. Michaela Smoláková
57. Ing. Martina Wolfová
58. Ing. Marek Kováč
59. Ing. Róbert Figmig
60. assoc. prof. Ing. Peter Mesároš, Phd.

IOP Conf. Series: Materials Science and Engineering **385** (2018) 011003 doi:10.1088/1757-899X/385/1/011003

Peer review statement

All papers published in this volume of *IOP Conference Series: Materials Science and Engineering* have been peer reviewed through processes administered by the proceedings Editors. Reviews were conducted by expert referees to the professional and scientific standards expected of a proceedings journal published by IOP Publishing.

Detection of reinforcement in the non-traditional building structures - historical statues

O Anton, L Vitek, V Hermankova, T Komarkova and P Cikrle

Department of Building Testing, Faculty of Civil Engineering, BUT in Brno, Veveri 95, Brno, Czech Republic

Abstract. During restoration and maintenance of historical buildings, statues and sculptures are often neglected elements in the pre-construction condition surveys, especially in case they are not part of the supporting structure but only an ornamental and aesthetic complement of the architectural whole. The condition of a number of such artworks requires restoration or static assessment from the viewpoint of safety and durability. In the case of statues made of stone, artificial stone or concrete, an inseparable part of the pre-construction condition survey is determining the method of their reinforcement with steel elements, which is, considering the often complex spatial aspects of their surface, relatively complicated. The paper offers some possibilities of its solution on three examples. The diagnostic methods were proposed based on the authors' workplace's long-term experience with the diagnostics of reinforced concrete structures. During the past 10 years, about 1,000 such diagnostics have been performed. These experiences were subsequently experimentally applied to several historical sculptures with great success.

1. Introduction

During restoration and maintenance of especially historical buildings, we relatively often face a situation where the structure includes statues or sculptures whose function is mostly not structural, but only aesthetic and architecturally decorative. The condition of a number of such artworks requires restoration or static assessment from the viewpoint of safety and durability. In the case of statues made of stone, artificial stone or concrete, an inseparable part of the pre-construction condition survey is determining the method of their reinforcement with steel elements, which is, considering the often complex spatial aspects of their surface, relatively complicated [1,2].

If we use an analogy from the pre-construction condition surveys of reinforced concrete structures, it is possible to use four different methods [1,2]:

- chopped probes,
- electromagnetic indicators,
- georadar,
- radiography.

In the case of artworks, the chopped probes method is unacceptable in most cases. The use of georadar or electromagnetic indicators is possible but only in the most elementary simple cases where the reinforcement is simple and the structure surface consists of large areas curved as little as possible (the probes of the radar and electromagnetic indicators are designed for use on flat surfaces of reinforced concrete structures). As regards electromagnetic indicators, critical moments of their use are their limited maximum depth and other aspects given by the principle of the method. An effective use of the electromagnetic indicator technology can be illustrated on an example of the reinforcement detection in the sculptures of the sculptural group called "Health" on the building of Zahradníkova 2-8 / Nerudova

11 in Brno. It is the building of a former Regional Sickness Fund from 1925, and the author of the sculptures was Myslbek's pupil, academic sculptor Václav Hynek Mach. In the sculptures made from artificial stone, we detected only sporadic reinforcement with several smooth reinforcing bars with a circular cross-section of 20 mm, which are insufficient in terms of statics, as is apparent in figure 1.

If the reinforcement of sculptures was more complicated, the method used would probably have failed to yield relevant results.

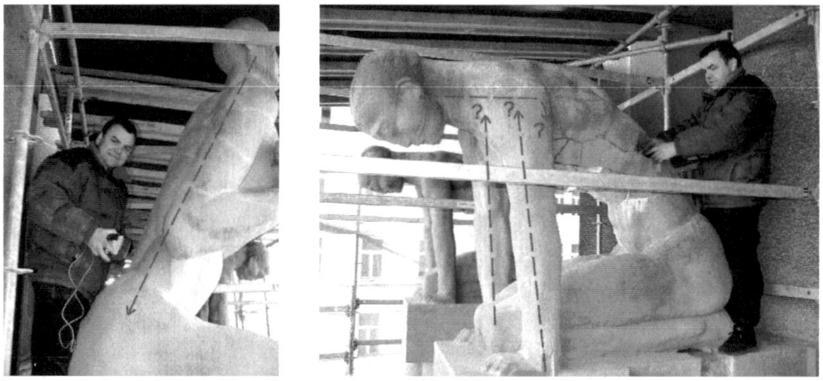

Figure 1. Measurement of the position of reinforcement in individual sculptures from artificial stone in the sculptural group called Health. The detection of reinforcement was relatively precise, considering the low number of reinforcing bars. Nevertheless, it was not possible to precisely determine the location and method of anchoring of individual reinforcing bars [3].

Detecting the reinforcement by means of the radiographic method offers substantially greater possibilities. This progressive method (described inter alia in ČSN 73 1376) is based on the utilization of the passage of X-rays, or gamma rays, through the material, where the radiation is attenuated according to the material's thickness and density. The passed-through radiation then hits the radiographic film (or one of the digital recording media), which, after development, shows blackening corresponding to the dose of radiation to which individual parts of the film were exposed. Any inhomogeneities in the irradiated area (metal reinforcement) are clearly displayed in the film by different blackening. The advantage of radiography is its visualization of the internal structure of the object and a possibility to identify even complicated types of reinforcement or holes, etc. [3,4,7].

As a radiation source, it is possible to use a radioactive emitter (Co60) for massive structures, or a technical X-ray for more subtle elements. As regards the Co60 emitter, it field use, especially in urban areas, is complicated, if not impossible, in the context of current safety regulations, and is therefore limited to closed irradiation facilities. As regards the X-ray, there is no problem to perform radiography in the field under certain conditions. The effective use of radiography with an X-ray is described in the following examples.

2. Survey of a part of the statue of Archangel Michael in Malešovice

The aim of the survey was to establish the existence or shape of reinforcement in the supplied fragment of the Baroque sandstone statue of Archangel Michael from Malešovice. It was the right hand of the statue broken under the shoulder. There were three other fractures on the fragment, in the elbow, in the middle of the forearm and in the wrist. In the past, these fractures had been repaired in an unknown manner, and at some places cement was added. The tests performed were to verify whether the statue fragment is reinforced at all, both at the fracture points, and outside them.

As a radiation source, we used a macrostructural Andrex CP 490 X-ray – 160 kV. The assessment was performed in the irradiation room of the Radiation Defectoscopy Centre at the Department of Building Testing of the Brno University of Technology. During irradiation, we used exposure parameters which guaranteed an ideal imaging of the potential metal parts in the statue fragment, and chose a higher focal distance (1,250 mm). Radiograms were exposed do Agfa Testix C7 films + Pb in darkroom packaging with the dimensions of 0.3 x 0.4 m placed in cassettes. The exposed films were developed in the laboratory using a standard procedure.

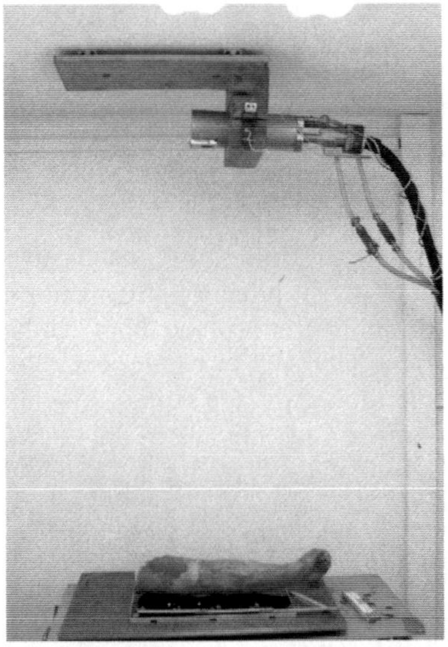

Figure 2. Radiography of the arm from the statue of Archangel Michael from Malešovice, above: x-ray tube of the Andrex CP 490 – 160 kV device, at the very bottom: cassette with an Agfa Testix C7 film.

The radiology revealed that the original reinforcement consists of one, probably forged, rod-like element with a diameter of approximately 13 mm. This element originally linked the arm with the corpus of the statue and ended at the elbow level, in the plane of the first repaired fracture. The image of this reinforcing element is well apparent in figure 3 and also in figure 5. The reinforcement is located in the cavity with a diameter of approximately 20 mm which is sealed with sulphur. An additional reinforcement was detected at the point of the repaired fracture in the elbow. The reinforcement consists of two steel reinforcing bars with a diameter of approximately 6 mm and a length of approx. 53 and 55 mm. According to the surface finish of the reinforcement apparent on the radiogram, it is probably a cut piece of the Kari net for concrete.

It is well apparent on the radiograms that reinforcements were inserted into the holes pre-drilled in both fragments, and the holes exceed the inserted reinforcement by approx. 11 - 17 mm. From the image of the pre-drilled holes it is possible to assume that they are either not filled at all, or that they are filled with a material with a substantially lower density than the volume mass of the arm's sandstone. The image of these reinforcing elements is well apparent in figure 3 and 4.

Another additional reinforcement was detected at the point of the repaired fracture in the middle of the forearm. The reinforcement consists of one steel reinforcing bar with a diameter of 6 mm and a length of 70 mm. According to the surface finish of the reinforcement apparent on the radiogram, it is probably a cut piece of the Kari net for concrete.

It is well apparent on the radiograms that reinforcement was inserted into the hole pre-drilled in both fragments, similarly to the additional reinforcement 1. The image of this reinforcing element is well apparent in figure 3 and 4.

IOP Conf. Series: Materials Science and Engineering **385** (2018) 012001 doi:10.1088/1757-899X/385/1/012001

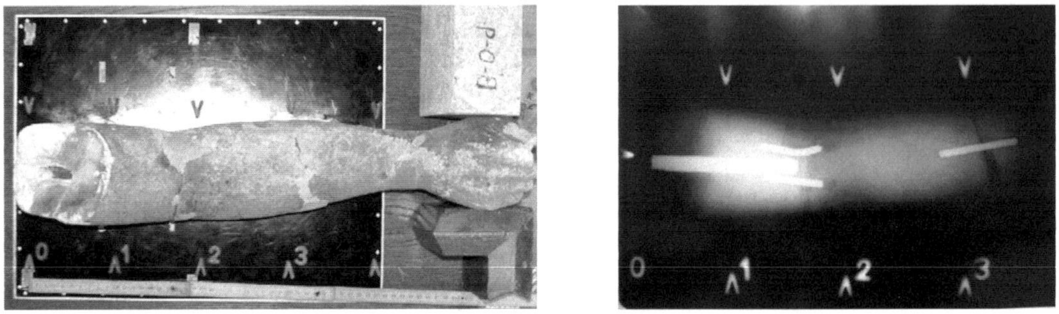

Figure 3. Radiography of the sandstone arm from shoulder to wrist in geometry 1, view from the x-ray tube (from the focus of irradiation), right picture: the resulting radiogram, well apparent metal connecting elements in both parts of the fracture (in the elbow and in the middle of the forearm) as well as the reinforcement ending in the elbow. The picture clearly shows also the pre-drilled holes for inserting the connecting elements in the fracture points. Also apparent is the hole in which the original reinforcement is placed, filled with sulphur.

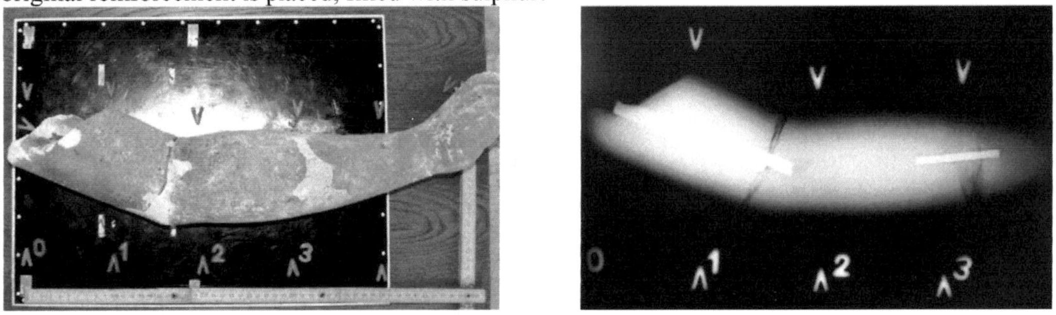

Figure 4. Radiography of the sandstone arm from shoulder to wrist in the control geometry 2, view from x-ray tube (from the focus of irradiation).

In the last fracture in the wrist of the arm, there is no reinforcement. Apparently, the joint is connected only by bonding.

As can be seen, radiography allowed for obtaining quite relevant data for an assessment of the damage to the statue and preparation of its restoration.

3. Survey of Pieta sculpture

The object of the survey was the sculpture of a musician from the scattering meadow of the cemetery in Přerov. As assumed, the sculpture was made of artificial stone or concrete composite, the aggregate used was not known. The sculpture was apparently reinforced with steel rods with square and circular cross-sections, which was obvious in parts that were damaged by exposure of the sculpture to weather. In these places, there were cracks or parts of the composite had fallen off, and the reinforcement was exposed. Surprisingly, even in these places, no serious corrosion of steel reinforcement was apparent. The aim of the research was to detect reinforcement in the sculpture (position and shape of reinforcing bars) and simultaneously to localize potential internal cavities in the object.

The sculpture was made from reinforced cement composite. The composite used was apparently a mixture of a cement binder and fine aggregate with an estimated fraction of 0 – 4 mm. The aggregate used was single-sized with a very high volume mass and high content of magnetic particles (a permanent magnet attached to the surface of the composite displayed high attraction).

After consulting a geologist, doc. RNDr. Pavel Pospíšil, Ph.D. from the Department of Geotechnics and Underground Engineering at VŠB TU – Ostrava, we identified the aggregate used, with high probability, as basalt originating from Bílčice area (active quarry). This is also supported by the high volume mass of the cement composite which, in spite of the porosity of the binder, has a wet volume mass of approx. 2700 kg/m^3.

The position of reinforcements was determined by rentgenography (radiography with X-ray as the source of radiation). As the source of x-ray radiation, we used a 300 kV technical X-ray called YXLON, the passed through and attenuated radiation was captured with memory films and the Dürr scanning system.

The assessed object was irradiated in various directions across it key parts. During the irradiation it became apparent that considering the extreme type of aggregate used with a high shielding capability, the widths of certain parts of the sculpture (back, pedestal) were on the very edge of, or behind, the possibilities of the X-ray (high width and simultaneously small differences in the surface mass at points with and without the reinforcement). That is why it was possible to irradiate, step by step, only small sections of the sculpture, always with a markedly different exposure parameters suitable for the imaging of reinforcement in the given section. At the same time it is necessary to say that the sculpture's shape was so complicated that the irradiation of certain parts was not possible – it was not possible to determine the axis Radiation source – Object – Memory film without interfering with another part of the object. Examples of irradiation configurations, examples of radiograms and the anticipated position of reinforcement are shown in the following figures.

Figure 5. Radiography of an object with such a complicated shape whose individual parts have various widths, requires a number of exposures with different exposure parameters and irradiation angles.

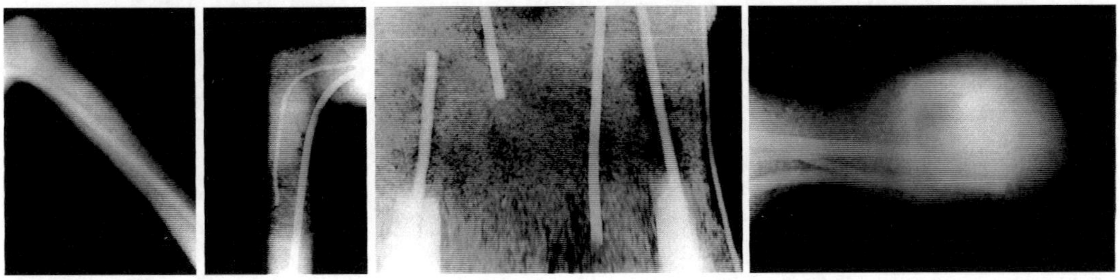

Figure 6. Examples of individual radiograms capturing reinforcement of individual sections of the sculpture.

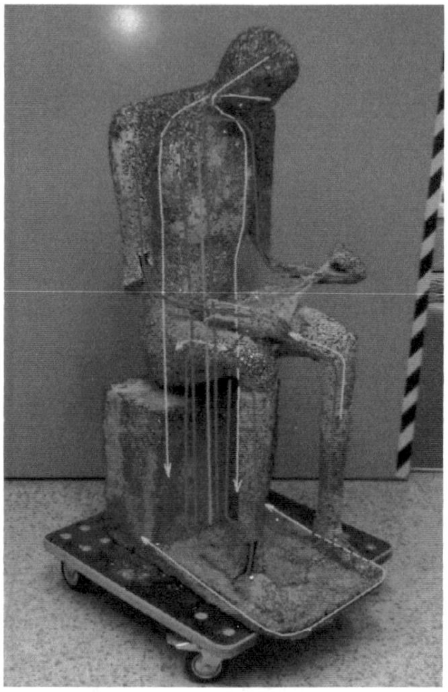

Figure 7. The anticipated pathways of the detected reinforcement in the sculpture. Red – rectangular cross-section of 10 x 10 mm, green – circular cross-section of Ø 6 mm, blue – reinforcement with an unknown cross-section with a width (diameter) of min. 15 mm. Reinforcement of the mandolin and fingers with reinforcing bars of Ø 6 mm is not captured in the picture.

4. Conclusion

On laboratory examples, the paper presented a method which is ideal in a number of cases for determining the position of steel reinforcement in historical statues and sculptures and which can be used, without much difficulty, also directly on the objects in the field, if they are sufficiently accessible from both sides. The surveys clearly showed that it is possible to apply the methods commonly used in the diagnostics of traditional reinforced concrete structures for the surveys of historical sculptures, the radiographic method with an X-ray in particular provides ideal results. The main problem halting the spread of this method for these kinds of applications is that the equipment needed (suitable sources of radiation) is only available to a very small number institutes, e.g. in the Czech Republic practically only in the workplace of the authors of the article.

Acknowledgement

This paper has been worked out under the project No. LO1408 "AdMaS UP - Advanced Materials, Structures and Technologies", supported by Ministry of Education, Youth and Sports under the „National Sustainability Programme I".

References

[1] Stefanidou M, Pachta V a Papayianni I 2015 *Constr. Build. Mater.* **93** 957-965
[2] Chastre C, Ludovico-Marques M 2018 *Handbook of Materials Failure Analysis* (New York: Elsevier) pp 255-294

[3] Anton O, Komarkova T and Hermankova V 2017 *Methods Appropriate For Determination Of The Prescribed Reinforcement Of The Elements Of Reinforced Concrete Structures - Nowadays Used Methods* (Praha: *CTU* Publishing House – production) (in Czech)

[4] Anton O, Komarkova T and Hermankova V 2017 *Conclusive Determination Of Compliance With The Prescribed Reinforcement Elements Of Concrete Structures - The Appropriate Methods And Their Capabilities* (Praha: CTU Publishing House – production) (in Czech)

[5] Anton O, Cikrle P and Hermankova V 2012 *Proc. Conf. on Testing and Quality in Construction* (Brno: Brno University of Technology) (in Czech)

[6] Hobst L and Anton O 2017 The Present State of Radiographic Evaluation of Reinforced Concrete Structures, *Proc. Conf. 10th seminar of Reconstruction of Concrete Structures*, (Bratislava: IRIS - Publishing and Printing s.r.o) (in Czech)

[7] Slaviková M, Krejci F, Zemlicka J, Pech M, Kotlik P, Jakubek J 2012 *Journal of Cultural Heritage* **13** 357-364

Development of new methods for determining the distribution of steel fibres in the hardened steel fibre reinforced concrete – possibilities of production and verification of test specimens

O Anton, T Komarkova, J Lanik and V Hermankova

Department of Building Testing, Faculty of Civil Engineering, VUT in Brno, Veveri 95, Brno, Czech Republic

Abstract Many research teams currently deal with the issue of determining the distribution of steel fibres in the hardened steel fibre reinforced concrete, both in terms of fibre concentration, and in terms of homogeneity of fibre distribution in the concrete. The basic element in the development of new non-destructive methods is the development of new principles and devices on adequate test specimens, in which precise spacial distribution of fibres, including the orientation of individual fibres, would be quite evident. The paper presents two technologically completely different possibilitites of dealing with this issue. X-ray radiography is a traditional and long-established method, but in the field of construction, the authors' workplace is one of the few in the Czech Republic able to use it. The method of creating transparent specimens is completely unique, as it has been developed and is used only at the authors' workplace.

1. Introduction

Steel fibre reinforced concrete is nowadays a preferred material, starting from heavy-duty industrial floors, and ending with e.g. supporting structures of traffic facilities in the tectonically active areas.

Despite the current progression of this material, the methods by which it is possible to determine the concentration of fibres in the finished structure and at the same time the homogeneity of its dispersion, lag behind. The existing normative regulations only deal with taking core samples from the structure and their subsequent crushing and determining the content of fibres in the specimen. No wonder that a number of research teams throughout the world are occupied with this issue and develop non-destructive methods based on various principles, which could be used in this area.

The department of the authors of this paper participates in the development and testing of experimental devices based on the principle of assessing the physical parameters describing the electromagnetic field. This field occurs in an area with electrically and electromagnetically conductive fibres in the tested composite after attaching an electromagnetic coil or permanent magnet. Although the basic principle of these newly designed devices is the same, the measuring methodologies themselves differ substantially and also their applicability is different.

An integral part of the development of new devices and methods is their testing. Testing requires test specimens with precisely defined properties. In this case it is necessary to use steel fibre reinforced concrete specimens of large dimensions - slabs. At the same time, however, it is necessary to have a precise idea of the concentration and distributional pattern (orientation) of fibres in any part of the test specimen. There are principally two ways of achieving this goal. Either to manufacture the specimen precisely according to the requirements, or to determine the required structural parameters in the finished test specimen. This paper does not deal with the manufacture of specimens with atypical structure which differ from the common structure of steel fibre reinforced concretes made in the building practice, e.g. atypical orientation of fibres etc. [1-4].

Figure 1. Compact measuring and recording NDT device for the evaluation of the impedance module / phase of the monitored component (steel fibers) of the composite material specimen.

2. Variant 1 - Manufacture of specimens with precisely determined parameters with the possibility of their control during manufacture

At the beginning of this section it is necessary to note that it is almost impossible to manufacture the required specimens with clearly defined parameters from concrete, if we have to be certain that the specimen does not deviate from the assumed and required limits in any of its parameters. Inhomogeneity of the dispersion of fibres in the laboratory specimens with a small volume is probable, it is caused both by a small volume of the prepared mixture, and e.g. by the edge effect in the proximity of walls of the mould etc. Even in the case of specially prepared specimens with atypical distribution of fibres, the use of concrete is inappropriate.

For the above mentioned cases, the authors of this paper developed a non-standard material, the structure of which simulates concrete, or steel fibre reinforced concrete, rather precisely, but which allows for corrections of the position of fibres, or detection of accidental inhomogeneities and problems which could degrade the test results of devices and methods.

The entry requirements on this material were as follows:

- Transparency to such an extent that it is possible to monitor the distribution of steel fibers with the considered thickness of the test slabs (within mm).
- Granularity structure similar to concrete which would guarantee similar parameters of dispersion and orientation of fibres as in real steel fibre reinforced concrete.
- Long workability time – high value of initial setting (in the order of hours) for the possibility of making corrections to the specimen structure.

Crucial for ensuring the required transparency is using a suitable binder and filler. At the same time, the binder should have a suitable consistency and the fractions of the filler should simulate, as precisely as possible, fine and coarse aggregates in the concrete mixture.

Clear crushed glass with suitable fractions, which is supplied to the glass art industry, turned out to be an ideal substitution for the aggregate. It became apparent that crushed glass with the required fractions and appropriate shape index is routinely available on the Czech market. The crushed glass used was a waste product from the production of the Glass Sphere s.r.o. company. From the supplied mixed crushed glass, individual fractions were separated at first, and subsequently, mixtures were prepared corresponding to the fraction proportions of the aggregate from the original concrete. Before that, all parts of larger fractions which did not have a suitable shape index (flat plates), were removed from the crushed glass [5].

Epoxy resin was chosen as a suitable binder. In the selection of epoxy resins, several factors were taken into consideration. The lowest possible viscosity guaranteeing an ideal penetration of the mixture, an appropriate, relatively long period of workability, minimum volume changes. From the currently available resins, we chose the all-pupose resin called Biresin L84.

Another issue was the possibility of an appropriate moulding of specimens. We used a silicon sheet which allowed for a good separation of the epoxy mixture from the plywood mould. Alternatively, it is possible to use a self-adhesive metal sheet and a separator but silicone turned out to be more suitable. Nevertheless, both alternatives guarantee surface smoothness and protection of the mould itself from temperature - during resin curing, an exothermic reaction occurs and the specimens are heated to approximately 90°C [6].

Figure 2. Separated fractions of crushed glass.

After the tests, we first mixed a dry mixture of crushed glass, and subsequently added fibres in an appropriate amount, which was the optimum procedure. The mixing was carried out manually with a visual check of fibre homogeneity. The dry mixture was subsequently poured into the mould. In the end, the mixture was poured over with Biresin L84. Thanks to the appropriately selected low viscosity, penetration into the whole specimen was guaranteed, and at the same time, thanks to the long workability time, the majority of air bubbles disappeared spontaneously. The composition of the mixture is given in table 1, examples of the specific amount of fibres for some of the test specimens are given in table 2.

Note: The bulk density of the mixed fractions of glass in the shaken state was 1600 kg/m^3, the porosity was 38.5%.

Table 1. Amount of glass and resin necessary for the proposed concrete mixture.

Material	Volume mass [kg/m^3]	Mass per 1m^3 of the mixture [kg]
Glass fraction 0 - 8	2600	960
Glass fraction 8 - 16	2600	640
Resin L84	700	269

Table 2. Amount of steel fibres necessary for individual mixtures.

Concentration (mass per 1m^3 of the mixture)	Amounts of steel fibres for V = 0.0135 m^3 (specimen with the dimensions of 300/300/150) FIBREX	TRI-TREG	DRAMIX
0.50% (39.25 kg/m^3)	0.53 kg = 9 104 pcs	0.53 kg = 1 628 pcs	0.53 kg = 2 307 pcs
0.75% (58.88 kg/m^3)	0.79 kg = 13 658 pcs	0.79 kg = 2 443 pcs	0.79 kg = 3 462 pcs
1.00% (78.50 kg/m^3)	1.06 kg = 18 209 pcs	1.06 kg = 3 257 pcs	1.06 kg = 4 615 pcs
1.25% (98.13 kg/m^3)	1.32 kg = 22 762 pcs	1.32 kg = 4 071 pcs	1.32 kg = 5 769 pcs
1.50% (117.75 kg/m^3)	1.59 kg = 27 312 pcs	1.59 kg = 4 885 pcs	1.59 kg = 6 923 pcs

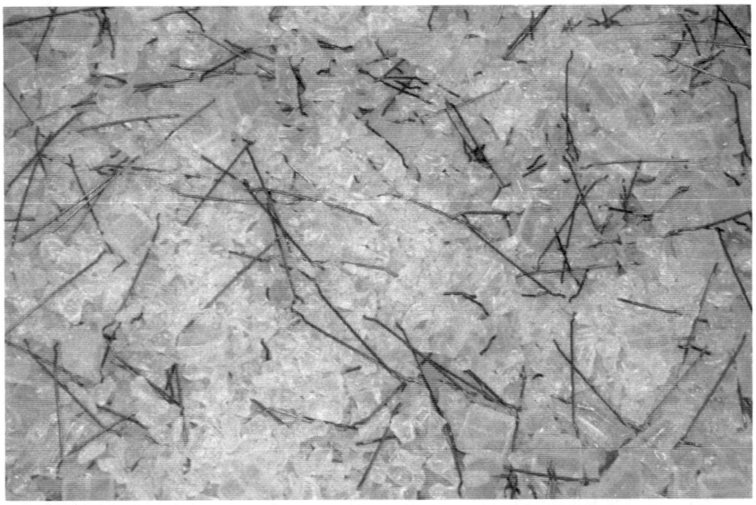

Figure 3. Mixture of a filler and fibres to be poured over with a binder.

Figure. 4: After pouring the epoxy resin over the filler, the specimen becomes transparent and it is still possible to make corrections and simultaneously to visually check the resulting structure of the specimen. The specimen will keep its transparency even after its hardening.

3. Variant 2 - Manufacture of standard specimens with a subsequent check of their structure without the possibility of influencing the manufacture

At the beginning of this section it is necessary to note that it is almost impossible to manufacture the required specimens with clearly defined parameters from concrete, if we have to be certain that the specimen does not deviate from the assumed and required limits in any of its parameters. Inhomogeneity of fibre distribution is always a certain risk, which is substantial in the case of manufacturing specimens with the dimensional limitations. Yet, due to potential inhomogeneities, the subsequent testing of the newly developed probes could be completely misleading.

It is therefore necessary to check the actual distribution of fibres in the manufactured specimen. In the case of specimens with a width of up to 150 (max. 200) mm, the ideal method of checking is radiography with X-ray, i.e. a method based on the passage and attenuation of x-ray radiation. In the case of limit thicknesses of concrete, it is necessary to have an X-ray device, ideally with an X-ray tube voltage of 300 kV and an electric recording of the image for accelerating the activity (compared to the classical radiographic films [7,8].

At the department of the authors of this paper, this measurement is done with the YXLON SMART 300 HP X-ray device, the passed and attenuated radiation is recorded with the memory film Dürr in combination with the CR35 NDT Plus scanner.

As an example, it is possible to present a radiographic inspection of the test steel fibre reinforced concrete slabs with the dimensions of 1000 x 1000 x 100 mm.

The focal distance chosen for the radiography of slabs was 1200 mm, the X-ray exposure parameters were determined according to the experimentally specified exposure nomogram and optimized in the following values:

Current on X-ray tube:	3 mA
Voltage on X-ray tube:	200 kV
Exposure time:	4.5 min

Due to the high dynamics of the blackening of memory films, small width of radiation beams and the influence of the dispersed radiation on the edges of the slabs, the individual pictures overlapped each other so that all the slab areas were evaluable.

When evaluating the images, it is appropriate to combine the classic and inverse display of the radiograms.

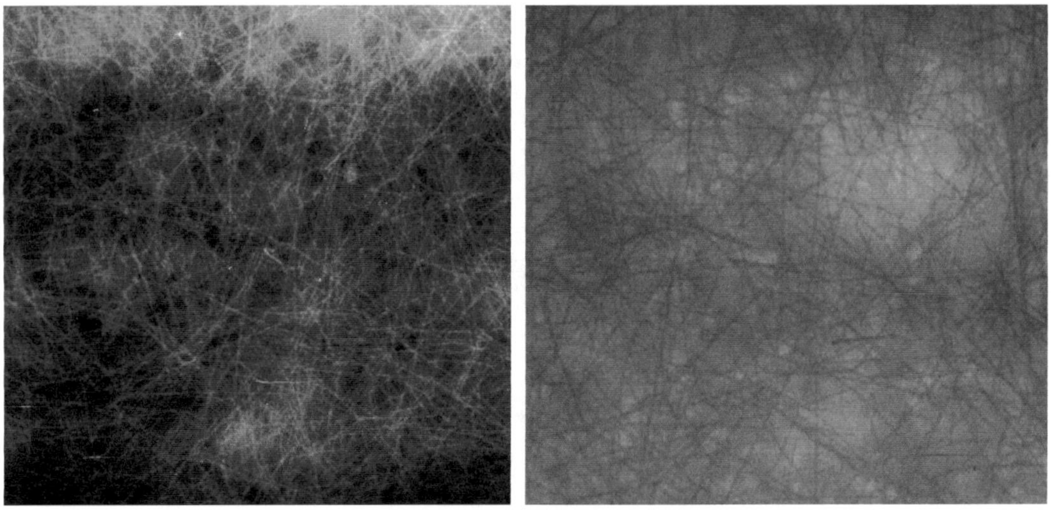

Figure 5. Example of a classic display of the radiogram where the blackening is proportionate to the radiation dose. Objects with a higher density are lighter, and on the right of the inverse variant of the radiogram, objects with a higher density are darker.

4. Conclusion

The described methods appear to be ideal for the preparation of test specimens for creating new diagnostic methods for the determination of fiber reinforced concrete in terms of fiber concentration and dispersion. This is based on several years of experience with their application in authors' workplace. Especially the method of using transparent specimens is absolutely unique and, to the best of authors' knowledge, not used anywhere else in the world. On the other hand, the x-ray radiography is a well-known method. However, it is also scarcely used for these purposes as the number of institutions engaging in construction diagnostics which are in possession of the equipment needed for this method is small.

Acknowledgement

This paper has been worked out under the project No. LO1408 "AdMaS UP - Advanced Materials, Structures and Technologies", supported by Ministry of Education, Youth and Sports under the „National Sustainability Programme I".

References

[1] Vassou V, Kettle R J 2005 *Insight - Non-Destructive Testing and Condition Monitoring* **47** 400-407
[2] Komarkova T, Fiala P, Steinbauer M, Roubal Z 2018 *Meas. Sci. Rev.* **18** 35-40
[3] Zezulova E, Komarkova T 2017 *Key Eng. Mat.* **755** 153-158
[4] Komarkova T 2016 *Key Eng. Mat.* **714** 179-185
[5] Anton O, Hermankova V, Hobst L 2012 *Conf. paper for the Construmat conference 2012(Zilina)* (Zilina: Stavebna fakulta, Zilinska univerzita)
[6] Hobst L, Anton O, Hermankova V, Vodicka J 2011 *Proc. Int. Conf. on Fibre concrete (Prague)* (Prague: CTU)
[7] Hobst L, Anton O 2017 *Proc. Conf. on Reconstruction of concrete structures* (Bratislava) (Bratislava: IRIS - Publishing and Printing)
[8] Hobst L, Anton O, Vodicka J, Scucka J 2013 Homogeneity detection of fibre-concrete structures by using radiographic technique *Nondestructive Testing Of Materials And Structures* ed O Guneş and Y Akkaya (Dordrecht: Springer) pp 323-328

Research of siliceous fly ash after denitrification utilisation in ternary binders

M Bambuchova and D Pytlik

Faculty of Civil Engineering, VSB – Technical University of Ostrava, Ludvika Podeste 1875/17, Ostrava – Poruba (70833), Czech Republic

Abstract. The requirement to reduce the nitrogen oxide emissions in the production of electricity and heat by combustion of mainly brown coal leads to the installation of NO_x reduction technologies. The presented paper focuses on a partial replacement of Portland cement with siliceous fly ash after denitrification by a selective non-catalytic reduction method (SNCR) where urgent urea or ammonia water solution is injected into the combustion chamber. As the third component of the ternary binder the limestone and/or recycled concrete powder was used. The activity index of compressive strengths and the workability of mortars were tested in accordance with EN 196-1.

1. Introduction

Production of Portland cement requires a great deal of material to be extracted from nature. Subsequent crushing, homogenization, drying of cement-based raw material and excavation of the raw material itself is a major environmental burden. During production of one ton of cement, nearly one ton of CO_2 is produced too, as well. For this reason, modern concrete technology of concrete focuses on the possibility of lowering cement content in the concrete by partially replacing it. Cement can be partially replaced by one or more components. This article is focused on ternary binders, where 25 % of the cement weight is partly replaced by two components. These ternary binders are a promising way of concrete technology [1-4]. For experiments in this paper fly ash from the power plant using SNCR method and limestone or recycled concrete powder were used. An optimum composition of ternary binder can modify hydration reaction in a positive way and it can bring a synergic effect. The synergic effect of all three components brings about to obtain better properties such as higher early strengths in comparison with binary binders as well as better durability [1-2,4,6].

Fly ash, in general, has several positive benefits in its use for the production of the concrete mixture. Especially, there is a positive impact on the rheology of cement paste due to its suitable grain morphology and also reduction of the hydration heat development. However, in 2016 legislation regulating the limits of NO_x came into force. For this reason, power plants and heating plants are taking steps towards new technologies that allow us to reduce the amount of nitrogen oxides in the flue gas. One of the technologies is selective non-catalytic reduction method (SNCR). SNCR denitrification technology consists in spraying the reduction solution into the flue gas stream in the 850 – 1100 °C temperatures. The effectiveness of this method also depends on a number of other things, though the chemical, physical and mechanical properties of the collected fly ash are changing. Such fly ash contains more compounds of ammonia than the usual one and its morphological structure is damaged by the heat shock when the reducing agent is injected.

2. Materials and experimental procedure

Some of the used fly ash properties after SNCR method are shown in table 1. To prepare the mortars, CEM I 42.5 R from Cement Hranice company was used. The composition of mortars is shown in table 2. Portion of the cement weight (25%) was replaced by fly ash with addition of limestone or recycled concrete powder. Sand normalized according to Czech standard ČSN EN 196 was used. Mortars were mixed in a laboratory mixer in accordance with EN 196-1. The workability was measured with a truncated cone, low base diameter 100 mm, upper base diameter 60 mm, height 60 mm. The cone flow was measured on a jolting table after 15 jolts in two perpendicular directions. Prepared mortars were put into molds (40 × 40 × 160 mm) and they were compacted. At the age of 24 hours, the specimens were demolded and 2 of them were tested for 24 hours strength. Other specimens were stored in water (with 20 ± 3 °C) until testing at the ages of 7 and 28 days. Every time the ternary mortars were made comparative, cement-plain specimens in accordance to EN 196-1 were also prepared.

Table 1. Some properties of fly ash.

	Date of withdrawal	pH ČSN ISO 10523 ČSN EN 12457-4	Ammonnia [NH_3] ČSN EN 12457-4 ČSN ISO 7150-1 [mg/kg]
Fly ash A	April 6, 2017	11.94	11.6
Fly ash B	January 29, 2018	12.08	19.8
Fly ash C	March 14, 2018	12.14	22.8

Table 2. Composition of mortars [kg/m^3].

Sign	0	6.5	12.5	18.7	25	CEM
CEM I 42.5 R	338	338	338	338	338	450
Fly ash	0	28	56	84	112.5	0
Recycled concrete powder P450 (or limestone)	112.5	84	56	28	0	0
Water	225	225	225	225	225	225
Normalized sand PG 1	450	450	450	450	450	450
Normalized sand PG 2	450	450	450	450	450	450
Normalized sand PG 3	450	450	450	450	450	450

2.1. Grain morphology of fly ash

The impact of SNCR method on fly ash was analyzed using electron scanning microscopy. A specimen of the fly ash which has been sprayed with a reducing agent and so-called pure specimen has been studied. However, when comparing fly ash after SCNR and without this method under a microscope, no significant difference in the fly ash structure was apparent (as it can be seen in figure 1 and figure 2). This is probably due to the low content of ammonia in fly ash. During mixing the fly ash after SCNR with cement and water, no release of ammonia into the air was observed. The power plant need to supply reducing agent in larger quantities for meeting the nitrogen oxide discharge limit is unlikely.

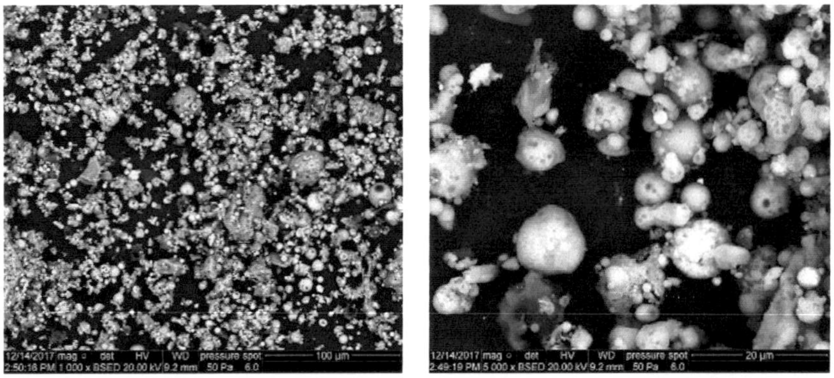

Figure 1. Grains of fly ash without SNCR method (magnified 1000x and 5000x).

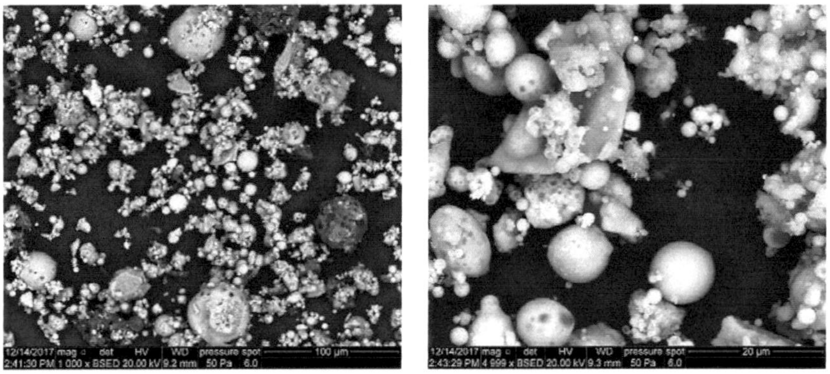

Figure 2. Grains of fly ash after SNCR method (magnified 1000× and 5000×).

3. Results and discussion

The presented results deal with the compressive strengths as well as workability of mortars with cement, fly ash after denitrification by the SNCR method in combination always with one more supplementary material – limestone or recycled concrete powder P450. Recycled concrete powder is made from concrete railway sleeper and contains 7.8 % of $CaCO_3$ and 18.8 mgCaO/g. Content of CaO includes $CaCO_3$ also.

3.1. Fly ash and limestone

At the age of 24 hours the highest compressive strengths are recorded for mixture with 25 % of fly ash after SNCR. The interesting thing about results are that there is not any significant synergy between fly ash and limestone at the age of 24 hours. Even at 7 days and 28 days the best strengths were recorded for ternary binder. Comparative cement mortar has the highest strengths – all activity indexes are lower than 100 % as we can see from figure 3.

The workability in figure 4 of each mixture improves with decreasing the proportion of fly ash in place of limestone. This is probably related to the higher absorptivity of water of the fly ash. Such as the fly ash after SNCR method, even the recycled concrete powder is another reason for worse workability.

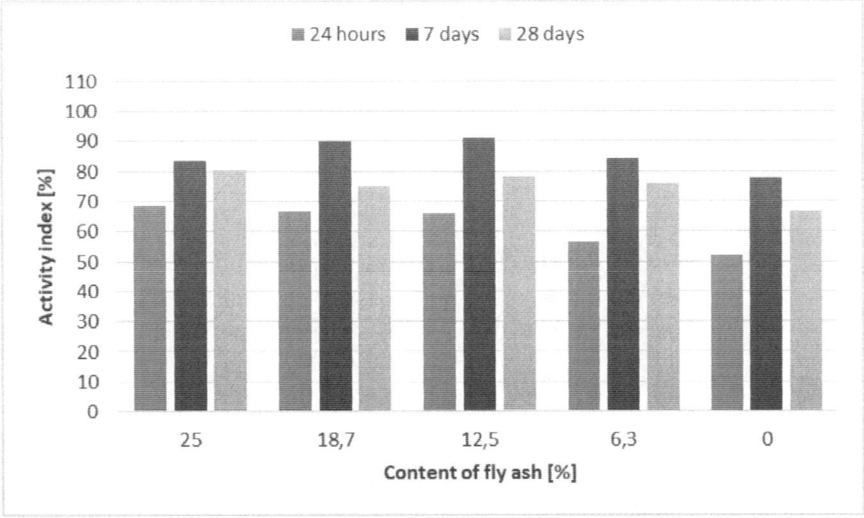

Figure 3. Activity index of compressive strength, 25 % of fly ash, and limestone.

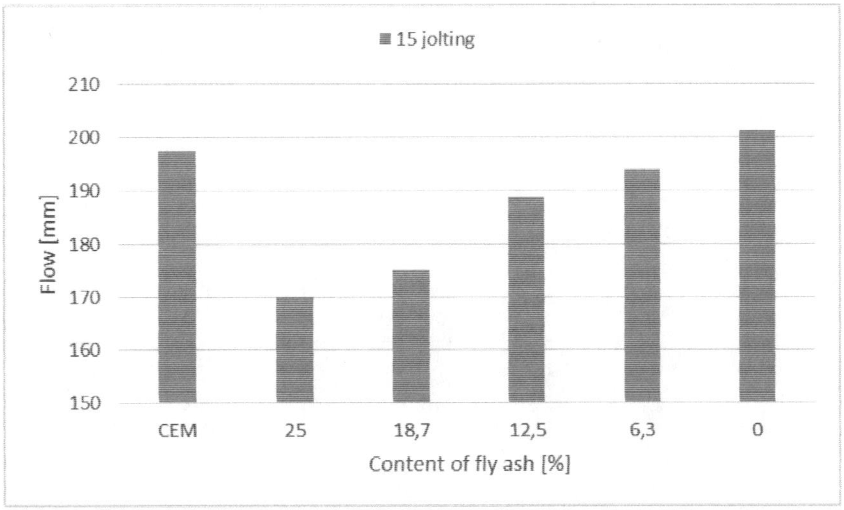

Figure 4. Workability of mortars, 25 % of fly ash, and limestone.

3.2. Fly ash and recycled concrete powder

For the combination of cement, fly ash and concrete powder the strong synergy at the age of 24 hours for the mortar with 6.3 % of fly ash can be seen in figure 5. At the age of 28 days the activity index is the highest for mortar with 25 % of fly ash. Comparative cement mortar has the highest strengths as well as mortars with limestone.

The workability in figure 6 is getting better with increasing the proportion of the recycled concrete powder. This fact, confirm high fly ash absorption of water. All workability considered, better workability for mortars with limestone was found. This is because finely ground limestone has a much lower absorption of water compared to the recycled concrete powder. Next research may focuses to use of plasticizer. This can improve the workability of mortars with fly ash after SNCR and recycled concrete powder.

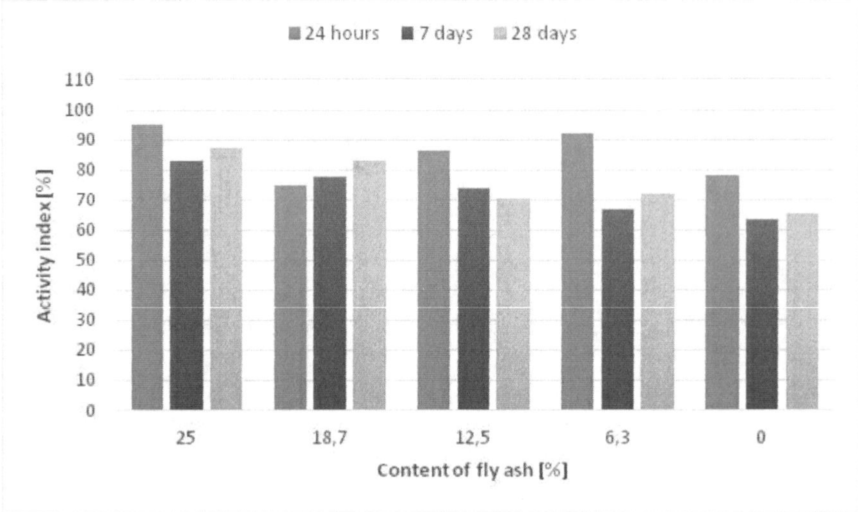

Figure 5. Activity index of compressive strength, 25 % of fly ash and recycled concrete powder P 450.

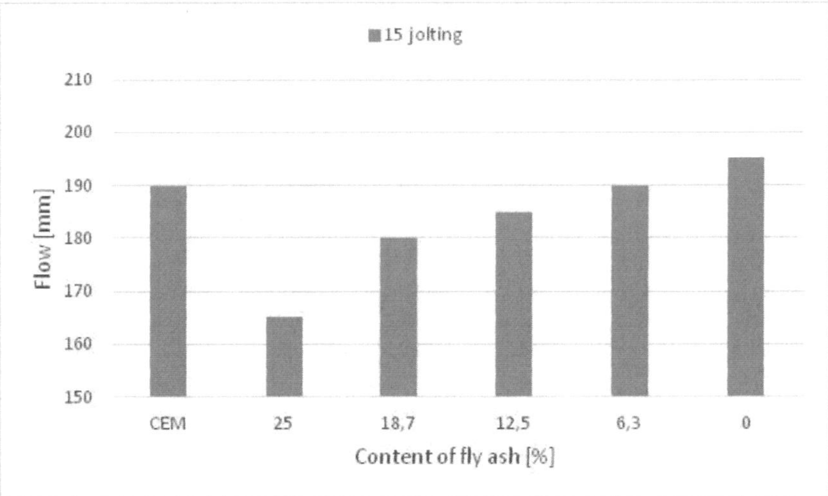

Figure 6. Workability of mortars, 25 % of fly ash, and recycled concrete powder P 450.

4. Conclusions

The presented results show the use of fly ash after SNCR method, limestone and recycled concrete powder in ternary binders. Fly ashes from a power plant are supplementary materials that are used successfully in concrete to improve the rheology of cement paste and to reduce hydration heat while reducing the total amount of cement in concrete. In conjunction, with one more supplement the binder is called ternary. Initial strengths and durability of ternary binders are typically better than of those with binary binders. However, the SNCR denitrification method can affect the fly ash but it is not know how. Fly ash, which was used, changes the properties of mortars. Especially workability as we can see above. This change leads to the worse rheology of cement paste. At the age of 28 days the activity index was similar for both types of mixtures.

Further research should focus on influencing the granulometry and other chemical and physical properties as on binary and ternary binders. Also, there should be research with other types of supplements.

It is above all the fact that fly ash properties are changing due to the changes in legislation and there is not enough information about how such fly ash will react in the cement paste. The same applies for binary and ternary binders.

Acknowledgments
This work was supported by the project TJ01000174 Research of siliceous fly ash after denitrification utilisation as a binder, as a blend to cement and alkali activated system granted by TAČR and by the project 734833 Utilization of secondary raw material in geopolymers production.

References

[1] Antoni M, Rossen J, Martirena F and Scrivener K 2012 *Cement Concrete Res.* **42** 1579–1589
[2] Bohac M, Palou M, Novotny R, Masilko J, Vsiansky D and Stanek T 2014 *Constr. Build. Mater.* **64** 333–341
[3] De Weerdt K, Kjellsen K O, Sellevold E and Justnes H 2011 *Cem. Concr. Comp.* **33** 30–38
[4] Sanchez Berriel S et al. 2016 *J. Clean. Prod.* **124** 361-369
[5] Bilek V 2013 *Cement Wapno Beton* **18** 343-352
[6] Decree No. 415/2012 Coll. on the permissible level of pollution and its determination and implementation of certain other provisions of the Air Protection Act as amended.

Effect of alkaline activator quantity and temperature of curing on the properties of alkali-activated brick dust

P Bayer and P Rovnanikova

Department of Chemistry, Faculty of Civil Engineering, Brno University of Technology, Veveri 331/95, Brno 602 00, Czech Republic

Abstract. The properties of alkaline activated brick dust mixed with different amounts of water glass with a silicate modulus (M) of 1.0 were studied in connection with the temperature of curing. The influence of the amount of alkaline solution and the temperature of curing on the microstructure and mechanical properties of hardened materials was studied. The flexural and compressive strengths of the hardened products, as well as their microstructure, were determined. Part of the samples was stored under laboratory conditions (20 °C) for 7 and 28 days, other set of samples was stored at 60 °C for 7 and 28 days. One set of samples was treated for 7 days at 60 °C and 21 days at 20 °C. The microstructure of the hardened products was studied by means of porosimetry and scanning electron microscopy.

1. Introduction

Much research has been devoted to alkali-activated materials, which have seen a great deal of development since the 1940´s. Alkali-activated materials based on different raw materials have been developed via many works of research [1-4]. Various (especially amorphous) aluminosilicates can be used as a precursor for alkali activation. Researchers are looking for new resources from which such materials can be developed. One of the possible materials capable of alkaline activation is the waste brick dust available from the production of calibrated thermal insulating bricks. In general, waste brick dust can be used as a substitute for cement in concrete, and for the alkaline activation of aluminosilicates as well due to the pozzolanic activity it exhibits [5,6]. The results of the alkali activation of aluminosilicates are dependent on many parameters, but mainly on the relationship between the precursor and the amount of alkaline activator, the composition of both, and the temperature of curing. Robayo et al. [7] described the effect of the SiO_2/Al_2O_3 and Na_2O/SiO_2 ratio on the compressive strength of alkali-activated red clay brick waste. The addition of soluble silica to the alkaline activator had a positive effect. Da Silva et al. [8] studied the effect of the SiO_2/Al_2O_3 ratio on the early-stage reaction kinetics of the geopolymerization of metakaolin at 40 °C using alkaline solution composed from sodium silicate and sodium hydroxide. The setting time increased with the increasing SiO_2/Al_2O_3 ratio of the initial mixture. Up to a certain limit, the SiO_2/Al_2O_3 ratio was also found to be responsible for the observed high-strength gains at later stages. Their investigation demonstrates that the properties of the resulting geopolymer systems can be drastically affected by minor changes in the available Si and Al concentrations during synthesis.

Most of the authors that have produced research in this area carried out the alkali activation of brick powder at an elevated temperature. They typically cured specimens at a temperature above 40 °C for different periods of time. Tuyan et al. [9] subjected alkali-activated mixtures to six different temperatures from 50 to 100 °C for four different durations, from 1 to 7 days. A gradual increase in compressive strength was observed in the mixtures cured at a temperature of up to 70 °C for all

periods. Reig et al. [10] investigated the properties and microstructure of alkali-activated red clay brick waste using a curing temperature of 65 °C for 7 days, along with alkaline solutions with different SiO_2/Na_2O ratios, and various water/binder ratios as well. Compressive strength increases till a SiO_2/Na_2O ratio of 1.46 is reached. Rovnaník [11] analyzed the effect of curing temperatures from 10 to 80 °C and time of curing on the compressive and flexural strengths, pore distribution and microstructure of alkali-activated metakaolin. The results have shown that the treatment of fresh mixture at elevated temperatures accelerates the development of the aforementioned strengths, but the 28-day mechanical properties are worse in comparison with results obtained at ambient temperature.
The present experimental study aims to assess the effect of the amount of alkaline solution, Si/Al, Si/Na and Al/Na ratios, and temperature of curing on mechanical properties, porosity and microstructure.

2. Materials and methodology

2.1. Characterization of raw materials

Waste powder from the manufacture of calibrated brick elements at the HELUZ Brickworks, v.o.s. factory located in Libochovice (Czech Republic) was used as a solid precursor for the purposes of this study. The chemical composition of the brick powder obtained via XRF analysis (PANalytical) is presented in table 1, while the mineralogical composition obtained via XRD analysis (SAXS - PANalytical Empyrean) is specified in table 2. The proportions of the present minerals were calculated via Rietveld analysis.

Table 1. Chemical composition of brick powder.

	SiO_2	Al_2O_3	Fe_2O_3	CaO	MgO	K_2O	Na_2O	P_2O_5	TiO_2	SO_3	Sum
Mass [%]	54.5	16.8	5.9	13.5	2.6	2.5	0.9	0.2	0.7	1.7	99.3

Table 2. Mineralogical composition of brick powder.

Mineral	Formula	[%]
Anorthite	$CaAl_2Si_2O_8$	13.4
Gehlenite	$Ca_2Al[AlSiO_7]$	2.6
Hematite	Fe_2O_3	2.7
Orthoclase	$KAlSi_3O_8$	8.9
Quartz	SiO_2	29.4
Illite 2M1	$K_{1.67}(Al_{3.29}Fe_{0.38}Mg0_{0.32})[(Si_{6.68}Al_{1.32})O_{20}(O)_4]$	19.7
Calcite	$CaCO_3$	1.5
Amorphous phase	-	21.8

The pozzolanic activity of the brick powder was determined using the modified Chapelle test. The result representing the amount of calcium hydroxide consumed in the reaction per 24 hours is 328 mg $Ca(OH)_2$ per 1 gram of brick powder. The particle size distribution of the brick powder with grains less than 1 mm in diameter is d_{10}= 1.753 μm, d_{50}=11.052 μm, and d_{90}=52.167 μm.

2.2. Composition of the tested mixtures

An alkaline solution (M_s=1.0) was prepared by mixing 70 g of sodium water glass (26.43% SiO_2, 16.61% Na_2O, and 56.96% H_2O) (Vodní sklo, a. s., Czech Republic), 9.5 g of NaOH, and 50 g of water. The mixture composition is presented in table 3. The Si/Al, Si/Na, and Al/Na ratios in the mixture are presented in table 4. While the Si/Al ratio slightly increases as the amount of alkaline solution grows, the Si/Na and Al/Na ratios markedly decrease in this case.

Table 3. Composition of mixtures.

Mixture	Brick powder [g]	Alkaline solution M_s=1.0 [g]	Water [g]
FL-151	200	30	87
FL-152	200	50	67
FL-153	200	70	47
FL-154	200	90	27

Table 4. Ratios of Si/Al, Si/Na, and Al/Na in the mixtures.

	FL-151	FL-152	FL-153	FL-154
Si/Al	2.86	2.93	3.01	3.08
Si/Na	9.43	6.56	5.09	4.19
Al/Na	3.29	2.24	1.69	1.36

3. Experimental procedure

The samples were treated as follows: two sets of samples were stored at the temperature 20 °C for 7 and 28 days, 2 sets of samples were stored at the temperature 60 °C for 7 and 28 days and one set of samples was treated for 7 days at 60 °C and 21 days at 20 °C.

The flexural and compressive strengths of 20×20×100 mm test specimens were determined using a hydraulic press (Tonindustrie, Germany). The porosity of the alkali-activated brick powder was assessed using a PoreSizer 9310 mercury porosimeter (Micromeritics). Cumulative pore volume was determined for samples cured at a temperature of 60 °C. Detailed microstructure images were taken via a MIRA3 XMU environmental scanning electron microscope (Tescan).

4. Results and discussion

4.1. Mechanical properties

The bulk densities grow from sample 151 to 154, and increase with the temperature and time of curing of individual samples. The bulk densities of the samples cured at 20 °C were determined to be between 1648 and 1825 kg·m^{-3}, while the heated samples had bulk densities from 1184 to 1613 kg.m^{-3}.

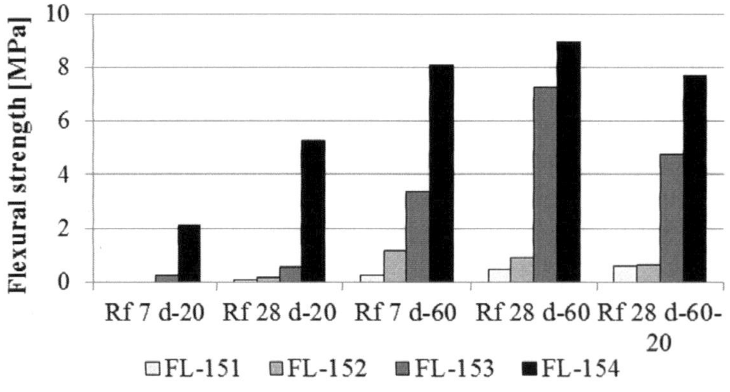

Figure 1. Flexural strengths.

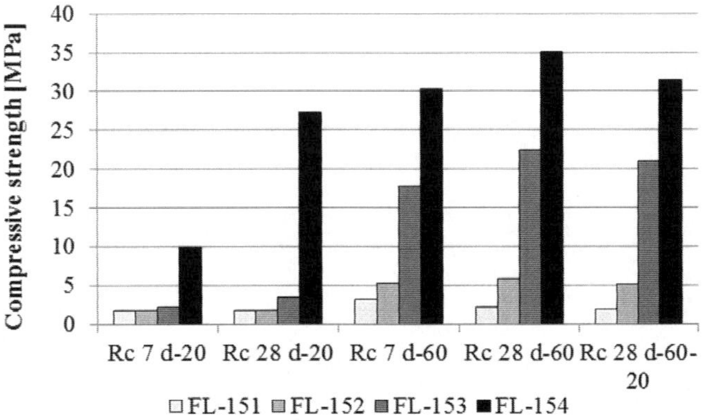

Figure 2. Compressive strengths.

The flexural strengths (Figure 1) increased with curing time at the same temperature, and with higher curing temperature as well. Significantly higher strengths were reached with higher proportions of alkaline solution (samples 153 and 154). Compressive strengths recorded the same development (Figure 2). Higher amounts of alkaline solution are related to the diminishing of Si/Na and Al/Na ratios, i.e. to the enhancement of sodium ion content, and the corresponding OH^- ion content.

4.2. Morphology of microstructure and pore structure

The microstructure of the samples differs depending on the composition of the fresh mixture. A higher amount of alkaline activator results in denser microstructure. Sample FL-151 has the least dense microstructure and the pores which are present are of a significant size for all curing temperature modes (Figure 3), and similar images were obtained for sample FL-152 (Figure 4). The microstructure of sample FL-153 (Figure 5) looks like that of the two previously named samples but the pore size is considerably smaller and little formations are visible on the surface. Sample FL-154 (Figure 6) features dense microstructure without pores, and only some cracks are visible; the images are quite different.

Pore structure was measured for all samples cured at a temperature of 60 °C for 28 days. The results for cumulative pore volume are presented in Figure 7. Pore volume decreases as the amount of alkaline solution in the mixture rises.

FL-151 – 28 d – 20 °C FL-151 – 28 d – 60 °C FL-151 – 28 d – 60 and 20 °C

Figure 3. Micrographs of FL-151 samples cured at different temperatures.

IOP Conf. Series: Materials Science and Engineering **385** (2018) 012004 doi:10.1088/1757-899X/385/1/012004

The sample with the lowest concentration of alkaline solution (FL-151) has the largest pores and greatest cumulative volume of them all. The sizes of the pores and their cumulative volume fall as the concentration of alkaline solution in the mixture increases.

FL-152 – 28 d – 20 °C FL-152 – 28 d – 60 °C FL-152 – 28 d – 60 and 20 °C

Figure 4. Micrographs of FL-152 samples cured at different temperatures.

FL-153 – 28 d – 20 °C FL-153 – 28 d – 60 °C FL-153 – 28 d – 60 and 20 °C

Figure 5. Micrographs of FL-153 samples cured at different temperatures.

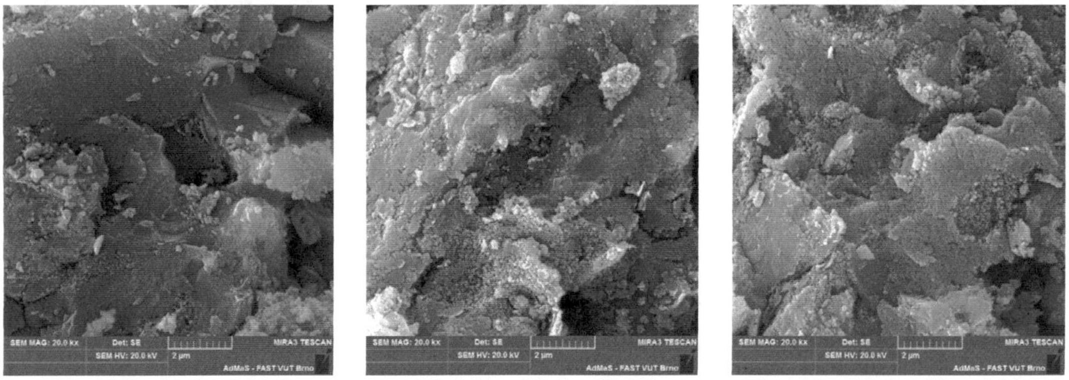

FL-154 – 28 d – 20 °C FL-154 – 28 d – 60 °C FL-154 – 28 d – 60 and 20 °C

Figure 6. Micrographs of FL-154 samples cured at different temperatures.

The sample with the highest concentration of alkaline solution (FL-154) contains the finest pores; their cumulative volume is the smallest of all the assessed samples. The results of porosimetric measurements, i.e. of pore size and cumulative pore volume, are in good agreement with the images of the samples in the micrographs above.

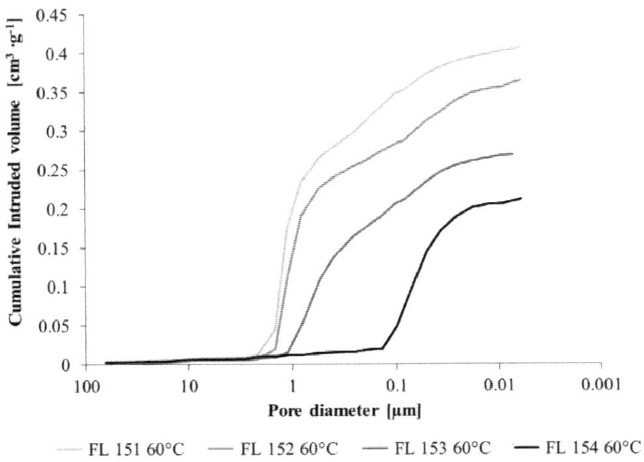

Figure 7. Cumulative intruded volume of pores.

5. Conclusion

This study has demonstrated the effect of the amount of alkaline solution and temperature on the microstructure and mechanical properties of alkali-activated brick powder. The following conclusions may be drawn from this study:

- Flexural and compressive strengths increase and microstructure is more comprehensive with growing Si/Al ratio.
- Decreasing Si/Na and Al/Na ratios mean a greater content of alkaline solution, resulting in a rise in the strengths and compactness of matter.
- The temperature of curing increases the strength of the hardened product. The most significant increase in strength was found in samples with the highest content of alkaline solution.
- Cumulative pore volume and pore diameter decrease with an increasing concentration of alkaline solution in the mixture.

Acknowledgments
This research was supported by the Czech Science Foundation under project No 16-02862S.

References

[1] Purdon A O 1940 *J. Soc. Chem. Ind.* **59** 191–202
[2] Davidovits J 2002 *Géopolymere (*Saint-Quentin, France) pp 1–16
[3] Kriven W M, Bell J L and Gordon M *Ceram. Transact.* 2003 **153** 227–250
[4] Xu H, Van Deventer J S J 2002 *Min. Eng.* **15** 1131–1139
[5] Baronio G, Binda L 1997 *Constr. Build. Mat.* **11** 41–46
[6] Navrátilová E, Rovnaníková P 2016 *Constr. Build. Mat.* **120** 530-539
[7] Robayo R A, Mulford A, Munera J and de Gutiérrez R M 2016 *Constr. Build. Mat.* **128** 163–169
[8] De Silva P, Sagoe-Crenstil K and Sirivivatnanon V 2007 *Cem. Concr. Res.* **37** 512–518
[9] Tuyan M, Andiç-Çakir Ö and Ramyar 2018 *Comp. Part B* **135** 242–252
[10] Reig L, Soriano L, Borrachero M V, Monzó J and Payá J. 2013 *Constr. Build. Mat.* **43** 98–106
Rovnaník P 2010 *Constr. Build. Mat.* **24** 1176–1183

Methods of testing the properties of fibre-cement composites

L Bodnarova[1], M Hosko[2] and P Martinec[2]

[1] Institute of Technology of Building Materials and Components, Faculty of Civil Engineering, Brno University of Technology, Veveri 95, 602 00 Brno, Czech Republic
[2] Cembrit a.s., Plant Sumperk, Pricna 26, 787 55 Sumperk, Czech Republic

Abstract. The paper deals with process of manufacturing the fibre-cement materials and methods of testing their climatic performance. Basic raw materials for production fibre-cement sheets are cement, cellulose fibers, reinforcing fibers - polyvinyl alcohol (PVA) fibers, filler, additives and water. Production follows the modified Hatschek's technology. The paper described the most important tests for climatic performance: Freeze-thaw test and Heat rain test. The fibre-cement profiled sheets were tested. The Freeze-thaw test includes assessment of frost resistance, precipitated salts efflorescence on the face-side surface, peeling of the coating on the edges, peeling of the coating on the face-side surface, longitudinal cracks in the various locations on the face-side, transverse cracks in the various locations on the face-side, shape deformation, swelling water between the layers of coating and material delamination on the edges. The frost resistance RL for fibre-cement profiled sheets after 100 freeze-thaw cycles was 0.95, when min. RL = 0.70. The fibre-cement profiled sheets were tested to 50 heat-rain cycles. After 50 cycles, the samples were checked for cracking (longitudinal, transverse and at the fixing points), delamination and other visible defects (especially the presence of formed drops on the underside of corrugated sheets). The tested modified recipe of fibre-cement profiled sheet met the Heat rain test requirements.

1. Introduction

Asbestos cement products have been produced in Bohemia and Moravia since 1910. Own patent for the production of this covering was filed by Ludwig Hatschek in 1901. Ludwik Hatschek himself called the products "Eternit", meaning "everlasting". The covering met with great success relatively quickly, because it was light and it was possible to use it even in harsh climates, where it was not possible to use burnt roofing. Asbestos was used to produce these sheets until the nineties. Asbestos fibers have been reported as harmful to health, therefore asbestos has been replaced by cellulose fibers, synthetic fibers and other additives. This has resulted in creation of an ecologically clean product. Since 1996 asbestos-cement technology has been being replaced by fibre-cement technology - no asbestos [1].

2. Composition of fibre-cement composites

Basic raw materials for production fibre-cement sheets are cement, cellulose, reinforcing fibers, filler, additives and water. Production follows the modified Hatschek's technology.

2.1. Cement

As the name suggests, the main raw material is cement. Choosing the kind of cement depends on the type of final product, on the curing process, on manufacturing facilities, and in some cases on water temperature. Availability and economical aspect both also play important roles in choosing the type of cement. Properties important when choosing type of cement:

- Development of hydration heat
- Beginning and the end of the cement hardening
- Particle size (Blaine)
- Final strength properties.

2.2. Cellulose fibers

Cellulose fibers are used as procedural fibers. These fibers absorb cement and other additives on the surface. For the production of fibre-cement sheets 2 types of cellulosic fibers are used. Bleached or unbleached pine fibers and sulphate fibers - high performance fibers. When fibre-cement sheets are manufactured, the distribution of cellulose fibers in water is important. This division is usually performed in the pulper during the pulping controlled time. Properties of the pulp fibers are molded in the disc mills. There the specific surface fibers and the technological parameters of the procedural fibers are improved. When modifying the properties of the pulp fibers, the fibers are larger, fibers are more flexible and fibers have larger external surface area of the fiber. External structure of fibers becomes broken and the fiber surface becomes hairy [1,2].

2.3. Polyvinyl alcohol fibers

Polyvinyl alcohol fibers (PVA) are used as reinforcing fibers. When using the PVA fibers in a concrete or cement composites several advantages over competing fibers of other materials can be observed. PVA fibers have high tensile strength, relatively high modulus of elasticity, good chemical compatibility with Portland cement, a good affinity with water and PVA fibers are without any health risks [3]. PVA fiber of a thickness of 0.014 mm with a tensile strength of 1600 MPa, modulus of elasticity 37 GPa and possible elongation of 7% are used for fibre-cement products. Growing concern about the dangers of asbestos inhalation brought the PVA fibers into the limelight as a substitute for asbestos fibers. Studies show that PVA fibers represent less risk than chrysotile fibers because they are too big to breathe in. PVA fibers do not cause any tissue reactions [4].

2.4. Fillers

The most commonly used fillers in the production of fibre-cement sheets are micro silica (specific surface area of 15000 - 35000 m^2/kg), ground limestone (specific surface area 300-400 m^2/kg) or another suitable micronized filler. [1] Besides applications of microparticles are tested nanoparticles with the use of in cement composites with cellulose fibers to influence the properties of the cement composites [5]. The chart below shows the proportional representation of the various components of fibre-cement composites.

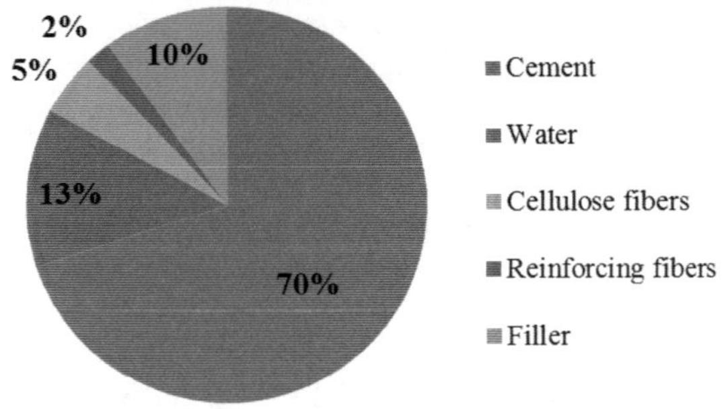

Figure 1. The proportional representation of the various components of fibre-cement composites.

3. Technology of production fibre-cement composites

The modified Hatschek's technology is used for the production of fibre-cement sheets. Hatschek machine for the production of fibre-cement was first developed in 1890 and it was logged on at the patent office by Ludwig Hatschek. The machine has been used in the same basic form till today, though modern Hatschek machines are much more productive, than the models at the start of the production. Hatschek process is particularly suitable for making flat and profiled sheets. This process is the process of filtering based on the drainage of the suspension of diluted through the cylindrical sieve. Thus, the primary layer consisting of fibers, cement, additives and fillers is separated on the sieve, and this layer is transferred to the endless filter carpet [6].

The fibrous component is mixed with cement and other additives in a turbomixer. From there it goes into the homogenizer where it regulates the density of the slurry. The suspension is transported into the production machine vats. In the vats, there are rotating sieves cylinders that transfer the fiber cement on a continuous layer. It is very important to maintain the constant and right level of suspension in the vat. If the level of the slurry moves, it could affect the thickness of each layer, and thus the final product. The Hatschek production machine has normally 3 or 4 manufacturing vats and contribution of each vat is about 0.3 mm. The individual layers are piled on the format roller, until the desired thickness of the product is reached.

The size of the holes on the sieve reaches values around 0.4 mm (400 microns) and a non-fibrous component used in the production are much smaller than the hole of the sieve, and that could cause their passage. Capturing the non-fibrous material therefore depends on the formation of the filtration layer of fibers on the surface of the sieve. Filter layer of fibers is created on the surface of the sieve at a short distance from the immersion of sieve, which is immersed into the suspension. Formation of the film continues on the sieve, but now it contains a lower proportion of the fibers and a higher proportion of non-fibrous materials. Subsequently, the film is dewatered and it is removed from the sieve by plotting on a continuous belt. Main drainage is performed by means of vacuum chambers, which are disposed on the upper side of the strip between the last production vat and a format cylinder. A vacuum chamber with a large amount of water is used to clean the belt.

The fibre-cement layer is transferred on the continuous belt and the belt is transferred on a format roller. After winding the desired thickness the material is cut and moves on the tow table. Winding the desired thickness of material on the format roller varies by product range in the range of 5-7 mm.
Fibre-cement material is cut to desired dimensions by steel disks, or there is cutting shapes, including holes and edges on the cutting press. All the parings, which represent approximately 10-20 % of total production, go to the dissolver and are reused in the production. The cutting material is transferred to the corrugator head, where it is formed to the corrugated roofing.

The individual sheets are located on the steel templates which are placed on carriages. The carriages are transported to pre-heating tunnel where they are hydrated over a period of 8-10 hours at about 60 °C. The steel templates are partly heated to prevent the flow of heat from the sheets during cure and achieve the required handling strength. Then the individual sheets are separated from the steel template and are stacked on wooden pallets. Each template is then brushed and sprayed with a thin layer of mineral oil in the oil machine. Application of oil ensures good separation of the steel plates and fibre-cement sheets after curing. After 10 hours, the sheets are moved on the wooden pallets and the pallets are transported into the ripening warehouse where the sheets under controlled conditions obtain final properties. Time for storing plates in the ripening warehouse is at least 7 days [1].

4. Methods for testing the properties of fibre-cement composites

Fibre-cement is an ideal building material for protecting the external shell of the building and it is used for roofing and facade cladding on new buildings and renovations. Today's fibre-cement covering is already incomparable with the one that was produced 30 to 40 years ago, the environmentally neutral raw material such as cement, water, ground limestone, microsilica, cellulose and synthetic fibers are used to produce fibre-cement roofing materials. Combining these materials makes it possible to form a non-combustible building material which is suitable for any type of building. Fibre-cement material combines the technical properties such as strength, durability, and weather resistance, easy workability, efficiency and environmental safety. It is resistant to water, frost, rot, corrosion, and can be processed by conventional tools on the construction site [1].

The following text expresses the most important tests (Freeze-thaw test and Heat rain test) for climatic performance. During a proposal of new or modified recipe must be verified the impact of changes in the recipe on the final properties of fibre-cement sheets.

4.1. Freeze-thaw test (EN 494+A1 Fibre-cement profiled sheets and fittings - Product specification and test methods)

The specimens were longitudinally cut from the middle of a complete sheet, with two complete corrugations. Twenty specimens were tested. Specimens were transversely cut to a length allowing a span of 12 times the height of corrugation.

The specimens were divided at random into two lots of 10. The first lot of 10 specimens was tested on the breaking load test (or to the bending moment test for short sheets). At the same time, the second lot of specimens was immersed in water at ambient temperature (> 5 °C) for 48 hours. Then the second lot of specimens was subjected to 100 freeze-thaw cycles:

- cooling (freeze) in the freezer which reach a temperature of (-20 ± 4) °C within 1 hour to 2 hour and holding this temperature for a further 1 hour
- heating (thawing) in the water bath which reach a temperature of (20 ± 4) °C within 1 hour to 2 hour and holding at this temperature for a further 1hour.

Then the breaking load test (or to the bending moment test for short sheets) was done. The following figure 2 and figure 3 show the results of freeze-thaw test of the fibre-cement profiled sheets.

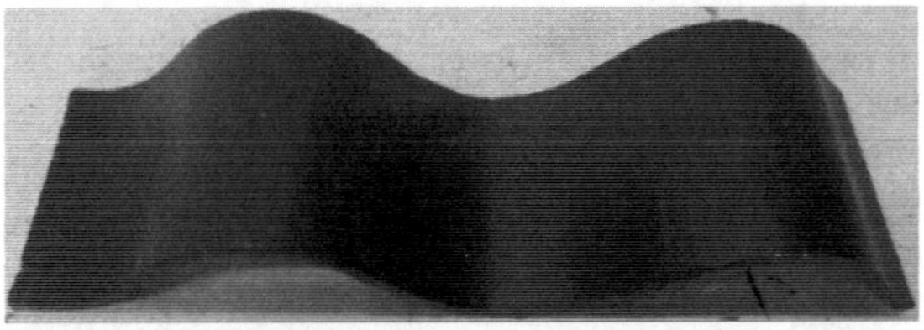

Figure 2. Sample of fibre-cement profiled sheet before freeze-thaw test.

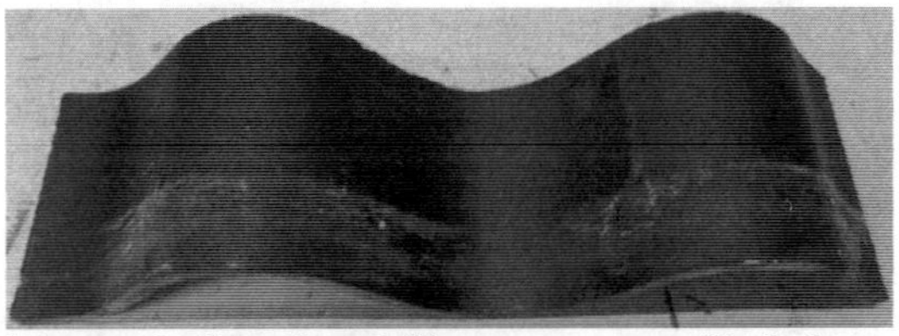

Figure 3. Sample of fibre-cement profiled sheet after 100 cycles of freeze-thaw.

For each of the two lots, the mean breaking load or bending moment and the standard deviation at the value was calculate. From these values is subsequently determined R_L ratio (see equation (1) – (3)). The ration R_L shall be not less than 0.70 after 100 freeze-thaw cycles [7]:

$$L_2 = X_2 - (0{,}58 \times s_2) \quad (1)$$

Which is the lower estimation of the mean breaking load or bending moment after freeze-thaw cycles at 95 % confidence level (second lot):

$$L_1 = X_1 + (0{,}58 \times s_1) \quad (2)$$

Which is the upper estimation of the mean breaking load or bending moment at 95 % confidence level of the reference lot (first lot):

$$R_L = \frac{L_2}{L_1} \quad (3)$$

The R_L ratio for fibre-cement profiled sheets (Cembrit, type A5, A6.5, A6, B7, B8) is min. 0.70 after 100 freeze-thaw cycles [1]. To evaluate the test for frost resistance is not only R_L ratio but also the assessment of following properties: Precipitated salts efflorescence on the face-side surface, peeling of the coating – on the edges, peeling of the coating – on the face-side surface, longitudinal cracks – in the various locations on the face-side, transverse cracks – in the various locations on the face-side, shape deformation, swelling - water between the layers of coating and material, delamination - on the edges [7]. The results of freeze-thaw test are of modified recipe shown in the table 1.

Table 1. The results of the freeze-thaw test of modified recipe.

	Reference	100 freeze-thaw cycles
Δ breaking load (kN/m)	4.77	4.74
L_1	4.82	-
L_2	-	4.62
R_L ratio		0.95

Together with the assessment of frost resistance (R_L) on the samples the peeling of paint on edges was assessed as it shown on the figure 2 and figure 3. In this test, there shall be no cracks on the paint leading to subsequent peeling of the surface treatment. The occurrence of efflorescence on the sample after 100 cycles is not a major problem because the main objective is to verify the durability of the surface treatment.

4.2. Heat rain (according to EN 494+A1 Fibre-cement profiled sheets and fittings - Product specification and test methods)

The specimens shall comprise a minimum of one full sized sheet with the underlapping and overlapping sheets/strips surrounding it a minimum of ½ sized sheets. 12 specimens for sheets of length equal to or less than 0.9 m or 9 specimens for longer sheets were tested. The specimens were stored for 7 days in a laboratory atmosphere. Each sheet were laid with overlaps at the four edges. The specimens were fixed on the frame according to regulations or, if no regulations exist, to manufacturer's instructions. One installation system, which the manufacturer regards as decisive for the sheets, shall be selected. The upper face of the sheets was tested to 50 cycles without interruption in accordance with table 2. Any visible cracks, delamination or other defects in the fibre-cement sheets shall not be of such degree as to affect their performance in use after 50 heat-rain cycles [7].

Table 2. Heat-rain cycle

Cycles	Duration
Wetting 2.5 $l/(m^2.min)$	2 h 50 min
Pause	10 min
Heating (70 ± 5^{0}C)	2 h 50 min
Pause	10 min
Total cycle	6 h

The figure 4 shows the comparison of results of heat-rain test of the fibre-cement profiled sheets after 4 cycles. On the left in blue border is modified recipe and on the right in red border is standard recipe. On the sample with modified recipe we can see the longitudinal and transverse microcracks, resulting in faster moisture penetration. From the figure 4 there is a clear difference in the rate of development of moisture traces. Therefore, it is important to assess not only the final results but also continuous results of individual samples. The presence of wet traces on samples is not a reason for non-compliance, but in no case shouldn't be formed drops on the underside of corrugated sheets. After 50 cycles, the samples are checked for cracking (longitudinal, transverse and at the fixing points), delamination and other visible defects [1].

During the 1st cycle: 75 minutes of sprinkling seq. – First occurrence of moisture traces on the left sample (thickness 6.5 mm) most probably due to the loose screw (still no problem of the fibre-cement material)

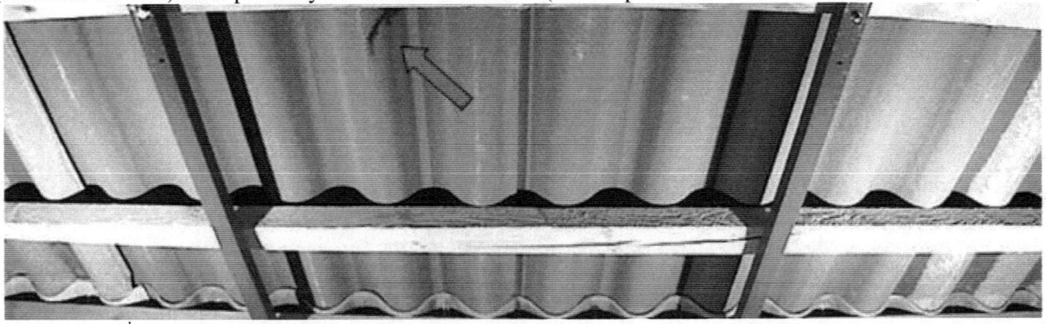

During the 2nd cycle: First occurrence of water soaked through material on the left sample (longitudinal cracks), big spot is commented in the previous picture

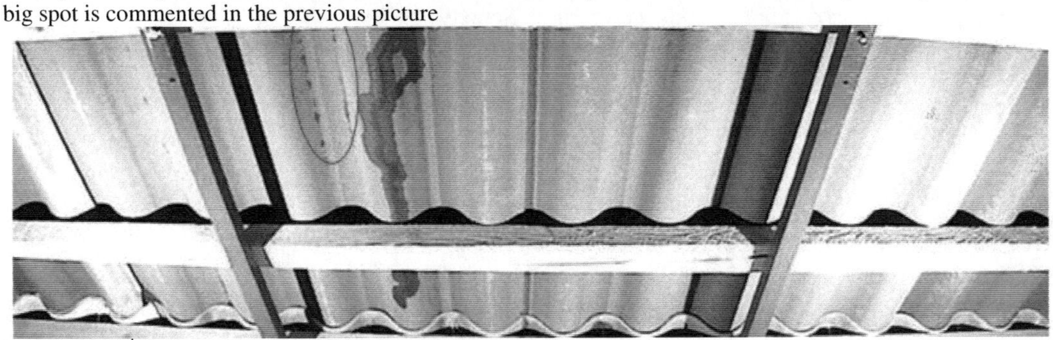

During the 4th cycle: First occurrence of transverse cracks on the left sample (with combination of longitudinal cracks also)

Figure 4. The comparison of the results of standard (left blue border) and modified (right red border) recipes after 4 cycles.

5. Conclusion

Fibre-cement products must meet the high durability, aesthetic and utility requirements. To fulfill all these requirements, it is necessary to use in the production of fibre-cement sheets suitable materials, which must meet the strict quality requirements. Fibre-cement products are made exclusively from natural materials environmentally friendly. Using purely natural materials, the products are fully recyclable (e.g. use in comminuted state as an admixture in concrete, or use as permanent formwork), and in line with the trend of sustainable development. From the viewpoint of achieving the right durability of these products is a key parameter the right technological process of production. The perfect manufacturing technology is important for strength and durability of fibre-cement sheets when thin layers of fibre-cement are wrapping themselves and before final maturation are exposed under intense pressure in the press. Great-value is the testing the properties of the fiber-cement sheet. We can assume the durability of these materials from the results of water absorption tests, freeze-thaw test and heat-rain tests. The combination of tests allows us to deduce further features such as resistance to rain and resistance to flowing fluids. To achieve the high and stable quality of fiber cement products, the technological process of production must be carefully observed. The quality of fiber cementitious material can be influenced by both the variability of feedstocks and the changes in cure and aging. Consistent control of key stages of production contributes to non-problem production process and quality of produced sheets.

Acknowledgements
This outcome has been achieved with the financial support of Project FAST-S-18-5410, supported by Brno University of Technology, and project GACR 18-25035S Study of effects of flowing fluids on cement composites wear and subsequent modelling of mechanical corrosion, supported by Czech Science Foundation.

References
[1] Internal materials provided by Cembrit 2018 *https://www.cembrit.com/* (online)
[2] Bodnarova L, Kostelanská K, Jankech J 2014 *Advanced Materials Research* **1054** 85 - 89
[3] Noushini A, Samali B, Vessala K 2013 *Constr. Build. Mater.* **49** 374-383
[4] Horikoshi T, Ogawa A, Saito T, Hoshiro H 2006 Properties of polyvinylalcohol fiber as reinforcing materials for cementitious composites *Int. Workshop on High Performance Fiber Reinforced Cementitious Composites in Structural Applications* (Paris: RILEM Publications) pp 145 - 153
[5] Santos S F, Rodrigues J A, Denzin T, Gustavo H, et al. 2014 *J. Mat. Sci.* **49** 7497-7506
[6] Cooke T 2002 *Proc. Int. Conf. on Inorganic-bonded wood and fiber composite materials (Sun Valley)* (Madison: Forest Products Research Society) pp 231-246
[7] EN 494+A1: 2016 *Fibre-cement profiled sheets and fittings - Product specification and test methods*

IOP Conf. Series: Materials Science and Engineering **385** (2018) 012006 doi:10.1088/1757-899X/385/1/012006

Study the effect of carbon nanoparticles in concrete

L Bodnarova and T Jarolim

Institute of Technology of Building Materials and Components, Faculty of Civil Engineering, Brno University of Technology, Veveri 95, 602 00 Brno, Czech Republic

Abstract. Carbon nanoparticles, whether in the form of nanotubes, fibers or graphenoids, can under suitable application improve mechanical as well as durability properties of cement composites. The use of carbon nanoparticles in concrete is limited by their relatively high cost but also by the difficulty of their dispersion in concrete. The experimental research was focused on the issue of dispersion of carbon nanotubes (CNT) and the effect of CNT in concrete. Concrete with different doses of multi-walled CNT (0.0015 % and 0.012 % from the weight of cement) was prepared. The dispersion of CNT, superplasticizing admixture based on polycarboxylate ether, and water was carried out using ultrasonic cavitator. The impact of various doses of CNT on the strength of concrete after 7 and 28 days and frost resistance of concrete after 100 cycles was observed. The best results were achieved with the addition of 0.012 % CNT of cement weight, representing a dose of 43.2 g per 1 m^3 of concrete. The increase in compressive strength was 11.9 %, 17.3 % in flexural strength, 15.2 % in the tensile splitting strength and 26.9 % is the tensile strength of concrete surface layers after 28 days compared to the reference samples without CNT. Changes in properties of concrete with the addition of CNT were observed also in the case of frost resistance of concrete (8 % increase in the coefficient of frost resistance of concrete with the addition of 0.012 % CNT of cement weight).

1. Introduction

The beginning of the new millennium is the era of nanomaterials and nanotechnologies. The term nanomaterials can be used to describe bulk materials, nanocrystalline layers or nanocrystals. Nanocrystals are grains of particle size in the order of 1–100 nm, representing only a few hundred atoms or molecules. Carbon nanomaterials are a significant direction of modern technology. Carbon nanotubes, fullerenes, and most recently also graphene, are nanomaterials that have interesting mechanical, physical and chemical properties. They can find use in various fields, such as medicine, electronics, but also in the construction industry.

Carbon is the basic building block of all organic compounds. It is stable in several molecular structures that we call allotropes. Among the most famous allotropes that occur in nature is graphite. Graphite can also be called amorphous carbon or fullerene. It is also a modification of carbon with a molecular – crystalline structure and is characterized by spherical formations with 60 carbon atoms. The more rare carbon allotropes include crystalline carbon, known as diamond [1].

Carbon nanotubes were discovered in 1991 by prof. Sumio Iijimo. A unique feature of these structures is their electronic, mechanical, optical and chemical properties [2].

There are different types of CNT that can be manufactured in different ways. The most common techniques for the production of carbon nanotubes today are: arc discharge, laser ablation, chemical excretion and flame synthesis. CNT cleaning can also be divided into several main techniques: oxidation, acid treatment, annealing, sonication, filtering and functionalization techniques [2].

Carbon nanotubes (CNT – carbon nanotube) are an allotropic modification of carbon. It is a substance composed solely of carbon atoms. CNT are elongated formations with a diameter of 1 to 100 nanometres and a length of 100 micrometres. Typically, two types are used: single-walled carbon nanotube (SWCNT) and multi-walled carbon nanotube (MWCNT) [3]. SWCNT have a diameter of 1.2 – 1.4 nm, MWCNT typically have a diameter of 1.2 – 20 nm, usually they have 5 – 20 layers spaced about 0.34 nm [4].

The development of cement-based materials is directed towards high compressive strength cement composites [5,6] and cement composites with special properties in extreme conditions [7]. However, such composites also exhibit a very brittle fracture. This results in composites sensitive to micro-cracks at an early stage of volume changes due to high autogenous stresses [8]. These properties of cement-based materials are serious drawbacks that limit design and affect long-term durability of structures. Carbon nanoparticles represent possibilities to reduce the undesirable brittleness of cement matrices. Compared to conventional fine fibers, the CNT exhibits significantly higher strength and stiffness, which should improve overall mechanical behaviour. It is also expected that the high CNT slenderness ratio will effectively stop the spread of micro-cracks. If the CNT is evenly dispersed, the spacing between the individual reinforcement fibers is reduced due to their diameter.

2. Aim of research work

With good application, the carbon nanoparticles improve mechanical and durability properties of cementitious composites. The so far published studies differ in terms of the recommended amount of specific type nanoparticles into concrete and in the question of a suitable method of applying nanoparticles to concrete.

The first stage of experimental research was therefore focused on the issue of dispersion of nanoparticles and the evaluation of nanoparticles dispersion in concrete. The correct dispersion of nanoparticles is often very problematic [9,10]. Commercial nanoparticles, although manufactured and supplied in sizes below 100 nm, are typically showing as large agglomerates of 1 to 100 in size. The agglomeration reduces the specific surface of nanoparticles, this preventing the full utilization of these materials in order to improve the desired properties of concrete and other cement mixtures. In addition, the formation and presence of agglomerates prevents proper dispersion of nanoparticles within the matrix, resulting in the formation of attenuated zones in the matrix of the final product. The ideal dispersion can be described as a state in which nanoparticles are completely separated from each other and there are no clumps or agglomerates found in them. The first step in properly dispersing nanoparticles should be to use sufficient energy to break down aggregates and agglomerates into smaller sizes. After breaking the agglomerates, the next step is to stabilize the fragments and prevent their re-agglomeration.

In the next phase of experimental work, concrete with various doses of carbon nanotubes (in the order of thousands of percent of the weight of cement) were produced. It was necessary to define the exact technological process of preparation of concrete with nanoparticles. Subsequently, the effect of different doses of nanotubes on concrete properties was monitored. Concrete compressive strength, flexural strength, tensile splitting strength, surface layer tensile strength of concrete and frost resistance of concrete were determined over 100 cycles.

3. Experimental part

3.1. Used materials

- Cement. Portland cement CEM I 42.5 R was used.
- Aggregates. Three fractions of aggregate were used: silica sand with a fraction of 0 – 4 mm, aggregates 4 – 8 mm and 8 – 16 mm biotic granodiorite.
- Water. The dispersion of carbon nanotubes was carried out in demineralized water. This water is stripped of all the minerals contained in the water by filtration. During the production of

concrete, drinking water was used as mixing water, which according to EN 1008 Mixing water for concrete meets the requirement for testing [11].

- Plasticizer. To improve the consistency of the concrete and to stabilize the stirred nanotubes during dispersion, the superplasticizing admixture COATEX ETHACRYL HF S.C. 40% was used. This admixture is based on polycarboxylate ether. Additionally, Sika ViscoCrete 4035 superplasticizing admixture was added to the concrete.
- Carbon nanotube. Carbon nanotubes manufactured by Yurui (Shanghai) Chemical Co., Ltd were used. There tubes were made using a 95 % purity chemical vapour deposition (CVD) method. The properties of carbon nanotubes are summarized in table 1.

Table 1. Selected properties of carbon nanotubes TNM7.

Properties	Value
Name	TNM7
Colour	Black
Type	MWCNT
Length	10 – 20 μm
Inside diameter	5 – 12 nm
Outside diameter	30 – 50 nm
Bulk density	0,22 $g{\cdot}cm^{-3}$
Specific weight	2,1 $g{\cdot}cm^{-3}$
Specific surface	> 60 $m^2{\cdot}g^{-1}$
Specific electrical conductivity	100 $s{\cdot}cm^{-1}$

3.2. *Mix design*

Carbon nanotubes as dispersed reinforcement are dosed up to 0.1 %. For this experiment, doses of 0.0015 % and 0.012 % by weight of the cement were selected. Each dose of nanotubes was dispersed in a solution containing the surfactant. Dosage of these substances is an empirical issue. It is reported that the amount of surfactants ranges from 10 % to 700 % by weight of CNT. In this case, a 400 % CNT weight dose was chosen, i.e. the ratio of surfactant: CNT = 4:1.

Determination of concrete properties was performed after 7 and 28 days. For each batch, a reference batch was created, which had the same dose of all raw materials, the only difference was it didn't contain carbon nanotubes. The recipe is shown in table 2.

Table 2. Composition of concrete per 1 m^3.

Materials		Amount per 1 m^3
CEM I 42.5 R		360 kg
	Sand 0-4 mm	840 kg
Aggregate	Coarse aggregate 4-8 mm	205 kg
	Coarse aggregate 8-16 mm	900 kg
Water		153 kg
Plasticizer Sika Viscocrete 4035 (SP)		2.5 kg (0.69 % z m_c)
Plasticizer Ethacryl HF		172.8 g (0.048 % z m_c)
CNT 0.0015		5.4 g (0.0015 % z m_c)
CNT 0.012		43.2 g (0.012 % z m_c)

3.3. *Preparation of suspense*

The first step was the weighing of the superplasticizing additive, carbon nanotubes and water. The dispersion was carried out using the BandelinSonopuls HD 3200 ultrasonic cavitator. The suspense was then mixed in a magnetic homogenizer and then dispensed into concrete.

3.4. Production of concrete
According to the recipe, individual raw materials were weighed. The laboratory mixer plate with forced circulation was first dosed with weighed aggregate, from the coarsest fraction to the finest. The aggregate was briefly mixed, then cement was added and once again the dry ingredients were mixed briefly for better homogenization. Additionally, water with superplasticizing additive was added. After sufficient stirring, the already dispersed CNT was poured into the mixture and subsequently thoroughly mixed again. The concrete slump test was carried out on fresh concrete. Test specimens with measurements of 100 × 100 × 400 mm were produced.

4. Results of experimental work and discussion

4.1. Consistency of fresh concrete
The slump test was performed just after the mixing of each concrete batch. The concrete recipe has been designed to the S3 consistency of concrete in the slump test. The results were 120 to 130 mm Slump [12].

4.2. Strength of concrete after 7 and 28 days
Compressive strength, flexural strength, tensile splitting strength and concrete surface layer tensile strength were determined on samples after 7 and 28 days. Tests were performed according to standards EN 12390–3, EN 12390–5, EN 12390–6, CSN 731318 [13-16]. The results are shown in the following table 3 and in figure 1.

Table 3. Concrete compressive strength, flexural strength, tensile splitting strength and concrete surface layer tensile strength after 7 and 28 days.

Concrete	Compressive strength [MPa]		Flexural strength [MPa]		Tensile splitting strength [MPa]		Concrete surface layer tensile strength [MPa]	
	7 days	28 days	7 days	28 days	7 days	28 days	7 days	28 days
REF	50.0	58.0	5.0	5.2	3.1	3.3	1.9	2.6
CNT0.0015	51.5	62.1	5.3	6.0	3.1	3.4	1.9	3.0
CNT0.012	52.4	64.9	5.5	6.1	3.2	3.8	2.0	3.3

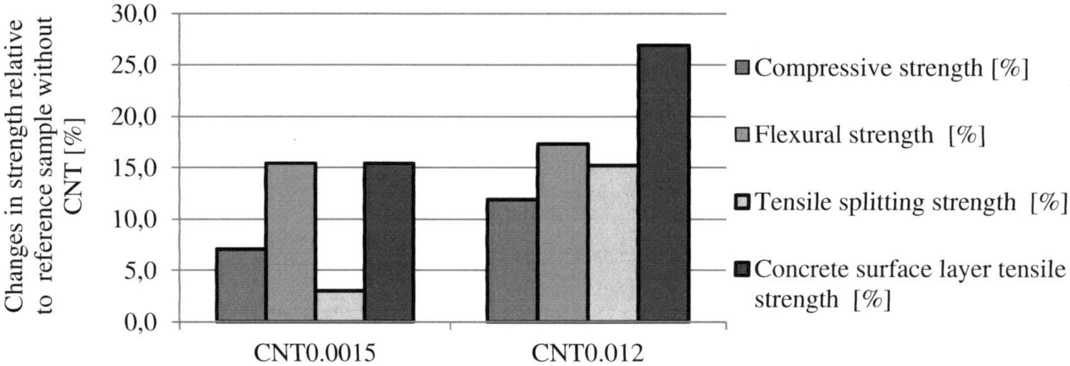

Figure 1. Changes in concrete strength after 28 days relative to the strength of the reference sample without the addition of CNT.

4.3. Frost resistance of concrete
The frost resistance of concrete was performed according to CSN 731322 [17]. Results are given in table 4.

Table 4. Flexural strength before and after 100 freezing cycles.

Concrete	Flexural strength [MPa]		Coefficient of frost
	No freezing	After 100 frost cycles	resistance
REF	7.1	7.0	98.6
CNT0.0015	7.2	7.2	100
CNT0.012	7.3	7.6	104.1

The best results were achieved with concrete with 0.012 % CNT addition. The increase in compressive strength compared to the reference samples was 11.9 % after 28 days.

Another observed characteristic was flexural bending strength. According to Chen and Chung, it has been found that flexural bending strength in samples using short carbon fibers can increase up to 85 %. In our experiments, the flexural strength of CNT0.0015 concrete has increased by 15.4 % and CNT0.012 has increased by 17.3 %.

After determining the tensile splitting strength after 28 days, there was an increase of 3 % for the CNT0.0015 mixture and 15.2 % for the CNT0.012 mixture. It is not possible to trace any CNT dose dependence on the strength characteristics of the monitored concrete from the results obtained in the tensile splitting strength test. It is, however, indisputable that the CNT test samples achieved the best results from the monitored sample set during the tensile splitting strength test, after 7 days by 3.2 % and after 28 days by 15.2 % compared to the reference samples.

Next test was to determine the concrete surface layer tensile strength, which determines the magnitude of the force applied perpendicularly to the observed sample at the time of the failure. The tensile strength test of the surface layers after 7 days showed increase only in the CNT0.012 recipe, 5.3 %. However, strengths after 28 days showed significant changes. Recipe CNT0.0015 showed an increase of 15.4 % compared to the reference samples. Recipe CNT0.012 showed an even larger increase of 26.9 %. Also interesting is the tensile strength of concrete surface layers in samples CNT0.0015, when a virtually 10 times smaller dose of CNT achieved an increase in strength by about 15 % compared to the reference samples.

Frost resistance is the ability of water-saturated concrete to withstand the effects of alternate defrosting and freezing. The test samples were stored for 25 days in a humid environment and then for 3 days in water. Subsequently, they were tested on 100 freezing cycles. For reference concrete, a decrease in strength was recorded after freezing, however, it met the condition that the flexural bending strength after freezing must be higher than 75 % of the strength of the non-frozen concrete. Consequently, the test reference concrete without the addition of CNT can be marked as frost-resistant. With recipes containing carbon nanotubes, it was possible to observe the positive influence of CNT in the area that is critical to the durability of all concrete structures in our climatic conditions. The frozen samples gained on the strengths compared to the non-frozen samples. This can be explained by the fact that the microstructure of cement stone is so dense that the effect of the alternating frost didn't manifest yet. Based on the results of the tests, it can be argued that carbon nanotubes improve the resistance of concrete to freezing.

5. Conclusion

The experiment was focused on the observation of different doses of CNT 0.0015 and 0.012% of cement weight on the observed characteristics of concrete – compressive strength of concrete, flexural strength, tensile splitting strength, surface layer strength of concrete and frost resistance of concrete. Observation of strength characteristics was performed at age 7 and 28 days. Frost resistance was determined after 100 cycles. The carbon nanotubes from Yurui (Shanghai) Chemical Co., Ltd. were used for the experiment. The best results were found in concrete with an addition of 0.012 % CNT of cement weight, representing a dose of 43.2 g per 1 m^3 of concrete. The results obtained with a dose of 0.0015 % CNT of cement weight (i.e. 5.4 g per 1 m^3 of concrete) were also positive. Changes in properties of concrete with the addition of CNT were observed also in the case of frost resistance of concrete (8 % increase in the coefficient of frost resistance of concrete with the addition of 0.012 %

CNT of cement weight). Specifically, the increase of resistance and durability of concrete with the addition of nanoparticles is an interesting area which we should focus more attention to and direct more research that way. Increasing the tensile strength of surface layers and frost resistance shows an increased resistance and durability of the concrete with the CNT addition. Follow-up experiments are therefore aimed at monitoring the resistance of concrete to abrasive action, both by mechanical influences and in particular by flowing fluids, clean water or water with solid particles.

Acknowledgment
This outcome has been achieved with the financial support of Project FAST-S-18-5410 and project GACR 18-25035S Study of effects of flowing fluids on cement composites wear and subsequent modelling of mechanical corrosion.

References

[1] Harris P J F, 2009 *Carbon Nanotube Science: Synthesis, Properties And Applications* (Cambridge: Centre for Advanced Microscopy, University of Reading, Cambridge University Press) p 179
[2] Ham H T, Choi Y S and Chung I J 2005 *J. Colloid Interf. Sci.* **286** 216–223
[3] Raki L, Beaudoin J J, Alizadeh R, Makar J M and Sato T 2010 *Material* **3** 918-942
[4] Aqel A, Kholoud M M, Ammar Reda A A and Al-Warthan A 2012 *Arab. J. Chem.* **5** 1-23
[5] Kolisko J, Hunka P, Rydval M and Kostelecka M 2013 *Proc. Int. Conf. on Central Europe towards Sustainable Building (Prague)* (Prague: GRADA PUBLISHING) pp 385-388
[6] Novakova I and Bodnarova L 2017 *Mater. Sci. Forum* **908** 164-170
[7] Stehlik M, Hermankova V and Vitek L 2015 *J. Civ. Eng. Manag.* **21** 177-184
[8] Holcapek O, Kotatkova K and Reiterman P 2018 *Advances in Civil Engineering* **2018** 1-10
[9] Foldyna J, Foldyna V and Zelenak M 2016 *Proceedia Engineer.* **149** 94-99
[10] Jarolim T, Labaj M, Hela R, Michnova K 2016 *Adv. Mater. Sci. Eng.* **2016** 1-6
[11] EN 1008: 2002 *Mixing water for concrete*
[12] EN 12350 – 2: 2009 *Testing fresh concrete* - Part 2: Slump-test
[13] EN 12390 – 3: 2009 *Testing hardened concrete* - Part 3: Compressive strength of test specimens
[14] EN 12390 – 5: 2009 *Testing hardened concrete* - Part 5: Flexural strength of test specimens
[15] EN 12390 – 6: 2009 *Testing hardened concrete* - Part 6: Tensile splitting strength of test specimens
[16] CSN 731318: 1987 Z1: 1994, Z2: 2003 *Determination of tensile strength of concrete*
[17] CSN 731322: 1968 Z1: 2003 *Determination of frost resistance of concrete*

IOP Conf. Series: Materials Science and Engineering **385** (2018) 012007 doi:10.1088/1757-899X/385/1/012007

Construction of precast segmental box girder bridge

P Bujnakova

Department of Structures and Bridges, University of Zilina, Faculty of Civil Engineering, Univerzitna 8215/1 Zilina, 010 26, Slovakia

Abstract. New road segmental bridge have been built near the city of Žilina, aligned on European highway corridor. The use of segmental concrete box girder was chosen as the flexible system and appropriate method in a municipal zone. The same parallel precast post-tensioned box girder structures were used with main spans of 60.5 m for total length of 1042 m. The balanced cantilever method, with the self-launching gantry was used for erection. The shorter end spans, adjacent to the abutment, had to be built by employing falsework. Generally, the early segmental bridges have not been realized without problems following to the structure collapse. In order to assure the safety under construction stage of bridges and subsequently in service, the data was collected during casting and erection of the bridges since August 2016 to November 2017. The experimental research was divided into the laboratory testing and continuous in-situ strain and temperature monitoring of the bridge superstructure. The paper illustrates fabrication and erection process of the bridge and some results of superstructure testing.

1. Introduction

Segmental construction is preferred in highway bridges for many reasons. One advantage of this technique is elimination of conventional falsework and temporary supports by adopting cantilever construction method. The major part of the work is performed in the precasting yard and erection process can start simultaneously with the foundation work. It is adaptable to almost any conceivable site conditions. Horizontal and vertical alignments, straight or curved bridges can be built without any difficulty. The long span structures have been built without any objection to traffic, over deep valleys or difficult terrain. Therefore, this method is used for large urban and suburban viaducts for most road and rail bridges all over the world. Nevertheless, there are some limitations due to the cost for precast yard and erection equipment. Generally, the method is more economical when number of segments exceeds 350 to 400 units per bridge [6]. Precast cantilever method is suitable for 40 to 150 m span length depending on erection method (span-by-span method, balanced cantilever method or progressive placement). The post-tensioned structure is made of individual precast units stressed together. The joints are either dry (in areas where climate permits) or made of a very thin epoxy resin (Europe, Asia). The joints are of negligible thickness without any structural function. To lock the segments together the shear keys are cast into the joint faces. They transmit shear forces and help in exact alignment of the segments. During the construction process, the structure changes in static system, in support, loading and magnitude of external prestressing. The whole section remains in compression under both dead and live load. An accurate evaluation of stresses and deformation in each construction stage is essential to maintain safety levels and to ensure the final required bridge alignment.

2. Bridge details

The segmental bridge is a part of the new Slovak motorway connection from Bratislava to the Ukrainian border forming the international connection E- 50, located in the urban area of village "Lietavská Lúčka" near the city Žilina. The bridge is composed of the two separate precast concrete structure. The total length of both structures is 1042 m. The similar segmental structure consists of a total of eighteen continuous spans with lengths 46.10 +15 x 60.50 + 49.80 + 32.80 m. The balanced cantilever method was accomplished utilizing the self-launching gantry for erection.

The roadway has a variable width from 11.75 to13.25 m between barriers and the total deck width varying from 14.25 to 15.75 m. The typical cross section consists of post-tensioned box girders of the constant depth 3.0 m for all the spans providing the most efficient section for casting. The span- to – depth ratio is corresponding to 20. The typical cross section is shown in figure 1. The constant web thickness 500 mm was determined by shear considerations, as tendon ducts internal to the concrete were present. Box girder webs are inclined. The local haunches are used at the intersection of the bottom slab and the webs to provide sufficient space for accommodating the required number of tendon ducts. The box segment has a bottom slab width of 6.5 m. The bottom flanges are 200 mm and 350 mm thick. The top slab thickness is variable, resulted from the limit deflection criteria under the live loading.

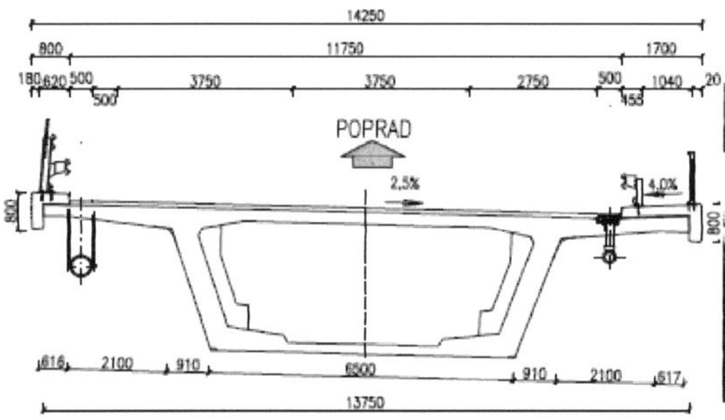

Figure 1. Bridge cross section.

Figure 2. Pier segment.

Two pairs of reinforced concrete column piers of square-shaped section provide vertical supports for spans at intermediate points. In addition to transferring the superstructure vertical loads to the foundations, the piers should resist the higher lateral actions caused by potential earthquake events in this low seismic area and consequently ductility aspects had to be provided by their design. The seat-type abutments were constructed separately from the bridge superstructure as the reinforced earth-retaining structure using multiple-layer strips from non-gradable fabrics to reinforce the fill material in the lateral direction. The face panels as slabs anchored by the strips are subjected to the lateral soil pressure. The bridge superstructure seats on the abutment stems through the pot bearings comprising plain elastomeric disks confined in shallow steel rings. Teflon sliding surfaces of expansion bearings can accommodate the translational movement. Keeper plates are used to retain the superstructure moving in presumed direction. The fixed bearings allowing only rotations, but restricting movements, are located at the top of the central piers. The placement of expansion joints within these long viaducts was necessary to accommodate the change in length of structure 560 mm due to creep, shrinkage, and thermal changes. The expansion joints are located at the centerline of the end abutments.

3. Fabrication and erection process

The segments were erected in existing stationary yard in Prefa Senec factory, allowing superior quality control. The short line match casting system was used and the rate of the segment production was four segments per line every 5 days. Total number of segments casted in stationary yard was 928 units. The precasting operation was divided to segment casting, geometry control, segment handling and storage. The control setting of the precast yard allowed production similar to an assembly line environment.

The box girder was designed as the most efficient and economical cross section for segmental bridges. The torsional stiffness of a box girder allows for temporary out-of-balance forces during the erection. The constant depth (3.0 m) and weight (60 t) of the segments was designed by the capacity of transport and placing equipment. The five type of segments were designed for each bridge. The three main types: abutment segments, pier segments, span segments and two other types of modified span segments - segments with anchor block for upper continuity cables and segments located near the pier. Pier and abutment segments are 1.65 m in length, the span segments are 2.2 m long. The span segments were designed of concrete class C 45/55 and pier segments of C 55/67. The pier segment was designed for both permanent and temporary conditions [9]. The stressing and grouting of the transverse tendons in pier segments were done in the storage yard before the shipping for erection. The three tendons has been designed with characteristic strength 1860 MPa consisting of 5 strands with diameter 15.7 mm. The longitudinal post-tensioning was designed as bonded tendons with low relaxation strands of 15.7 mm and characteristic strength 1860 MPa.

The multiple shear keys were designed to provide locations for joints of adjacent segments and to carry shear load, figure 2. They were formed across the full width of the web and in the top and bottom flanges. The typical segment was made of reinforcement cage prefixed using jigs, figure 4. The web form was supported by a frame connected to the casting cell foundation.

The new segment was match-cast against the previous one. When the new segment had hardened, it was moved to the match- cast position to serve as a front bulkhead of the casting cell for the next segment, figure 3. Control points were located over the webs to eliminate any influence of the top slab movement. The center points establish theoretical centerline. The successive segments were cast against the adjoining segment in the correct relative orientation with each other starting from the first segment away from the pier. To obtain a vertical curve of the bridge, the match-cast segment had first to be translated and given a rotation in the vertical plane. To obtain a horizontal bend, the conjugate unit was given a rotation in the horizontal plane. All these adjustments of the conjugate unit had to be combined for obtaining the desired geometry of the bridge. The balanced cantilever method requires additional geometry control to account for the temporary construction loads on the structure, than an additional camber was built into the segments.

Figure 3. Precasting bed.

Figure 4. Reinforcement cage.

The typical cycle of erection consists of placing segments, installing and stressing post-tensioning tendons and launching the gantry to its next position.

The 60-ton segments were precast and delivered by trailers. The transportation may produce excessive stresses, which were considered. Segments were lifted into place by a self-launching gantry placed on the previously completed portion of the deck (figure 5). In order to assure the safety under construction of bridges. The self-launching gantry has a length slightly longer than the typical span. The construction started from the piers cantilevering out to both sides to minimize the negative moments. The each phase was tied to the previous ones by the post-tensioning tendons, incorporated into the permanent structure. Cantilever tendons are designed as internally bonded within the top flange.

During the erection of the cantilever, the centre leg rests on the pier while the rear leg reposes on the cantilever tip of the previously erected span, which must resist the corresponding reaction. Prior to launching, the back spans had to be made continuous. Then, the centre leg was moved to the forward cantilever tip, which had to resist the weight of the gantry plus the weight of the pier segment. This stage controls the design of the gantry, which must be made as light as possible, and of the cantilever. The use of epoxy joint requires a perfect fit between the ends of adjacent segments. The joints were made of 2 mm thin epoxy resin, which provides water tightness and durability at the joints and did not alter the match-cast geometry.

Figure 5. Balanced cantilever construction with launching gantry.

The mid- span closures between cantilevers were constructed by casting in- situ joint of concrete class C 35/45 and approximately 0.6 m in length. Grouting procedure provides corrosion protection to the prestressing steel and develops bond between the prestressing steel and surrounding concrete.

The end spans adjacent to the abutment had to be built on falsework. It helps to balance moments in different spans under operating loads and provides positive support reactions on the abutments. In order to allow the movement necessary for right joints, the jacks with capacity of 2000 kN were placed between the bottom of the segments and the falsework.

The several actions had to be considered in checking verification. Firstly a possible out- of- balance one segment on the cantilever. Then, the presence of a stressing platform, live loading on one side of 1.5 kN/m^2, wind loading acting during the construction and the possibility of one cantilever having a 2.5% higher dead weight than the other [1-2, 8].

4. The measurement and monitoring of the bridge structure

The monitoring of the bridges is able to detect reliably structural changes or damage that is not visually achieved [4]. Structural changes may result from other causes such as sudden settlement of foundation, ground movement, excessive traffic or failure of post-tensioning tendons [5]. To avoid the unpredictable structural changes both bridge superstructures were monitored. The program was divided into following parts:

- Laboratory tests of mechanical properties.
- Strain and temperature recording during the erection.
- The static and dynamic load tests.
- Long- term monitoring of bridge performance in service.

IOP Conf. Series: Materials Science and Engineering **385** (2018) 012007 doi:10.1088/1757-899X/385/1/012007

The data have been collected continuously during the fabrication and erection process in 2016 and 2017. The magnitude of stresses was evaluated at all the stages of construction. The specimens were tested in Accredited Laboratory of Civil Engineering at University of Žilina.

4.1. Verification of mechanical properties

Uniform quality and higher strength of concrete are essential for segmental construction. It is recommended the ideal concrete mixes as near as practical zero slump [7]. The zero slump concrete allows for direct stripping of the unhardened product after filling and vibrating the mold and subsequent transport to a place with defined curing condition.

The main purpose of the laboratory tests was to verify the real values of basic mechanical properties of concrete such as compressive strength, modulus of elasticity and long-term concrete strain development over time due to creep and shrinkage, figure 6. The compressive strength as the common performance property was measured by breaking cube concrete specimens of 150 mm and cylindrical concrete specimens of 150×300 mm size in the laboratory compression testing machine and calculated from failure load divided by the cross-sectional resisting area. The specimens were tested in curing room with temperature of 20 ± 2°C and relative humidity of 80 % [10]. Three specimens from each batch of selected casting segments were tested. The results from the cast cubes and cylinders validated existing concrete strength, as well as adequacy of curing and protection measures [3]. Measured material properties were used in computing model for non-linear analysis to calculate stresses and deflection of the bridge structure.

a) compressive strength test b) creep test c) strain gauge

Figure 6. Testing procedure and tools.

The next test method measured the load-induced time-dependent creep strain on molded concrete samples subjected to sustained longitudinal compressive load (figure 6b). Creeping testing machine consists of main unit using disk springs specifically selected to maintain constant load over the range of deformation probable during creep tests and measured using externally mounted mechanical gage points. For strain time dependent variation, the seventh and sixteenth mid-spans, as well as the two neighboring piers sections of the left bridge, were implemented with sensors during the fabrication and erection process. The vibrating wire strain gauges, suitable for long term readings were placed on the concrete deck and on the girder web. Four strain gauges according to figure 6c were placed at the intersection of the slabs and the webs of cross sections and spaced along the length of each girder. Since there were two selected segments in each span, a total of 16 strain gauges were placed on the left bridge (span no. 7 and no. 16) and also on the right bridge (span no. 2 and no. 9). The histories of strain development were recorded and structural response computed from the measured readings. At the same time, temperature measurements were made on the specified 32 different locations across the centre of the span and pier sections. The data acquisition for each site investigation took about six hours. Figure 7 illustrates the normal stresses variation in the mid-section of the sixteenth span of the left bridge during the construction stages (in the stage of cantilever, Pnk), before loading (PZS) and in loading test (ZS). The prediction stress distribution and obtained values in bottom and top slab of the cross section achieved acceptable agreement.

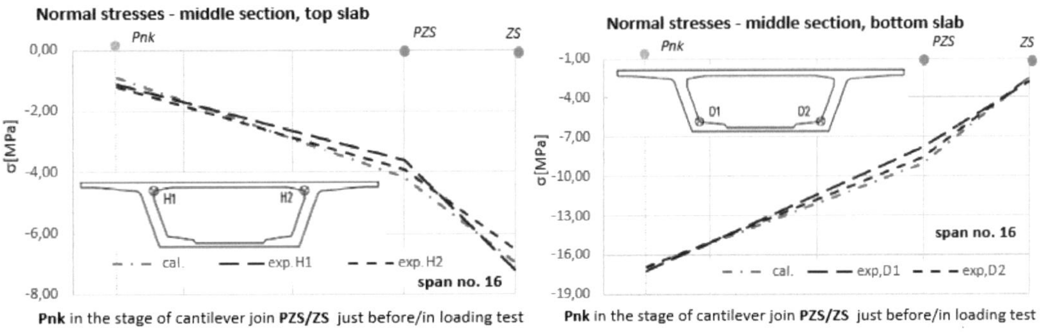

Figure 7. Normal stresses variation (span no. 16).

5. Conclusions

Prestressed box girder bridges typically exhibit small deflections during the first years of service. Global stresses and deformations of segmental box-girder bridges due to the effects of dead load and post-tensioning, as well as the long-term effect of creep can be predicted during the design process by the use of a computer analysis program. But the real values are dependent, to a large extent, on the method of the structure construction, the age of the segments when post-tensioned and the time when other loads are applied [3]. Real more correct behaviour, especially of major structures can be identified only by field measurements. Monitoring system is a way for understanding the complex behaviour of bridges, especially early structural changes in service condition. The results of this study could help in design, construction process and especially in bridge maintenance of the precast segmental bridges in further years.

Acknowledgements
This research work was supported by the Slovak Grant Agency under contract No. 1/0336/15 and No. 1/0343/18.

References

[1] Bujnak J 2015 *Design Review Of The Right Bridge SO 209 At The Highway Construction D1 Hricovske Podhradie – Lietavská Lucka* (Zilina: University of Zilina)
[2] Bujnak J 2015 *Design Review Of The Left Bridge SO 209 At The Highway Construction D1 Hricovske Podhradie – Lietavska Lucka (*Zilina: University of Zilina)
[3] Bujnak J and Bujnakova P 2018 *Monitoring Report Of The Bridge SO 209 At The Highway Construction D1 Hricovske Podhradie – Lietavska Lucka (*Zilina: University of Zilina)
[4] Massicotte B and Picard A 1994 *PCI Journal* **39** 66- 80
[5] Woodward R J 1989 *Transp. Res. Rec.* **1211** 38- 59
[6] Design guide: 2003 *Prestressed concrete bridges built using the cantilever method* (Sétra: FR)
[7] Guide to good practice, FIB report: 2017 *Precast Segmental Bridges* p 183
[8] STN EN 1991-2: 2006 Eurocode 1: *Actions on Structures*. Part 2: Traffic Loads on Bridges. (UNMS, Bratislava, SK)
[9] STN EN 1992-2: 2012 Eurocode 2*: Design of Concrete Bridges.* Design and Detailing Rules. (UNMS, Bratislava, SK)
[10] STN EN 206 +A1: 2017 Concrete. *Specification, performance, production and conformity.* (UNMS, Bratislava, SK)

IOP Conf. Series: Materials Science and Engineering **385** (2018) 012008 doi:10.1088/1757-899X/385/1/012008

Experimental determination of the initial compressive strength of concrete using a rebound test hammer

P Cikrle, D Kocab and P Misak

Faculty of Civil Engineering, Brno University of Technology, Veveri 331/95, 602 00 Brno, Czech Republic

Abstract. This paper deals with rebound hammer testing of concrete in the early stage of ageing. Six types of concrete were produced for the purpose of the presented experiment and differed in the water/cement ratio and were otherwise identical as to the input components. SilverSchmidt PC L test hammer was used to determine the rebound coefficient Q and compressive strength of cube test specimens. The aim of the experiment was to verify the possibility of creating a conversion relationship between the determined parameter of the used non-destructive method and the real compressive strength of concrete in the first days of ageing of concrete. Based on the conducted experiment, it may be stated that the SilverSchmidt PC L test hammer with a mushroom plunger is a suitable instrument for determining the initial compressive strength of concrete, starting from the value of about 5 N/mm^2. An important conclusion is the fact that it is possible to create a regression model with relatively high accuracy, in this case approximately ± 3 N/mm^2.

1. Introduction

Rebound hammer test of concrete is one of the most widely used non-destructive methods (NDT) in the construction industry since the measurement is relatively simple and the character of the method is practically non-destructive [1,2]. It is essentially a surface test of a concrete element with measurements and evaluation of a prescribed number of local failures or elastic reactions caused by the impact.

The rebound hammer test method is in practice used principally in two basic ways. The first is the use of a rebound test hammer to diagnose older structures with the main purpose being the determination of the strength class of the tested concrete. In case of such structures, however, it is usually necessary to remove not only the plaster, but also the surface layer of concrete to prevent possible negative influence of the rebound hammer measurement due to carbonation. Furthermore, it is also necessary to supplement the non-destructive measurements with compressive testing of core samples taken from the tested concrete. The second way of using the rebound hammer test method is more advantageous and focuses on quality control of concrete of new constructions, especially those which have a smooth surface and which therefore eliminate the need of grinding. Concrete is tested at sufficient age, at least 28 days of age, and the compressive strength estimation is best carried out using a conversion relationship (i.e. the regression dependence between the NDT parameter and the actually determined compressive strength) created especially for the tested type of concrete.

In addition to quality control of already aged concrete, new types of hardness testers can also provide estimates of the relatively low, so-called formwork-removal, strength of non-aged concrete, e.g. at the age of several hours or a few days after concreting the construction element. This paper deals with the issue of creating the conversion relationship to estimate the initial strengths of concrete and to determine the appropriate moment for removing the formwork from the structure.

2. Types of rebound test hammer

A number of rebound hammers have been developed over the past few years but the ones used the most are the Schmidt type test hammers. There are several types of rebound test hammers that differ from each other in the energy of the impact, the size and the shape of the plunger, or the mechanical design. The usability of the basic rebound test hammers with respect to compressive strength and the thickness of the concrete layer is presented in table 1 [3], which also includes new data relating to new digital instruments SilverSchmidt.

Table 1. Range of use of rebound test hammers [3].

Used test hammers	Cube strength of concrete [N/mm^2]	The smallest concrete layer thickness [mm]	Smallest grouting width [mm]
Original Schmidt N	17 – 60	100	30
Original Schmidt L	13 – 50	60	30
Schmidt M	25 – 60	200	not used
SilverSchmidt PC N	10 – 100	100	30
SilverSchmidt PC L	5 – 30	60	30

The rebound test hammer SilverSchmidt is a modern variant of the original Schmidt N test hammer. The rebound coefficient Q is the test indicator and is calculated from the equation (1):

$$Q = 100 \cdot \frac{energyrecovered}{energyinput} \tag{1}$$

The SilverSchmidt test hammer uses optical sensors to measures the speed of the impact as well as of the rebound immediately before and after the impact while calculating the amount of energy that can be recovered. The Q value allows to extend the range of standard conversion relationships for concrete strength at both ends of the scale. The strength range of the SilverSchmidt test hammer is from 10 N/mm^2 to 100 N/mm^2, which allows testing of modern high strength concrete. When using the mushroom plunger for SilverSchmidt PC L (see figure 1), even a low compressive strength from 5 N/mm^2 can be determined. The described experiment deals particularly with concrete tests with this mushroom plunger.

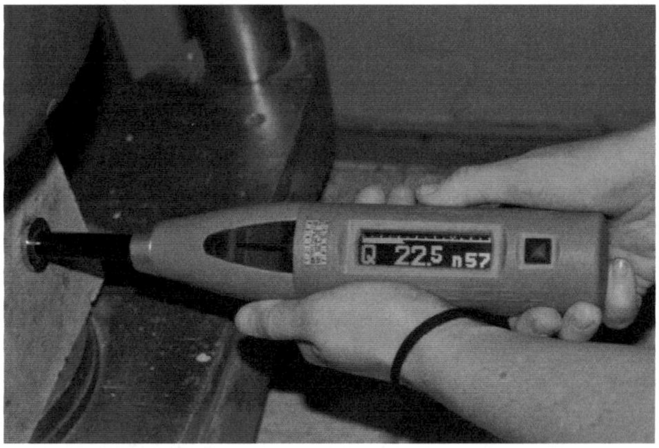

Figure 1. Test hammer SilverSchmidt PC L with the mushroom plunger [1].

3. Experiment

Conversion relationships can be created in numerous ways. The given experiment followed the principles of the standard ČSN 73 1370 [4] and therefore non-destructively and destructively tested 150 mm concrete cubes.

4. Material and test specimens

Concrete cubes based on six concrete formulas were created for the experiment and differed only in the amount of cement, water and admixtures, and therefore differed only in the water/cement ratio. The production of the six types of concrete, designated letters A to F, used the same components – cement CEM I 42.5 from the cement plant Mokrá, 0/4 mm sand from the Bratčice area, crushed natural aggregate with the grain size 4/8 mm and 8/16 mm from Olbramovice, plasticiser SikaViscoCrete 4035 and air-entraining agent Sika LPS A 94. Concrete A and B contained the plasticiser in the amount of 0.25% of the cement weight and concrete C and D contained the plasticiser in the amount of 0.50% of the cement weight. These two types of concrete were not air-entrained. Concrete designated E and F contained the same amount of plasticiser as concrete C and D, but contained also the air-entraining agent. The amount of used water was always adjusted according to the amount of other ingredients so that all six types of concrete were similarly workable. The formulas for all the concrete types are given in table 2 and table 3 presents the determined basic parameters of the concrete in the fresh state, including the water/cement ratio. The properties of fresh concrete were determined in accordance with the standards EN 12350 [5].

Table 2. Composition of concrete used in the experiment.

	kg / 1 m^3 of fresh concrete					
Component	A	B	C	D	E	F
CEM I 42.5 R (Mokrá)	350	400	350	400	350	400
Sand 0/4 mm (Bratčice)	875	825	875	825	875	825
Aggregate 4/8 mm (Olbramovice)	185	185	185	185	185	185
Aggregate 8/16 mm (Olbramovice)	695	695	695	695	695	695
Water – mixing	165	165	160	160	150	150
Water – in aggregate	14	14	14	14	14	14
Water – total	179	179	174	174	164	164
Plasticiser (Sika ViscoCrete 4035)	0.88	1.00	1.75	2.00	1.75	2.00
Air-entraining agent (Sika LPSA 94)	-	-	-	-	0.75	0.75

Table 3. Properties of the concrete in the fresh state.

	Concrete					
Property	A	B	C	D	E	F
Flow table test [mm]	350	370	340	330	330	320
Slump-test [mm]	50	60	70	50	60	50
Air content [%]	2.8	3.2	3.5	3.0	6.2	5.7
Density [kg/m^3]	2300	2300	2270	2300	2190	2260
Water/cement ratio (w_{mixing}/cement)	0.47	0.41	0.46	0.40	0.43	0.38

In addition to the cubes that were designed for the purposes of the described experiment, additional test specimens were produced for control tests at the age of 28 days. These specimens were removed from the moulds after 24 hours and placed into a water bath with the temperature of (20 ± 2) °C. Table 4 presents the results of the tests of hardened concrete at the age of 28 days that were carried out in accordance with the standards EN 12390 [6].

Table 4. Properties of concrete in hardened state – average values with sample standard deviation in the brackets.

Property	Concrete A	B	C	D	E	F
Density [kg/m^3]	2330 (4)	2360 (2)	2340 (16)	2320 (6)	2280 (31)	2320 (16)
Compressive strength [N/mm^2]	50.8 (0.2)	56.4 (2.2)	52.2 (4.8)	57.2 (1.7)	53.7 (0.9)	63.7 (1.4)
Flexural strength [N/mm^2]	4.9 (0.1)	5.6 (0.4)	5.6 (0.2)	5.7 (0.1)	5.3 (0.1)	6.7 (0.3)

The described experiment used 6 concrete cubes from each concrete type produced in laboratories of the Faculty of Civil Engineering BUT, totalling 36 test specimens. Concrete was placed in polyurethane moulds, where it aged until the time of demoulding and testing. The filled moulds were covered with a lid of the same material and stored under standard laboratory conditions. The aim of the experiment was to determine the conversion relationship for the initial compressive strength of concrete, also known as "formwork-removal strength" – which is the compressive strength of concrete (usually 5 to 10 N/mm^2) defined by the manufacturer of the concrete as the limit when it is possible to remove the formwork from the concrete element.

5. Testing method

The test cubes of each concrete type were tested sequentially on the second day after pouring concrete – the age of the concrete was therefore always in the range from 20 to 36 hours. After removing the cubes from the moulds, they were immediately subjected to weight and dimension measurements. Each cube was then clamped to compression testing machine and 10 values of the Q coefficient were determined with the test hammer SilverSchmidt PC L with the mushroom plunger – each time in a series of five on the two opposite sides of the test cube. Compressive strength was then determined on the test specimens.

6. Results and discussion

The test results from the rebound test hammer on the individual specimens were first assessed separately. After verifying the normality of the data, the Grubbs test at the significance level of 0.05 was used to identify the outlying values. These values were removed from the experiment.

Figure 2 presents a regression relationship created from all non-outlying Q values. A linear regression model in the form of a straight line $f_c = b_0 + b_1 \cdot Q$ seems to be the most appropriate means, where b_0 and b_1 are the sought after regression coefficients. In addition to the linear regression model, two regression bands are also presented in the graph. The narrower band, the so-called confidence band, defines the limits in which we can expect the obtained curve (linear model) with 95% confidence. This band is determined by interval estimates of regression coefficients. The wider band, the so-called predictive band, indicates the limits in which we can expect the f_c value estimated on the basis of the obtained Q value. The prediction band forms a 95% interval estimation of the mean value of f_c determined based on the observation of the Q value. This regression model allows us to estimate the compressive strength with a 95% confidence and accuracy of approximately ± 5 N/mm^2.

The graph in figure 3 presents a regression model created from the average Q values obtained from individual specimens. This model allows us to estimate the compressive strength of concrete with 95% confidence and accuracy of approximately ± 3 N/mm^2.

The results of the conducted regression analysis indicate that due to the high variability of the measured Q values, the model compiled from the average Q values obtained always on one specimen / test point seems to be a more appropriate model for practical purposes. Such a regression model can be used to estimate the initial compressive strength with the accuracy of ± 3 N/mm^2 and 95% confidence. Assuming the initial compressive strength of 10 N/mm^2, it can be stated on the basis of the created model that the minimum Q value to guarantee 95% confidence is for the used concrete 22.

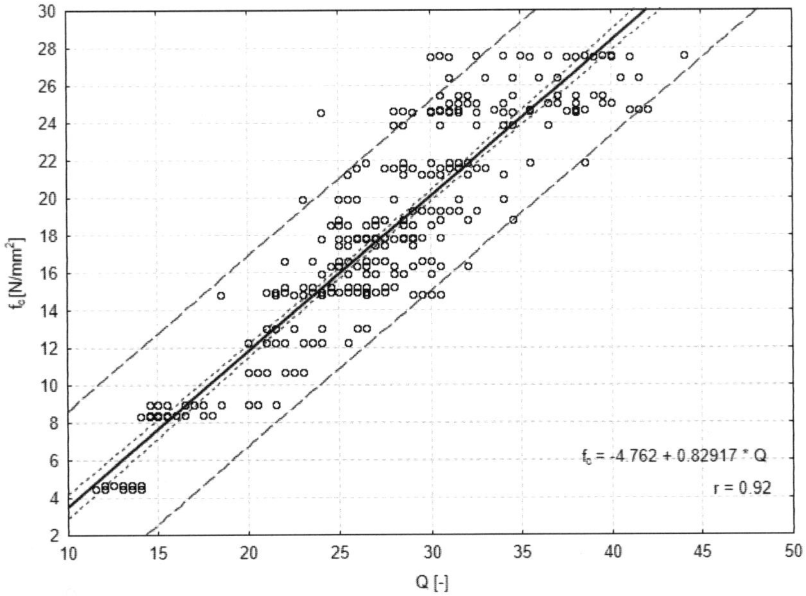

Figure 2. Linear regression model of dependence of the rebound value Q and compressive strength f_c; the confidence and predictive regression bands are dashed; r - correlation coefficient.

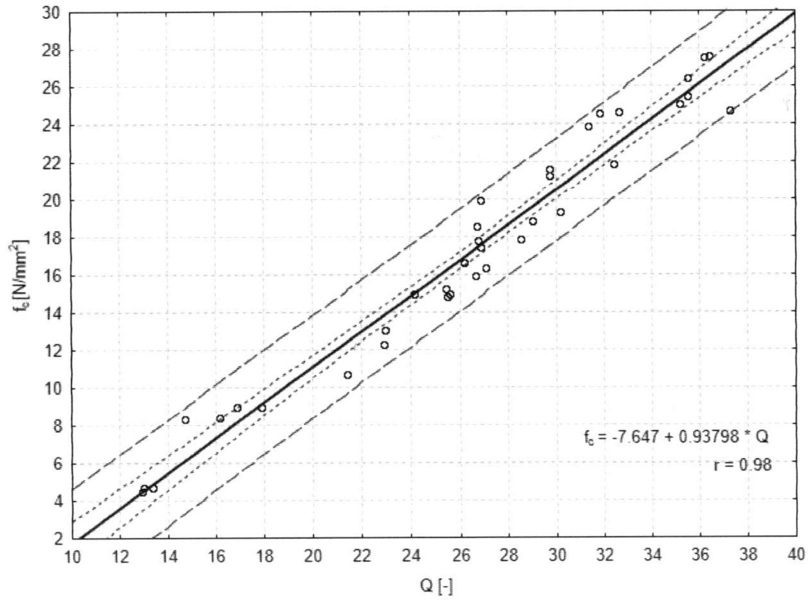

Figure 3. Linear regression model of the relationship of the mean rebound Q values and compressive strength f_c; the confidence and predictive regression bands are dashed; r - correlation coefficient.

It is important to mention that the established conversion relationship may not work on concrete of different composition. However, the evaluation process revealed that the regression model may not only apply to a single concrete formula or a single concrete strength class. If the same components are used for the production of concrete, the decisive factor is neither the amount of cement, nor the amount of water, nor the amount of admixtures that the fresh concrete contains. It would seem appropriate to supplement the conducted experiment with concrete tests that will use concrete of varying amounts of sand and varying aggregates grain sizes. This experiment employed the same amount of coarse aggregate in all six different concrete types.

7. Conclusions

Based on the conducted experiment, it may be stated that the SilverSchmidt PC L test hammer with a mushroom plunger is a suitable instrument for determining the initial compressive strength of concrete, starting from the value of about 5 N/mm^2. The advantage of the rebound hammer test method is the simplicity of the measurements and the non-destructive (non-invasive) character of the test when using the selected instrument since its large mushroom plunger leaves only a slight trace of the impact on relatively slightly hardened concrete. In addition, the test can be repeated on the same place over time and thus provide an overview of the real course of ageing of concrete directly in the construction. An important conclusion is the fact that it is possible to create a regression model with relatively high accuracy (in this case approximately $\pm$ 3 N/mm^2) for more concrete formulas. However, the use of the same input materials is essential.

Acknowledgement

This paper has been worked out under the project No. LO1408 “AdMaS UP - Advanced Materials, Structures and Technologies”, supported by Ministry of Education, Youth and Sports under the “National Sustainability Programme I”.

References

[1] Cikrle P and Anton O 2015 *Beton TKS* **3** 3-7 (in Czech)
[2] Hannachi S and Nacer Guetteche M 2014 *Proc. Int. Conf. on Architecture And Civil Engineering (Dubai)* pp 118-127
[3] CSN 73 1371: 2011 *Non-destructive testing of concrete – Determination of compressive strength by hardness testing methods (in Czech)*
[4] CSN 73 1370: 2011 *Non-destructive testing of concrete – Common regalutions (in Czech)*
[5] EN 12350: 2009 *Testing fresh concrete*
[6] EN 12390: 2009 *Testing hardened concrete*

IOP Conf. Series: Materials Science and Engineering **385** (2018) 012009 doi:10.1088/1757-899X/385/1/012009

Determining reinforcement coverage using an electromagnetic rebar detector

P Cikrle, O Anton, A Kinclova and D Kocab

Faculty of Civil Engineering, Brno University of Technology, Veveri 331/95, 602 00 Brno, Czech Republic

Abstract. Correct reinforcement is an essential part of constructing reinforced concrete structures. Errors in positioning, diameter, or coverage may result in defects, which is why it is necessary to make sure the reinforcement is implemented well. Even though electromagnetic detection is the most common non-destructive method for locating reinforcement in reinforced concrete, science literary sources hardly discuss it. The main reason is a lack of graphic test results – thus far the most common research outcomes have been photographs of lines drawn on the concrete surface or additions to technical drawings of the reinforcement. Only modern detectors expanded the possibilities of the method's application including graphic measurement records and a more accurate estimation of reinforcement diameter including corrections for adjacent reinforcement spacing. The paper focuses on the principle of measurement and practical examples of determining the position and coverage of reinforcement in reinforced-concrete structures using a state-of-the-art electromagnetic rebar detector – a Profometer PM-630. Aside from data recording and statistical evaluation during coverage measurement, modern instruments also allow the data to be exported and visualised.

1. Introduction

Originally, the method exploited the magnetic properties of the material being detected. Older instruments responded to changes in the magnetic resistance in a magnetic circuit [1,2]. The stability of measurement was affected by the dependence of the magnetic properties of the detection coil's core on temperature, as well as interference of other magnetic fields including Earth's magnetic field. The contemporary method operates on the principle of pulse induction [3].

Only very sensitive instruments can be used for this kind of measurement. Here, the most important fact is whether a reinforcement bar is solitary, or if there are any other rebars in its vicinity (whether parallel or perpendicular). Measurements made on a solitary rebar return straightforward results. Based on a change in the indicator (needle motion, numerical value) coverage is determined by means of a fitted curve. Newer instruments allow the user to enter the rebar diameter and directly obtain the coverage value.

A rebar is considered solitary if the indicator response is not affected by adjacent reinforcement (whether parallel or perpendicular) by more than 5 %. The minimum distance of adjacent reinforcement is determined by measurements using a model concrete member [4]. If this distance is not kept, the instrument returns a value of coverage which is smaller than the actual one (several rebars deep inside the member behave like a single rebar closer to the surface). During the determination of reinforcement coverage in a member with dense reinforcement and known distribution it is possible to set correction for coverage using a calibration measurement performed on a model member with reinforcement which is identical to that of the member being measured.

However, the diameter of reinforcement is not always known, in spite of the fact that it is an important parameter. For instance, if coverage is evaluated using a curve for reinforcement of Ø 16 mm in diameter, but the real diameter is Ø 20 mm, the resulting coverage would be lower than the actual value. However, as long as the diameter can be at least approximately estimated or non-destructively measured, the distortion of concrete cover measurement is relatively low.

2. Determining the position of reinforcement bends in a bridge cross beam in Ostrava

To give an example of accurate determination of reinforcement cover, we can describe the diagnostics of an end cross beam of a bridge in Ostrava - Mariánské Hory (see figure 1). The impulse for the diagnostics was given by the city planner. The end beam suffered some transverse cracking and thus required remedial measures to seal the cracks and prevent their further propagation. To stabilise the cross beam structurally, it was post-tensioned in the longitudinal direction. However, this required core drilling at a diameter of 57 mm in order to allow the installation of post-tensioning cables and thus introducing additional reinforcement to the structure. The drilling posed a risk of damage to the cross beam's main reinforcement (of 32 mm in diameter), which was a serious problem. The goal of the diagnostics was to ascertain whether the position of the rebars and especially the bends corresponds accurately to the project documentation. Only after the positions of the bends were identified, the post-tensioning ducts could be drilled.

Figure 1. A view of the bridge in Ostrava - Marianske Hory; the deck is terminated by a massive cross beam.

The proposal intended to perform core drills through the beam in places where it lacks three rebars (numbers 4, 6, 8) out of a total of eleven main reinforcement bars with a diameter of Ø 32 mm (see figure 2). This is because the rebars identified as 4, 6, and 8 bend at a distance of approx. 1.9 m from the end of the beam. The position of the bends in main reinforcement as well as the ducts for post-tensioning cables can be seen in figure 3.

The inclined drills were to be performed downward from the end faces of the cross beam and exit through its bottom at an approximate distance of 1.85 m from either end, i.e. in close proximity to the bends in the main reinforcement. This reinforcement was not to be been damaged by the drilling; according to the construction design, the drills would miss the reinforcement by a mere 50 mm.

Figure 2. Three rebars (4, 6, 8) bent an incline, remaining 8 bent by 90° straight upwards from the bottom.

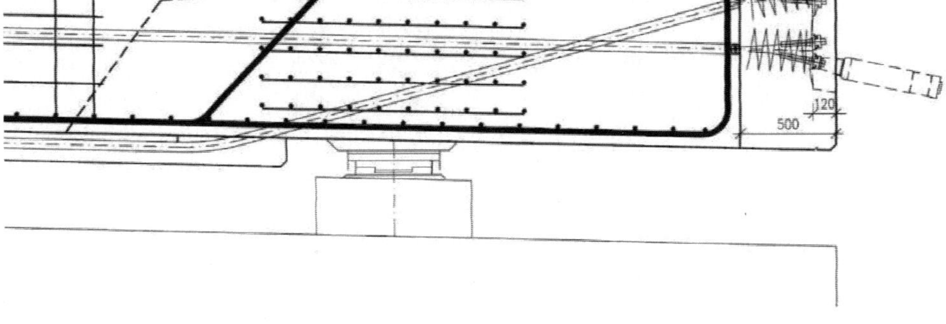

Figure 3. Positioning of main reinforcement and the proposed post-tensioning cables for the structural stabilization of the bridge.

Thanks to sufficient distance between the bridge's bearing bed and the cross beam, i.e. 0.48 m, it was fairly simple to perform measurements on the bottom face of the beam using a new-generation rebar detector Profometer PM-630. The measurement took the form of several linear scans across the cross beam at a varying distance.

Even though the concrete cover was 50 mm, it was possible to locate each bar on the bottom of the beam as they were spaced at a great enough distance of 100 mm to 150 mm.

The Profometer PM-630 retains data from each linear scan in its memory, allowing access to it after the measurement. A total of 7 linear scans were performed at a distance of 1.71 m, 1.77 m, 1.83 m, 1.89 m, 1.95 m, and 2.01 m from the end so as to capture the area from the left as well as from the right from where the bends were expected to be located. Figure 4 shows one of the linear scans performed in the transverse direction at a distance of 1.89 from the end of the beam. In this place the 3 "red" bars are covered by up to 15 to 20 mm more concrete than the other 8 bars.

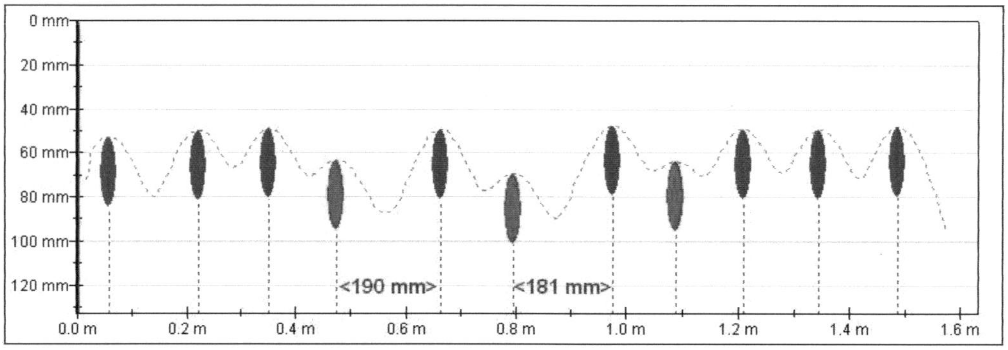

Figure 4. Visualisation of the line scan performed by the Profometer PM-630 in the transverse direction and distance of 1.89 from the end of the beam. The red rebars bend and are covered by up to 15 to 20 mm more concrete.

The results of coverage measurements from all 7 scans were later compiled into a table and plotted in a graph. Table 1 shows the values of concrete coverage in mm for each of the 11 main reinforcement bars. Figure 5 shows an accurate plot of reinforcement coverage in place where the bars bend as well as the location of the drill.

Table 1. Cover depth of all 11 reinforcement bars measured along 7 lines using the electromagnetic rebar detector Profometer PM-630.

	Coverage [mm] along lines from the end of the beam						
Reinforcement No. / Position	1.71 m	1.77 m	1.83 m	1.89 m	1.95 m	2.01 m	2.07 m
1	52.6	52.5	52.6	52.8	53.2	53.6	55.3
2	49.7	49.6	49.4	49.8	50.0	49.8	49.7
3	48.9	48.7	48.8	49.1	49.1	49.4	49.3
4	108.0	92.0	76.7	63.1	56.3	52.3	52.3
5	49.7	49.4	49.5	49.7	49.6	49.9	49.8
6	116.0	99.0	84.0	69.5	60.7	52.7	52.6
7	47.4	47.2	47.4	47.6	47.6	47.8	47.8
8	109.5	95.0	78.0	64.0	56.6	50.8	50.7
9	49.2	48.9	49.1	49.3	49.5	48.9	49.0
10	50.0	49.4	49.5	49.4	49.1	49.5	49.3
11	48.5	48.2	48.0	48.4	48.4	48.3	48.1

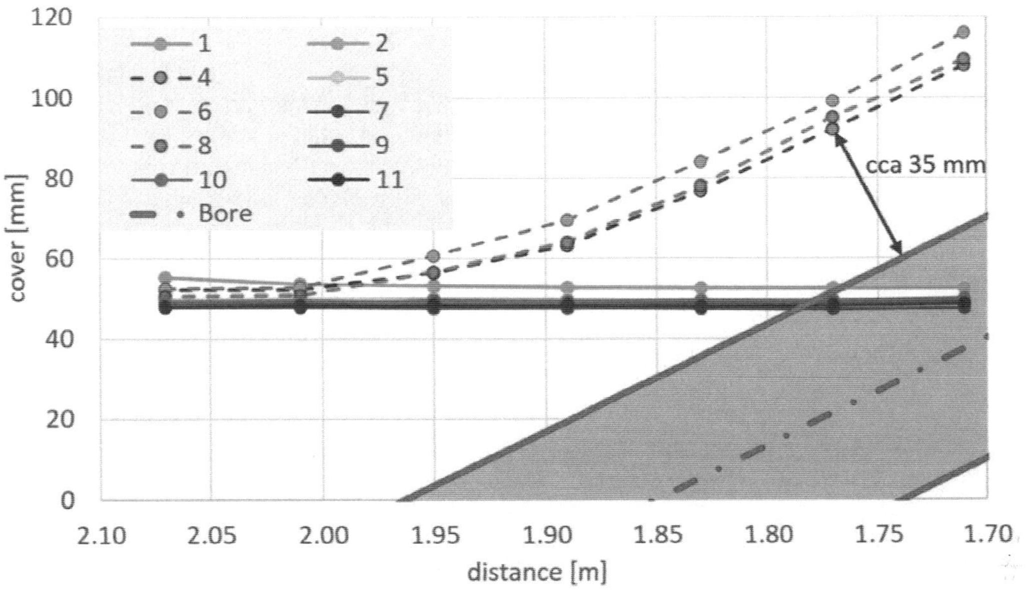

Figure 5. A plot showing the thickness of concrete cover for each of the reinforcement bars; the post-tensioning duct is shown in red.

The diagram in figure 5 shows that the edge of the drill is in fact 35 mm from the surface of the reinforcement, which ensures sufficient cover. Given the axial spacing of 100 - 150 mm there was a chance that adjacent reinforcement might affect the accuracy of measurement. This is why a laboratory measurement was performed with a model incorporating three Ø 32 mm bars, two of which were straight and the third, middle one, was bent. The model measurement showed that the adjacent reinforcement has the greater an effect on measurement, the deeper the bend is. As long as the cover of all the bars was 50 mm, the measurement was very accurate with a deviation of only ± 1 mm. Measurements in place where the bend was 75 mm deep showed a deviation of -3 mm (adjacent reinforcement reduces the depth of cover), and at a depth of 100 mm, the deviation was -10 mm. This means that the reinforcement is actually located a little deeper than the instrument shows. However, the location of the beginning of the bend has been located very accurately [5].

3. Conclusion

Methods for locating reinforcement have recently seen substantial progress. This also applies to the Profometer PM-630, which is the first instrument currently available on the market to allow visualising a cross-section through a steel-reinforced element, indicating the number of bars, coverage, spacing, and diameter of reinforcement. The paper presents a specific example of a fairly sophisticated determination of the precise location of bends in main reinforcement, which allowed effective post-tensioning of the damaged cross beam of the road bridge.

Acknowledgement

This paper has been worked out under the project No. LO1408 "AdMaS UP - Advanced Materials, Structures and Technologies", supported by Ministry of Education, Youth and Sports under the "National Sustainability Programme I", and with the support of specific research project No. FAST-S-18-4824 at Brno University of Technology.

References

[1] Balayssac J P and Garnier V 2017 *Non-destructive Testing and Evaluation of Civil Engineering Structures* (London: Elsevier) p 376

[2] Cikrle P, Anton O and Hermankova V 2014 *Proc. Conf. on Testing and Quality in Construction (Brno)* (Brno: VUT) pp 27-36 (in Czech)
[3] Cikrle P and Anton O 2015 *Beton TKS* **3** 3-7 (in Czech)
[4] Kocab D, Cikrle P, Pospichal O and Hronová P 2009 *Proc. Conf. on Testing and Quality in Construction (Brno)* (Brno: VUT) pp 493-498 (in Czech)
[5] Cikrle P 2016 *The Importance Of Non-Destructive Diagnostic Methods For The Assessment Of Reinforced Concrete Structures* (Brno: VUT) p 174 (in Czech)

IOP Conf. Series: Materials Science and Engineering **385** (2018) 012010 doi:10.1088/1757-899X/385/1/012010

The influence of binder dilution on the properties of lightweight composites

A Daubnerova and S Uncik

Department of Material Engineering, Slovak University of Technology, Faculty of Civil Engineering, Radlinskeho 11, 810 05, Bratislava

Abstract. In the experiment was monitored the influence of thermal-technical properties by binder amount. As binder was used organic based glue 4,4′MDI (Conipur 360). At the same time was verified possible dilution of binder with water. The binder was diluted in amounts of 5, 10, 15, 20, 25 and 30%. However, in the case of 25 and 30% has not been achieved complete dilution. Recycled plastic EVA (ethyl vinyl acetate) was used as filler.

1. Introduction

The biggest global problem in last years is recycling of produced waste. Slovak republic produced in 2015 more than 10 million tons of waste [1]. Paper, clothes, glass, various metal and tires are recycled every day. Plastics also belongs to recycles, which constitute 1/3 from global waste production [1-3]. Plastics are used in all parts of an industry. It is due to their low weight, long life and nontoxicity [4]. Recycling process of plastics is pretty simple because of their workability. But what to do then with recycles? One solution is to use it in civil engineering. Plastic recycles are used as filler of lightweight concrete (e.g. polystyrene concrete). Potential usefulness of plastics as filler of lightweight concrete is explained in next paragraphs.

2. Materials

As filler was used recycled EVA plastic (figure 1.). Plastic was used firstly as the part of sneakers to absorb walking impacts. Plastic was processed to fraction 4/8 after use. Advantage of this plastic is sunlight resistance, aging and water resistance, nontoxicity [4]. It is used also to produce child toys, bags or bottles. In the experiment will be demonstrated good thermal-technical properties of lightweight concrete with EVA plastic used as filler.

Figure 1. Eva plastic.

As binder was used organic based adhesive – 4,4´MDI (methylene – diphenyl – diisocyanate) with the business name Conipur 360 (figure 2.). 4,4´MDI is mostly produced diisocyanate of the industry. In 2016 was made more than 6 million tons of it [5]. It is the main component of polyurethane. 4,4´MDI has good thermal insulation properties. Compared to other organic cyanates, MDI has a relatively low human toxicity, but it is an allergen and sensitizer. Handling MDI requires strict engineering controls and personal protective equipment [6].

In the experiment was monitored the effect of binder dilution with water to selected sample properties – bulk density, thermal conductivity coefficient and tension at defined deformation.

Figure 2. Organic based adhesive 4,4´MDI.

3. Samples preparation

Mixtures components were portioned by the volume. Ideal amount of components were defined in previous experiment as 0,4L of binder and 8L of filler [7]. All of made mixtures were filled to forms 100x100x100mm. There were made 7 different mixtures (table 1.) with constant amount of filler. From every mixture were made 6 samples. Binder amount was changed because of binder dilution with water in ratio 5, 10, 15, 20, 25 and 30% of the binder amount. In 30% water dilution was not achieved homogenous solution (approximately 1/20 of water was not diluted with the binder). One mixture was made without water dilution (reference mixture – Z1). Samples were removed from the mold after 24 hours and stored in humid environment (humidity is more than 90%).

Table 1. Mixture proportions.

Mixture	Filler [L]	Binder [L]		
	EVA	4,4´MDI	Water amount	Dilution [%]
Z1	4,00	0,20	0,00	0
Z2	4,00	0,19	0,01	5
Z3	4,00	0,18	0,02	10
Z4	4,00	0,17	0,03	15
Z5	4,00	0,16	0,04	20
Z6	4,00	0,15	0,05	25
Z7	4,00	0,14	0,06	30

4. Test methods

4.1. Bulk density

Bulk density was classified as ratio between weight and volume of made samples. Weight was measured by scales accuracy of 0,01g.

4.2. Thermal-technical properties

Thermal-technical properties were monitored by ISOMET 2114 with surface prob. For this experiment was evaluated only thermal conductivity coefficient. However, appliance is capable to monitor thermal temperature coefficient and specific volumetric capacity.

4.3. Tension at defined deformation

The compressive strength was specified as tension at defined deformation, because in lightweight concrete based on foam plastics does not happen fragile fracture. The measurement was performed at samples 100x100x100mm in the hydraulic press. Monitored deformations were defined as 2,5, 5, 7,5 and 10% in previous research.

5. Results

5.1. Bulk density

Measured values of bulk density are shown in table 2.

Table 2. Measured values of bulk density.

Mixture	Dilution [%]	Dimensions [mm]			Mass [g]	Bulk density [kg/m^3]
		a_1	a_2	h		
Z1	0	98,95	99,81	100,70	211,68	210
Z2	5	99,38	99,67	101,16	211,04	210
Z3	10	99,46	98,65	95,93	204,19	220
Z4	15	99,06	99,38	101,51	196,51	200
Z5	20	99,97	101,44	101,23	207,01	200
Z6	25	99,78	99,49	100,70	201,56	200
Z7	30	99,28	99,81	100,88	200,52	200

5.2. Thermal-technical properties

Measured values of thermal conductivity coefficient are shown in table 3.

Table 3. Measured values of thermal conductivity coefficient.

Mixture	Mass [g]	Dilution [%]	Thermal conductivity coefficient [W/m.K]
Z1	211,68	0	0,0706
Z2	211,04	5	0,0691
Z3	204,19	10	0,0700
Z4	196,51	15	0,0691
Z5	207,01	20	0,0689
Z6	201,56	25	0,0675
Z7	200,52	30	0,0690

5.3. Tension at defined deformation

Measured values of tension at defined deformation are shown in table 4.

Table 4. Measured values of tension at defined deformation.

Mixture	Dilution [%]	Tension at defined deformation [MPa]			
		Deformation [%]			
		2,5	5	7,5	10
Z1	0	2,22	2,40	2,54	2,64
Z2	5	2,22	2,40	2,51	2,60
Z3	10	2,15	2,36	2,49	2,59
Z4	15	2,14	2,27	2,37	2,46
Z5	20	2,17	2,29	2,40	2,49
Z6	25	2,14	2,27	2,35	2,46
Z7	30	2,14	2,28	2,39	2,46

6. Conclusion

Binder was not completely diluted in 30% dilution. It was not possible to make a homogenous solution (approximately 1/20 of water was not diluted with the binder). With the decreasing amount of binder, the bulk density of samples was decreasing. Values of bulk density were measured between 200 – 220 kg/m^3. The highest bulk density (220 kg/m^3) was measured on samples with 10% water dilution. With the decreasing amount of binder, the thermal conductivity coefficient was decreasing. The lowest value (0,0675 W/m.K) reached the mixture with 25% dilution of binder. Amount reduction of binder caused reduction of tension at defined deformation. The lowest values had sample with 25% water dilution, but values were very similar. Therefore the water dilution has no effect on tension at defined deformation. After these measurements is possible to state, that the replacement of binder seems like potential.

Acknowledgments

To Slovak Research and Development Agency within the project APVV-15-0681 "Investigation of Hydrothermal and Mechanical Properties Linked to the Consistency and Structure of Porous Building Materials, based on Mathematical Modeling." , and to MESRSSR – Ministry of Education, Science, Research and Sport of the Slovak Republic, for financial support by the operational program Research and Development with the project: University Science Park of STU in Bratislava, ITMS 26240220084, co-financed by the European Regional Development Fund.

References

[1] Schabjuk T 2018 Did you know how much waste we produced in 2015? (in Czech)
[2] Rousekova I 1985 *Plastic Materials* (Bratislava: SVST in Bratislava) p 236
[3] Waste separation http://www.triedenieodpadu.sk/plasty.php (online) (in Slovak)
[4] Kovacic L and Bina J 1974 *Plastics, Propertie, Production, Use* (Bratislava: Alfa) p 339
[5] International Agency for Research on Cancer 1999 *IARC Monographs on the Evaluation of Carcinogenic Risks to Humans* **71** 1049-1058
[6] Almaguer D 2006 *National Institute for Occupational Safety and Health* 42
[7] Dragomirova J 2017 *Utilization of recycled EVA plastic for the production of heat insulating polymer concrete* (Bratislava: Slovak University of Technology in Bratislava) Diploma thesis

The evaluation of the effect of crystallization additives on long term durability of self-compacting concrete

A Dufka and T Melichar

Brno University of Technology, Faculty of Civil Engineering, Veveri 331/95, 602 00 Brno, Czech Republic

Abstract. The article deals with the influence of crystallisation additives on the life of self-compacting concrete (so-called SCC concrete) in case of exposure to aggressive gases. The effect of crystallization additives on the properties of such modified composites is generally tested during exposure of liquid media, in this case the attention is focused on influence of selected types of aggressive gases (CO_2 and SO_2). The impact of individual types of aggressive gases is assessed on the basis of a set of physico-mechanical and physico-chemical analyses.

1. Introduction

In common concrete, as well as self-compacting concrete, one way to limit the adverse influence of aggressive substances from the external environment is to use additives for 'secondary crystallization' (hereinafter referred to as 'crystallization materials'). The article analyzes the effect of these types of additives on the long life of SCC concrete in selected types of aggressive gases. The results obtained after the annual and one-year exposure of SCC concrete in aggressive environments are presented.

The use of crystallization materials results in slowing down the degradation of concrete and thus increasing its service life. By contrast to conventional solutions (bitumen strips, plastic foils, etc.), crystallization materials cause changes to the internal structure of concrete.

Self-compacting concrete (SCC) is a high-performance type of concrete. It is a type of concrete that shows both sufficient mobility and resistance to segregation while it is still fresh. Self-compacting concrete is able to fill the form and to flow past even closely spaced reinforcing bars. The fact that SCC does not need to be compacted leads to the relatively higher homogeneity of the resulting concrete structure [3].

The interaction of crystallization additives in the capillary porous structure of 'standard' concrete types has been studied by many authors [1, 2]. In the case of SCC and with respect to the fact that they are new and highly progressive materials, the situation is different. The effects of crystallization additives on the capillary porous structure of self-compacting concrete represent an area that has not yet been explored properly.

The effect of aggressive environments on the developed SCC concrete was evaluated by a set of physico-mechanical and physico-chemical analyses.

2. Methods and experiments

The methodical procedure used for conducting experiments is described in the following text. The composition of the SCC followed on from its previous development at the Institute of Technology of Building Materials and Components, as well as the knowledge mentioned in [3]. For the composition of SCC, refer to the table below:

Table 1. Composition of the SCC concrete.

Component	Component amount [in kg per m^3 of concrete]
Cement CEM I 42.5R	390
Aggregate 0-4 mm	545
Aggregate 4-8 mm	850
Aggregate 8-16 mm	210
Fly ash	180
Water	200
Plasticizer Visocrete 1035 (0.9% of cement weight)	3.50
Xypex Admix C-1000NF (1% of cement weight)	3.90

Additionally, concrete samples of the same composition, but without the crystallization additive, were made. The samples were subjected to a test that would allow a correct assessment of the effect of the additive on the SCC properties of both fresh and hardened concrete. In particular, the following determinations were made:

- Determination of the consistency of fresh SCC concrete – L-box test carried out according to ČSN EN 12350-10,
- Determination of water absorption according to ČSN 731357,
- Determination of capillary attraction according to ČSN 731357-2.

To determine the effect of aggressive gases on SCC concrete modified with crystallization additive and without it, the test samples were divided into several sets, namely:

- Set I – the test specimens of this set were left in laboratory conditions ($t=20\pm2^0C$, $\varphi=50\pm10\%$) throughout the experiment, i.e. for one year. Test samples of this set were considered as reference,
- Set II – the test samples were exposed for 360 days to sulfur dioxide, the concentration of 3%, $t=20\pm2^0C$, $\varphi=50\pm10\%$,
- Set III – test specimens were exposed for 360 days to carbon dioxide, the concentration of 3%, $t=20\pm2^0C$, $\varphi=50\pm10\%$.

The samples were placed in the test environment at 28 days of age. The concentration of aggressive media was monitored and maintained at a constant level throughout the experiment. To determine the effect of aggressive media on SCC concrete modified with crystallization additive and without it, the following set of experiments took place:

- Determination of tensile strength of surface layers of concrete was performed according to ČSN 731318,
- Determination of compressive strength of concrete was made according to ČSN EN 12390-3.

Microstructure analysis using electron scanning microscope, differential thermal, X-ray diffraction analysis and SEM microscopy. Value of pH was also evaluated. These assays were performed according to the methodology described in [4].

3. Results

3.1. Determination of the characteristics of fresh concrete

For the results of the determination of the consistency of the tested SCC, refer to table 2. Values shown represent the arithmetic mean of three measurements.

Table 2. Workability of SCC.

Mix designation		Investigated parameter		
		T_{500} [s]	Slump-flow [mm]	L-box, flow rate
SCC without crystallization additive	Values	1.7	680	0.84
	Classification	VS1	SF2	PL2
SCC with crystallization additive	Values	1.8	675	0.82
	Classification	VS1	SF2	PL2

It has been proven that the modification of SCC concrete by the crystallization additive has no significant influence on the workability of fresh concrete.

3.2. Characteristics of the capillary porous structure of concrete
The effects of the crystallization additive on the characteristics of the capillary porous structure of SCC were assessed by determining absorptivity and capillarity. The results are summarized in Table 3. The values shown represent the mean of six measurements.

Table 3. Absorptivity of SCC.

Mix designation	Absorptivity [%]
SCC without crystallization additive	5.9
SCC with crystallization additive	3.7

These results document the positive effect of the crystallization additive on the capillary-pore structure, respectively on the sorption properties of concrete. This aspect is highly perspective in terms of the concrete's resistance to aggressive environments.

3.3. Determination of the effects of aggressive environments on SCC
As already mentioned above, the effects of selected types of aggressive environments on tested SCC was investigated by both physico-mechanical and physico-chemical analyses. The results of the determinations of strength characteristics (i.e. tensile strength of surface layers and compressive strength) are listed in the following tables. For the sake of clarity, the results that were obtained after a one year exposure in aggressive environments and then after two years exposure in these environments, are shown below. The values shown represent a mean of six measurements (figures 1-4).

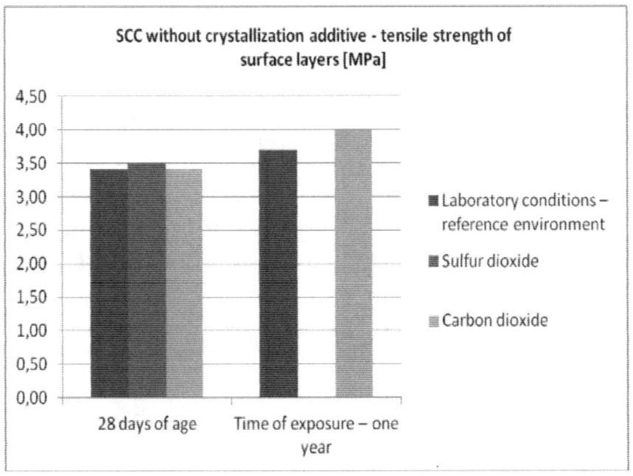

Figure 1. Tensile strength of surface layers - SCC without crystallization additive.

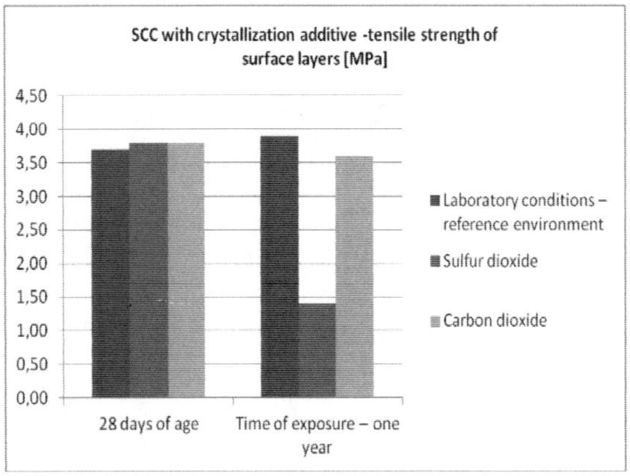

Figure 2. Tensile strength of surface layers - SCC with crystallization additive.

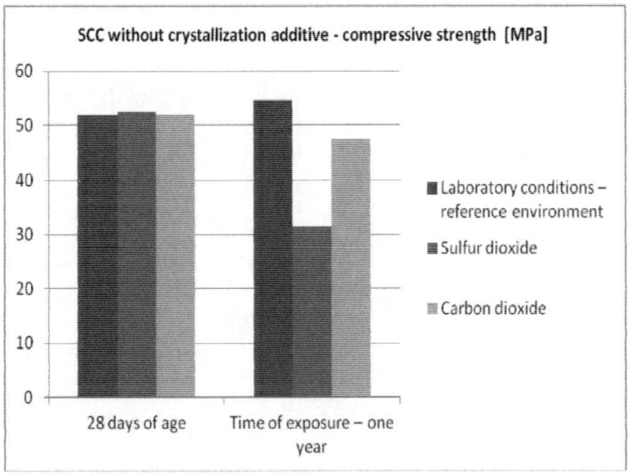

Figure 3. Compressive strength - SCC without crystallization additive.

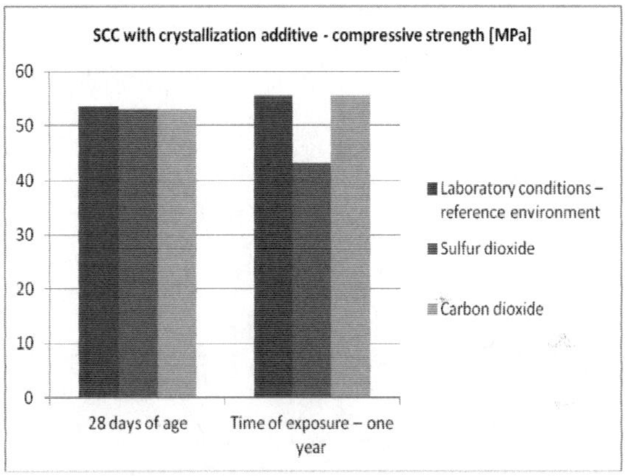

Figure 4. Compressive strength - SCC with crystallization additive.

The knowledge obtained during the testing of strength parameters has been further extended and supplemented by X-ray diffraction analysis, differential thermal analysis and by pH determination. The findings obtained by these methods are captured in the following tables.

Table 4. Results of X-ray diffraction analysis

Mix designation	Exposure conditions	Identified mineral 28 days of age	Time of exposure – one year
SCC without crystallization additive	Laboratory conditions – reference environment	Calcite, calciumhydrosilicate, portlandite, monosulfate, quartz, feldspar	Calcite, calciumhydrosilicate, traces of portlandite, monosulfate, quartz, feldspar
	Sulfur dioxide	Calcite, calciumhydrosilicate, portlandite, quartz, feldspar	Calcite, calciumhydrosilicate, gypsum, ettringit, quartz, feldspar
	Carbon dioxide	Calcite, calciumhydrosilicate, portlandite, monosulfate, quartz, feldspar	Calcite, traces of calciumhydrosilicate, monosulfate, quartz, feldspar
SCC with crystallization additive	Laboratory conditions – reference environment	Calcite, calciumhydrosilicate, portlandite, monosulfate, quartz, feldspar	Calcite, calciumhydrosilicate, , portlandite, carbonate complexmonosulfate, quartz, feldspar
	Sulfur dioxide	Calcite, calciumhydrosilicate, portlandite, monosulfate, quartz, feldspar	Calcite, calciumhydrosilicate, traces of portlandite, monosulfate, quartz, feldspar
	Carbon dioxide	Calcite, calciumhydrosilicate, portlandite, monosulfate, quartz, feldspar	Calcite, calciumhydrosilicate, monosulfate, quartz, feldspar

Table 5. Results of pH determination

Mix designation	Exposure conditions	pH value 28 days of age	Time of exposure – one year
SCC without crystallization additive	Laboratory conditions – reference environment	12.1	11.9
	Sulfur dioxide	12.0	9.8
	Carbon dioxide	12.1	10.7
SCC with crystallization additive	Laboratory conditions – reference environment	12.2	12.1
	Sulfur dioxide	12.1	10.4
	Carbon dioxide	12.1	11.9

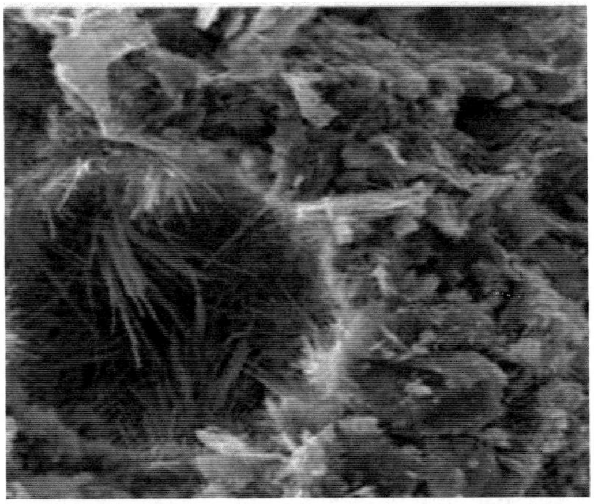

Figure 5. View of microstructure of SCC concrete without crystallization additive after exposure to sulfur dioxide. The massive formation of ettringite neoplasms in the microstructure of concrete is evident.

Figure 6. View of microstructure of SCC concrete modified with crystallization additive after exposure to sulfur dioxide. Clearly, the presence of the crystallization agent has significantly reduced the occurrence of corrosive neoplasms.

4. Conclusions

Negative effect of selected types of aggressive gases is particularly more noticeable on the tensile strength of the surface layers of concrete, less significantly in the case of compressive strength. The fact that the surface layers of concrete are primarily interfered has also been proved by physico-chemical analyzes. Of the tested types of aggressive gases, the most dynamic effect is caused by the sulfur dioxide. One year long exposure to this environment caused in the SCC concrete sample without the crystallization additive in virtually complete destruction of the surface layers of the concrete, the depth of concrete distortion was approximately 5 mm. The one year exposure in the sulfur dioxide resulted in a decrease in compressive strength of approximately 40%. The deterioration of mechanical properties of SCC concrete without crystallization additives was also confirmed by physical-chemical analyses. By these analyses was showed the occurrence of corrosion neoplasms, especially in the surface layers of concrete (formation of gypsum etc).

In the case of SCC concrete modified by the crystallization additive, the decrease of these parameters was significantly lower. The tensile strength of the surface layers in this case decreased by about 60% due to the one year exposure to sulfur dioxide. Decrease in compressive strength after one year long exposure did not exceed 20%. For concrete modified by the crystallization additive, the incidence of corrosion neoplasms in their microstructure was relatively low.

The impact of carbon dioxide on the properties of SCC concrete was lower than that of sulfur dioxide. For example, in the case of concrete tensile strength of the surface layers, SCC concrete without a crystallization additive there has been a relative increase. This effect is due to the formation of calcium carbonate crystals which compact the matrix structure of the surface layers. In the case of modified concrete, this effect was not detected. This observation, together with the results of physico-chemical analyses, indicates that the crystallization additives limit the penetration of carbon dioxide into the concrete structure.

Another aspect, which documents the positive effect of crystallization additives on the durability of self-compacting concrete, is parameter of pH value. It has been shown that for concrete modified by additives the decrease of concrete alkalinity is reduced. This is particularly positive in terms of the ability of the concrete to protect reinforcement against corrosion.

It can be stated that the crystallization additives positively affect the resistance of self-compacting concrete not only to the action of liquids but also to the action of aggressive gases.

Ackowledgements
This paper has been worked out with financial support of project GAČR 14-25504S "Research of Behaviour of Inorganic Matrix Composites Exposed to Extreme Conditions" and under the project No. LO1408 "AdMaS UP - Advanced Materials, Structures and Technologies", supported by Ministry of Education, Youth and Sports under the "National Sustainability Programme I".

References
[1] Teng L W, Huang R, Chen J, Cheng A and Hsu H M 2014 *Materials Basel* **7** 399-412
[2] Tang S W, Yao Y, Andrade C, Li Z J 2015 *Cem. Concr. Res.* **78** 143-154
[3] Collepardi M 2009 *Modern Concrete* (Prague: Information Centre CKAIT) p 342
[4] Matousek M, Drochytka R 1998 *Atmospheric Corrosion Of Concrete* (Prague: IKAS) p 171
[5] Hela R 2007 *Concrete Technology II* (Brno: Brno University of Technology) p 148
[6] Anil K 1999 *Cem. Concr. Compos.* **21** 235
[7] Paraic C R and O'Connor A 2016 *Contr. Build.Mater.* **120** 504-513

The microstructure of a cement composite based on a secondary raw materials

P Figala and R Drochytka

Faculty of Civil Engineering, VUT Brno, 331/95 Veveri, Brno 602 00, Czech Republic

Abstract. Waste management and its use as a secondary raw material is a matter that is being given increasing attention. Such raw materials also include waste products from the energy industry or municipal waste disposed of from the recycling process. This study explores the use of secondary raw materials in specific cement–based cement composite as a partial binder substitute. Selected secondary raw materials were pre–treated by grinding to a specific surface of 300 and 500 $m^2 \cdot kg^{-1}$. As a reference material, a commercially available mixture was selected to produce cement composites intended for the environment with high physical–mechanical demands. Cement was substituted by secondary raw materials at a rate of 20 and 40% in the reference recipe. Basic physical–mechanical characteristics such as compressive strength and tensile strength at bending were determined on standard 160x40x40 mm prisms. Strengths were determined after 3 and 28 days of ageing. In addition, the influence of binder substitute by the admixtures on the microstructure of cement stone was monitored by sample scanning using an RTG tomography. The positive effect of a part of the binder substitution by selected secondary raw materials on microstructure was proved in all tested samples. From the measured data, it is evident that the higher specific surface of the pre–treated admixtures did not have a significant positive effect either on strengths or the microstructure.

1. Introduction

One of the traditional secondary raw materials used in the construction industry is slag. It is an inorganic waste generated on grates or in furnaces by the combustion of coal, flammable slate and other solid fuels. Slag is used in building materials due to the pozzolanic character, but its greater use is limited by the large inhomogeneity and frequent occurrence of more pollutants. The use of slag as aggregate in concrete is influenced by several factors. The quality of slag is influenced by the rate of mechanical degrading pollution of organic or inorganic origin, quality and type of default fuel, combustion method, amount of fuel combustible residues, free CaO and MgO content, sulphur compounds and soluble salts [1][2].

The time and method of its deposition in temporary landfills also influences the characteristics of the slag. Prior to the use of slag, it is recommended to store it for a minimum of six months on free heaps where leaching of soluble salts occurs, which would otherwise cause blistering in concrete. Further, there is oxidation and leaching of sulphur compounds. Standard ČSN EN 450 provides a number of other criteria for slag as an admixture for concrete. For example, there is a requirement for a high SiO_2 content. In addition to the amorphous phase, the slag also contains mullite, anorite and melilite and lime clay rocks such as magnetite or gypsum. [3] Finally, the usability of the slag is limited by the maximum content of combustible substances (10%), the chemical character of the slag extract (pH in the range of 4–8) or by the specific activity (<150 $Bq \cdot kg^{-1}$) [1].

The use of silicate raw materials of amorphous character as building materials has also been the subject of studies for decades. In the alternative raw materials for concrete a whole range of glass types (solid substances with a high content of amorphous SiO_2) have a place [4]. The most commonly used is glass recyclate, especially container glass from bottles, but also glass from glasses or glass panes, both in clear and coloured form. It is usually made from 50% siliceous sand, 10% Na_2CO_3 (soda), 12% $CaCO_3$ (limestone), 18% crushed culled of waste glass and 4% other substances [5]. Not everything can be used for recycling, which is a major environmental problem around the world. For example, only glass sheets made of building demolition in the Czech Republic produce approximately 40 thousand tons per year.[6] Limiting the recyclability of waste glass causes different colouring, impurities, etc. The most widely produced glass type is soda–lime glass, accounting for almost 80% of all glass waste (63% of white coloured glass) [7] [8].

When using glass recyclate in cement composites, it is necessary to consider the possible occurrence of an alkali–silica reaction (ASR) between the used glass (source of active SiO_2) and alkalis. Due to the amorphous structure of glass, significant pozzolanic activity is also expected, which leads to the consumption of portlandite during the formation of neologisms. These are very similar to CSH phases and result from the hydration of Portland cement. These neologisms envelop the reaction edges of the glass grains and reduce the monosulfate level. This phenomenon reduces the risk of ASR occurrence and is closely related to the fineness of glass recyclate grinding. Various studies state different limits of glass grading. In general any risk of ASR is eliminated if the average size of particles is below 300 μm [7] [8] [9].

2. Material composition

2.1. Characteristics of secondary raw materials

Thermal power plant slag (further Slag or EOS) was delivered in a fraction of 0–8 mm. Grounded recyclate of container glass was derived from controlled production and was delivered in a fraction 0–32 mm. Pre–treatment of secondary raw materials for research purposes began by adjusting them to a 0–4 mm fraction using a jaw crusher. In the next step, the secondary raw materials were adjusted to predetermined specific surfaces: variant A = 300 $m^2 \cdot kg^{-1}$ (± 50 $m^2 \cdot kg^{-1}$) and B = 500 $m^2 \cdot kg^{-1}$ (± 50 $m^2 \cdot kg^{-1}$). The grinding drum for discontinuous grinding was selected for grinding. For minimization of the variables the mill rotation speed and emptying times were identical in all cases. For all samples of secondary raw materials, the following physical characteristics were verified density with a helium pycnometer, specific surface with a permeable method according to Blain's device and granulometry using laser diffraction analysis (table 2).

Table 1. Physical characteristics of pre–treated secondary raw materials.

Raw material	Designation	Density ($kg \cdot m^{-3}$)	Variant	Specific surface ($m^2 \cdot kg^{-1}$)
Slag	EOS	2610	A	298
			B	530
Waste glass	Glass	2560	A	300
			B	541

Table 2. Granulometry of pre–treated secondary raw materials.

Raw material	Variant	%>0,063 mm	0,01>%> 0,006 mm	%< 0,01 mm
Slag	A	42.83	55.63	1.48
	B	2.44	91.49	6.07
Waste glass	A	44.55	54.04	1.41
	B	12.47	77.89	9.64

At the same time, pre–treated grain microscopy was monitored by SEM analysis (Figure 1–3).

Figure 1. Microscopic images of cement particles (CEM I).

Figure 2. Microscopic images of slag particles (EOS).

Figure 3. Microscopic images of waste glass particles (Glass).

2.2. Test recipes

As a reference substance (hereafter referred to as PH3), a high–strength floor screed from the renowned building materials manufacturer was chosen for research purposes (table 3). Substance PH3 was delivered as a dry homogeneous mixture and its material composition was based on a commercially available product. For the purpose of research, the manufacturer reduced the amount of CEM I 42.5 R Portland cement in dry mixture to 50%, so the other 50% could be substituted by secondary raw material. The reference dry mixture (hereinafter referred to as REF) was composed of the PH3 substance and the missing binder and was homogenized with a vertical screw homogenizer. The resulting composition of substance REF was based on the manufacturers prescribed ratio: 5:1:1 (PH3:CEMI42.5R:water). The water coefficient for the reference substance was determined by the manufacturer for 0.17 kg of water/1 kg of dry mixture with 100% dose of cement. The methodology of the dry mixture homogenization as well as the methodology of production and storage of prisms, which is not excluded by the technical sheets of the PH3 screed, was determined for research purposes. The standardized beams of dimensions 160×40×40 mm were selected as ideal test bodies for research purposes.

Table 3. Approximate material composition of the reference substance.

Material	The proportion in the mixture (%)
Portland cement CEM I	37–46
Coarse aggregate 4–8 mm	14–20
Fine aggregate 0–4 mm	24–38
Pozzolanic admixture	3.5–4.4
Ground limestone	3–5
Mixture of additives	1.15–1.95

The other test recipes were based on the modified composition of the reference recipe. The binder was partially substituted by grounded slag (recipe marked with EOS) and glass recyclate (marking with Glass) in a 20 and 40% rate of the total Portland cement CEM I 42.5 R dose. Overall, eight recipes were designed and manufactured containing two different admixtures, which were modified to two finenesses of grinding. Individually modified mixtures weighing 20 kg were mixed with a screw homogenizer and the homogenization time was set to 30 minutes to ensure perfect mixing.

In the next step, the optimal water coefficient was determined for each test recipe by the consistency test on a jogging table (CSN EN 1015–3). The water coefficient was chosen so that the consistency of the fresh substance was always the equivalent of REF. Subsequently, standardized prisms of 160×40×40 mm were made and stored in a laboratory environment under PE foil and treated regularly with water. Test recipes has been marked as a follow-up to the secondary raw material used with a definite specific surface.

3. Evaluation of experimental testing

When designing the quantity of mixing water for individual recipes, it was found that for all recipes with the addition of secondary raw materials, the consumption of mixed water was reduced while maintaining the reference consistency. The major influence was the different water absorption of cement and alternative raw materials; the higher portion of the glass phase in the slag and in the glass was also important. Despite the lower water coefficient and the similar consistency set on the jogging table, the workability of substances with grounded glass was significantly better than the reference substance or the substances with slag. The effect was not only water absorption of glass but also a suitable form of the grains leading to a simpler compaction of the substance. The used screed was characterized by a considerable development of the hydrating temperatures after the beginning of the hydrating processes. This phenomenon was probably caused by the hydrating of the contained aluminous cement. Due to the significant development of hydrating heat and a very rapid increase of the initial strengths, it was suggested that the microstructure of the cement stone and the influence of the secondary raw materials on the development of the hydrating cracks will be studied on samples of selected recipes.

3.1. Physical–mechanical characteristics

Compressive strength is the most significant material characteristic of cement composites. The production of test bodies and their testing complied with the standard CSN EN 13892–2, resp. standard for the testing of screed systems. The effect of the cement substitution rate by individual secondary raw materials on short–term strength after three days maturation and on long–term strength after 28 days maturation was monitored.

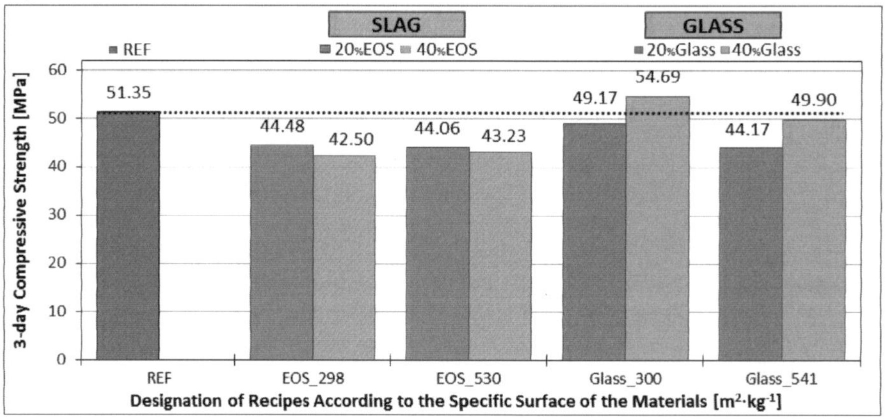

Figure 4. Compressive strength after three days of maturation.

Determination of early compressive strengths has shown that it is possible to replace larger amounts of cement in the screed while achieving similar or even higher strengths. These strengths (figure 4) were achieved with the use of grounded glass at a recipe of 40% Glass300.

For long–term compressive strengths (figure 5), the dependence between the lower specific surface of secondary raw materials (300 $m^2 \cdot kg^{-1}$) and higher compressive strengths was proven. The highest strengths after 28 days of maturation were achieved with 20%EOS298 recipes (113% of reference strength) and 40%Glass300 (112% of reference strength).

Long-term compressive strengths were determined also after 90 days of maturation and the typical pozzolanic trend of strength growth was confirmed. But the indicative properties of the specific material are after 28 days of maturation, so the results after 90 days are not essential and are not part of this article.

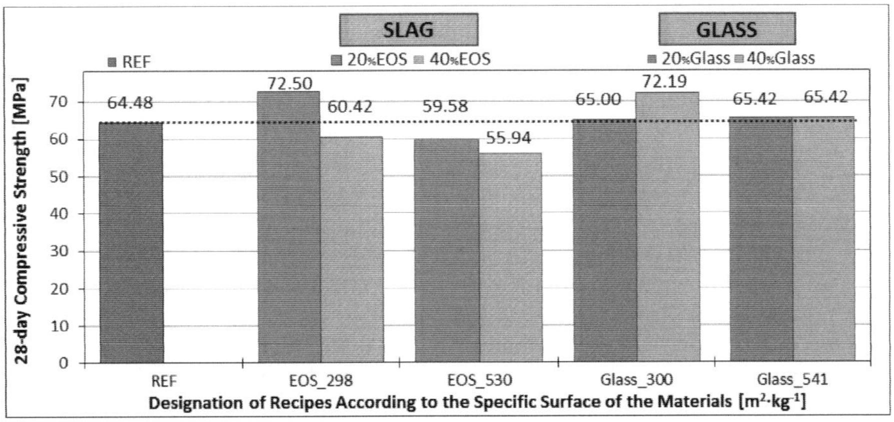

Figure 5. Compressive strength after 28 days of maturation.

3.2. Microstructure

The internal microstructure of prisms was studied using an X–ray (RTG) tomography. It is a method based on the radiography of the sample from a variety of angles in one plane, resulting in several hundred projections. These are subsequently reconstructed, and the output is a surface section of the examined object. Based on the evaluation of the strengths determined for the individual test recipes, two representative recipes for the study of the cement stone microstructure were selected.

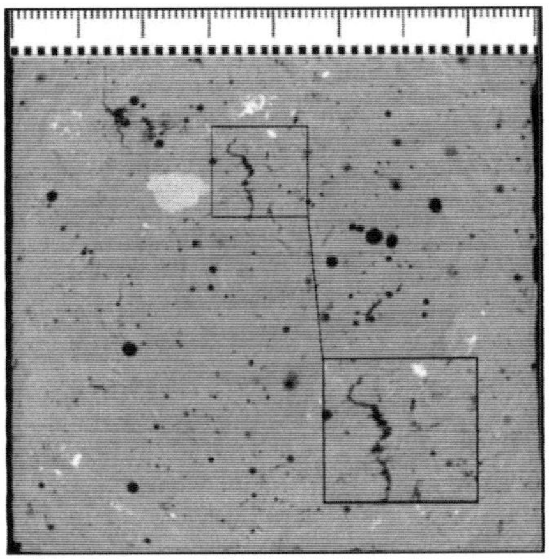

Figure 6. A tomographic image of the test body structure REF.

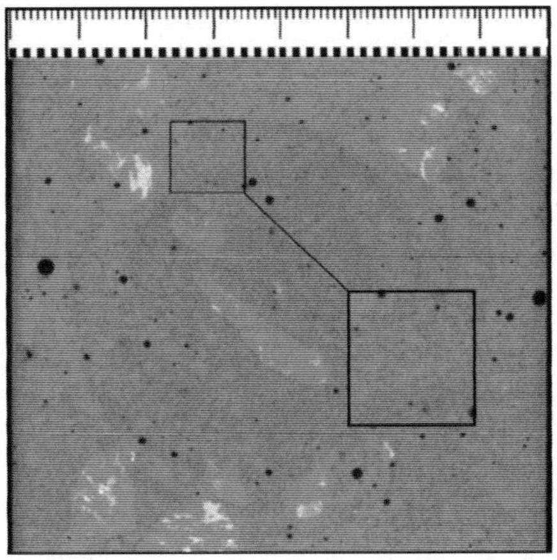

Figure 7. A tomographic image of the test body structure 40%Glass300.

There were a large number of shrinkage cracks in the REF reference screed (Figure), as demonstrated by scanning the prisms with a tomography. This confirmed a research concern, namely that due to the high hydrating temperatures developed at the early stage of the PH3 screed maturation, a disruption of the cement stone integrity occurs. By studying the microstructure of recipe bodies 40%Glass300 (Figure 7), a significant contribution of grounded glass as a raw material for the floor screed has been proven. The premise that the glass calms down the hydrating processes was confirmed. According to images of the screed internal structure, almost no shrinkage cracks occurred. Simultaneously, the number of larger cracks occurring by the processing of the fresh screed was significantly reduced in the prisms.

4. Conclusion

Each figure should have a brief caption describing it and, if necessary, a key to interpret the various lines and symbols on the figure.

Based on the results of experimental research it can be stated that:

- Substitution of the binder with finely grounded slag and glass has a positive effect on the workability of all modified substances regardless of the substitution rate or the admixtures fineness of grinding.
- The dependence between the lower specific surface of used secondary materials and the higher compressive strengths has been proven. The effect of cement substitution by secondary raw materials on compressive strength increased with maturation time and was more pronounced in milled glass recipes due to the higher pozzolanic activity of the admixture. Higher compressive strengths of up to 12% compared to the reference recipe were achieved.
- Due to the time and economic demands of concrete admixtures grinding, the optimal specific surface was chosen as 300 $m^2 \cdot kg^{-1}$. With this fineness of grinding, there is also no risk of an alkali–silica reaction occurring in the grounded glass.
- The lower the specific surface of the secondary raw materials leads to the higher strengths from the insights of the characteristic properties of the reference screed. These secondary raw materials were less reactive and contributed to slowing down hydration processes and the development of smaller hydration cracks.
- By studying of the screed microstructure with a partial substitution rate of cement by secondary raw materials, the positive influence of all alternative raw materials was proven. The best results were achieved with a higher rate of cement substitution with grounded sodium–calcium glass. The use of grounded glass recyclate of a specific surface of 300 $m^2 \cdot kg^{-1}$ influenced the microstructure of the screed very positively.

Acknowledgments

This paper has been worked out with financial support from the budget under the project FAST–S–18–5406, “Study of advanced silicate composites structure with emphasis on the characterization of secondary raw materials influence” and under the project No. LO1408 "AdMaS UP – Advanced Materials, Structures and Technologies", supported by Ministry of Education, Youth and Sports under the “National Sustainability Programme I”.

References

[1] Pytlik P 2000 *Concrete Technology* (Brno: VUTIUM) p 346
[2] Drochytka R and Matulova P 2006 *Lightweight Building Materials: Modul M01* (Brno: VUT Brno) p 203
[3] Smarzewski P and Barnat–Hunek D 2016 *Constr. Build. Mater.* **121** 9–17
[4] Melichar T, Dufka A and Bydzovsky J 2017 *Materiali in tehnologije* **51** 243–249
[5] Radejova L 2013 *Development of new industrial floors using secondary raw materials* Dissertation thesis (Brno: VUT Brno) p 169
[6] Secondary raw materials policy of the Czech Republic approved by the Government of the Czech Republic on 15.9.2014 vol 1 (Praha: MPO)
[7] Shi C and Zheng K 2007 *Resources, Conservation and Recycling* **52** 234–247
[8] Lee H, Hanif A and Usman M 2018 *J. Clean. Prod.* **170** 683–693
[9] Zheng K 2016 *Cement Concrete Comp.* **67** 30–38

IOP Conf. Series: Materials Science and Engineering **385** (2018) 012013 doi:10.1088/1757-899X/385/1/012013

Influence of moisture content of samples on the depth of penetration of water under pressure

R Figmig and M Kovac

Department of Material Engineering, Institute of Environmental Engineering, Faculty of Civil Engineering, Technical University of Kosice, Vysokoskolska 4, 042 00 Kosice, Slovakia

Abstract. The article deals with the impact of samples moisture content on the results of the depth penetration of water under pressure according to EN 12390-8, considering the fact, that technical standard does not prescribe definite condition of the samples prior testing. To obtain results of the water penetration depth, samples from eight different batches (strength classes from C 12/15 to C 55/67) were prepared. The experiment simulated two boundary conditions: saturated samples were tested immediately after taking from the water. This procedure of samples treatment was marked as the reference (REF). The second one: oven dried samples were tested after cooling to room temperature. This procedure of samples treatment was marked as the comparative (COMP). The results showed an increased tendency in the penetration depth when the samples were oven dried compared to the reference method. There is also discussed the substance, aim and relevance of the technical standards in connection with practical use.

1. Introduction

Durability is one of the essential concrete properties [1]. Basic durability parameter determined on the hardened concrete is watertightness that defines the resistance of concrete against penetration of water [2]. Watertightness is in the close relationship with the permeability of concrete and plays important role in the durability of concrete in sense, that prevent the concrete structure from penetration groundwater under pressure, that may contain chemical compounds, causing its deterioration [3-4].

At present, the penetration test according to EN standard [5] is required to declare the watertightness of concrete. For this parameter, an achievement the depth of penetration of water under pressure must be lower or equal to 50 mm of the highest measured result from the set of testing samples. Watertightness is an obligatory parameter of environmental exposure classes XF2, XF3, XF4, XA1, XA2 and XA3 in the Slovak Republic [6]. More strictly requirements are defined in the Czech Republic, where the obligatory depth of water penetration limits are 50 mm for classes XC3, XC4, XD1, XD2, XF1, XA1, 35mm for XF2 and XA2 and 20mm for XD3, XF3, XF4 and XA3 [7].

Based on mentioned above, it can be declared, that unified and definite complex testing procedure for the determination of a depth of penetration of water under pressure as classification criteria of watertightness should be included in the harmonized technical standard. It ensures precision (reproductivity, repeatability) and comparability of measuring. According to the authors' opinion, the technical standard should prescribe methods of testing concrete (but not only), that simulate the real behavior of concrete in the environment. From the authors' experience, some technical standards determine procedures and requirements, which are not relative to industrial practice. As examples can be noted optimal grain curves of aggregate according to STN EN 206+A1 [8], which do not take into account the aggregate properties as a source, type, grain shape etc. [9]; water/binder ratio requirements (criticized e.g. by prof. Hela on the Concrete 2016 conference, Oct. 2016, Štrbské Pleso, Slovakia; also [10]); or procedure determination of water needed for standard consistency of cement paste, test by EN 196- 3 [11], which determines the temperature of testing cement at 20°C, what is temperature, which is rarely obtained when delivered from cement plant, what subsequently caused increased water need.

Another example of uncertainty is the standard EN 12390-8 which does not prescribe the conditions of samples prior the testing exactly. The common practice in performing the test is, testing the samples taken out of the water 24 hours before the test. However, because not exact standard description, there is the risk that non-uniform state of samples will affect the results of depth of penetration. In the following experiment, the influence of moisture content of the samples on the results of the depth of penetration of water under pressure is demonstrated. The purpose of the experiment is to point out the differences in the results regarding the extreme values of specimens' saturation.

2. Materials and methods

For the experiment, 8 samplings (divided into the 1st and 2nd sampling performed in the different interval) from the random and independent batches of various concrete strength classes of standard cement concrete were performed. The sampling covered various strength classes of concrete to obtain various results in the depth of water penetration. Concrete samples were divided in 4 groups regarding the strength class. Concretes being in the 1st and 2nd group were intended as common ready-mixed concretes (strength class from C 12/15 to C 25/30). Concretes being in the 3rd and 4th group were intended as watertight precast concrete (strength class from C 35/45 to C 55/67). Considering the purpose of the experiment, it is not necessary to give the particular mix design. The samples (cubes 150 x 150 x 150 mm) were taken from 1.0 m^3, respectively 1.5 m^3 batch of concrete. For each strength class 6 samples were taken in the 1st and also in the 2nd sampling. For each of 2 testing methods, 3 samples from each concrete batch were tested. Sampling, preparation, and curing were performed according to EN technical standards as standard procedures. Boundary conditions of samples moisture content were intended as fully saturated and oven dried on the other hand.

First set of samples (of minimal age 28 days) was tested right after samples were taken from the water. This procedure was marked as reference method (REF). This method is currently almost the same compared with the usual one in laboratory praxis. Usual praxis resides in taking samples from the water 1 day before the water penetration test. Second set of samples was tested in the oven dried conditions. Samples were dried at 105 °C ± 5 °C until constant mass of samples was achieved. This procedure was marked as comparative (COMP). Comparative procedure of samples treatment represents an opposite state of the samples moistness state compared with the reference one.

The depth of penetration of water under pressure as such was performed according to standard EN 12390-8 [5] (the water under a pressure of 500 ± 50 kPa was applied for 72 ± 2 h). No deviations were observed during performing the water penetration tests.

3. Results

In this chapter, the results of determining the depth of penetration of water under pressure are given. In Table 1 and Figure 1, results of samples from 1st sampling are given and in Table 2 and Figure 2 results of samples from 2nd sampling are given. In both, 1st and also 2nd sampling the decreasing tendency in depth of water penetration along with increasing strength class of concrete was achieved what is in accordance with theoretical assumption. All samples from 1st sampling treated by REF procedure complied the requirement for water tightness and so, achieved the lower depth of water penetration than 50 mm.

Samples from 1st sampling treated by COMP procedure achieved multiple higher depth of water penetration compared with the samples treated by REF procedure. Only samples from 3rd group complied the requirement for water tightness and so, achieved the lower depth of water penetration than 50 mm. Using comparative testing method (COMP), multiple increments of the depth of water penetration was observed in all tested samples. The same behavior was observed during testing the samples from 2nd sampling. In this case, all samples (excluded 1st group) treated by REF procedure complied the requirement for water tightness. If more strict requirements were considered according to [7], none of tested concrete samples did comply requirements for water tightness of concrete. Samples from 2nd sampling treated by COMP procedure also achieved multiply higher depth of water penetration compared with the samples treated by REF procedure.

Differences between selected testing methods are very high and they point out the importance of uniform standard procedure and sample preparation. Significant differences in the results of determination the depth of water penetration under pressure are illustrated also in Figure 3.

Table 1. Results of determining the depth of penetration of water under a pressure of samples from 1st sampling.

Testing method	1st sampling C 12/15	C 25/30	C 40/50	C 55/67
	25	8	3	5
REF [mm]	25	10	5	8
	45	12	10	9
	75	20	0	15
COMP [mm]	82	35	30	54
	85	95	35	60

Table 2. Results of determining the depth of penetration of water under a pressure of samples from 2nd sampling.

Testing method	2nd sampling C 16/20	C 25/30	C 35/45	C 55/67
	35	40	22	10
REF [mm]	48	41	23	14
	53	42	39	21
	114	65	64	52
COMP [mm]	118	84	95	54
	123	95	106	55

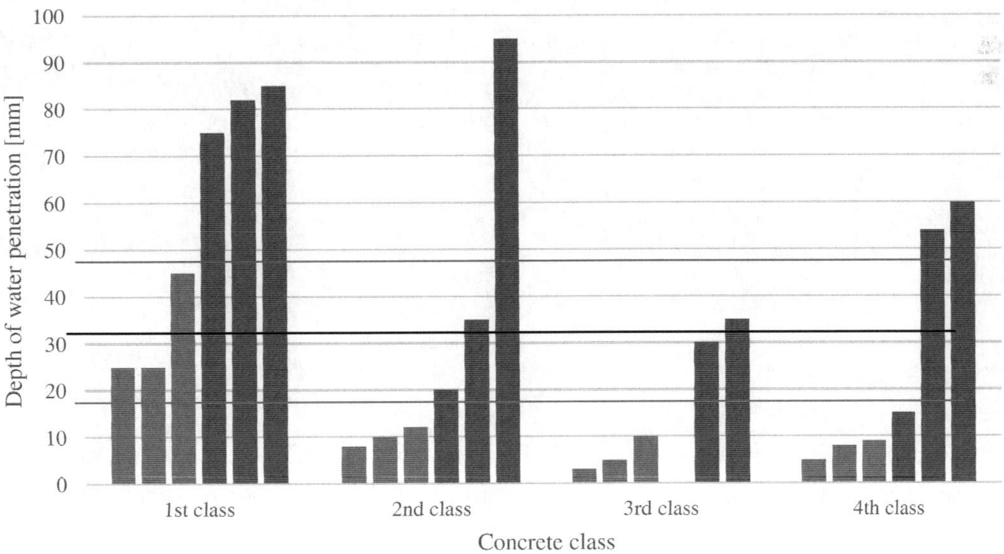

Figure 1. Depth of penetration of water; 1st sampling, Blue: REF, Red: COMP, Limits for maximal depth of water penetration according to [7].

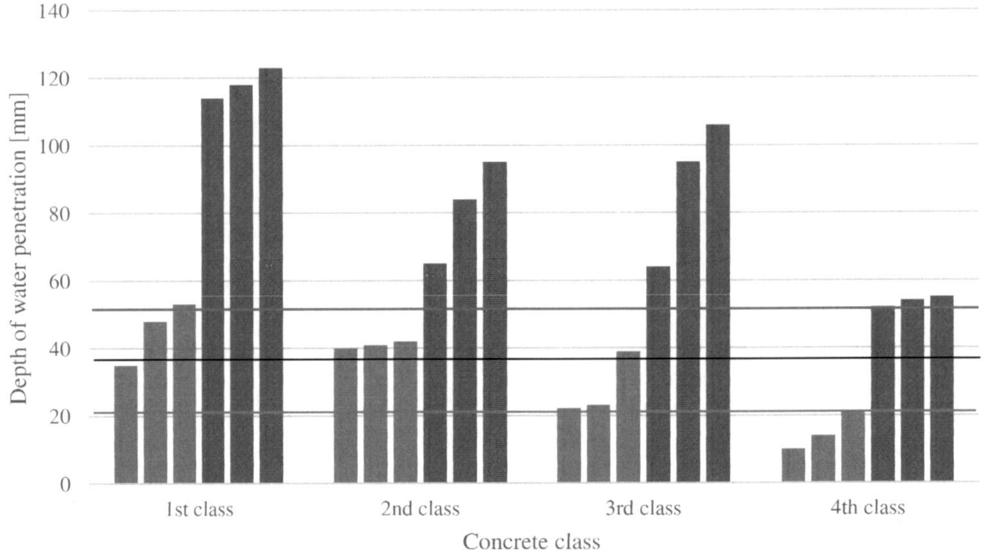

Figure 2. Depth of penetration of water; 2st sampling, Blue: REF, Red: COMP, Limits for maximal depth of water penetration according to [7].

Figure 3. Splitted samples after water penetration test, red line marks the maximum depth of water penetration, samples treated by COMP procedure are on the left, and samples treated by REF procedure are on the right.

4. Conclusion

Based on the results, it can be deduced, that the moisture content of sample has a crucial impact on the result of a watertightness test. Results obtained by testing samples treated by REF procedure are several times lower than that of obtained by samples treated by COMP procedure. In the case, when this parameter is demanding, moisture content has an impact on results in the sense of meeting or failing to meet the standard requirements. The query is, which of the state, saturated or dried, simulates the real concrete condition and environment the best. The saturated state is different considering various pore-capillary systems in connection with different concrete mix designs. Absolutely dryness will rarely be reached in real condition. Based on the authors´ experience and also in according to [12], oven-drying causes change in concrete micro-structure (micro-cracking, particular heat material degradation) what leads to higher permeability. The solution of described issue may lay in measuring the real casting concrete moisture condition in particular environment. Then the specimens should be prepared (cured) in relation to those results.

In authors' opinion it would be more suitable to express the results by maximal value of depth of water penetration (covering extreme value of a set of samples) and also by average value of depth of water penetration (covering particular samples deviation, which can be caused by unequal specimens making).

Acknowledgment
This work was supported by the Slovak Grant Agency for Science [1/0648/17] and by the project of Slovak Scientific Grant Agency VEGA [grant number 1/0524/18].

References

[1] Neville A M 1995 *Properties of Concrete* (London: Longman) p 482-497
[2] Schlumpf J, Bicher B and Schwoon O 2012 *Sika Concrete Handbook* (Zürich: Sika Services AG) p 126-129
[3] Zhang H 2011 *Building Materials in Civil Engineering* (Sawstone: Woodhead Publishing) p 113
[4] Zongjin L 2011 *Advanced Concrete Technology* (Hoboken: John Wiley & Sons) p 217-221
[5] EN 12390-8: 2011 *Testing hardened concrete* Part 8: Depth of penetration of water under pressure
[6] STN EN 206/NA: 2015 *Concrete. Specification, performance, production and conformity*
[7] ČSN P 73 2404: 2016 *Concrete - Specification, performance, production and conformity - Additional information*
[8] STN EN 206+A1: 2017 *Concrete. Specification, performance, production and conformity*
[9] Figmig R, Kováč M and Sicakova A 2017 *The influence of grain aggregate on the consistency of freshness concrete* (Bratislava: SAVT) p 35-44 (in Slovak)
[10] Wasserman R, Katz A and Bentur A 2009 *Mater. Struct.* **42** 973-982
[11] EN 196-3: 2017 *Methods of testing cement - Part 3: Determination of setting times and soundness*
[12] Hager I 2013 *Bull. Pol. Ac.: Tech.* **61** 1-10

IOP Conf. Series: Materials Science and Engineering **385** (2018) 012014 doi:10.1088/1757-899X/385/1/012014

Study of hydration process of ternesite clinker

M Fridrichova, D Gazdic, K Dvorak, J Mokra and K Kulisek

Faculty of Civil Engineering, Brno University of Technology, Veveri 331/95, 602 00 Brno, Czech Republic

Abstract. Currently ternesite, $Ca_5(SiO_4)_2SO_4$, is used in the cement production, is one of two key phases of calcium sulfoaluminate cement. Some investigators claim that this phase is nearly inert or very low reactive and it hydrates only after later age of hydration. In order to improve knowledge in this field, process of hydration of neat ternesite clinker under specific conditions of exposure was observed in this study. Ternesite was prepared by firing of raw meal consisting of high calcium limestone, microsilica and calcium sulfate hydrate. The resulting ternesite clinker was subsequently hydrated in four storage environments. Kinetics of hydration process was monitored in water and in environment of saturated carbonic acid, at two different temperatures of 5 and 40 °C. Mineralogical composition of hydrated samples was analyzed by X-Ray diffraction analysis (XRD) and by differential thermal analysis (DTA) in particular ages of hydration. Based on the results, it can be claimed that rate of decomposition of ternesite was related to conditions of storage. The rate of decomposition was significantly accelerated by the environment of carbonic acid and also by the elevated temperature. Presence of calcium carbonates in mineralogical forms of calcite, vaterite and aragonite, coupled with gypsum and thaumasite was identified.

1. Introduction

Ternesite, also known as the sulfo-spurrite [1], is a colorless to light blue mineral with a composition of $Ca_5(SiO_4)_2(SO_4)$, shorthand $C_5S_2\bar{S}$. It can be found in xonotlits rich in calcium [2]. This mineral is orthorhombic and without cleavage, creates long prismatic crystals in radial structure [3].

Ternesite occurs in cement kiln coating in small quantities where is green due to addition of iron ions [3,4]. Shen et al. [5] claims that ternesite is formed as an intermediate phase approximately at temperature of 900 °C, authors [3, 6] state temperature 1100 °C. To create ternesite, respectively stabilize it at temperature higher than 1200 °C (up to 1250 °C), a mineralizing agent has to be taken into consideration during the firing – phosphates or fluorides, or ternesite has to be produced in two steps terminated by slow cooling to 800 °C [6]. That is for the reason ternesite is formed mainly at temperature range between 1200 °C and 950 °C [4]. Ternesite is released into decomposition products belite and anhydrite [1] at temperatures higher than 1200°C, Ukrainczyk et al. [7] state the temperature range 1200-1280 °C. Irran et al. [3] claims decomposition occurs at temperature higher 1150 °C or higher than 1300 °C in open system (platinum crucibles) respectively.

In general, ternesite is considered to be almost inert or to react only at later stages of hydration – is not hydraulically active [1,2,4,8-10]. But it immediately reacts with water in the presence of alumina ions (e. g. coming from blast furnace slag, metakaolin, anhydrite) and forms solid framework [9,11].

In recent years, ternesite is mainly utilized in cement industry due to reduction of CO_2 emissions. Ternesite is a main phase in ternesite calcium-sulphoaluminate clinker/cement and an additive in calcium-sulphoaluminate (ferrite) (belite) clinker/cement, CSA(F)(B) [5,6,12]. Researchers from Heidelberg Technology Centre in Germany registered patents for the Belite-Calciumsulfoaluminate-

Ternesite (BCT) clinker [2,6,8,10]. Not only belite, but also ternesite is the main reactive phase in this clinker [1, 4]. BCT clinker reacts with water to give hydration products, such as stratlingite, monosulfoaluminate, silicon hydrogarnets and C-S-H phases [6]. Calcium-sulfoaluminate (CSA) cements are fired at lower temperatures, around 1250 °C, and generate up to 30 % less emissions of CO_2 than ordinary Portland cements, which are fired at temperatures of approximately 1450 °C. By-products such as fly ash or blast furnace slag can be used as source of aluminum instead of expensive natural bauxite. Moreover, these cements are easily to grind, therefore the energy demand for grinding and milling processes is reduced [12].

The aim of this work was synthesis of ternesite and subsequent monitoring of the hydration process to improve knowledge in this field. In order to follow the progress of hydration ability, hydration of ternesite in four storage environments, namely in storage in water and in the presence of carbonic acid and at low and higher temperature in both cases was observed.

2. Methodology

The experiment was based on the study of hydration of ternesite clinker in four different storage environments. Raw meal for ternesite was synthesized from three components: high calcium limestone – calcium carbonate, $CaCO_3$ (purity 99.00 %), gypsum – industry by-product from flue gas desulfurization, $CaSO_4{\cdot}2H_2O$, (purity 95.51 %) and microsilica, SiO_2, (purity 98.00 %). Raw materials were dried to a constant weight (limestone and microsilica at temperature of 105°C, gypsum at 40 °C) using circulating oven BINDER ED APT line II with forced circulation. The dosage of respective oxides was based on stoichiometric proportions of particular components in ternesite, $5CaO{\cdot}2SiO_2{\cdot}SO_3$, namely 5 mol of CaO, 2 mol of microsilica and 1 mol of calcium sulfate hydrate, specifically (taking purities into consideration). The dosage was performed with accuracy of ± 0.01g. Subsequently, the raw materials were blended with isopropylalcohol into consistency of suspension which was crushed and homogenized in a planetary mill for 20 minutes by 300 rpm using FRITSCH Pulverisette 6 with stainless steel grinding bodies. The resulting slurry was completely dried at temperature of 40 °C to a constant weight. Approximately 25 g of dry powder mix was poured into 5 platinum crucibles and manually compacted. The firing process was performed in a laboratory furnace CLASIC 2018S CLARE 4.0 at temperature of 1150 °C at heating rate of 10 °C/min, soak 1 hour and next quenched by air flow. Ternesite clinker was ground in the planetary mill for 10 minutes by 500 rpm before the hydration. The storage environments of exposure were chosen with respect to possible processes, as follows. One half of samples was placed into the environment of saturated carbonic acid with the aim to support supposed formation of thaumasite, into whose structure is built the carbonate unit. One half of these samples was put into CO_2 saturated water and the second half of these samples into pure water.

Hydration in water was executed that ground ternesite was poured into a bottle of volume of 1l which was filled by water of temperature of around 7°C. Hydration in the presence of saturated carbonic acid was executed in a pressure bottle and water was saturated by carbon dioxide. All the samples were placed into the storage environments up to 90 days:

- Saturated carbonic acid, temperature 5 °C – CO_2/5°C.
- Saturated carbonic acid, temperature 40 °C – CO_2/40°C.
- Water, temperature 5 °C – w/5°C.
- Water, temperature 40 °C – w/40°C.

Mineralogical composition was regularly characterized by XRD using diffractometer Empyrean PANalytical with Cu K-alpha cathode (λ = 1.54184 Å) and PIXcel3D detector. The testing was supplemented by quantitative TG/DTA analysis using TGA/DTA851 Mettler Toledo (the heating rate 10 °C/min, open air conditions). Samples of hydrated ternesite clinker were ground in a XRD-Mill McCrone (Micronising mill) before testing. The measurement of XRD analysis was carried out under conditions, as follows: generator settings were 45 kV and 40 mA; scan range 5–80°, step size of 0.0131°, 255 active channels. The XRD patterns were analyzed by software HighScore Plus with PDF-2 powder diffraction files database.

3. Results

Figure 1 to 4 show the XRD patterns of the hydrated ternesite clinker exposed to the individual storage environments.

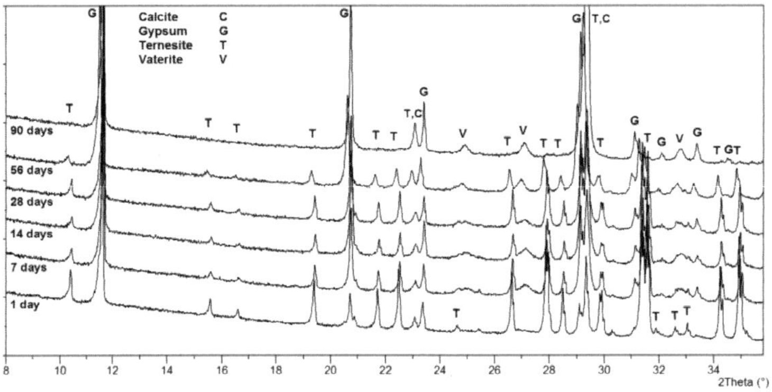

Figure 1. XRD patterns of hydrated ternesite clinker, CO_2/5°C.

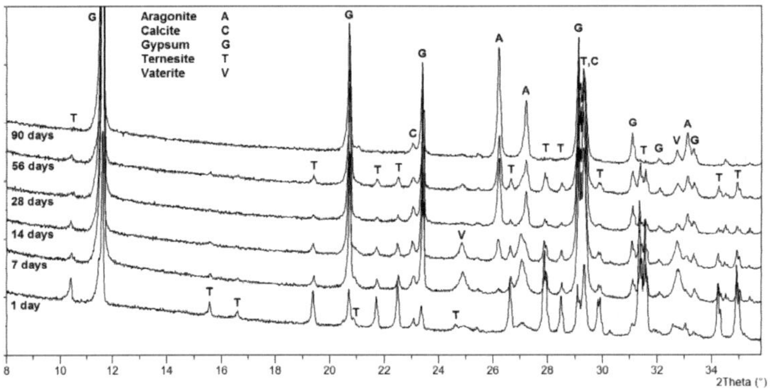

Figure 2. XRD patterns of hydrated ternesite clinker, CO_2/40°C.

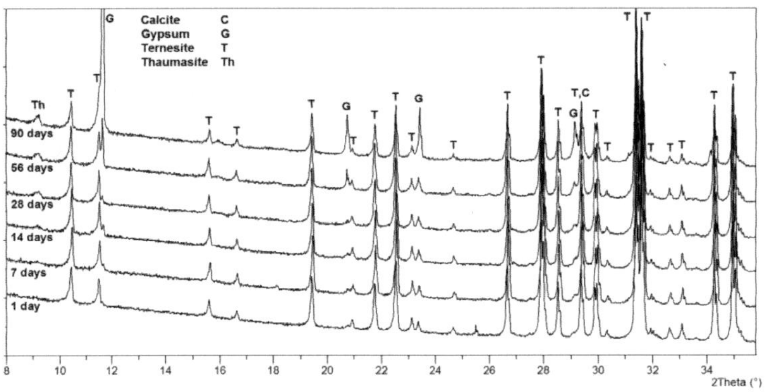

Figure 3. XRD patterns of hydrated ternesite clinker, w/5°C.

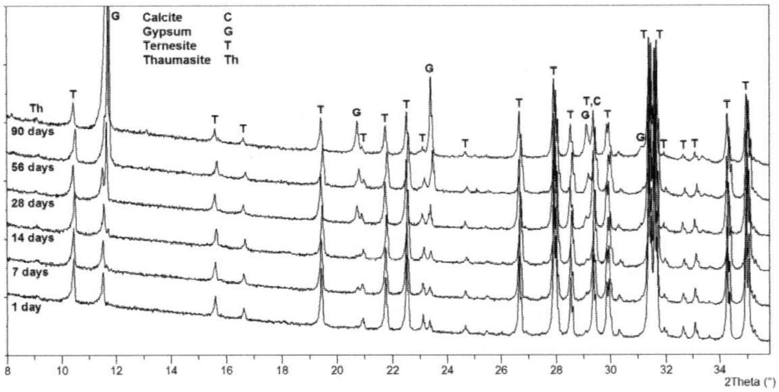

Figure 4. XRD patterns of hydrated ternesite clinker, w/40°C.

Table 1 presents the minerals identified in the XRD patterns of hydrated ternesite clinker.

Table 1. List of identified minerals in samples.

Mineral	Marking	ICDD-PDF-card
Aragonite, $CaCO_3$	A	00-041-1475
Calcite, $CaCO_3$	C	00-005-0586
Gypsum, $CaSO_4 \cdot 2H_2O$	G	00-033-0311
Ternesite, $5CaO \cdot 2SiO_2 \cdot SO_3$	T	00-049-1807
Thaumasite, $CaSiO_3 \cdot CaCO_3 \cdot CaSO_4 \cdot 15H_2O$	Th	00-046-1360
Vaterite, $CaCO_3$	V	01-072-0506

TG and DTA curves for individual samples of clinker in specific ages of hydration were obtained by differential thermal analysis. From the course of curves, presence of calcium carbonate in the form of calcite and mildly crystalline phases of vaterite and aragonite was identified, gypsum and thaumasite were also found. Content of gypsum and also the sum of CaO coming from calcium hydroxide and carbonate (marked as $\sum CaO$) were calculated. From the quantification, the figure 5 and 6 graphically provide the results depending on the age of hydration and the storage environment.

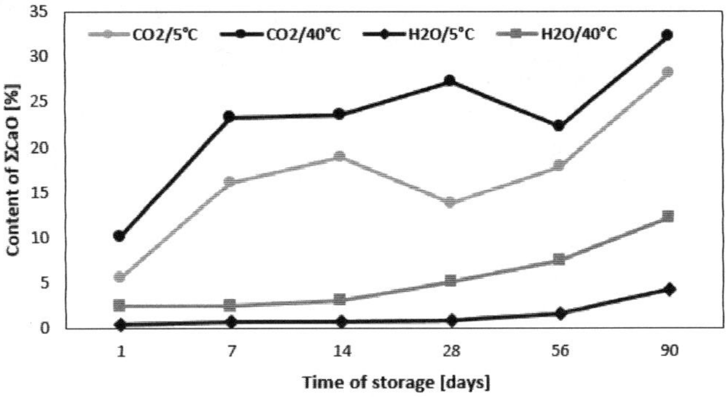

Figure 5. Content of $\sum CaO$.

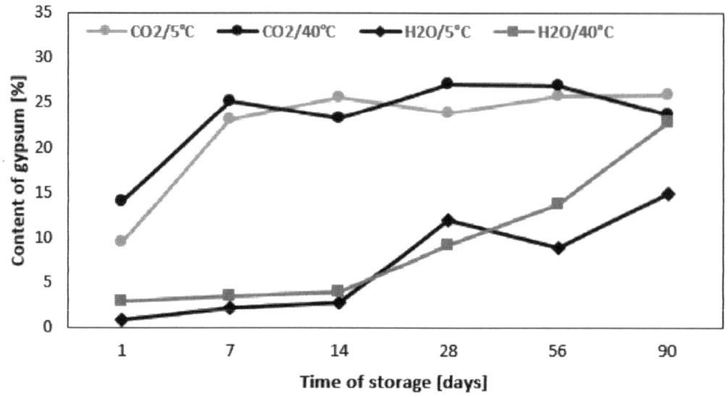

Figure 6. Content of gypsum.

4. Discussion

It is quoted in the literature [1,2,4,8-10] that ternesite is nearly inert when mixed with water, or reacts only at later stages of hydration. According to the experiment performed on the clear ternesite clinker using XRD, decomposition of ternesite already at an age of 1 day of hydration was observed. In the environment of carbonic acid, the rate of decomposition of ternesite was very high. Gypsum, calcite or vaterite and aragonite were identified as the products of decomposition. No new products of hydration of ternesite with water were observed in this environment. Assumed formation of thaumasite was not recorded. Regarding the temperature of storage, the rate of decomposition at the temperature of 40 °C was greater than at 5 °C. No more lines of ternesite were found at an age of 90 days of hydration, complete decomposition occurred.

The decomposition of ternesite in the water took place substantially slowly in comparison to the former conditions. As a consequence of low rate of decomposition of ternesite clinker, gypsum and calcite were formed very gently, the diffraction lines of vaterite and aragonite were not observed. Except that, basal diffraction line of silicate mineral thaumasite, $CaSiO_3 \cdot CaCO_3 \cdot CaSO_4 \cdot 15H_2O$, was observed in samples at an age of 14 days. This process was well apparent for samples exposed to the temperature of 5 °C, this line was marginal for samples from the temperature of 40 °C. Authors [14, 15] describe the formation of thaumasite and the influence of the temperature on its formation. They claim the formation of thaumasite requires temperatures to 15 °C, ideally from 0 to 5 °C.

In accordance with XRD evaluation, the second biggest rate of decomposition of ternesite stored in the environment CO_2/40°C of all samples was observed. On the contrary, the amount of $\sum$CaO was very low in the water storage, namely the smallest value - in accordance with the formation of thaumasite - was calculated for the samples from the environment w/5°C.

In course of time, amount of gypsum in hydrated clinker was increasing in all the observed storage environments. However, gypsum as the product of decomposition of clinker stored in the presence of carbonic acid was formed very rapidly and in a raised amount, conversely, when stored in a pure water, gypsum was formed slowly and to a markedly lesser extent.

Based on the results from DTA and TG curves of ternesite clinker after an age of 90 days of hydration, it was possible to observe which also mean quantify the mineral thaumasite. Its content was calculated as a value of around 3.2 %.

5. Conclusion

This paper investigated hydration process of ternesite. For the purpose of studying it, the raw meal was prepared firstly, optimized burning regime was employed for the firing of the raw meal subsequently. The resulting ternesite clinker was hydrated and exposed in four specific environments, namely in water and in the presence of saturated carbonic acid, at the low and elevated temperature for both cases mentioned. The progress of hydration process was monitored using XRD and DTA analysis. Slow to complete decomposition of ternesite was observed within the time period studied. The rate of decomposition depended on the conditions of storage. The highest rate of decomposition was determined for the ternesite clinker from the storage environment of saturated carbonic acid at elevated temperature, conversely, the lowest rate of decomposition of the clinker was observed in the storage in water at low temperature. When the ternesite was exposed to water environment, the formation of thaumasite was discovered, which was firstly assumed for the environment of the carbonic CO_2. It can be stated that minerals having great rate of consolidation (i.e. rate of creation and crystallization) are preferentially formed by very high rate of decomposition of the ternesite clinker. By contrast, the slow decomposition of ternesite enables consolidation of slowly formed thaumasite.

Acknowledgment
This work was financially supported by project number: GA17-24954S "The Conditions of Thermodynamic Stability and Transformation of AFt Phases" and project No. LO1408 "AdMaS UP - Advanced Materials, Structures and Technologies", supported by Ministry of Education, Youth and Sports under the „National Sustainability Programme I".

References
[1] Ludwig H M and Zhang W 2015 *Cement Concrete Res.* **78** 24–37
[2] Dienemann W 2013 *Int. Cong. 7th VDZ (Düsseldorf)* (Leimen:Heldelberg Cement AG technology Center) pp 24-27
[3] Irran E, Tillmanns E, Hentschel G 1997 *Mineralogy and Petrology* **6** 121-132
[4] Dominguez R, Gomez-Millan J, Alvarez M, de Aza S, Contreras L and de Aza A H 2010 *J. Am. Ceram. Soc.* **30** 1879–1885
[5] Shen Y, Qian J, Chai J and Fan Y 2014 *Cem. Concr. Comp.* **48** 67–74
[6] HeidelbergCement *New type of clinker significantly reduces CO_2 emissions Ternesite clinker (Germany)*
[7] Ukrainczyk N, Mihelj N F and Sipusic J 2013 *Chem. Biochem. Eng. Q.* **27** 83–93
[8] HeidelbergCement 2013 ZKG Int. **5**
[9] Beretka J, de Vito B, Santoro L, Sherman N and Valenti G L 1993 *Cement Concrete Res.* **23** 1205-1214
[10] Bullerjahn F and Ben Haha M 2013 *Cement International* **11**
[11] Gartner E and Hirao H 2015 *Cement Concrete Res.* **78** 126–142
[12] Schneider M 2015 *Cement Concrete Res.* **78** 14–23

IOP Conf. Series: Materials Science and Engineering **385** (2018) 012015 doi:10.1088/1757-899X/385/1/012015

Effects of expansive additive on cement composite properties

V Gregorova, Z Stefunkova and M Ledererova

Department of Material Engineering, Faculty of Civil Engineering, Slovak University of Technology in Bratislava, Radlinskeho 11, 810 05 Bratislava, Slovakia

Abstract. Shrinkage is a due to chemical and physical processes in concrete. The results of these are volume changes that may cause cracking, deformation, decrease in strength and especially durability of a construction. The amount of water in concrete, the fine particles content, the dimension of construction, the method of curing of the fresh concrete and environmental conditions (temperature, humidity) are some factors influencing the shrinkage of concrete. This article deals with reduction the shrinkage of cement composites by addition of expansive cement Denka CSA and synthetic gypsum. The additives were used at a dosage of 10% of the original weight of cement. The obtained results showed different impact of these additives to the volume changes and the compressive strength of cement composites. The Denka CSA additive had a positive effect to monitored properties of cement mortars. It caused the lowest shrinkage; slight increasing of long-term compressive strength and decreasing of bulk densities after 2, 28 and 90 days of hardening.

1. Introduction

The first mention of composite materials dates back to the ancient world. The composite material is a material consisting of two or more components with different properties. These components give the resulting material novel properties, which is not in itself any of its components. These materials mainly consist of a filler and binder. The binders may be different, but most often the cement composites are based on Portland cement clinker. Nowadays, when modern materials are produced, concrete due to its mechanical, physical, chemical and structural properties belongs to the most used building materials. In the design and realization of concrete elements and structures it needs to be focused on the volume changes occurring in the concrete. They have a significant effect on later properties. These changes are: shrinkage, swelling, creeping and volume changes from the temperature effect. Inherent characteristic of any concrete and other cement-based composites is its shrinkage.

Shrinkage of cement composites can be characterized as a consequence of reducing the moisture content of the porous structure. Dependence mainly affects the capillary forces exerted in the composite. These forces compress the cement composite from all sides, resulting in its shrinkage [1].

Shrinkage generally depends on: the dosage of water in the concrete, the fine particle content, dimension of construction, the method of fresh concrete curing, the age of the concrete and the environmental conditions (temperature, relative humidity, pressure and air velocity) [2]. The shrinkage of concrete can be divided into 2 main stages: the early and the long term ages. The early stage is commonly defined as the first day (within the first 24 hours), while the concrete is setting and starting to harden. After hardening of concrete, the water which has not been consumed by cement hydration leaves the pore system, in case there is not a balance between the moisture content in the composite and the surrounding environment. While drying, capillary forces are formed in the pore system together with the surface tension of water causing reduction of the pore and material volume. This is known as drying shrinkage [2-3]. The next more relevant type of shrinkage is autogenous shrinkage.

There are many ways of reducing the shrinkage of cement composites. The one of the ways is suitable reinforcement of concrete. Different types of dispersed fibers can be used to eliminating the effects of shrinkage of self-leveling cement screeds [3-5]. The next option is to use such an additive inorganic material as the fly ash and the shrinkage reducing agent (SRA) [6-7]. This SRA is a kind of chemical substance that helps reducing the surface tension by dropping the capillary stress and, hence, reducing the shrinkage of the paste in terms of both the chemical reaction and the external environment. The expansive additives generally help reducing the shrinkage due to many different factors, for example, the enlargement from water absorption and the forming of the pore and crystal structure in the process of hydration reaction [8-9].

Calcium sulfoaluminate cement (CSA cement for short) was developed in Japan. This special cement distinguishes itself from Portland cement by a high-speed bonding, fast strength development, and a shrinkage reduction. CSA cement has been used for decades as a binder in concrete for bridges, airport start- and runways, concrete road-repair, and many other applications where a quick reuse is required. Nowadays, CSA cement is also used in dry mortars for self-levelling floors, levelling compounds, casting mortars, tile adhesives, grouts, etc. [10-11]. It is produced by crushing a sintered compound mainly consisting of limestone, gypsum, and bauxite (calcium oxide, calcium sulfate and aluminum oxide) in adequate amounts. The chemical composition of sintered CSA clinker mainly consists of hauyne, free lime and free gypsum. Hydration of these minerals forms ettringite ($3CaO.Al_2O_3.3CaSO_4.32H_2O$), a very small crystal with dimensions of several microns (figure 1). In the cement paste hardening process, this colloid-like crystal becomes burr-like within minute gel voids. It thus reduces hardening shrinkage and dry shrinkage of the gel, and furthermore it acts to expand the gel. Therefore, the presence of ettringite increases the density of the concrete matrix, reduces dry shrinkage, and induces compressive stress into a concrete structure under a restrained condition. Through the above mechanism, CSA reduces dry shrinkage cracking in concrete and increases the water-tightness of concrete structures [12].

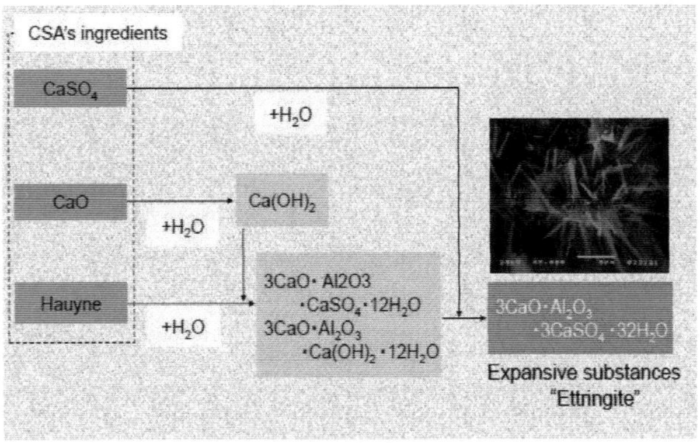

Figure 1. The hydration system of DENKA CSA [12].

2. Materials and mixture design

The experimental study was focused on the reduction of shrinkage of cement mortars by using the expansive cement Denka CSA and synthetic gypsum.

Cement CEM II A/S 42.5 R, in according with STN EN 197-1 [13], with following properties: compressive strength after 2 days 21.6 MPa, after 28 days 49.7 MPa, initial setting time 200 min, setting time 250 min and soundness (expansion) by Le Chatelier 7.2 mm.

Natural quarried aggregate divided by sieve analysis on fraction 0/1, 1/2 and 2/4 mm.

Fly ash is characterized as the gray powder with apparent density from 700 to 900 $kg.m^{-3}$; activity index after 28 days min 75%; activity index after 90 days min 85% and other characteristics declared

IOP Conf. Series: Materials Science and Engineering **385** (2018) 012015 doi:10.1088/1757-899X/385/1/012015

by producer. The chemical compositions of fly ash (wt. %) are: SiO_2 (38.76); Al_2O_3 (16.08); Fe_2O_3 (3.21); CaO (21.28); MgO (1.98); TiO_2 (0.51); SO_3 (13.99).

Water supplied by the public water supply company was used in the mixtures. All cement mortars had w/c ratio = 0.55.

The agent *MS 4-12,* based on modified polycarboxylates, was used for reference mortar (labeled REF). It is characterized by density at 20 °C 1060 $kg.m^{-3}$ ± 20 $kg.m^{-3}$; pH value at 20°C 6,0 ±2.0; dry matter content 20% ± 1%; chloride content max 0.10% by weight of agent. Furthermore, *BR stabilizer 101* with density 1026 $kg.m^{-3}$, 3.6% dry weight and pH 9.5 was used.

Expansive cement *DENKA CSA* is Calcium sulfoaluminate cement (CSA) cement produced from a special clinker and milled to the particle size of cement. It is off-white powder with specific density ca. 2860 $kg.m^{-3}$; loose apparent density from 800 to 900 $kg.m^{-3}$; tapped apparent density ca. 1500 $kg.m^{-3}$ and specific surface by Blaine ca. 370 $m^2.g^{-1}$. Synthetic gypsum or calcium sulfate (gypsum) is the end product of flue gas desulphurization. It represents chemically very pure material with a purity of 97-98% which is mainly used in the manufacture of cement and plaster. Further application is in the manufacture of plaster mixtures, aerated concrete and as a setting regulator. It is characterized by the moisture content of 10%; pH value of 6 – 9 and purity of 95%.

Figure 2. The expansive cement Denka CSA.

Figure 3. The synthetic gypsum.

2.1. Mixture design

The design of cement mortar composition was based on commercially produced self-leveling cement screeds. Reference mortar was modified by expansive additive at a dosage 10% of weight of cement. The proportions of cement mortars are given in table 1.

Table 1. The overview of cement mortars mixing ratios to cubic meter.

	Reference mortar ($kg.m^{-3}$)	Modified mortar ($kg.m^{-3}$)
Cement	243.15	218.84
Expansive additive		
- Denka CSA	-	24.32
- Synthetic gypsum	-	24.32
Fly ash	154.00	154.00
Aggregate		
- fraction 0/1	518.72	518.72
- fraction 1/2	207.49	207.49
- fraction 2/4	311.23	311.23
Plasticizing agent		
- MS 4-12	6.00	6.00
- BR stabilizer 101	6.24	6.24
Water	218.84	218.84

3. Measuring methods

Cement mortars were mixed in a normalized laboratory mixer. After that, the properties of the fresh mortars were determined: consistency of the mortar (determined by a flow test on the Haegermann flow table) and the bulk density. The test samples to determine compressive strength were cylindrical with diameter and heights of 30 mm and were cured in humid environment. The compressive strength of hardened mortars was measured after 2, 28 and 90 days. The test samples to determine shrinkage were beams of 40 x 40 x 160 mm with glass contacts in front. They were cured for 24 hours in humid environment. After this time, samples were unmolded and the initial length measurements were done. Subsequently, the samples were kept in a laboratory environment (temperature 20 ° C ± 5 ° C, relative air humidity φ = ca 50%). The drying shrinkage of cement mortars was determined as loss of length compared to the initial length determined after 24 hours of curing in moist environment. Shrinkage was measured after 2, 4, 7, 14, 28, 56 and 90 days of curing in the laboratory environment. The length measurement was done using the Graf-Kaufmann device. The weight loss was measured on the same samples used for the shrinkage measurements.

4. Results and discussion

4.1. Consistency and bulk density

The measured values of the consistency and the bulk density of fresh cement mortars for each mixture are given in table 2.

Modifications of cement mortars with expansion additives caused a slight worsened consistency compared to the reference sample (REF). Flow diameter ranged from 233.5 to 244 mm.

The expansive additives had a different impact on the bulk density of fresh mortars. Replacing the cement with expansive cement Denka CSA (labeled PC) led to a slight increase in bulk density to value 2140 kg.m^{-3} compared to the reference mortar. Using the synthetic gypsum (sample labeled PE) had an opposite effect; reducing bulk density to value 2000 kg.m^{-3}.

Table 2. Properties of fresh cement mortars.

	Bulk density (kg.m^{-3})	Consistency (mm)
REF	2120	282.5
PC	2140	244.0
PE	2000	233.5

4.2. Shrinkage and weight loss

Measured values of shrinkage, respectively swelling of hardened cement mortars are given in figure 4. Using the expansive additives had a various impact to the volume changes of the hardened cement mortars. Shrinkage of the reference cement mortar and mortar with synthetic gypsum additive was approximately the same during 14 days. The Denka CSA additive caused swelling of the mortars in the first days. In other monitoring days, the gradual shrinkage of samples was already observed. The total value of shrinkage of the reference mortar after 90 days was ca. 0.74‰; mortars modified by additive Denka CSA ca. 0.68‰ and synthetic gypsum ca. 0.79‰.

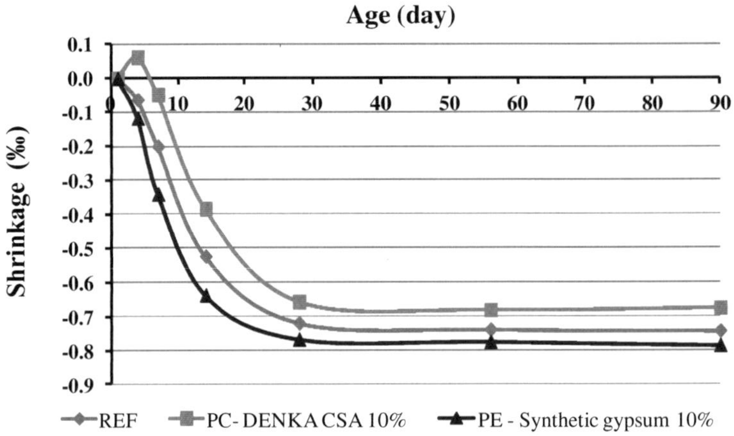

Figure 4. Shrinkage of the reference sample and samples with 10% doses of expansive additive.

4.3. Compressive strength

Application of expansive additives had impact on the physical properties of hardened cement mortars, which can be seen in table 3.

Table 3. The results of hardened cement mortars.

	Bulk density ($kg.m^{-3}$)			Compressive strength (MPa)		
	2 days	28 days	90 days	2 days	28 days	90 days
REF	2160	2010	1990	9.85	24.38	26.00
PC	2050	1910	1870	11.02	23.09	27.92
PE	2030	1710	1690	6.11	15.94	20.06

Using additive Denka CSA resulted in only a slight increase in the initial (2-day) and the long term (28 and 90 days) compressive strength of cement mortars compared to the reference mortar. The values of compressive strength in the monitoring period ranged from 11.02 MPa to 27.92 MPa. The values of bulk density of cement mortars were less than the reference mortar after 2, 28 and 90 days of hardening.

Cement mortar modified by synthetic gypsum additive showed significantly the lower strength characteristics compared to the reference and the second modified mortar. The compressive strength values ranged from 6.11 MPa to 20.06 MPa in the same monitoring period. The bulk density values achieved the lowest values compared to other studied cement mortars.

5. Conclusion

Effect of using expansive additives to cement mortars were studied and recorded in this research. The achieved results are summarized hereinafter:

- Application of the expansive cement Denka CSA led to reduce the shrinkage of cement mortars and the complete elimination of shrinkage in the first days of hardening. The significant benefit is also the increase of the initial and the long term compressive strength. The values of compressive strength ranged from 11.02 MPa to 27.92 MPa after 2, 28 and 90 days of hardening. The dosage of CSA cement did not have significant impact on the bulk density.

These values are very similar in comparison to values of samples modified by synthetic gypsum additive.

- Synthetic gypsum additive had negative effect on the properties of cement mortars. It caused increasing the shrinkage (ca. 0.74‰ after 90 days of hardening) and also a significant the compressive strength reduction (from 6.11 MPa to 20.06 MPa after 2, 28 and 90 days of hardening). The positive effect of this additive was reflected in the decreasing values of bulk densities compared to other studied cement mortars. The reference mortar´s bulk density values after 2, 28 and 90 days of hardening ranged from 2160 $kg.m^{-3}$ to 1990 $kg.m^{-3}$; the cement mortar values modified by Denka CSA cement ranged from 2050 $kg.m^{-3}$ to 1870 $kg.m^{-3}$ and cement mortars with synthetics gypsum additive ranged from 2030 $kg.m^{-3}$ to 1690 $kg.m^{-3}$.

Acknowledgment
This article was created with the support of the Ministry of Education, Science, Research and Sport of the Slovak Republic within the Research and Development Operational Program for the project "University Science Park of STU Bratislava", ITM S 26240220084, co-funded by the European Regional Development Fund and by the Slovak Research and Development Agency under the contract No. APVV-15-0681.

References

[1] Adamek J, Jurankova V and Michalko O 2001 *Monitoring of Sealing Technology Quality of Concrete by Determination of its Water and Gas Permeability* (Ostrava: VSB - Technical University of Ostrava) p 218-223
[2] Bajza A and Rousekova I 2006 *The Concrete Technology* (Bratislava: Jaga Group) p 190
[3] Dubek M, Bederka M and Grinc M 2017 *Czech Journal of Civil Engineering* **3** 40-45
[4] Dubek M, Makys P, Petro M and Briatka P 2017 *Detecting Fibers In The Cross Sections Of Steel Fiber-Reinforced Concrete* (London: Taylor&Francis Group) p 383-388
[5] Dubek M, Makys P, Dubek S and Petro M 2018 *J. Civ. Eng. Manag.* **24** 183-192
[6] Collepardi M 2005 *Cem. Concr. Compos.* **27** 704-708
[7] Matulová K and Uncik S *2017 Effect Of Shrinkage-Reducing Admixture On The Properties Of Cement Composites* (London: Taylor & Francis Group) p 329-334
[8] Sarapon T and Komsan M 2017 *Adv. Mater. Sci. Eng.* **2017** 8917957
[9] Winnenfeld F and Lothenbach B 2010 *Cem. Concr. Res.* **40** 1239-1247
[10] Caltra Nederland BV 2016 *The advantages of C.S.A. cement* Report
[11] Xiaolu G, Huisheng Shi, Wenpei H and KaiWu 2014 *Cem. Concr. Compos.* **46** 26-31
[12] Suhara K 2008 http://cement08.in2p3.fr/Presentation%20Workshop%20pour%20site%20web/Tuesday%2014/PM/SUHARA.pdf (online)
[13] STN EN 197-1: 2012 *Cement Part 1: Composition, specifications and conformity criteria for common cements*

Optimization of heavy weight concrete composition and process of prefabrication for prefabricated shielding cladding tiles

R Hela[1], D Jancarikova[1], P Louda[1,2] and J Jobanek[3]

[1] Faculty of Civil Engineering, Brno University of Technology, AdMaS Centre, Purkynova 139, Brno 61200, Czech Republic
[2] Prefa Brno, a.s., Kulkova 10, Brno 615 00, Czech Republic
[3] Faculty of Civil Engineering, Brno University of Technology, Veveri 331/95, Brno 602 00, Czech Republic

Abstract. The aim of experimental research was to verify the possibility of the prefabrication of the thin shielding cladding tiles from heavy weight concrete with fine. The minimal required density of the concrete was 3800 kg/m^3 and the maximal particle size was 2 mm. Because of the though over technology of the complicated locks in the touch of tiles was necessary to achieve consistency around 210 mm flow with the Chryso cone test (modified slum flow test). To obtain this consistency without danger of the heavy weight aggregate segregation the superplasticizer was used. In first step was determined optimal amount of mixing water, in the second step we tested different frequency of vibration to achieve maximal consolidation and as a last step to define composition of mixture we tested 3 types of accelerators of concrete hardening to shorten time of hardening and to facilitate more effective production. As the optimal frequency of vibration for compacting process was determined 75 Hz. The best accelerator of concrete hardening for optimalized mixture was MasterSert XS100 from BASF producer.

1. Introduction

Heavy weight concrete is concrete with the density above 2600 kg/m^3 in the dry state [1]. High density is achieved by the using heavy weight aggregate (with density higher than 3000 kg/m^3) or in the special cases by the using barium sulphate cement [2-4]. Heavy weight concrete is used as the shielding concrete for the protection again the radiation in the nuclear powerplants, the hospitals and in the research centres. Another way of using is to load the construction for example in the case of the machines subjected by the dynamic forces. As a heavy aggregate are used the steel bullets, the artificial corundum or the natural aggregate barite or the ironstones such as the magnetite and the hematite. The heavy aggregate complicates mixing of the fresh concrete - the lower volume of fresh concrete in the mixing machine because of the high weight, worse homogenisation and the tendency to the segregation.

To produce the thin shielding cladding tiles there had to be designed the heavy weight concrete with the fine aggregate - heavy weight mortar. Requests for mixture were following:

- Water/binder ratio as low as possible but with the requested consistency. High water/binder ratio can lead to the segregation and lower density.
- Using of the superplasticizers with the best coworking with the accelerator of hardening.
- Higher content of the cement because of the higher specific surface of the fine aggregate.

- Minimalization of the air content.
- As effective compaction of the concrete as possible.

The compaction of the concrete is technological process which aim is to achieve maximal compactness and minimal porosity. Balance between inner and outer forces is destroyed during the compaction. Inner forces are capillary forces, surface tension, aggregate friction and physically-chemical bonds. Outer forces are gravitational force and loads affected concrete. Quality of concrete (mortar) can be influenced in the compaction process by:

- Separation of fresh concrete.
- Segregation - separation of mixture components according to their density.
- Bleeding.

In more than 90 % of the compaction processes is used vibration. By vibration are particle moved and inner forces are changed. With higher speed of vibration, the speed gradient of particles is higher too. The viscosity of cement paste is lower and mixture is compacted.

On Figure 1 we can see influence of frequency on time of compaction.

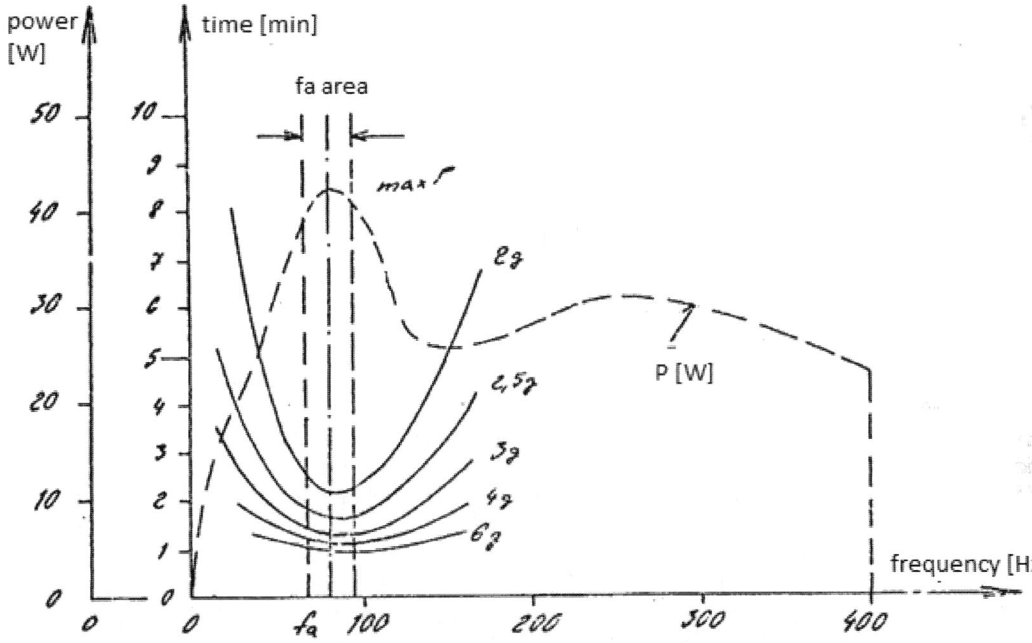

Figure 1. Influence of frequency on time of compaction [5].

The time of compaction is increasing till the fa (absorption frequency) and then is decreasing. By the tests was found that each concrete mixture has different absorption frequency.

The aim of the experimental research was to verify the possibility of the prefabrication heavy weight concrete with the fine aggregate for the manufacturing thin shielding cladding tiles. Required density of the concrete was 3800 kg/m^3 and the maximal particle size was 2 mm. Because of the though over the technology of the complicated locks in the touch of the tiles was necessary to achieve the consistency around 210 mm by slump test. To obtain this consistency without danger of the heavy weight aggregate segregation different types of the superplasticizers was tested. Accelerators of the hardening was tested to shorten time of hardening and to facilitate more effective production. Optimal frequency of vibration was determined to achieve maximal density.

IOP Conf. Series: Materials Science and Engineering **385** (2018) 012016 doi:10.1088/1757-899X/385/1/012016

2. Materials and Methods

For design of reference concrete mixture were used cement CEM I 52.5 R, as aggregate was used magnetite in fraction 0-2 mm imported from Sweden, silica fume, stone filler, superplasticizer Glenium 430 from BASF producer and accelerator of hardening Master set SX 100 from BASF producer.

In first step was looking for optimal amount of water to achieve maximal density, in second step we found optimal frequency of vibration to get best compaction and as a last step we tested another two types of accelerator of concrete hardening.

2.1. Magnetite

Magnetite imported from Sweden in fraction 0-2 mm was used. Density of aggregate is 5120 kg/m^3 in natural state and 5060 kg/m^3 in dry state. Particle size distribution is shown on Figure 2.

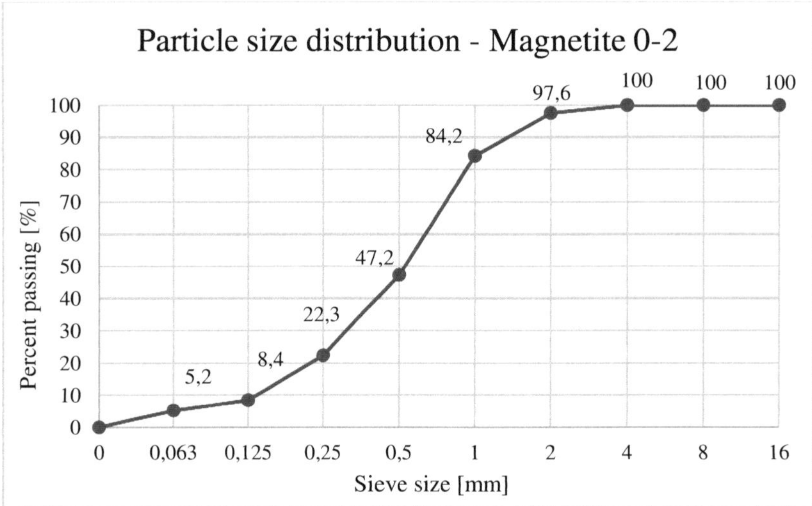

Figure 2. Particle size distribution - magnetite 0-2.

2.2. Amount of water

As assumed before, we looked for optimal dosage of water to achieve maximal density. In reference mixture 7 dosages of water was tested. Composition of reference mixture is shown in Table 1. Tested amount of water and corresponded density is shown in Table 2.

Table 1. Composition of reference mixture.

Component	amount for 1 m^3 of concrete
Cement CEM I 52,5 R from Cement Hranice producer	490 kg
Fine aggregate 0-2 mm - sweden magnetite	3240 kg
Silica fume	20 kg
Stone filler - Želešice	50 kg
Superplasticizer BASF Glenium 430	10 kg
Accelerator of concrete hardening BASF master set XS 100	21.2 kg

Table 2. Tested amount of water and correspondent density of concrete.

amount of water for 1 m^3 of conrete [l]	140	145	150	155	160	165	170
density of fresh concrete [$kg \cdot m^{-3}$]	4117	4192	4212	4151	4134	4126	4087

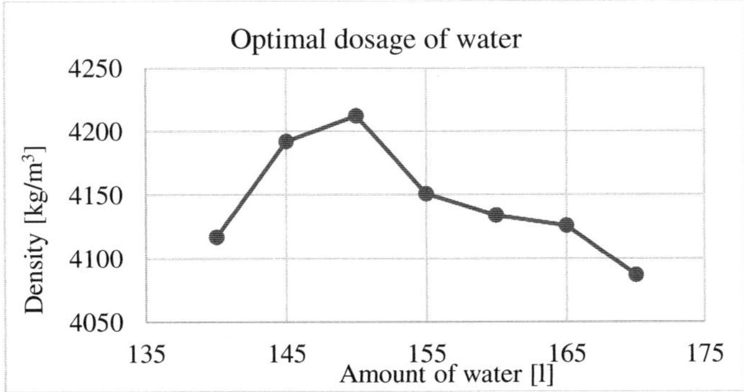

Figure 3. Influence of amount of water to density of fresh concrete.

As optimal dosage of water was determined amount of 150 l per 1 m^3 of concrete.

2.3. Frequency of vibration
The second step was determination of optimal frequency of vibration. To reference mixture from table 1 was added 150 l of water. To achieve smoother surface of concrete 40 kg of stone filler Želešice per 1 m^3 of concrete was added to reference mixture. This mixture was vibrated on vibrating table with possibility of changing frequency from Wacker-Werke producer. 4 frequencies were tested. Frequency, speed of motor rotation and density of fresh concrete are shown in Table 3.

Table 3. Optimization of compaction process.

speed of motor rotation [min^{-1}]	frequency of vibration [Hz]	density of fresh concrete [$kg \cdot m^{-3}$]
4000	66.7	4069
4300	71.7	4079
4500	75.0	4197
4700	78.3	4114

As optimal frequency for compaction process was determined frequency 75 Hz.

2.4. Accelerators of concrete hardening
Three types of accelerators of concrete hardening were tested. MasterSet XS 100 from BASF producer, Betodur A5 from STACHEMA producer and Fast Kick 111 from MC Bauchemie producer. The accelerators of hardening did not influence consistency of fresh concrete.

2.4.1. Betodur A5. Betodur A5 is chloride free admixture accelerating hardening of concrete. Density is 1450 kg/m^3.

2.4.2. MasterSet XS 100. MasterSet XS 100 is ready to use liquid accelerating admixture for use in concrete and portland cement. MasterSet XS 100 admixture brings down setting times in general and specially under cold weather, leading to early high and ultimate increased strengths. In addition, it reduces bleed and segregation while improves workability. MasterSet AC 100 does not contain any added chloride ions ensuring that the product does not contribute to the corrosion of reinforcing steel.

In fact, studies indicate MasterSet XS 100 helps in reduction of chloride attack on concrete. Density is 1260 kg/m^3.

2.4.3. Fast Kick 111 - MC Bauchemie. With MC-FastKick, MC has developed a new functional group of accelerators as a further development of the BASF-patented concept of stabilising CS phases with polycarboxylate ethers. MC-FastKick not only kick-starts the hardening of the concrete but also results in significant plasticisation, providing the mix with very good processing, flow and application properties, including high early strength values. These accelerators thus greatly reduce cycle times in precast operations and will also facilitate shorter hardening times during wintertime construction. MC-FastKick has a positive influence on crystal growth in the calcium-silicate-hydrate phases within the concrete. This results in a significant increase in the crystallisation rate and thus substantially higher early strength values – but without negatively influencing the other concrete properties.

Composition of compared mixtures is shown in Table 4. Dosage of accelerators were according to producers recommendation. Compressive strength was determined in age 6 hours, 24 hours and 28 days.

Table 4. Composition of mixtures with different types of concrete hardening accelerator.

	Component		Note	Amount for 1 m^3 of concrete
	Cement CEM I 52,5 R from Cement Hranice producer			490 kg
	Fine aggregate 0-2 mm - sweden magnetite			3240 kg
	Silica fume			20 kg
	Stone filler - Želešice			90 kg
	Superplasticizer BASF Glenium 430			10 kg
Mixture A	Accelerator of concrete hardening	BASF MasterSet XS 100	0.4 % m_c	19.6 kg
Mixture B		Stachema CZ Betodur A5	0.3 % m_c	14.7 kg
Mixture C		MC Bauchemie Fast Kick 111	0.2 % m_c	9.8 kg

Table 5. Time evolution of compressive strength of concretes with different accelerator of concrete hardening from different producers.

Mixture	Accelerator of concrete hardening	Compressive strength [N/mm^2]		
		6 hours	24 hours	28 days
A	MasterSet XS 100	6.0	37.7	66.6
B	Betodur A5	3.2	38.3	63.1
C	Fast Kick 111	1.5	39.1	59.5

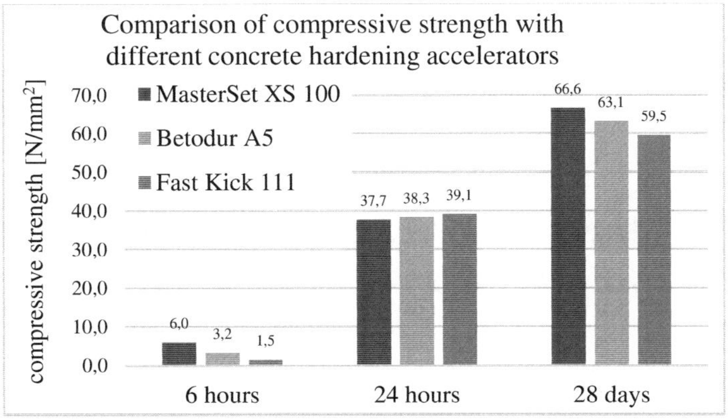

Figure 4. Comparison of compressive strength.

Average density of fresh concretes was 4100 kg/m^3 and in age of 28 days and in natural humidity 4040 kg/m^3.

3. Conclusion

Optimal composition of the heavy weight concrete for prefabrication thin shielding cladding tiles was determined. With 490 kg of cement, 3240 kg of magnetite, 20 kg of silica fume, 90 kg of stone filler, 10 kg of superplasticizer Glenium 430 and 19.6 kg of accelerator of concrete hardening MasterSet XS 100 we achieved concrete with the density over 4000 kg/m^3, compressive strength in 6 hours age 6 N/mm^2, in 24 hours 37.7 N/mm^2 and in age of 28 days 66.6 N/mm^2. Consistency of the mixture fulfil requested 220 mm flow by Chryso cone test. Due to the low water cement ratio we achieved smooth surface of the concrete without the pores. For the process of the compaction the optimal frequency of vibration was determined as 75 Hz. Due to used accelerator of the concrete hardening we got compressive strength in 24 hours around 38.0 N/mm^2 which means rapid reusing moulds during manufacturing.

Acknowledgment

This research was supported by project MPO FV10680 funded by Czech Ministry of Industry and Trade.

Reference

[1] Ouda A S 2015 *Prog. Nucl. Energy* **11** 328-338
[2] Azreen N M 2018 *Contr. Build.Mater.* **172** 370-377
[3] Horszczaruk E 2015 *Procedia Eng.* **108** 39-46
[4] Pillai C S, Santhakumar A R, Poonguzhali A, Pujar M G, Kumar J A, Preetha R, Mudali U K 2016 *Contr. Build.Mater.* **126** 453-465
[5] Hela R 2005 *Technology of Building Components - Studying Materials* (Brno: Brno University of Technology) p 115

IOP Conf. Series: Materials Science and Engineering **385** (2018) 012017 doi:10.1088/1757-899X/385/1/012017

Optimization of thermal insulation of external walls of wooden frame constructions

K E Hellova and A Struharova

Department of Material Engineering, Faculty of Civil Engineering, Slovak University of Technology in Bratislava, Radlinskeho 11, 810 05 Bratislava, Slovakia

Abstract. The article deals with the optimisations of external structures and points to energy-savings and requirements of low energy construction. It defines and evaluates the problems of required thermal-technical characteristics achievement set by standard for external structures of low energy and energy passive houses. At the same time, it describes in detail different variants of their optimisation.

1. Introduction

Work goal is to optimise chosen external wall composition of wooden frame construction from the viewpoint of thermal insulation. There was compared most used mineral wool with additional thermal insulation and without additional thermal insulation in the partition. The parameters and specific characteristics were modified by optimisation, and that exposed the behaviour of such external structure to the ordinary user [1].

Concerning the thermal-technical characteristics, the external structure optimization can improve its parameters, and at the same time reduce the built-up area. Material optimisation of wood-based walls from the viewpoint of thermal technical characteristics (heat conductivity coefficient (C-value), phase displacement and condensation) for low-energy houses (LEH) and energy-passive houses (EPH) [2].

Optimisation criteria are:

- Minimal value of heat conductivity coefficient U for LEH = 0.22 W/m^2K.
- Minimal value of heat conductivity coefficient U for EPH = 0.15 w/m^2K.
- Diffuse open structure (without vapour barrier) for natural transfer of vapour, so-called, structure breathing.
- Phase displacement if is the bigger, the better the structure is (recommended value is 14 hours).
- Critical surface temperature for the formation of $T_{si,80}$ (the formation of molds) and $T_{si,100}$ (the formation of condensation) - corresponding to 80% or 100% relative humidity in close to the surface of the building structure, depends on the temperature and relative humidity of the interior the air [3].
- Thriftiness to human and environment with using of natural materials, that can be recycled or easy disposable.
- Production and usage of carbon dioxid during construction must be more produced than consumed.
- Sustainable usage of raw materials – favour the ecological and renewable materials, recycle the waste and re-use it during construction [4].

2. Methods

The simulation software "Teplo 2010" [5], that evaluates and optimises the final thermal-technical parameters in accordance with Standard STN 730540-2, was applied for the optimisation of external wall composition from the viewpoint of thermal-technical characteristics.

3. Achieved results of optimization

Optimisation of designed EW (external wall) composition is mainly focused on usage main and additional TI (thermal insulation) on the interior side (EKO 2A – EKO 2 B) and their combination to achieve minimal values of thermal-technical characteristics in accordance with STN 730540-2 for LEH and EPH [4].

By optimisation of particular EW composition were improved foremost the values of heat conductivity coefficient U (see table 1 from standard), phase displacement and water vapour condensation in the structure. The bigger the phase displacement is, the longer structure withstands the high temperatures and prevents overheating.

Recommended value of phase displacement is 14 hours. Recommended value of condensed vapour in the structure per year is maximum 0.5 kg.m^{-2}. year^{-1} [4].

Table 1. Heat conductivity coefficient through structure from standard in W.m^{-2}. K^{-1}[4].

	Minimal value U_{min}	Standard value U_N	Reco mmended value U r1 (after 2015)	Target reco mmended value U_{r2} (after 2020)
EW and sloped roof over living space with slope > 45°	0.46	0.32	0.22	0.15

4. Mineral TI ISOVER Uni – properties

Table 2. Thermal-technical characteristics of used TI.

Material	Bulk density ρ [kg.m^{-3}]	K value λ [W.m^{-1}.K^{-1}]	Specific thermal capacity c [J.kg^{-1}.K^{-1}]	Diffuse resistance factor μ [-]
Mineral Wool ISOVER Uni	45	0.033	940	1

4.1. EW Composition – EKO 2 A EKO 2

A is external wall, having ISOVER Uni as main TI, while the EW have no any additional TI in the partition what you can see in table 3.

Table 3. Structure composition (from the interior side).

Nr.	Name	T [m]	λ [W.m^{-1}.K^{-1}]	C [J.kg^{-1}.K^{-1}]	ρ [kg.m^{-3}]	μ [-]
1	Plasterboard	0.0125	0.2200	1060.0	750.0	9.0
2	Closed air pocket	0.0500	0.2940	1010.0	1.2	0.2
3	OSB 3 EKO	0.0150	0.1400	1700.0	650.0	150.0
4	ISOVER Uni	0.1400	0.0390	940.0	17.0	1.0
5	STEICO PRO	0.0600	0.0480	2100.0	230.0	5.0
6	Mineral adhesive	0.0060	0.8400	920.0	500.0	18.0
7	Weber	0.0030	0.8600	920.0	1700.0	50.0

By evaluation, it is important to care about the minimal required and calculated interior surface temperatures shown in the figure 1. The calculated values related critical surface temperature for the formation of $T_{si,80}$ (the formation of molds) and $T_{si,100}$ (the formation of condensation) - corresponding to 80% or 100% relative humidity in close to the surface of the building structure, depends on the temperature and relative humidity of the interior the air [3].

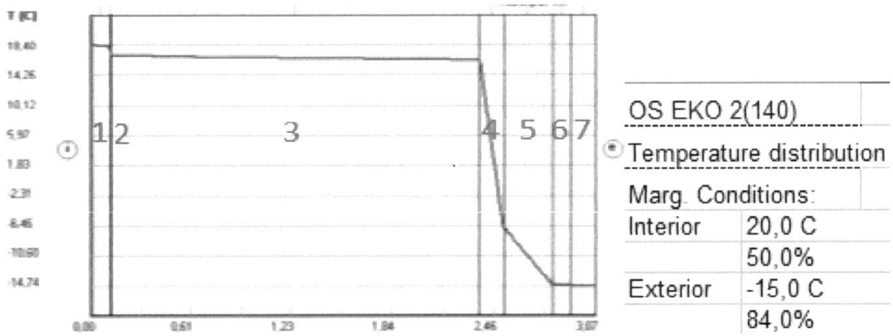

Figure 1. Temperature distribution in the standard section in the EW structure – EKO 2 A [5].

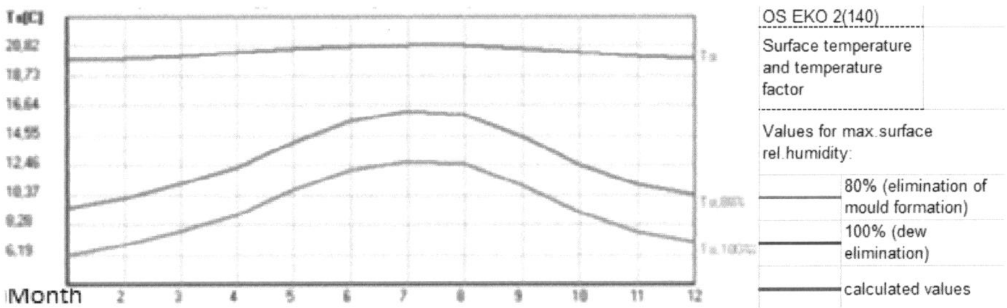

Figure 2. Minimal required and calculated interior surface temperature of the EW – EKO 2 A [5].

Table 4. Final values of the EW structure optimisation – EKO 2 A [5].

Used TI material		ISOVER UNI			
Main TI thickness [mm]		120	140	200	240
Total structure thickness [mm]		266.5	286.5	346.5	386.5
Structure surface temperature [°C]		18.7	18.23	18.58	18.74
Heat conductivity coefficient U [$W.m^{-2}.K^{-1}$]		0.227	0.207	0.165	0.146
Structure resistance R [$m^2.K.W^{-1}$]		4.24	4.67	5.89	6.67
Phase displacement Ψ [h]		6.4	6.5	7.2	7.6
Vapour condensation in the structure [$kg.m^{-2}.rok^{-1}$]	G_{λ}	0.050	0.050	0.050	0.049
	G_V	7.585	7.569	7.533	7.516
Structure suitability concerning the U value according to STN 730540-2:Z1/2016	EPH U_{min}=0.15	fail	fail	fail	fulfil
	LEH U_{min}=0.22	fail	fulfil	fulfil	fulfil

Designed basic composition of external structure with overall thickness of 286.5mm and main thermal insulation ISOVER UNI thickness of 140mm, has been optimised in main insulation thicknesses of 120mm, 200mm and 240mm to achieve required values of heat conductivity coefficient U for LEH and EPH in accordance with standard STN 730540-2. Minimal required values are defined in table 1. By using main TI of 140mm thickness in the EW composition, the final values of heat conductivity coefficient U are not sufficient for construction of LEH or EPH [6].

Main TI of 140mm thickness in the EW composition has sufficient values of heat conductivity coefficient U for construction of LEH, however insufficient for the construction of EPH.
By using the main insulation of 200mm thickness in the EW composition, the final values of heat conductivity coefficient U are favourable for the LEH, however still insufficient for the construction of EPH. Main TI of 240mm thickness in the EW composition causes the fulfilment of required values of heat conductivity coefficient U both for construction of LEH as well as EPH.

By using the main insulation of 240mm thickness in the EPH structure, the phase displacement in the structure has the value of 7.6 hours. Concerning the LEH structure with the main TI thickness of 140mm, it has 6.5 hours. With the thickness of 200mm, phase displacement value is 7.2 hours and with the thickness of 240mm, it is 7.6 hours [4].

4.2. EW Composition
EKO 2 B EKO 2 B is external wall, having ISOVER Uni as main TI, while the EW have additional TI in the partition what you can see in table 5.

Table 5. Structure composition (from the interior side).

Nr.	Name	T [m]	λ [W.m^{-1}.K^{-1}]	C [J.kg^{-1}.K^{-1}]	ρ [kg.m^{-3}]	μ [-]
1	Plasterboard	0.0125	0.2200	1060.0	750.0	9.0
2	ISOVER Uni	0.0400	0.0390	940.0	17.0	1.0
3	OSB 3 EKO	0.0150	0.1400	1700.0	650.0	150.0
4	ISOVER Uni	0.1400	0.0390	940.0	17.0	1.0
5	STEICO PRO	0.0600	0.0480	2100.0	230.0	5.0
6	Mineral adhesive	0.0060	0.8400	920.0	500.0	18.0
7	Weber	0.0030	0.8600	920.0	1700.0	50.0

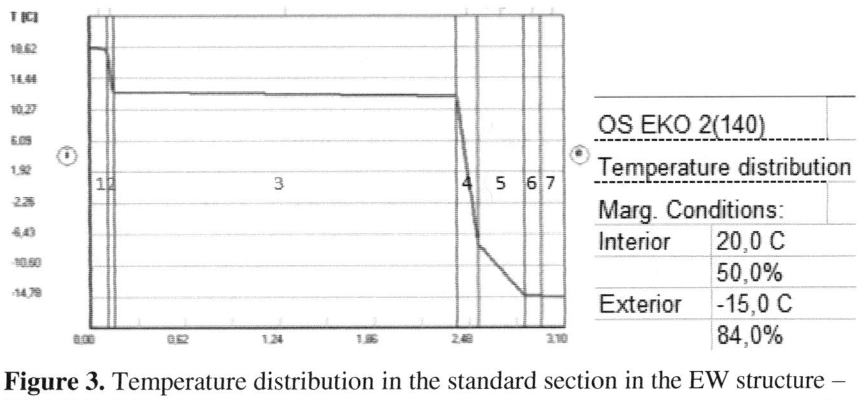

Figure 3. Temperature distribution in the standard section in the EW structure – EKO 2 B [5].

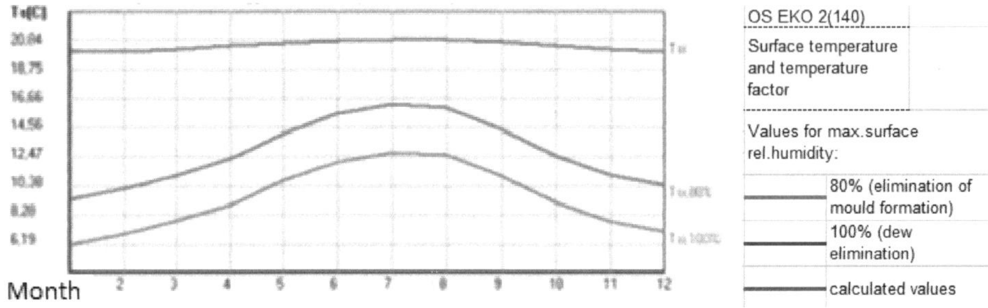

Figure 4. Minimal required and calculated interior surface temperature of the EW – EKO 2 B [5].

Minimal required values are defined in table 1 [4]. By using main TI of 120mm thickness in the EW composition, the final values of heat conductivity coefficient U are not sufficient for construction of EPH, but favourable for construction of LEH.

Main TI of 140mm thickness in the EW composition also has sufficient values of heat conductivity coefficient U for construction of LEH, however still insufficient for the construction of EPH. However, by using the main insulation of 200mm thickness in the EW composition, the final values of heat conductivity coefficient U are favourable for both construction types, the LEH and the EPH.

Increasing the thickness of main insulation up to 240mm in the EW composition causes the improvement of required values of heat conductivity coefficient U compared with previous results, and are favourable for both construction types LEH as well as EPH [5].

Table 6. Final values of the EW structure optimisation – EKO 2 B [5].

Used TI material		ISOVER UNI			
Main TI thickness [mm]		120	140	200	240
Total structure thickness [mm]		256.5	276.5	336.5	376.5
Structure surface temperature [°C]		18.33	18.45	18.72	18.85
Heat conductivity coefficient U [W.m^{-2}.K^{-1}]		0.196	0.181	0.149	0.134
Structure resistance R [m^{2}.K.W^{-1}]		4.94	4.35	6.54	7.29
Phase displacement Ψ [h]		8.1	8.4	9.0	9.5
Vapour condensation in the structure [kg.m^{-2}.rok^{-1}]	G_{λ}	0.050	0.050	0.049	0.049
	G_V	7.560	7.547	7.519	7.505
Structure suitability concerning the U value according to STN 730540-2:Z1/2016	EPH U_{min}=0,15	fail	fail	fulfil	fulfil
	LEH U_{min}=0,22	fulfil	fulfil	fulfil	fulfil

By using the main insulation of 200mm thickness in the EPH structure, the phase displacement in the structure reaches value 9.0 hours and increasing the main insulation to the thickness of 240mm raises the phase displacement in the structure to 9.5 hours.

For LEH constructions and at the main insulation thickness of 120mm, the phase displacement in the structure is 8.1 hours, at the thickness of 140mm, the phase displacement in the structure is 8.4 hours. At the thickness of 200mm, the phase displacement in the structure is 9.0 hours and at the thickness of 240mm, the phase displacement value in the structure reaches up to 9.5 hours [5].

5. Results and Discussion

Optimising the designed composition of external structure for the construction of LEH and EPH imposed a gradual improvement of monitored thermal-technical values. Final parameters can be compared reciprocally, either by using only one type of TI material or various TI types.

Optimisations ran same in all cases concerning the thickness of main thermal insulation of 120mm, 200mm and 240mm.

In the EW composition EKO 2 A and EKO 2 B, the mineral insulation ISOVER Uni was used. In case A, it was applied solely as main TI within the structure and the structure contained 50mm thick air pocket under the plasterboard on the interior side. Values of heat conductivity coefficient U for all optimisations were favourable for EPH at TI thickness of 240mm with value of 0.149 W/m^2K, and for LEH at thicknesses from 140 to 200mm with values from 0.210 to 0.149 W/m^2K.

Optimising the TI to 140mm, 200mm and 240mm in the structure caused the change of thermal-technical characteristics and improvement of required values. Values of heat conductivity coefficient U were favourable for EPH only at TI thickness of 240mm (U=0.146 W/m^2. K). Also the values of heat conductivity coefficient U were favourable for LEH only at higher TI thicknesses 200mm (U = 0.165 W/m^2K) and 240mm (U = 0.146 W/m^2K).

In case B, ISOVER Uni was applied as main TI, but the air pocket was filled with same TI material of 40mm thickness. Final parameters achieve better values of heat conductivity coefficient U in connection with the increase of main TI thickness. Parameters of heat conductivity coefficient U for EPH are favourable for thicker TI, specifically for 200mm thickness (U = 0.149 W/m^2K) and 240mm thickness (U = 0.134 W/m^2K). Parameters for LEH are fulfilled in all cases of optimisation for all thicknesses with U values varying from 0.134 to 0.196 W/m^2K [5].

6. Conclusion

Optimisation consisted from the thicknesses and use of additional TI in plasterboard partition walls on the interior side of structure. By comparing the observed thermal-technical characteristics of the external structure, we can estimate specific behaviour of particular structure compositions in connection to the required thermal-technical characteristics set in standard STN 730540-2. From achieved results, it is clear that every TI material acts different in the same structure composition. Fulfilling the required thermal-technical values for LEH and EPH, it is necessary to choose well the TI material type and thickness. Low energy construction demands a high quality of construction works as well as high quality of used materials. For these constructions, it is important to choose suitable TI material, that fulfils the required thermal-technical parameters, has impact on reduction of built-up area material consumption and is environmentally friendly.

Acknowledgment

This article was created with the support of the Ministry of Education, Science, Research and Sport of the Slovak Republic within the Research and Development Operational Programme for the project "University Science Park of STU Bratislava", ITMS 26240220084, co-funded by the European Regional Development Fund.

References

[1] Havirova Z 2005 *Woodenhouse* (Czech Republic: ERA) p 104

IOP Conf. Series: Materials Science and Engineering **385** (2018) 012017 doi:10.1088/1757-899X/385/1/012017

[2] Berggren B and Wall M 2012 *Proc. Int. Conf. on Moisture Conditions in Exterior Walls for Net Zero Energy Buildings in Cold Climate* (Calgary: ASHREA)
[3] Husek P 2008 *Wooden Frame Construction From The Point Of View Of Thermal Moisture Behaviour* (Brno: Brno University of Technology Press) p 10-108
[4] STN 73 0540-2:2016 *Thermal protection of buildings. Thermal performance of buildings and components Part 2: Functional requirements*
[5] Teplo 2010, simulation software (in Czech)
[6] Dudas J and Jochim S 2008 *Structural Wood Materials* (Zvolen: Technical University in Zvolen Press) p 155
[7] Hens H 2010 *Applied Building Physics–Boundary Conditions, Building Performance and Material Properties* (Berlin: Ernst & Sohn Verlag)
[8] Glass S V and Zelinka S L 2010 *Moisture Relations and Physical Properties of Wood* (U.S.A Wisconsin: Department of Agriculture, Forest Products Laboratory)

IOP Conf. Series: Materials Science and Engineering **385** (2018) 012018 doi:10.1088/1757-899X/385/1/012018

Possibilities and pitfalls of revitalizing the timber structure of an industrial building from the 19th century from the viewpoint of diagnosis

V Hermankova, O Anton, P Cikrle, T Komarkova and M Stehlik

Department of Building Testing, Faculty of Civil Engineering, BUT in Brno, Veveri 95, Brno, Czech Republic

Abstract. A typical feature of contemporary cities is the permanent shortage of land suitable for residential development inside the city. This leads to attempts in recent years to re-use the so-called Brownfields which have the appearance of the former industrial areas, the buildings of which are mostly far beyond the estimated lifetime. Apart from the standard procedure, i.e. demolition and new construction, it is possible to come across efforts, in recent years, to renovate and newly utilize the existing industrial buildings, some of which have a historical value and considerable design potential for an alternative use. The paper presents a case where an original five-story warehouse building from the mid-19th century should be prospectively used for residential purposes, while maintaining and utilizing, for aesthetic purposes, the timber ceiling structures. The paper describes the procedure of diagnosis of a timber structure and its results, suggesting potential risks of similar renovations where the object of interest has not been used and maintained for a long period of time.

1. Introduction

As part of the gradual revitalization of the former premises of textile production in Brno, designers consider utilizing one of the original buildings from the 19th century, probably a warehouse, for luxury housing. The building has five floors and its vertical supporting structures are made from quality massive brickwork, according to which the building was dated to around 1860. Architects and designers were attracted by its timber ceiling structures, supported inside with ornamental cast-iron pillars. The designer plans to use the timber ceilings, highly aesthetic at first sight, as a visual design element in the future interior. Within the scope of the planning, the Department of Building Testing, Faculty of Civil Engineering, BUT in Brno carried out a diagnosis of the building, and this paper focuses on the performed diagnosis of the mentioned timber structures and its results.

2. Scope of diagnosis and selected methods

On every floor, 1 or 2 testing sites were identified on beams or girders and 1 testing site on the beam heads. The diagnosis itself was performed by a proven combination of testing methods. They are:

- visual inspection,
- measurement of moisture content,
- resistance driving of the pin.

Figure 1. Timber ceiling structure supported with cast-iron pillars.

2.1. Visual inspection

The basis of a diagnosis of timber built into structures is a visual inspection. It is used to obtain information about the properties and condition of the material. The inspection of the condition of elements includes determining the type of timber used, timber defects, and detecting surface biotic and abiotic damage to timber [1].

The visual inspection reveals not only information about the condition of timber but also about technological procedures used, additional interventions into the structure and about an approximate age of the timber structure [1].

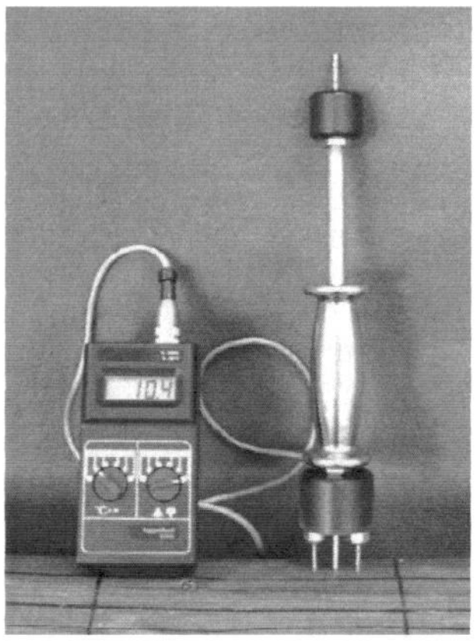

Figure 2. Device for measuring the moisture content of timber - Hygrotest 6500.

2.2. Measurement of moisture content

Timber is a hygroscopic material and has the ability to change its moisture content according to the humidity of its surrounding environment. The increased moisture content of timber influences both the risk of biotic degradation of timber caused by wood-decaying fungi and timber boring insects, and its physico-mechanical properties [1].

The CSN EN 335-1 [2] defines a potential occurrence of biological agents depending on the moisture content of solid wood. Generally, it is possible to say that if the moisture content of solid wood permanently exceeds 20%, timber is susceptible to be affected by wood-decaying fungi and timber boring insects [1].

The increased content of water can influence the survey results and at the same time, it can provide guidance e.g. in the identification of water leakage. The timber moisture content was measured by means of the Hygrotest 6500 device with an impact probe (figure 2). This method of measurement is regulated by CSN EN 13183-2 [3].

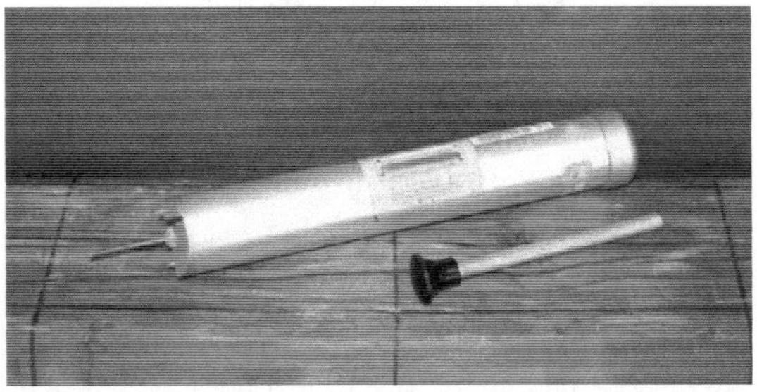

Figure 3. A device for measuring the depth of pin penetration - Pilodyn 6J.

2.3. Resistance driving of the pin

Resistance driving of the pin by means of the Pilodyn (Proceq) device is a method for quick estimate of the density of timber. Measurement with a Pilodyn device can be classified as semi-destructive testing because damage to the tested material is very low, almost negligible [4].

For the purposes of this research, a Pilodyn 6J model was used (figure 3), a simple mechanical device allowing to measure the depth of penetration of the pin with a diameter of 2.5 mm and a length of 40 mm driven into the timber with a constant driving force of 6 J. Using a dynamic impact of a calibrated shot, which is responsible for the penetration of the pin into the surface of the material, it is possible to measure the depth of pin penetration. The maximum depth of pin penetration is 40 mm.

When using Pilodyn, it is necessary to keep the direction of penetration strictly in the radial direction, due to a regular alternation of spring and summer parts of annual rings. When Pilodyn is used in the tangential direction, the pin can be potentially driven into one part of the annual ring only, and the measurement results can be significantly distorted. When measuring in the radial direction and with a deflection of less than 30°, the variability changes less than by 10% [5].

Density is one of the most important physical characteristics of timber, most mechanical properties of timber are positively correlated with density. To determine the density of structural timber on the basis of measuring the moisture content and penetration depth of the impact pin into timber, the following equation (1) can be used [6]:

$$\rho_{12} = 0.727987 - 0.027102 \cdot t_p \cdot [1 - 0{,}007 \cdot (w - 12)] \qquad (1)$$

where ρ_{12} - is the timber density with a moisture content of 12% [g/cm^3];
t_p - penetration depth of the impact pin into timber with known moisture content [mm];
w - timber moisture content at the time of measurement [%].

From the density, it possible to determine the flexural strength using regression dependencies [7]. The results of non-destructive testing of timber are always only indicative. In case they should be made more precise, it would be necessary to perform conclusive tests. Conclusive tests are performed according to ČSN EN 384 "Structural timber - Determination of characteristic values of mechanical properties and density" and ČSN EN 408 "Timber structures – Structural timber and glued laminated timber – Determination of some physical and mechanical properties" and are designed for a direct determination of some physico-mechanical properties.

Figure 4. The girder is connected with splayed heading joints above some of the pillars.

3. Results and detected defects

On the basis of the visual inspection, we determined the structural concept of the beamed ceiling:
On the cast iron pillars, there are headpieces and girders. The headpiece is connected to the girder by a pair of steel bolts with washers. The headpiece from hard boradleaf timber (probably oak) is carved, the girder, probably from fir timber, is beveled. Above some pillars, the girder is continuous, above the others, it is connected with butt joints or splayed heading joints. The girder head sits in the air pocket in the facade masonry and lies probably on an oak base. Perpendicularly to the girders, there are joists (ceiling beams), which are partially embedded into the girders (cogging) at some places. The joists, probably from oak timber, are full-edged, some of them are beveled up to approximately 1/3. The joist head sits in the pocket in the central wall. Perpendicularly to the joists, planks are nailed down with butt joints, and they represent the ceiling and the floor at the same time [8].

Other conclusions of the visual inspection:

- In many places of the ceiling, the outer parts of headpieces with steel bolts had been removed in the past, and this impaired the functionality of the pillar – headpiece – girder whole (figures 5 and 6).
- There are large shrinkage cracks in the girders (figure 7).
- In many places of the ceiling, parts of beams were cut out, and in the floors, there are holes for the facilities (figure 8).
- At the location of the elevator shaft, parts of beams were removed without replacement (figure 9).
- Removal of parts of beams, and highly inappropriate prostheisng performed (figure 10).

Figure 5. The outer parts of the saddle removed.

Figure 6. Non-functional steel bolt.

IOP Conf. Series: Materials Science and Engineering **385** (2018) 012018 doi:10.1088/1757-899X/385/1/012018

Figure 7. Shrinkage crack in the girder.

Figure 8. Weakening of the supporting beam.

Figure 9. Removal of beams without replacement.

Figure 10. Incorrectly performed prosthesing.

- Biotic damage to timber in the ceiling structures

Most of the elements in the ceiling structure had no visible biotic damage (they do not have traces of decay or affection by timber boring insects). The exceptions are, however, heads of three girders which are visibly damaged by decay and timber boring insects to a varying extent (figures 11 and 12). The actual damage can have a far greater extent because this damage typically spreads through the centre of the element first and only then gets to the surface of the structure.

Figure 11. Biotic damage to the girder.

Figure 12. Biotic damage to the girder.

The moisture content determined by Hygrotest 6500 and the density determined by Pilodyn were recorded for each test point in the table 1 and subsequently, an overall assessment was performed.

Table 1. Example of the record.

Measurement - beam			1	2	3	Average
Moisture content	ω	%	12.9	13.0	13.2	13.0
Pilodyn	t_p	mm	14.0	14.0	14.0	14.0
Estimate of timber density with 12% moisture content						
ρ_{12} =	351	kg/m^3				
Estimate of average flexural strength of timber with 12% moisture content						
f_m =	32.5	MPa				

4. Evaluation of the survey and recommendations

4.1. Moisture content of timber in the ceiling structures
The values of measured moisture content ranged from 9.3% to 16.2%. This means that all the elements of the ceiling structure have a value of moisture corresponding to the standard for timber built in the structure. This is also the reason why the detected biotic damage does not spread further.

4.2. Strength of timber in the ceiling structures
The estimate of average flexural strength of timber with 12% moisture content ranged from 18.8 MPa to 47.8 MPa.

4.3. Biotic damage to timber in the girder heads
Biotic damage to timber detected in the girder heads is a fundamental problem. Considering the fact that it was one third of the probes on the girder heads, it cannot be excluded that the same damage occurs in other (non-tested) elements. That is why we recommend to gradually expose all the heads and if biotic damage is found, to exchange them.

4.4. Removed headpieces above pillars
In a number of places, the outer parts of headpieces between pillars and girders, including steel bolts, were removed in the past. If ceiling structures are maintained, we recommend their adequate functional replacement.

4.5. Missing parts of girders and beams
In many places of the ceilings, end sections of girders (elevator shaft) and end sections or other sections of beams were removed in the past, and they were either not replaced or replaced in an inappropriate manner. We recommend adequate replacement of the missing parts of the elements.

5. Conclusion
The results of the pre-construction condition survey of the timber structures of ceilings in the building suggested considerable risks and potential complications of an attempt to newly utilize old industrial buildings which had not been maintained for a long period of time. What the architects considered at first sight to be a beautiful, monumental timber structure combined with cast-iron pillars with magnificent design, became rather a nightmare after a detailed diagnosis. The survey revealed places with significant decay (beam heads) and damaged supporting elements above pillars, which, due to the removal of parts of headpieces, also lost part of their stability. At the same time, ceilings are affected by a typical problem of the industrial objects used for a long time, where, due to multiple changes to

the function of the building and the facilities installed therein, various secondary openings in the structure were created, non-professional replacement of parts of elements was performed, etc. All this brings a new insight into a potential restoration, where simple maintenance of the existing timber structure without changes is not quite realistic.

Acknowledgement
This paper has been worked out under the project No. LO1408 "AdMaS UP - Advanced Materials, Structures and Technologies", supported by Ministry of Education, Youth and Sports under the „National Sustainability Programme I".

References

[1] Hermankova V, Kloiber M, Tippner J and Anton O 2011 *Proc. Conf. on Testing and Quality in Construction (Brno)* vol 10 (Brno: VUT) pp 79-96 (in Czech)
[2] CSN EN 335-1: 1995 *Durability of wood and wood-based products – Definition of use classes – Part 1: Application to solid wood* (in Czech)
[3] CSN EN 13183-2: 2002 *Moisture content of a piece of sawn timber - Part 2: Estimation by electrical resistance method* (in Czech)
[4] Kloiber M, Tippner J, Hermankova V and Stainbruch J 2012 *Proc. Int. Conf. on Structural Analysis of Historical Construction* (Wroclaw) pp 2035-2043
[5] Kuklik P 2007 *Determining the Properties of Materials When Evaluating Existing Structures* (Prague: CVUT) p 10 (in Czech)
[6] Ronca P and Gubana A 1998 *Const. Build. Mater.* **12** 233-243
[7] Kuklik P 2005 *Wooden Structures* (Prague: CKAIT publication) (in Czech)
[8] Hermankova V and Anton O 2018 *Final report on diagnostics of wooden construction OSEVA Prizova Brno* (in Czech)

IOP Conf. Series: Materials Science and Engineering **385** (2018) 012019 doi:10.1088/1757-899X/385/1/012019

Monitoring the long-term durability and microstructure of polymer concrete patching materials with high content of secondary raw materials using advanced methods

J Hodul, L Meszarosova and R Drochytka

Institute of Technology of Building Materials and Components, Faculty of Civil Engineering, Brno University of Technology, AdMaS Centre, Veveri 331/95, Brno 602 00, Czech Republic

Abstract. For commercially produced polymer systems, only primary raw materials are used as fillers. These fillers are predominantly pure silica sand of various grading. The substitution of primary materials with secondary raw materials, and their efficient use in the manufacture of polymeric repairing substances, is an indisputable advantage of the use of secondary raw materials. There is currently no general methodology for monitoring the long-term durability and microstructure of repairing substances on a polymeric basis. In this article, long-term durability was assessed by subjecting developed substances with a high content of secondary raw materials (60% fly ash, 75% glass recyclate) to freezing cycles and subsequent cohesion testing. Additionally, through exposure of a concrete element reprofiled by the developed substance, the weather effects for approximately one year. The microstructure of polymeric repairing substances cannot be observed by the same methods used for silicate substances. With regard to the type of tested substance and the filler used, advanced CT tomography and high resolution optical microscopy were chosen for monitoring the microstructure. Based on the results of the tested substances cohesion with the underlying concrete following the freezing cycles, and the evaluation of the taken images, it can be stated that the test repairing substances have excellent long-term durability and can be used in different applications. Furthermore, it was confirmed that replacing the primary filler with the secondary raw material does not reduce the long-term durability of the polymeric repairing substances on an epoxy basis. It was confirmed that only slight degradation occurs due to the photooxidation of epoxy resin by UV radiation, but this does not have a significant effect on physical and mechanical properties.

1. Introduction

In recent years, the use of polymer substances in the field of repairing substances has become widespread in the building industry. All industrially manufactured repairing substances use primary raw materials. The use of a polymer binder, together with secondary raw materials, is therefore an important topic.

Vipulanadan and colleagues concluded that the optimal amount of polymer binder in a mixture ranges from 14% to 16% [1-3]. Regarding the use of secondary raw materials in the epoxy matrix, Singl and Chawl traced a positive contribution alongside the use of fly ash. They found that with an increasing addition of fly ash (in the range of 30% to 54%), the strength of the resulting composite increased [1, 4]. Fly ash is a by-product of coal combustion in thermal power plants and is rich in silicon dioxide (SiO_2), iron oxide (Fe_2O_3), and aluminium oxide (Al_2O_3) [5]. Its use in polymer matrix (comprising polyethylene terephthalate, polyester, or polyoxomethylene composites) was addressed by

Chand, Li, Koszkul, Kwiatkowski and Baheti [1,6-8]. This paper follows up their research and its partial aim is to explore the possibilities of using fly ash after denitrification of flue gas process and waste container glass.

2. Materials

Following the work of Singla and Chawl [4], the possibility of using fly ash in the epoxy matrix was investigated. Compared to the work of these researchers, the reprofiling material was, from an economic and ecological point of view, filled with the maximum possible quantity of secondary raw materials as it concerns the workability and preservation of important properties.

2.1. Composition of the polymer concrete patching material (PM)

The tested polymeric repairing substances contained a bisphenol A-based epoxy resin (P-A) as a binder. The polymer binder is composed of component A, which is an epoxy monomer (epoxy resin), and hardening agent B based on aliphatic and aromatic amines. Two substances, PM-P Premium) and PM-E (Economy), were tested for long-term durability and microstructure. The composition of these substances is shown in table 1.

Table 1. The composition of tested polymer concrete patching materials.

Formulation	Binder	Filler
PM-P	40 % P-A	60% filter fly ash devalued by the denitrification process
PM-E	25 % P-A	75% crushed transparent glass recyclate (0-1.5 mm)

2.2. Properties of input raw materials

A two-component, low viscosity solvent-free epoxy-based substance with a density of 1.1 kg/l was used as a binder. Its workability ranged from 15 minutes to 20 minutes. The modulus of elasticity is 3000 N/mm^2, elongation at break is 4% to 7%, and compressive strength ranges from 80 MPa to 90 MPa. This resin is primarily used for repairing mechanically demanding parts, and to repair locations exposed to thermal and chemical stress.

The fly ash, which was influenced by the denitrification of the flue gas was used as a filler in one material (PM-P). The fly ash used was generated by the combustion of pulverised coal in a thermal power plant at high-temperature, in the north of Bohemia. Due to the new technology of flue gas denitrification, ammonium ions are present in fly ash created in the power plant. In the second substance, glass recyclate was used as a filler. These are grounded glass fragments of transparent container glass, collected during waste separation.

Table 2. Chemical composition of the fly ash (filler to PM-P material).

Parameter	SiO_2	CaO	Total organic carbon (TOC)	Ammonium ions	Sulphates
Amount	53.40%	4.13%	5.20%	3.79 mg/kg dry	0.25%

Table 3. Chemical composition of the glass recyclate (filler to PM-E material).

Parameter	SiO_2	Na_2O	CaO	MgO	Al_2O_3	Fe_2O_3	K_2O	BaO
Amount	74.80%	11.91%	7.69%	4.27%	0.55%	0.17%	0.19%	0.09%

3. Methods

For cured epoxy repairing polymer concrete patching mortar, the effect of cyclic temperature change (cyclic freezing and defrosting) was monitored over a total of 300 hours. The microstructure was also monitored by means of an optical microscope and computer tomography (CT).

3.1. Determination of adhesion following freezing cycles
Samples coated with a thin layer of tested substances were first saturated with water and subsequently subjected to cyclic temperature changes in the range of -15°C to 20°C. One cycle lasted six hours. During this time, four hours were employed for cooling to -15°C and two hours for heating to 20°C. The substances were exposed to 50 freezing cycles.

3.2. Determination of the microstructure
Monitoring of the microstructure of the tested polymeric repairing substances was performed by computed tomography (CT) and a high resolution optical microscope.

3.2.1. Optical microscope. The developed polymer concrete patching mortars (PM-P, PM-E) were monitored by an optical microscope for effects of weathering. Samples of individual substances were taken and subsequently observed with an optical microscope.

3.2.2. CT Tomography. Using special computed tomography (CT) designed for building materials, it was possible not only to observe the microstructure of the test samples, but also to consider the internal structure and distribution of pores and filler. In our case, whether the influence of weathering on the reprofilated part of the concrete curb degraded the polymeric repairing substance was monitored. As a tomographic sample, a cut-off section of the reprofilated corner of the concrete curb was used (see figure 1 and figure 2).

Figure 1. Sample of the PM-E patching mortar for CT Tomography.

Figure 2. Sample of the PM-P patching mortar for CT Tomography.

The observation primarily focused on structural failures, particularly crack formation, the separation of the repairing substance from the underlay, the irregular dispersion of filler, and the inhomogeneity of the developed polymer patching mortar.

4. Results and Discussion

4.1. Determination of the adhesion after freezing cycles
The adherence results of the tested repairing mortars after 50 freezing cycles (T50) are shown in table 4. The samples examined showed a breakage in the concrete underlay, which means that the adhesion of the repairing material to the underlay exceeded the cohesion of the underlying concrete. Highest cohesion was recorded for the PM-P substance, namely 2.2 MPa. It can therefore be concluded that developed substances are able to withstand temperature cycles and low temperatures, while

maintaining high adhesion to the concrete underlay. The use of secondary raw materials as fillers did not affect the cohesion of substances following cyclic temperature changes.

Table 4. Results of the adhesion after 50 freezing cycles (T50).

Type of patching mortar	Stress at breakage before T50 [MPa]	Stress at breakage after T50 [MPa]	Place of breakage
PM-P	2.20	1.28	Sub-concrete
PM-E	1.67	1.48	Sub-concrete
PM-REF	1.98	1.02	Sub-concrete

4.2. Determination of the microstructure

4.2.1. Optical microscope. From figure 3 and figure 4, it is evident how the epoxy resin, due to its low viscosity, flows into the porous structure of the underlay concrete and thus creates a strong connection at the contact zone. This is very important to ensure good adhesion of the reprofiled mortar to the underlay. The PM-P sample in figure 4 is not sufficiently transparent for viewing in transmitted light, but the interface is visible very well, even in incident reflected light. The figure shows the PM substance in the left lower part, and it is obvious that it is very well grounded in the underlying concrete. In the microscopic observation of the PM-E sample surface (figure 3), it is possible, with transmitted light, to observe in detail the interface of the existing concrete underlay (at the top of the figure), and the polymer concrete patching mortar, due to its high transparency. Furthermore, it is possible to observe the perfect anchoring of the reprofiling substance into the concrete underlay.

Figure 3. Connection detail of the PM-E reprofiling substance to the concrete base (at the top of the figure) (100x).

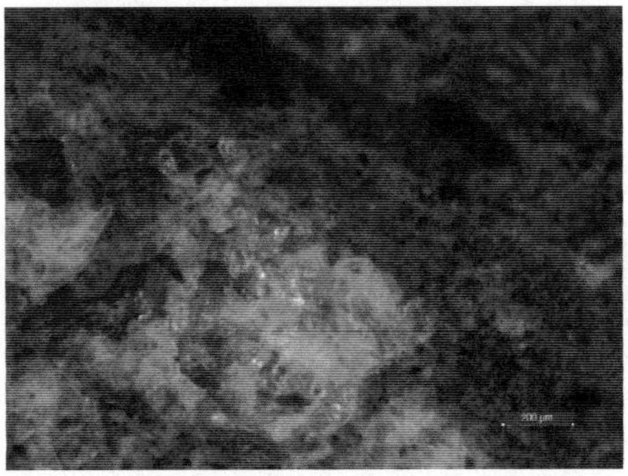

Figure 4. Connection detail of the PM-P reprofiling substance (in the lower left part) to the concrete substrate (100x).

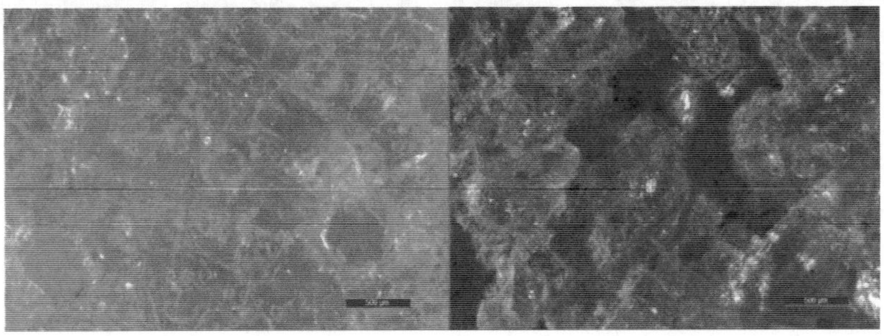

Figure 5. Surface detail of PM-E sample – left sample intact by sunlight (UV rays), UV affected surface can be seen in the right image (50x).

Figure 5 clearly shows the perfect coating of glass shards in an epoxy matrix, which is slightly yellowish. The micrograph in the figure 5 right shows the degradation effect of UV radiation, which affected the applied substance as a result of the exposure of the reprofilated curb to weather effects. Due to the effects of solar radiation, the epoxy matrix decreased slightly as a result of photooxidation (macromolecule breakdown by UV radiation).

Figure 6 on the left shows the microstructure of the PM-P substance surface, where it is possible to observe the uniform distribution of fly ash grains (cenospheres) in the epoxy matrix. In figure 6 on the right, it is possible to observe the apparent change in the surface of the substance due to photooxidation; this is a sample that has been subjected to weathering for 300 hours.

From the microscopic analysis, overall, the results indicate that the effect of atmospheric influences does not produce micro-cracks or the separation of substances from the concrete underlay.

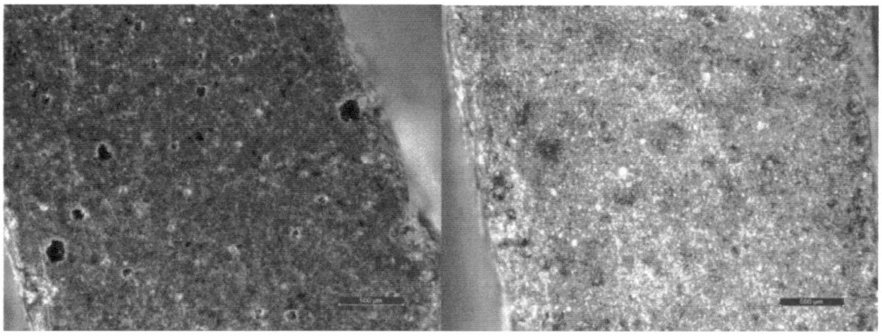

Figure 6. Surface detail of PM-P sample – left sample intact by sunlight (UV rays), UV affected surface by photooxidation can be seen in the right image (50x).

4.2.2. CT Tomography. From a computer tomography (CT) observation (see figure 7 and figure 8), it can be concluded that the patching mortar is compact in both variants on overall cross-section, shows no cracks, tears or higher porosity or separation of the repairing substance from the underlay, compared to the comparative REF samples. Due to the appropriate viscosity, the substance is firmly anchored to the underlay. Additionally, the PM-P substance had a denser and more compact structure compared to the PM-E substance, which can be said to have higher strength and better long-term durability.

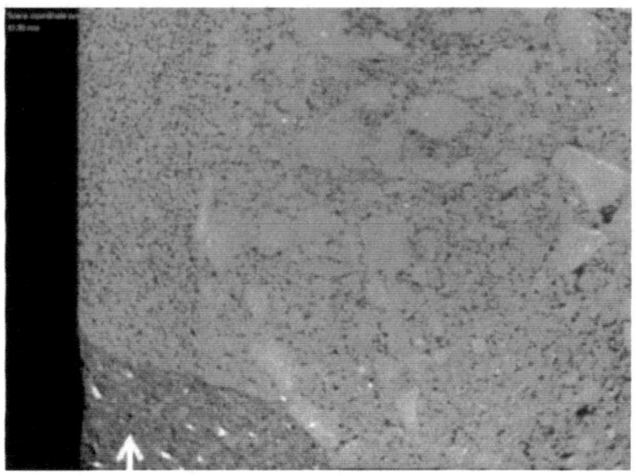

Figure 7. Internal structure of the PM-E substance (indicated by arrow) and detail of connection to the concrete curb.

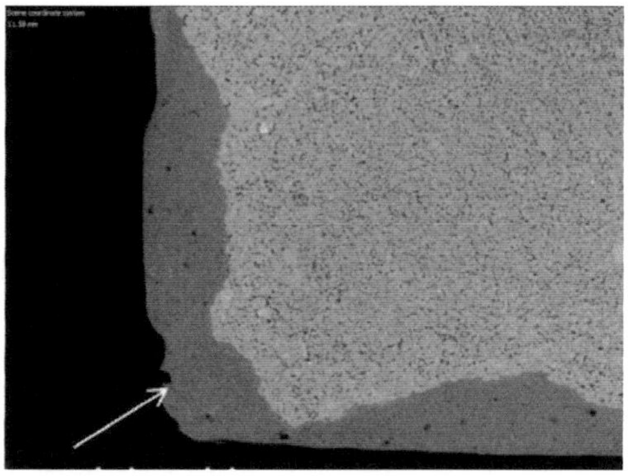

Figure 8. Internal structure of the PM-P substance (indicated by arrow) and detail of connection to the concrete curb.

5. Conclusion

In this paper, the long-term exposure influence of two developed polymer concrete patching mortar samples to atmospheric influences, and the monitoring of their microstructure using progressive analytical methods was investigated. Approximately one year later, the surface was slightly degraded by UV radiation (photooxidation). Based on the visual assessment, slight cracks and wrinkles was also observed. In the case of the substance with waste container glass as a filler, the surface damage was greater than for the substance filled with high temperature fly ash contaminated with ammonium ions. This greater surface disturbance was likely due to the transparency of the glass filler, which allowed for easier penetration of the sun's rays, and more intense UV exposure to the epoxy matrix. It is known that strong UV radiation can weaken chemical bonds in epoxy resin and thus accelerate aging and degradation of the material due to photooxidation. However, this effect did not impact physical and mechanical properties. The substance was subsequently subjected to a microstructure analysis by CT tomography and using an optical microscope. There was no evidence of any substance breakage or loss of cohesion with the sub-concrete in either case. A significant benefit of these patching mortars is the use of a large amount of secondary raw material in the form of fillers. Newly developed polymer concrete patching mortars indicate, in addition to existing commercially produced substances, better physical and mechanical parameters.

Acknowledgments

This work was supported by the project FAST-J-18-5343 “Development of highly resistant repair composite for the reconstruction of cast basalt objects using secondary raw materials”.

References

[1] Uygunoglu T, Gunes I and Brostow W 2015 *Mat. Res.* **18** 1188–1196
[2] Vipulanandan C and Dharmarajan N 1987 *Cem. Concr. Res.* **17** 219–230
[3] Vipulanandan C, Dharmarajan N and Ching E 2006 *Mater. Struct.* **21** 268–277
[4] Singla M and Chawla V 2010 *JMMCE* **9** 199–210
[5] Baheti V, Militky J, Mishra R and Behera B K 2016 *Composites Part B* **85** 268–276
[6] Li Y White D J and Peyton R L 1998 *Resour. Conserv. Recycl.* **24** 87–93
[7] Koszkul J and Kwiatkowski D 2004 *J. Mater. Process. Technol.* **157** 360–363
[8] Purohit R Sahu P Rana R S Parashar V and Sharma S 2017 *Mater. Today* **4** 3102–3109

Analytical methods for the determination of the input material quality for gypsum wallboard production

M Holub[1], A Estokova[2], S Demcak[2] and C Stone[1]

[1] Laboratory of Excellent Research, Faculty of Civil Engineering, Technical University of Kosice, Park Komenskeho 10/A, 042 00, Kosice, Slovakia
[2] Institute of Environmental Engineering, Faculty of Civil Engineering, Technical University of Kosice, Vysokoskolska 4, 042 00, Kosice, Slovakia

Abstract. Determination of the input material quality is crucial for calcination and for gypsum wallboard production itself. The article focused on the practical use of analytical methods in building practice. It presents procedures on how to determine the content of calcium sulphate in the raw material as well as the results from real measurements of the raw material samples, further calculation and evaluation. Together four samples from three different sources located near the potential gypsum wallboard production line were analysed using X-ray Diffraction (XRD) and Thermogravimetry (TGA).

1. Introduction

Wide use of gypsum wallboard in the construction of buildings (residential and commercial) is due its relatively easy installation with minimal finishing, which saves time and financial resources.

Gypsum wallboard consists of a hardened gypsum – containing a core covered with paper. It creates a sandwich structure which is commonly manufactured by placing an aqueous slurry (comprised of calcined gypsum) between two sheets of paper. Hardening of the core slurry is caused by rehydration of the calcined gypsum (calcium sulphate dihydrate). Hardening of the final product is controlled by heat treatment in a dryer which removes redundant water from the wallboard.

The manufacturing process begins with the addition of pre-generated foam to the slurry mix in order to decrease the final weight of the board. Foam is generated in a foam generator where an appropriate foaming agent is mixed with air and water.

Paper located on a long moving belt creates a carrier for the foamed gypsum slurry flowing from the mixer. Slurry is placed on top of the first paper, covered by a second paper, thus the sandwich is complete and passes through a forming station. A forming station is setup to finalize the required thickness of the gypsum board which can be changed. The width of the board is also controlled by a forming station, but it is rarely changed because of the standard dimension of 1200 mm [1].

The setting of the whole mix is controlled by adding admixtures in order to facilitate the continuous operation of the plant. Setting of the slurry is fast and begins immediately after it passes through the forming station. Then, as the belt moves, the board passes on the belt several meters to provide sufficient time for hardening and subsequent cutting. As mentioned before, the final step in gypsum wallboard production is drying so the boards are moving continually into a board dryer where they acquire their final properties. For simplification, the whole process is depicted in the figure 1.

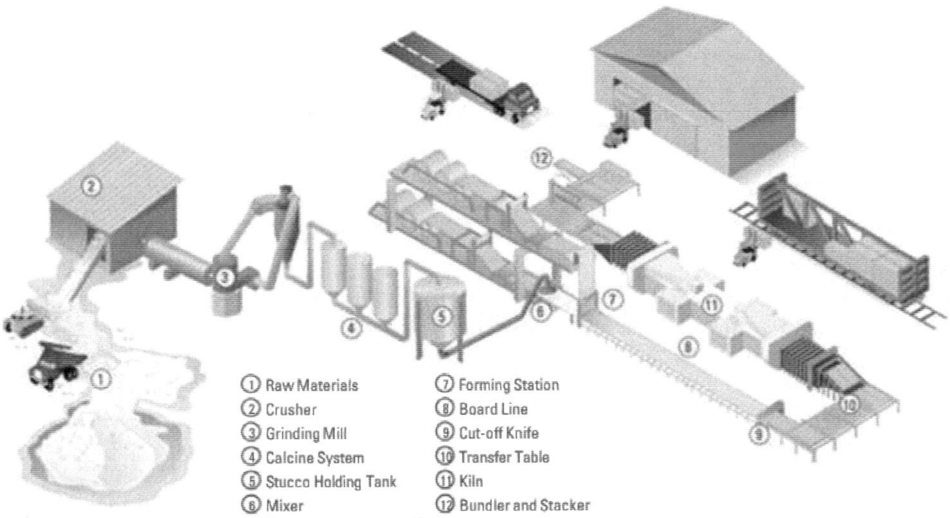

Figure 1. Typical process of gypsum wallboard manufacturing [2].

As mentioned previously the major ingredient consists of calcined gypsum, i.e. stucco, which is produced form natural rock. It is also possible to use a synthetic material (e.g. from desulphurisation of flue gases) with high content of gypsum.

The process of the stucco manufacture is in both cases similar. In the first stage the input material is put into a kiln to remove the moisture from the rock. As soon as the excess water is removed the material is ready for grinding to a required fraction. Next step consists from calcination, while the ground gypsum is put into a calcination unit (rotary kettle), where following chemical reaction takes place:

$$CaSO_4.2\,H_2O + heat \rightarrow CaSO_4.\,0.5H_2O + 1.5\,H_2O \uparrow \quad (1)$$

The hardening of stucco represents a reverse reaction after mixing of stucco with an appropriate amount of water connected with heat release. The quality of the input material is crucial for the calcination and also for the production of the gypsum wallboard itself [3].

According to the economics, in each case of gypsum wallboards plant design, the first and most important fact is proximity to the source of input rock or synthetic material with appropriate quality. This appropriate quality is defined by the content of gypsum (purity) and is, as mentioned above, key for the manufacturing of wallboard. High levels of purity are desired because lower masses of wallboard can be produced. That means the higher the purity the lower the mass of wallboard produced. From this fact it is needed to find an input material that contains at least 80% gypsum. Generally, the purity of natural rock varies usually from 80 – 96% according to local conditions. The purity of synthetic material is more than 95% and another benefit is its stable composition compared to natural sources.

The article is focused on the practical use of analytical methods for the determination of the input material quality for gypsum wallboard production. It presents a procedure on how to determine the content of calcium sulphate in the raw material as well as results from the real measurements of the raw material samples (suitable/unsuitable). Together four samples from three different sources were analysed using X-ray Diffraction (XRD) and Thermogravimetry (TGA).

2. Materials and methods

The content of the gypsum in the rock or synthetic material can be determined by several analytical methods including Differential Scanning Calorimetry/Thermal Gravimetry (DSC/TGA), X-Ray Fluorescence Spectroscopy, X-Ray Diffraction, SO_3 analysis and also their combination [4]. As mentioned before results from XRD and DSC/TG are presented. XRD was used to define the quality of the samples and quantitative analyses were performed by TG. Four samples from different locations (near the location of potential gypsum wallboard plant) were analysed. Three of them were from the Slovak Republic and one was from Poland. Two samples were collected from the same location but from different layers in the mining locality.

Qualitative analyses of the samples were carried out with an X-Ray Diffraction (XRD) using diffractometer Bruker D2 Phaser (Bruker AXS, GmbH, Germany) in Bragg-Brentano geometry (configuration Theta-2Theta), using the 1.54060 Å CuKα radiation, Ni Kβ filters and scintillation detector at a voltage of 30 kV and 10 mA current. The XRD patterns were processed using the software Diffrac.EVA v. 2.1. The ICDD PDF database (ICDD PDF – 2 Release 2009) was utilized for the phase identification.

The thermogravimetric analysis experiments and differential scanning calorimetric measurements (DSC) were carried out by using an instrument (STA 449F3 Jupiter, Netzsch, Germany). The record of mass loss in response to the temperature was collected to determine both the TG and derivative thermogravimetric (DTG) curves. Thermal decomposition of the samples was monitored from an ambient temperature to 300°C at a constant heating rate of 10°C/min under a high-purity nitrogen atmosphere of 100 mL/min.

3. Results

Thermogravimetry can be a good analytical method for the determination of gypsum rock purity because of losing crystalline water bonded in calcium sulphate dihydrate during the heating. As the dihydrate is heated the elimination of the water takes place in two steps. In the first step dihydrate is transformed to a form of hemihydrate (see equation 1) and then to the anhydrous form. It is also possible to separate the determination of both forms, but it is slightly complicated because the temperature ranges of these reactions are close together. For testing of gypsum content in natural rock this approach is not necessary and it is only needed to know the total mass loss to anhydrous form. According to stoichiometry, complete dehydration of 100% pure material is equal to mass loss of 20.9% and it can be divided into two steps (as mentioned before): 15.7% mass loss from dihydrate to hemihydrate and subsequent 5.2% mass loss to anhydride [5]. The assumption is valid only in the case, that the natural rock does not contain any other compound in the hydrated form. That is the reason why it is needed to use another analytical method for quality determination (e.g. XRD).
The results of thermogravimetry and further calculation are summarized in table 1.

Table 1. Result of thermogravimetry measurements.

Sample	Mass Loss [%]	Gypsum content [%]
DN	18.99	90.86
GOS1	5.53	26.46
GOS2	18.13	86.75
SH	3.25	15.55

It is clearly visible that two of the samples with low content of gypsum are not applicable for potential gypsum wallboard production. On the other hand, two have a high potential and can be used as a source for stucco production.

For the comparison of different samples (applicable and not applicable) representative results from thermogravimetry measurement are depicted in the figure 2 and 3. According the DSC curve (figure 2) it is possible to observe two separate reactions in the dehydration of the sample. First is located in the temperature range approx. from 120°C to 150°C (transformation to hemihydrate) followed by another one to complete dehydration around 200°C.

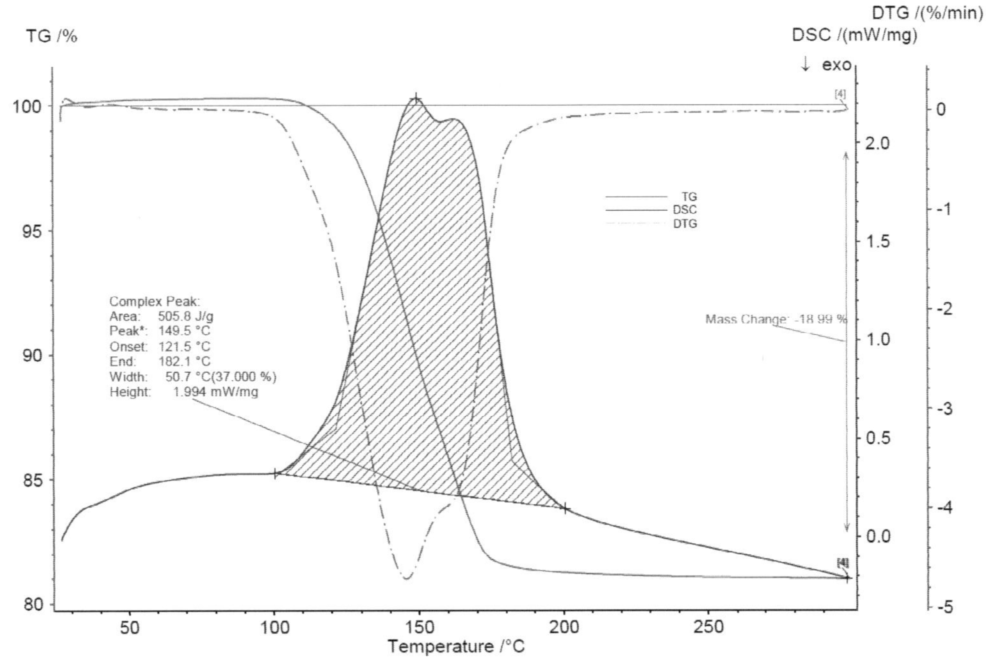

Figure 2. Thermogravimetric analysis of rock material (Sample DN - high content of gypsum).

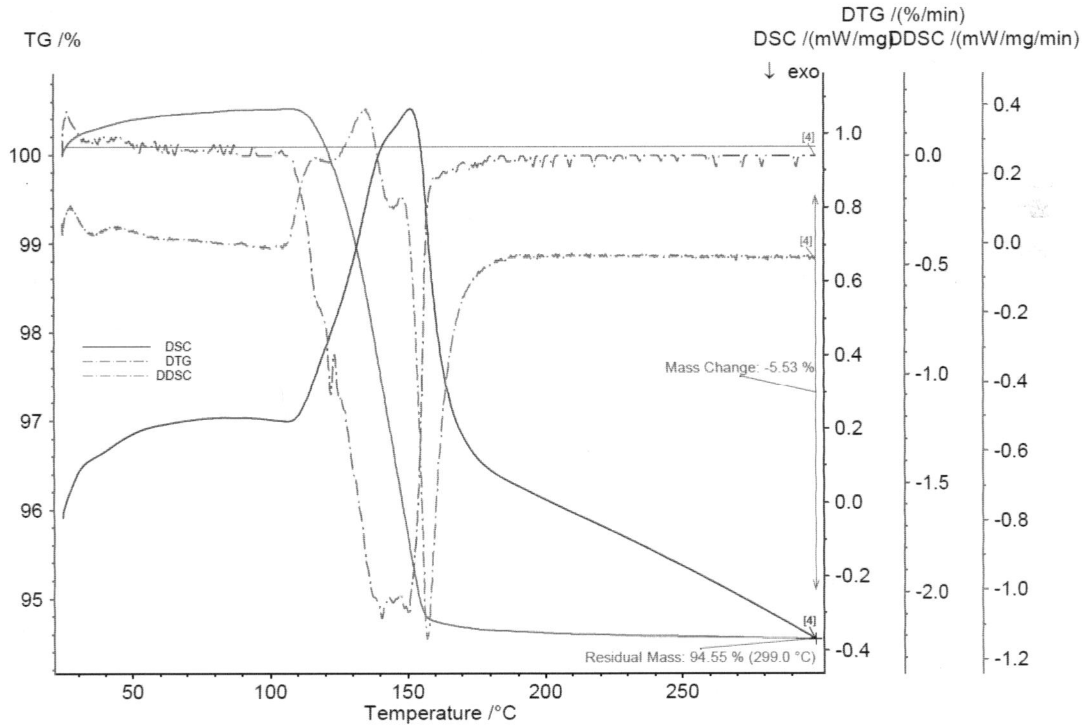

Figure 3. Thermogravimetric analysis of rock material (Sample GOS1 - low content of gypsum).

The GOS samples were collected from the same locality with different results of gypsum content. Probably it is caused by heterogeneous structure of the deposit and the content of gypsum is varying through the layers. According the XRD, the GOS1 sample contains additional mineral phases e.g. calcite, dolomite or quartz (see figure 4).

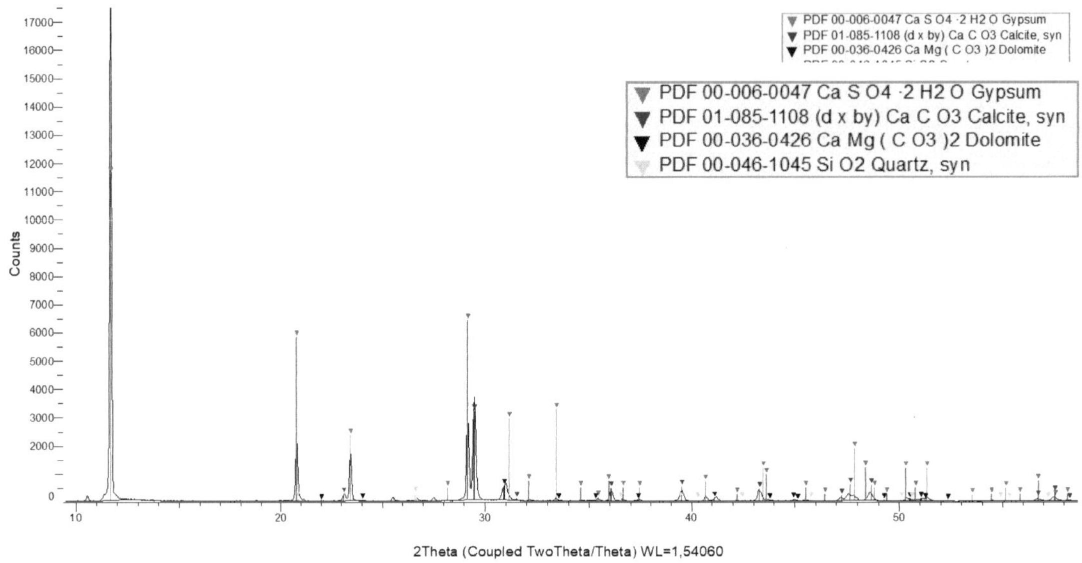

Figure 4. XRD analysis of rock material (Sample GOS1 - low content of gypsum).

To confirm the thermogravimetric analysis of DN sample, XRD pattern is also presented (figure 5). From the pattern it is evident that the sample contains only one hydrated compound which is represented by the calcium sulphate dihydrate. No others phases were observed.

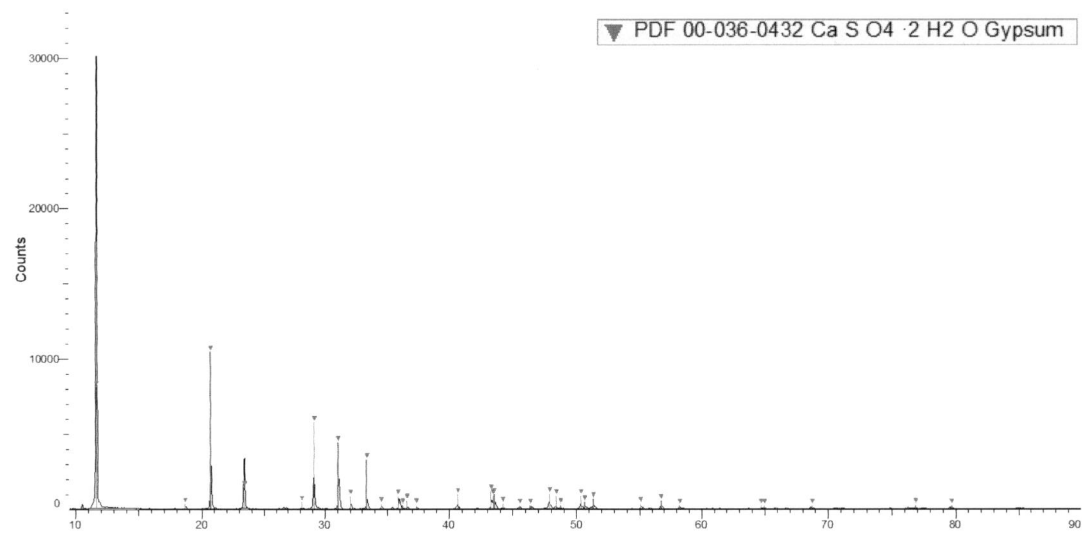

Figure 5. XRD analysis of rock material (Sample DN - high content of gypsum).

4. Conclusion

Based on the presented results, it can be stated, that the combination of XRD and TG can be useful for fast and precise determination of calcium sulphate dihydrate in natural rock material. It can also be used for the quality measurement of the material in the production itself. According to analyses, two of four analysed samples proved to be appropriate for use in gypsum wallboard production (content of $CaSO_4.2H_2O$ more than 85%). Another two samples exhibit inadequate properties because of impurities content and must be excluded from the process of production planning.

Acknowledgment

This work has been supported by the Slovak Grant Agency for Science (Grant No. 1/0563/15).

References

[1] Savoly A and Elko D 2015 GEO Specialty chemicals, INC. *Gypsum wallboard chemical additives*

[2] Georgia-Pacific Gypsum LLC 2010 Typical process of manufacturing gypsum board – Gypsum board production line (Scheme)

[3] Lootens D, Ampudia A, Hampel Ch and Mueller M 2012 Laboratory scaling of gypsum board production *ZKG International* **1** pp.1-6

[4] Henkels P J and Gynor J C 1996 *American Chemical Society Division of Fuel Chemistry* **41** 570-574

[5] Mettler-Toledo AG *Thermal Analysis Application No. UC 264* Analytical 2010 Determination of calcium sulfate dihydrate and hemihydrate in cement

Variability in cement properties - influence on bleeding of cement paste

D Jancarikova[1], R Hela[1], D Netsvet[2] and T Perina[3]

[1] Brno University of Technology, Faculty of Civil Engineering, AdMaS Centre, Purkynova 139, Brno 612 00, Czech Republic
[2] Belgorod State Technological University named after V.G. Shoukhov, Ulitsa Kostyukova, 46, Belgorod 308012, Russia
[3] Brno University of Technology, Faculty of Civil Engineering, Veveri 331, Brno 602 00, Czech Republic

Abstract. The aim of this study is to find relation between cement properties and tendency of cement paste (concrete) to bleeding. We tested 19 cement samples collected from different batches. On cements were tested granulometry, specific surface, strengths, alkali content, content of C_3A modifications, setting times and bleeding. Tendency to bleeding of all 19 samples was tested in 3 types of cement paste mixtures. One reference mix and two mixtures with superplasticizers from different producers. Results show, that we can see partly relation between specific surface area and tendency to bleeding. With lower specific surface area, the tendency to bleeding is higher. While the content of alkali seems to have small influence. Where the higher content of alkali means higher tendency to bleeding. On the other hand, these results are not conclusive and it is required further testing on different parameters such as content of sulphur and type of gypsum added to clinker during grinding.

1. Introduction

Cement is one of the most worldwide used binder in building materials. In European Union is cement under control of standard EN 197 and tested according to EN 196. This standard controls strength, chemical composition, setting times, fineness, heat of hydration and amount of water soluble chromium. If all these properties are in allowed range, cement should have constant behavior in concrete mixture. Nevertheless, sometimes there are differences in concrete behavior on situ. Even if the composition of concrete mixtures is constant, workability is different and we can see signs of concrete bleeding in various intensities.

Bleeding is a phenomenon where water in a freshly mixed cement-based composite is drained to the surface when solid component of the mixture consolidate in a form [1]. Bleeding influences the quality of fresh concrete and durability after the concrete is hardened [2]. There were several bleed channels and pores on the surface of freshly placed concrete, which bleeded too much [3]. It is known that concretes which exhibit less bleeding, have a stronger surface [4].

During the investigations, it was observed that concrete mixtures containing high cement contents reduce bleeding [5]. When cements having high fineness were used, the quality of bleeding and the rate of bleeding were decreased. It was also found that cements having a high C_3A content reduces bleeding [6].

Topic for our research came from a concrete producer. They have a problem with instability of concrete behavior on situ. Even if they use the same composition of mixture, concretes with cements from different batches have different behavior - sometimes there were signs of bleeding. In cooperation with cement producer we took 19 samples of cements from consecutive batches in cement plant. Samples were determined on granulometry, alkali content, C_3A content and its modifications, setting time, strengths and tendency to bleeding.

2. Materials and Methods

Tendency to bleeding was tested on 3 types of cement paste mixtures. In these mixtures were used all 19 samples of cement. All tests were done in laboratory room with temperature around 22 °C. First mixture - REFERENCE - plain cement paste, second mixture - with superplasticizer Dynamon SX 14, third mixture - with superplasticizer Stachement 2489

2.1. Tested cements

19 samples of cement CEM I 42.5 R were taken from consecutive batches in the time span from September to November 2017. C_3A and alkali content in cement is show in table 1.

Table 1. C_3A and alkali content.

Sample no.	C3A			alkali		
	cubic	ortho-rhombic	total	K_2O	Na_2O	Na_2O eq.
1505	3.75	2.14	5.89	0.96	0.14	0.77
1535	3.86	2.13	5.99	0.96	0.14	0.77
1553	3.37	2.46	5.83	0.97	0.14	0.78
1580	3.17	2.81	5.98	0.94	0.15	0.77
1601	2.86	3.38	6.24	0.91	0.15	0.75
1630	3.02	3.40	6.42	0.94	0.15	0.77
1651	2.48	2.76	5.24	1.00	0.17	0.83
1675	2.32	3.46	5.78	0.99	0.17	0.82
1692	2.51	3.18	5.69	1.00	0.15	0.81
1721	2.23	3.54	5.77	0.97	0.17	0.81
1738	2.54	2.94	5.48	0.95	0.16	0.79
1767	2.52	3.12	5.64	0.96	0.15	0.78
1785	2.81	2.20	5.01	0.95	0.18	0.81
1810	2.36	3.73	6.09	0.95	0.16	0.79
1827	2.34	3.16	5.50	0.99	0.16	0.81
1850	2.71	3.81	6.52	0.97	0.16	0.80
1867	2.92	3.02	5.94	0.97	0.16	0.80
1889	2.52	3.48	6.00	0.93	0.15	0.76
1905	2.35	3.92	6.27	0.93	0.16	0.77

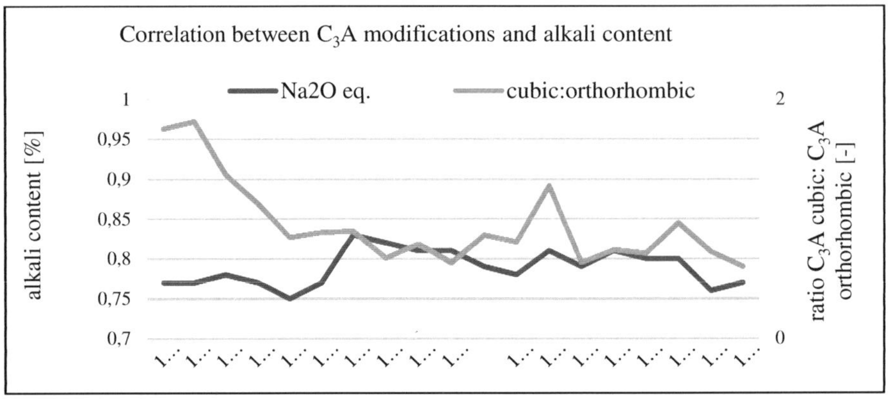

Figure 1. Correlation between content of C_3A modifications and content of alkali.

Total content of C_3A ranges between 5.0 and 6.5 % and the ratio between cubic and orthorhombic modification is changing from 0.6 to 1.8. As it can be seen on Figure 1 with higher content of alkali is lower ratio C_3A cubic / C_3A orthorhombic.

2.2. Superplasticizer

For preparation testing cement pastes were used two types of superplasticizers. Dynamon SX 14 from MAPEI producer which is a totally formaldehyde-free water solution of acrylic polymers that can effectively disperse cement granules with secondary components that significantly improve the cohesion and the pumpability of concrete. Second admixture was Stachement 2489 from Stachema producer. Stachement 2489 is polycarboxylate based admixture with high plasticizing effect especially used for ready-mix concrete.

3. Results and discussion

3.1. Granulometry and specific surface area

Cement granulometry was determined by laser particle size analyser Mastersizer 2000. Results of the analysis are shown in Figure 2 - 4. As the second parameter of particle size distribution was determined the middle particle size - Figure 5-6. Specific surface area was determined by automatic analyser according to CSN EN 196-6 - Permeable method (Blaine).

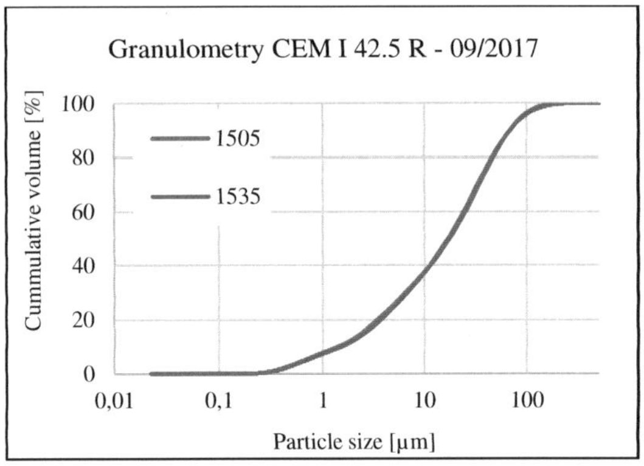

Figure 2. Granulometry CEM I 42.5 R-September.

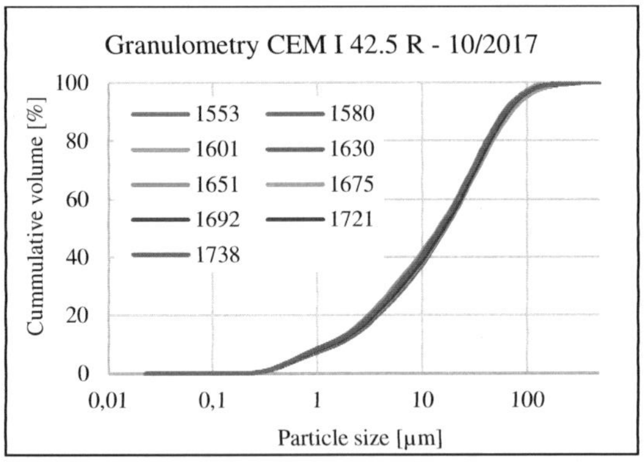

Figure 3. Granulometry CEM I 42.5 R-October.

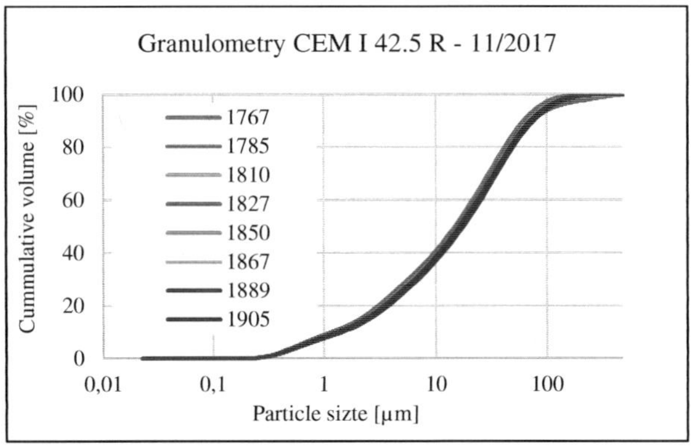

Figure 4. Granulometry CEM I 42.5 R – November.

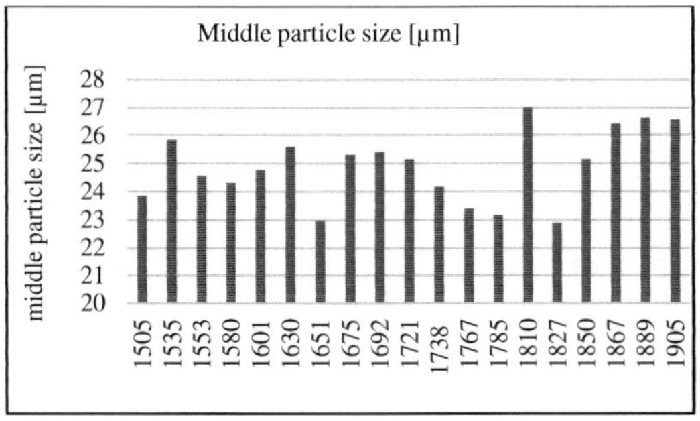

Figure 5. Middle particle size - CEM I 42.5 R.

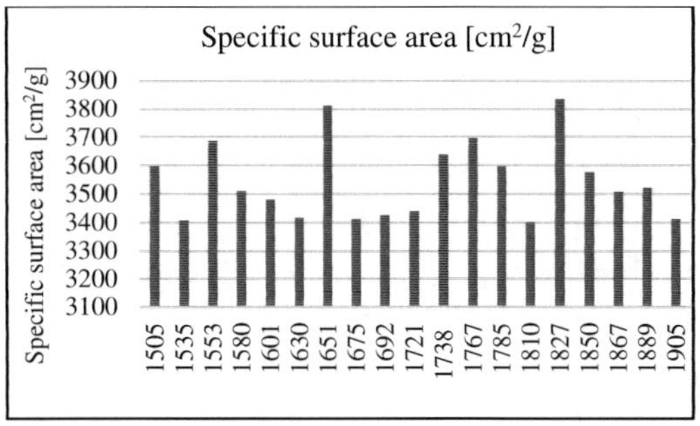

Figure 6. Specific surface area - CEM I 42.5 R.

As it can be seen in Figure 2 - Figure 4 granulometry of cements is quite stable. Even if granulometry of cement seems to be constant, differences are in middle particle size as can be seen in Figure. Value of middle particle size is in range 23 - 27 µm. Specific surface area is variable too and is in the range of 3400 - 3820 cm^2/g.

3.2. Setting time and strengths

Setting time was determined according to CSN EN 196-3. Tensile strength and compressive strength were determined according to CSN EN 196-1. Results are shown in table 2.

Sample no.	Setting time [min]	Tensile strength [N/mm^2]		Compressive strength [N/mm^2]		Sample no.	Setting time [min]	Tensile strength [N/mm^2]		Compressive strength [N/mm^2]	
		2 days	28 days	2 days	28 days			2 days	28 days	2 days	28 days
1505	196	6.0	9.3	34.0	60.7	1738	227	5.6	9.0	32.5	58.7
1535	216	5.6	8.8	30.8	57.2	1767	196	5.9	8.8	32.8	59.7
1553	197	5.8	8.7	33.5	59.8	1785	220	5.4	8.7	31.0	57.8
1580	219	5.5	8.6	29.6	59.0	1810	246	5.4	9.2	29.7	57.4
1601	229	5.6	8.8	30.5	59.4	1827	211	6.2	9.1	35.3	59.8
1630	235	5.6	9.1	30.7	58.4	1850	215	5.8	8.6	32.2	58.6
1651	191	5.9	9.4	34.3	59.9	1867	241	5.7	8.6	32.4	58.7
1675	225	5.4	8.7	30.7	56.6	1889	246	5.7	8.8	32.9	59.1

1692	231	5.4	9.2	29.4	56.8	1905	236	5.7	8.8	30.4	56.6
1721	251	5.3	8.9	30.6	56.8						

Whereas the cement is supposed to have compressive strength in 28 days 42.5 N/mm^2, the range of 56.6 - 60.7 N/mm^2 is enough. Setting time is in the range of 191 - 251 minutes. This variability is wide but still in allowed limits.

Table 2. Setting time, tensile strength and compressive strength.

Sample no.	Setting time [min]	Tensile strength [N/mm^2]		Compressive strength [N/mm^2]		Sample no.	Setting time [min]	Tensile strength [N/mm^2]		Compressive strength [N/mm^2]	
		2 days	28 days	2 days	28 days			2 days	28 days	2 days	28 days
1505	196	6.0	9.3	34.0	60.7	1738	227	5.6	9.0	32.5	58.7
1535	216	5.6	8.8	30.8	57.2	1767	196	5.9	8.8	32.8	59.7
1553	197	5.8	8.7	33.5	59.8	1785	220	5.4	8.7	31.0	57.8
1580	219	5.5	8.6	29.6	59.0	1810	246	5.4	9.2	29.7	57.4
1601	229	5.6	8.8	30.5	59.4	1827	211	6.2	9.1	35.3	59.8
1630	235	5.6	9.1	30.7	58.4	1850	215	5.8	8.6	32.2	58.6
1651	191	5.9	9.4	34.3	59.9	1867	241	5.7	8.6	32.4	58.7
1675	225	5.4	8.7	30.7	56.6	1889	246	5.7	8.8	32.9	59.1
1692	231	5.4	9.2	29.4	56.8	1905	236	5.7	8.8	30.4	56.6
1721	251	5.3	8.9	30.6	56.8						

3.3. *Bleeding*

Tendency to bleeding was determined on cement pastes. Compositions of mixtures are in

	name of mixture		
	Reference	DYNAMON SX 14	STACHEMENT 2489
cement CEM I 42,5 R	450.0 g	500.0 g	500.0 g
Water	292.5 g	175.0 g	175.0 g
superplasticizer Dynamon SX 14	-	2.5 g	-
superplasticizer Stachement 2489	-	-	2.5 g

Cement was put into water and measuring of time started. Cement paste was mixed for 1 minute by plastic spatula. Mixed paste was put into 500 ml glass volume cylinder and the high of level was noted. After 15 and 90 minutes was determined level of separated water. After 90 minutes water from the top pipetted out with small amount of cement. After drying we got real amount of separated water - intensity of bleeding. Separated water was recounted according to real amount of cement paste. Amount of separated water is indicator of cement paste tendency to bleeding.

Table 3. Composition of cement paste mixtures for bleeding determination.

	name of mixture		
	Reference	DYNAMON SX 14	STACHEMENT 2489
cement CEM I 42,5 R	450.0 g	500.0 g	500.0 g

Water	292.5 g	175.0 g	175.0 g
superplasticizer Dynamon SX 14	-	2.5 g	-
superplasticizer Stachement 2489	-	-	2.5 g

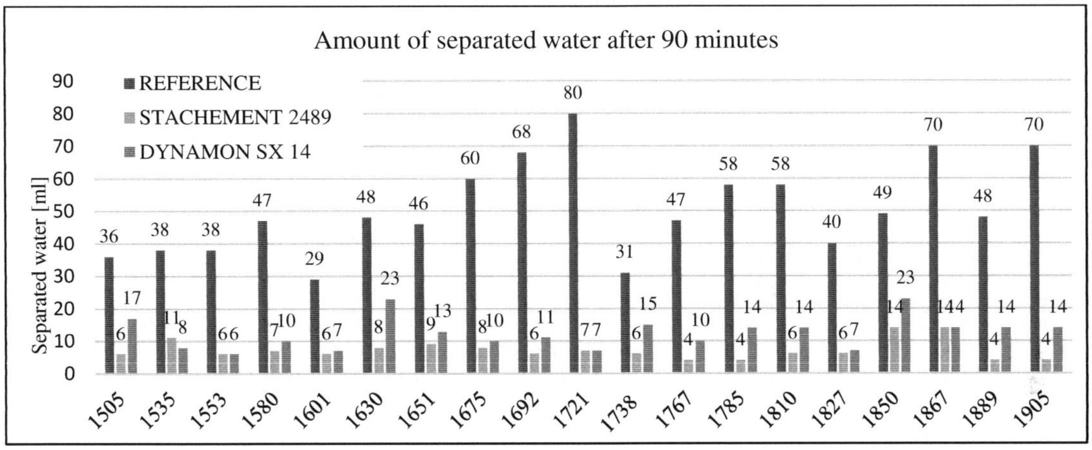

Figure 7. Amount of separated water after 90 minutes.

In 15 minutes, the amount of separated water is around 7 ml without superplasticizer and around 3 ml with admixture. Difference between with/without admixture is more visible after 90 minutes. Whereas the mixtures without superplasticizer have up to 80 ml of separated water, the mixtures with plasticizer have less than 23 ml. It is apparent, that the variability in amount of separated water is quite wide. With admixture adding, the variation is smaller. Of course, this fact is caused by reducing water-cement ratio. Nevertheless, this can be the manner how to reduce variability in the concrete tendency to bleeding.

The comparisons of cement properties and the amount of separated water from reference mixture after 90 minutes are shown on the following Figures.

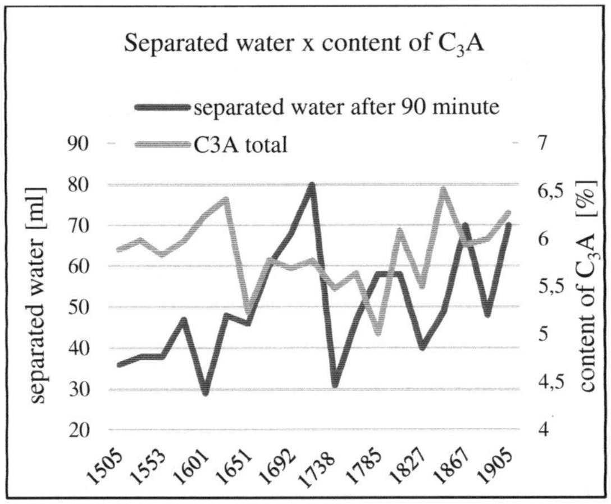

Figure 8. Relation between content of C3A and amount of separated water.

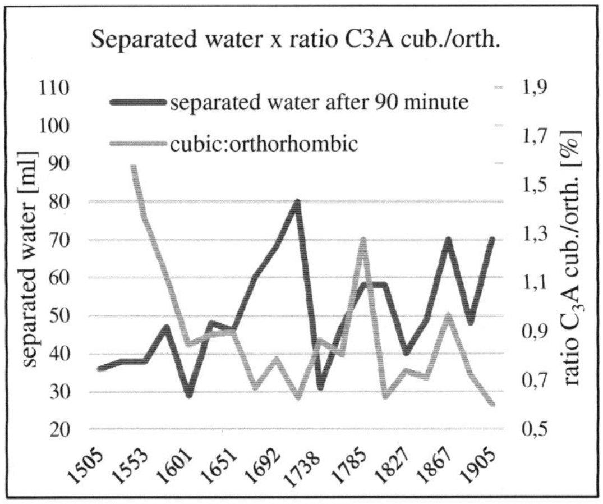

Figure 9. Relation between ratio of C_3A modification and amount of separated water.

IOP Conf. Series: Materials Science and Engineering **385** (2018) 012021 doi:10.1088/1757-899X/385/1/012021

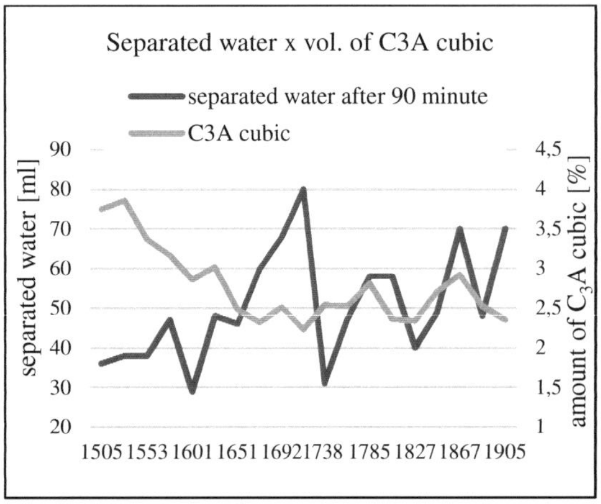

Figure 10. Relation between content of C_3A cubic and amount of separated water.

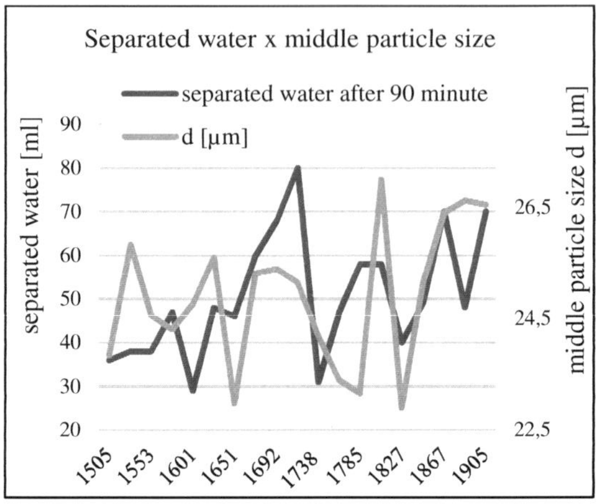

Figure 11. Relation between middle particle size and amount of separated water.

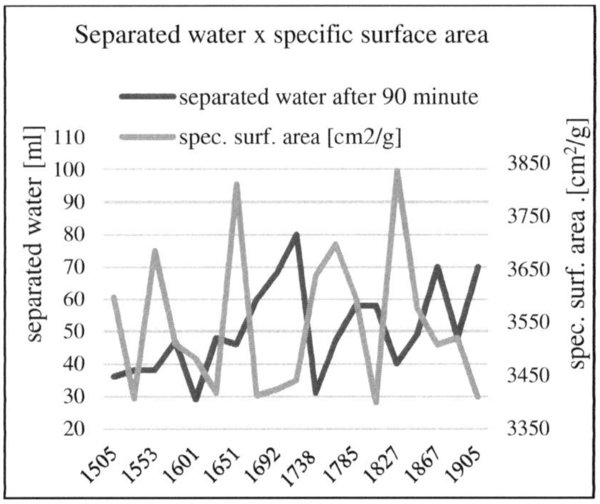

Figure 12. Relation between specific surface area of cement and amount of separated water.

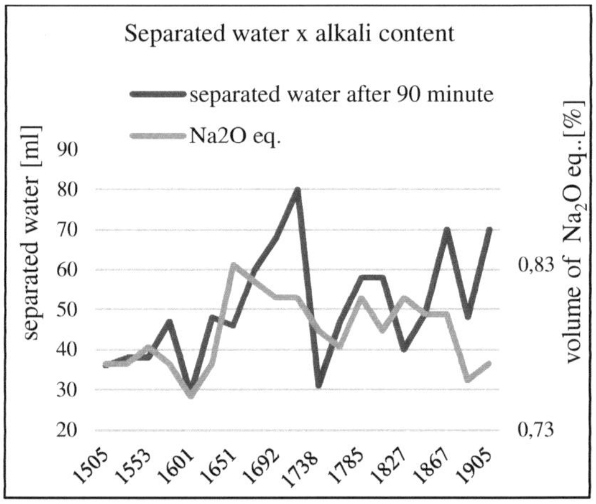

Figure 13. Relation between content of alkali and amount of separated water.

As it can be seen on Figure 8, the total content of C_3A is not valid parameter to describe tendency of cement paste to bleeding. Same situation is visible on Figure 10 where is compared the content of C_3A in cubic modification with the amount of separated water. On samples 1785 - 1905 (cements produced in November) can be seen the correlation between content of C_3A cubic and amount of separated water. With increasing amount of these C_3A modifications is visible increasing amount of separated water. On Figure 9 is amount of separated water compared with the ratio of cubic and orthorhombic modification of C_3A. Higher ration leads to higher tendency to bleeding. These results do not match the assumption, that higher amount of orthorhombic modification of C_3A increases tendency to bleeding. Comparison on figure 12 is between specific surface area of cements and amount of separated water. Here we can see demonstrable the dependence of tendency to bleeding on specific surface. With higher specific surface area of cement the tendency to bleeding is lower. On Figure 13 where is compared the amount of separated water with content of alkali is visible the dependence of these two parameters. Tendency to bleeding is higher with higher amount of alkali.

4. Conclusion

As mentioned in previous chapter, major influence on tendency to bleeding looks to have specific surface area or middle particle size. With higher specific surface tendency to bleeding decrease. Partial influence has ratio between content of C_3A cubic and orthorhombic. Higher amount of orthorhombic C_3A leads to higher tendency to bleeding. Addition of superplasticizer helps decrease tendency to bleeding, because of lower water/cement ratio. Whereas the results are not final another testing is requested. As another parameter, the content of sulphur and type of gypsum added to clinker during the grinding will be taken into consideration.

Acknowledgment

This research was supported by project MPO FV20019 funded by Czech Ministry of Industry and Trade.

References

[1] Powers T C 1968 *The properties of Fresh Concrete* (New York: J Wiley&Sons)
[2] Yim H J, Kim J H and Kwak H-G 2014 *Cement Concrete Res.* **57** 61-69
[3] Neville A M 1981 *Properties of Concrete* (Essek: Longman)
[4] Arslan M 2000 *Magazine of ZKU Karabuk TEF* 59-65
[5] Ozkul M H and Oztekin E 1994 *Proc. Int. Conf. on Concrete Congress* (Instanbul: the Turkish Chamber of Civil Engineering) p 163–174
[6] Topcu I B and Elgun V B 2004 *Cement Concrete Res.* **34** 275-281

Concrete based on full natural aggregate replacement by glass household waste

J Junak and N Junakova

Institute of Environmental Engineering, Faculty of Civil Engineering, Technical University of Kosice, Vysokoskolska 4, Kosice 04200, Slovakia

Abstract. The amount of municipal waste produced is increasing every year. Waste preventing is the best way, but it is not always possible. There is a need to look for new ways of municipal waste management, such as material recovery of waste in the construction industry. This paper shows the possibilities of using glass waste in the production of concrete. Three fractions of glass bottle cullet - 0/4, 4/8 and 8/16mm, as a full-scale replacement of natural aggregates were used. The results of the compressive strengths of more than 50 MPa and the bending compressive strength at more than 4.4 MPa, show considerable potential in this direction of use. In addition, at half of the prepared samples, cement was replaced by fly ash from biomass combustion. The negative effect of alkali-silica reaction of glass did not appear in our experiment.

1. Introduction

In the Slovak Republic, 10.5 mil. tonnes of wastes were created in 2015. Compared to 2014, the year-on-year increase in total waste generation in 2015 is almost 16%. In the production of waste according to the classification of economic activities, the largest producer of waste is industrial production, which accounts for about 38% of the total waste production, followed by the section of construction with about 24% share [1].

With regard to municipal waste (MW) production, in Slovakia about 1.8 million tonnes of municipal wastes were generated in Slovakia in 2015, or 348 kg per capita. Compared to 2014, this represents an increase of 17 kg per capita, but in a Europe-wide comparison, Slovakia is still among the countries with the lowest MW output. Negative is the fact that the dominant activity in the management of MW was landfilling. The share of landfilled municipal waste in total MW management was 69%, which represents a year-on-year increase of 3%. Municipal waste recycling, calculated according to Decision 2011/753/EU, reached only 20.11% in 2015. Reducing landfilling rates is a prerequisite for more efficient use of material resources in our country [2].

The main objective of Slovakia's waste management by 2020 is to minimize the negative effects of the generation and management of waste on human health and the environment. In order to achieve the stated objectives, a stronger enforcement and compliance with the binding waste management hierarchy will be avoided, with the aim of increasing waste recycling, in particular for municipal waste and construction and demolition waste, in line with the requirements of the Waste Framework Directive [3-4].

It is the Construction industry that appears to be an appropriate sector that can increase the share of the use of waste - both industrial and municipal waste. Many studies in Europe and in the world have shown that filler and binder in concrete can be successfully replaced by different materials: hemp hurds and cellulosic fibers [5-8], slag [9-12], fly ash [13-19], recycled brick and concrete aggregate [20-22], polyurethane foam [23], crushed glass [24-26], but by bottom sediment also [27-29].

Glass waste is produced in many forms, including packaging of container glass (bottles, jars), flat glass (windows, windscreens), bulb glass (light globes), cathode ray tube glass (TV screens and monitors), all of which have a limited life in the form they are produced and need to be reused/recycled. Current collection methods for glass products are quite limited, therefore, only a small fraction of the solid waste can be recycled directly to the primary market - the bottling and container industry [30-32].

The term fly ash was first used in the electrical power industry in 1930, the first comprehensive data on its use in concrete in North America were reported in 1937 [33]. Coal fly ash is generated by fluid combustion of black or brown coal in thermal power plant. The coal fly ash produced in thermal power plants and heat plants is waste and therefore needs to be disposed or store. This is, however, quite costly and also involves a considerable ecological burden. The coal fly ash, with its finenest increases the fine particle content of the concrete, which contributes favourably to the improvement of the processability of fresh concrete. However, due to the need for content of tap water, it is necessary to use a plasticizer in the manufacture of such concrete. The coal fly ash has a beneficial effect on the course of hydration heat in concrete and prevents cracks from excessive shrinkage. A higher percentage of coal fly ash in concrete partially reduces the rate of growth and final strength of concrete, nevertheless this type of concrete is suitable for many design solutions, even for special applications [34-37].

This paper is devoted to the study of the possibility of using glass waste as a full replacement of natural aggregate for concrete production. The following properties were observed: real density, compressive strength and flexural strength. On prepared beams with dimension 100x100x400 mm, compressive strength and flexural strength were after 14, 28, 90, 180, 365, 730 and 1095 days tested.

2. Material and methods

The materials used in our experiment were: Portland cement, fly ash, glass bottles cullet and natural aggregate.

In our study were used two different types of binders. The main binder was Portland cement CEM I 42.5 N with specific weight 3100 kg/m^3 [38]. As a substitute for cement, the fly ash from a biomass burning power plant (Kosice, Slovakia) with an average particle size of 12.038 μm and molar ratio of SiO_2 /Al_2O_3 = 3.88 was used. The chemical analysis of the fly ash is as follows: MgO (6.33%), Al_2O_3 (1.37%), SiO_2 (5.32%), P_2O_5 (3.09%), SO_3 (7.26%), K_2O (12.26%), CaO (47.33%).

Two different types of fillers were used in the preparation of concrete composites. Natural aggregate - three fractions (0/4 mm, 4/8 mm and 8/16 mm) from company RRH Slovakia were used like basic filler [39]. Further, three fraction of glass bottles cullet (GBC), 0/4 mm, 4/8 mm and 8/16 mm were used as natural aggregate full replacement in concrete mixtures. GBC was obtained from mixed coloured glass bottles, which were manually crushed in the laboratory and then sieved.

To prepare concrete mixtures drinking water was used [40].

Table 1 shows basic composition of concrete to 1m^3.

Table 1. Reference concrete mix proportion per cubic meter.

Composition	1 m^3
Cement (kg)	350
Aggregates 0/4 mm (kg)	955
Aggregates 4/8 mm (kg)	210
Aggregates 8/16 mm (kg)	710
Water (l)	198

For research the possibilities of crushed colour glass bottles cullet (GBC) utilization as a natural aggregate full replacement in concrete, ten different mixtures (samples G0 - G4 and GF0 - GF4) at w/c ratio of 0.6 were prepared. Two of these mixtures was comparative (samples G0 and GF0). Samples GF0 - GF4 were prepared with 25 wt.% cement replacement by fly ash without any treatment. Table 2 shows composition of samples based on a natural aggregate replacement by mixed coloured glass bottle culets.

Table 2. Composition of samples based on natural aggregate replacement by crushed coloured glass bottles.

Sample	Cement	Fly ash	Natural aggregate			H_2O	GBC		
			0/4	4/8	8/16		0/4	4/8	8/16
G0	●	-	●	●	●	●	-	-	-
G1	●	-	●	●	-	●	-	-	●
G2	●	-	●	-	●	●	-	●	-
G3	●	-	-	●	●	●	●	-	-
G4	●	-	-	-	-	●	●	●	●
GF0	●	●	●	●	●	●	-	-	-
GF1	●	●	●	●	-	●	-	-	●
GF2	●	●	●	-	●	●	-	●	-
GF3	●	●	-	●	●	●	●	-	-
GF4	●	●	-	-	-	●	●	●	●

Experimental mixtures were processed in the laboratory mixer with a horizontal rotary drum with a capacity of 150 l. Compounding process was chosen as follows: to a mixing drum are tipped aggregate in the order of fractions 8/16 mm, 4/8 mm, 0/4 mm. After a careful mixing of the all components concrete mixture were placed into cleaned metal forms. Thus prepared forms were then over 15s compacted on a vibrating table. After filling, prism forms were labelled and placed on a flat surface next 48 hours. After 48 hours, the prism were removed from the forms and then placed in a water bath. In order to realize the experiment program was made prism with dimensions of 100x100x400 mm. All of these concrete mixtures were tested in compression and flexural strength with strength machine ELE 2000 after 7, 28, 90, 180 and 360 days of curing according to the standard [41].

3. Results and discussion

Figure 1 shows the results of real density. The real density has been monitored during period three years.

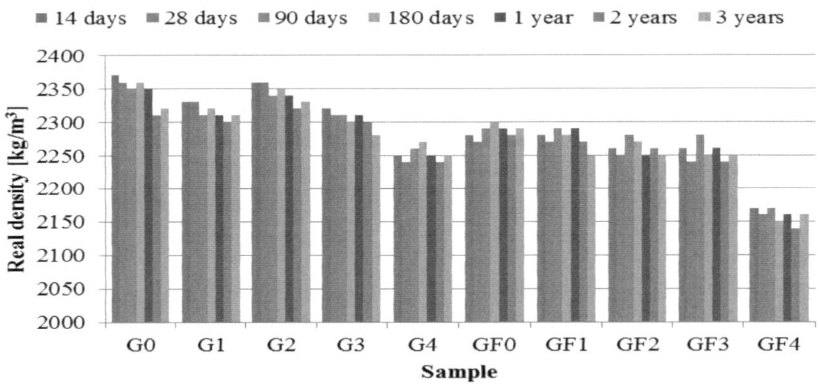

Figure 1. Real density of samples prepared with crushed coloured glass bottles.

The development of density value all samples is not unambiguous. The highest real density was obtained by the comparison sample (G0). On the contrary, the lowest density sample contained the glass in all three fractions and, in addition, was prepared with 25% cement replacement by fly ash. Values of real density ranged between 2370 kg.m^{-3} (sample G0) and 2140 kg.m^{-3} (GF4).

The results of compressive strength are shown in figure 2. Similar to real density, the highest compressive strengths after 3 years (except for the comparison sample) reached sample G2. The achieved strength of 51.3 MPa is at the level of concrete strength class C 40/50 according [41]. The lowest compressive strengths reached sample GF4 (29.8 MPa after 3 years; C 20/25), which contained no single fraction of natural aggregate and the standard binder CEM I 42.5 N was replaced with fly ash (25%). The substitution of natural stone with glass bottel cullet meant a decrease in the strength by more than 8 MPa in the G2 sample (13% decrease) and by 17 MPa in the G4 sample (28% decrease), compared to the comparison sample. Sample G2 contained glass waste in place of natural aggregate fraction 4/8 mm. Composites containing fly ash and glass waste reached values of compressive strength lower on average 18%, in comparison with samples prepared with glass waste, only.

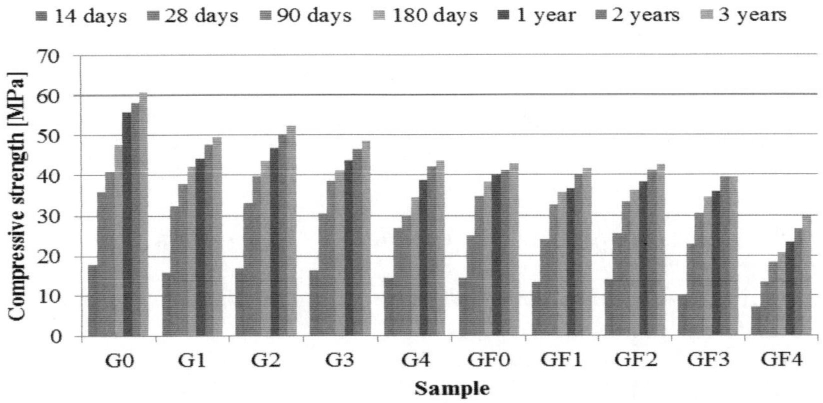

Figure 2. The running of the compression strengths of the experimental samples.

Flexture strength of experimental samples (figure 3) were tested on samples after compressive strength test. The obtained values of strength ranged between 5.6 MPa and 2.4 MPa.

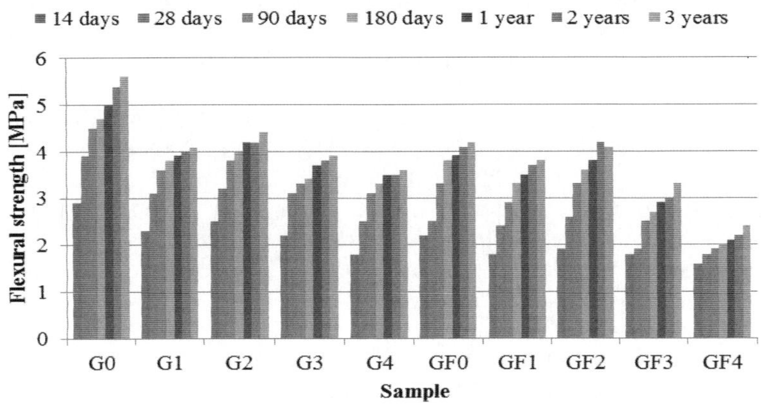

Figure 3. Flexural strength of samples based on glass waste as a filer.

If the comparative sample is not taken into account, the highest values of flexural strengths (4.4 MPa) have been achieved by samples containing glass waste fraction 4/8 mm - this applies equally to the sample containing the fly ash as a binder. The decrease of flextural strength of fly ash concrete samples is significantly lower than the compressive strength, is at the level of 10%.

The increase of the each strengths (compressive and flexural) was more noticeable at the beginning of the period under review (28 days), and then the strengths was increased slowly. Such development of concrete strengths is no exception, especially when we used "N" cement as a binder and the samples were standardly treated (water bath, 20°C) [42].

4. Conclusion

The aim of the paper was to show a new dimension of the reuse of household waste - glass bottle cullet. The results of real density, compressive and flexural strength indicate great potential for use in construction industry. Replacement of natural aggregate by glass waste was partially reduced: real density (5%), compressive strength (13%) and flexure strength (10%). However, generated waste will be used to save the natural raw materials and free place on landfill site. Moreover, if fly ash as a binder in the samples is used, greenhouse gases production will be significantly reduced. In the end, reached properties of the prepared samples, in particular the compressive strength of nearly 50 MPa predict this concrete for versatile use.

Acknowledgments

This research has been carried out within the Slovak Cultural and Education Grant Agency (contract No. 073TUKE-4/2018).

References

[1] European Environment Agency 2016 *Overview of national waste prevention programmes in Europe:* Slovakia
[2] OECD 2017 *Making the Slovak Republic a more resource efficient economy:* OECD environment policy paper no. 7
[3] Slovak republic: 2015 *Waste*, Act 79/2015
[4] European Environment Agency 2016 *Chapter 36: Waste production and management*
[5] Stevulova N, Cigasova J, Sicakova A and Junak J 2013 *Chem. Eng. Trans.* **35** 589-594
[6] Cigasova J, Stevulova N, Schwartzova I, and Junak J 2014 **37** *Chem. Eng. Trans.*685-690
[7] Stevulova N, Schwartzova I, Hospodarova V and Junak J 2016 *Chem. Eng. Trans.* **50** 367-372
[8] Cigasova J, Stevulova N, Sicakova A and Junak J 2013 *Chem. Eng. Trans.* **32** 1615-1620
[9] Vaclavik V, Dvorsky T, Simicek V, Ondova M, Valicek J, Kusnerova M and Gola L 2016 *Sol. St. Phen.* **244** 77-87
[10] Junak J, Stevulova N and Ondova M 2013 *13th Int. Mul. Sci. Geo. SGEM 2013 (Albena)* vol 1 (Sofia: STEF92 technology Ltd.) pp 393-399
[11] Baricova D, Pribulova A, Demeter P, Bulko B and Rosova A 2012 **51** *Metalurgija* 465-468
[12] Stevulova N, Vaclavik V, Junak J, Grul R and Bacikova M 2008 8th *Int. Mul. Sci. Geo. SGEM 2008 (Albena)* vol 1 (Sofia: STEF92 Technology Ltd.) pp 193–200
[13] Brenek A, Vaclavik V, Dvorsky T, Daxner J, Dirner V, Bendova M, Harnicarova M and Valicek J 2015 *Advanced Structured Materials* **70** 177-188
[14] Junak J and Stevulova N 2011 11th *Int. Mul. Sci. Geo. SGEM 2015 (Albena)* vol 3 (Sofia: STEF92 Technology Ltd.) pp 567–572
[15] Stevulova N and Junak J 2014 *Chem. listy* **108** 620-623
[16] Junak J, Stevulova N and Ondova M 2014 *Pollack Periodica* **9** 95-104
[17] Xu G, Shi X 2018 *Resour Conserv Recycl.* **136** 95-109
[18] Junak J and Stevulova N 2008 *Chem. listy* **102** 882-883
[19] Yu Z, Ni C, Tang M and Shen X 2018 *Constr. Build. Mater.* **175** 458-466
[20] Junak J and Sicakova A 2017 *Buildings* **8** Article number 2
[21] Ryu H S, Kim D M, Shin S H, Lim S M and Park W J 2018 *IJCSM* **12** Article number 19
[22] Junak J and Sicakova A 2017 *IOP Conf. Ser.: Earth and Env. Sci.* **92** Article number 012023
[23] Dvorsky T, Vaclavik V, Simicek V and Brenek A 2015 *Inzynieria Mineralna* **2** 51-56
[24] Yang S, Cui H, Poon, C S 2018 *Cement Concrete Comp.* **90** 266-276

[25] Junak J 2016 *Solid State Phenomena* **244** 102-107
[26] Hajimohammadi A, Ngo T, Kashani A 2018 *Constr. Build. Mater.* **171** 223-231
[27] Junakova N, Junak J 2017 *Procedia Eng.* **180** 1292-1297
[28] Vitola L, Sahmenko G, Erdmane D, Bumanis G and Bajare D 2017 *IOP Conf. Ser.: Mater. Sci. Eng.* **251** Article number 012038
[29] Junakova N and Junak J 2017 *Sustainability* **9** Article number 852
[30] Soliman N A 2017 *Constr. Build. Mater.* **145** 243-252
[31] Junak J, Stevulova N and Junakova N 2017 *Int. Conf. Sust. Mater. Sci. Tech.* 140-144
[32] Liu S, Shu W, Tang W, Hu N and Wei J 2015 *Materials* **8** 6849-6862
[33] Davis R E, Carlson R W, Kelly J W and Davis H E 1937 *J. Am. Concr. Inst.* **33** 577-612
[34] Velandia D F, Lynsdale C J, Provis J L and Ramirez F 2018 *Cement Concrete Comp.* **91** 11-20
[35] Verian K P and Behnood A 2018 *Cement Concrete Comp.* **90** 27-41
[36] Nuaklong P, Sata V, Wongsa A, Srinavin K and Chindaprasirt P 2018 *Constr Build Mater* **174** 244-252
[37] Ondova M, Stevulova N and Junak J 2013 *13th Int. Mul. Sci. Geo. SGEM 2013 (Albena)* vol 1 (Sofia: STEF92 Technology Ltd.) pp 605–612
[38] STN EN 197-1: 2012 *Cement. Part 1: Composition, specifications and conformity criteria for common cements*
[39] STN EN 12 620 *Aggregates for concrete*
[40] STN EN 1008: 2003 *Mixing water concrete. Specification for sampling, testing and assessing the suitability of water, including water recovered from processes in the concrete industy, as mixing water for concrete*
[41] STN EN 206: 2017 *Concrete. Specification, performance, production and conformity*
[42] Kim J K, Moon Y H and Eo S H 1998 *Cem. Conr. Res* **28** 1761-1773

IOP Conf. Series: Materials Science and Engineering **385** (2018) 012023 doi:10.1088/1757-899X/385/1/012023

The effect of sediment treatment on strength and water absorption capacity of sediment based concrete

N Junakova and J Junak

Institute of Environmental Engineering, Faculty of Civil Engineering, Technical University of Kosice, Vysokoskolska 4, Kosice 04200, Slovakia

Abstract. The paper presents the possibility of the Ružín (Slovakia) reservoir sediments' reuse as a 40% replacement of Portland cement in concrete. Treatment of sediments consists of sediment dry milling, adding of fly ash from local biomass incinerator plant into the dry milling process due to the low CaO sediment content and sediment's mechano-chemical activation using sodium hydroxide in the form of solid and solution as a chemical activator of pozzolanic properties. The results show the possibility of sediment reuse in building materials. However, treatment of sediments with the addition of fly ash doesn't significantly affect the development of compressive strength. Sodium hydroxide is not an effective pozzolanic activator for sediments. The absorptivity of composites increases with the fly ash and sodium hydroxide addition. Also the NaOH application form (as a solid or solution) influences the total absorption water capacity. Pouring the mixture with 5M NaOH increases the absorptivity of the hardened composites.

1. Introduction

River and reservoir sediment is a naturally occurring element in many water bodies. They are significant environmental, socioeconomic and geomorphological source. Most of them originate through erosion processes and are further transported in a hydrographic network [1]. The transported particles silt watercourses, settle in water reservoirs, where they cause different environmental [2-3] and water management problems [4]. Worldwide, reservoir sedimentation occurs at a rate of about 0.8 percent per year [5]. Most of the existing reservoirs in the world could be completely silted up in 200 to 300 years from now, with large reaches of river system permanently lost [5]. The sediment extraction is an essential part of maintaining the flow and functionality of watercourses and water structures. But the problem is how to manage with the extracted sediments [6,7]. Because sediments refer to the conglomerate of organic and inorganic materials that can be carried away by water [8], the sediment grain size classes ranging from rock to clay can be applied for a lot of different purposes [9]. Due to the perpetual availability of the sediments from the reservoirs and their properties, conversion of dredged sediments into secondary raw material as lightweight aggregate [10-12] or bricks and tiles [13-15] creates a permanent disposal solution that is cost predictable over the long-term and cost competitive with alternative options [16]. An unconventional idea not only in the world but also in the Slovak Republic in the sustainable management of river and bottom sediments is the use of sediments and remediation products as a secondary raw material in the production of concrete [7,17-19].

In this paper, the possibility of the Ružín (Slovakia) reservoir sediments' reuse as a partial replacement of binder in concrete is discussed. The compressive strength and water absorption capacity of sediment based composites are studied.

2. Material and methods

A laboratory-scale study was performed using the sediments extracted from the Ružín reservoir, located in the east of Slovakia. Based on the initial physico-chemical analysis of sediments (particle size distribution, mineralogical and chemical composition) [19], they were used as a partial (40%) cement replacement in concrete. The choice of 40% binder (Portland cement) substitution with sediments resulted from the literature [20]. Pontikes and Snellings [20] state, that supplementary cementitious materials containing less CaO can be blended into cements only up to 40% total weight without significant loss in binding properties.

The blended concrete specimens were prepared with modified sediments (fraction below 0.063 mm) as a binder as follows. With respect to the coarse-grained texture of the original (extracted) sediments from Ružín, mechanical modification of the sediments by dry milling in a laboratory vibrating mill for 3 minutes was performed. Due to the low CaO content in the sediment, fly ash from local biomass incinerator plant was added into the dry milling process in a ratio of 1:1 (sediment : fly ash). Finally, mechano-chemical activation of the sediment consisted of sodium hydroxide addition as a chemical activator of pozzolanic properties [20-21] into the sediment, and sediment and fly ash mixture. Granulated NaOH was added to a dry milling process at the weight ratio of 2.6:1 (sediment : NaOH); and 1:1:0.8 (sediment : fly ash : NaOH) to achieve 5 M NaOH solution. Sodium hydroxide activator was also used in the form of a pre-prepared 5M solution, which was added into the blended mixture instead of water. Portland cement (CEM I 42.5 N) and two different fractions of natural aggregates (0-4 mm and 4-8 mm) were used as other raw materials. Table 1 shows the compositions of the studied mixtures. A precise mixing procedure was followed according to standard STN EN 206 [22]. Control mixture (SM0) was also made only with Portland cement to compare the effect of sediment as a binder replacement on compressive strength. Double set of mixtures were prepared.

Table 1. Composition of mixtures.

Mixture	Binder (kg/m^3)		Filler-Natural Aggregate (kg/m^3)		Water (kg/m^3)
	CEM I	Sediment	Fraction 0-4 mm	Fraction 4-8 mm	
Control SM0	350	-	1123	717	193
SRM1	210	140 (milled)	1123	717	193
SRM2	210	140 (milled with NaOH)	1123	717	193
SRM3	210	140 (milled with FA)	1123	717	193
SRM4	210	140 (milled with FA and NaOH)	1123	717	193
SRM5	210	140 (milled and poured with NaOH)	1123	717	193
SRM6	210	140 (milled with FA and poured with NaOH)	1123	717	193

Note: FA – fly ash

Three prisms (40 mm × 40 mm × 160 mm) from each set were prepared for mechanical strength and water absorption capacity tests per studied age.

The compressive strengths of hardened mixtures were determined after 28 and 90 days of curing in accordance with standard STN EN 12390-3 [23]. The mechanical tests were carried out with an ELE ADR 2000 (ELE International Ltd, Leighton Buzzard, Bedfordshire, UK) testing machine after 28 and 90 days of curing in water.

The total absorption water capacity, as an important indicator of concrete durability, was another tested parameter in hardened specimens. It was determined according to STN 73 1316 [24]. The test consisted of two major steps: drying the specimens and their saturating. Specimens were dried in ventilated oven at a temperature of 105±5°C until the difference in mass was less than 0.1% (m_d).

Then specimens were immersed in water for 24 hours to obtain saturated mass (m_s). The water absorption by immersion (N) was expressed as the water uptake relative to the dry mass using this formula:

$$N = \frac{m_s - m_d}{m_d} * 100\ (\%) \quad (1)$$

3. Results and discussion

The compressive strength of the prepared sediment-based composites as a filler and control mixture after 28 and 90 days of curing are shown in table 2. All data are presented as the mean ± standard deviation of the six determinations made on a double set of three prisms.

Table 2. Composition of mixtures.

Composite	Compressive strengths (MPa)	
	28 d	90 d
SM0-control	38.3	47.2
SRM1	20.3	20.7
SRM2	4.6	6.1
SRM3	22.7	22.0
SRM4	8.4	10.3
SRM5	6.3	9.6
SRM6	8.2	10.9

The average compressive strength of composites prepared with milled sediment as a 40% partial cement replacement was at about 20 MPa after 28 and 90 days of hardening. However, a higher percentage of cement replacement by sediment resulted in a decrease of compressive strengths compared to a control mixture by half.

Mechanical activation of the sediment by its milling together with fly ash (with the purpose of adding CaO into the mixture) increases the compressive strength of prepared composites after 28 days of hardening by more than 10%, with the strength values achieved at 23 MPa. The addition of fly ash did not affect the development of compressive strength after 90 days of curing. The composite' compressive strength was at the same level with the addition of ash (SRM3) and without its addition (SRM1) after 90 days of hardening. The expected increase in compressive strength due to the addition of fly ash to the mixture has not been demonstrated [25]. It can be due to the low contents of SiO_2 and Al_2O_3 in fly ash that are essential for the formation of C-S-H / C-A-H gels important for strength development.

The last group of composites was prepared with mechano-chemical activated sediments by milling with the presence of NaOH activator in solid form and in the form of 5M solution. These composites achieved the lowest compressive strengths of all studied mixtures (see table 2). Higher initial compressive strengths achieved blends with addition of fly ash (SRM4 and SRM6). It has also been observed that the composites prepared with NaOH in the form of solution achieved higher strengths compared to the granulated NaOH application in the milling process. The results have shown that the method of the activator application (granulated NaOH vs NaOH solution) may influence the development of compressive strengths. However, significantly lower compression strengths of SRM2 and SRM4-SRM6 composites compared to others indicate that the conversion degree to the C-S-H gels in the composite system is too low due to the high concentration of hydroxyl ions in the system (pH of fresh blended mixtures were 13.8). This causes the shift of equilibrium towards the left side in the hydration process of C_3S and C_2S [26-27].

In composites, the total absorption water capacity was also studied (figure 1). Results have shown that the addition of fly ash to the mixture increases the total absorption water capacity of the hardened composites. Examples are SRM1 and SRM3, where addition of ash into the mixture has increased the absorption capacity of the hardened sample by more than 20%.

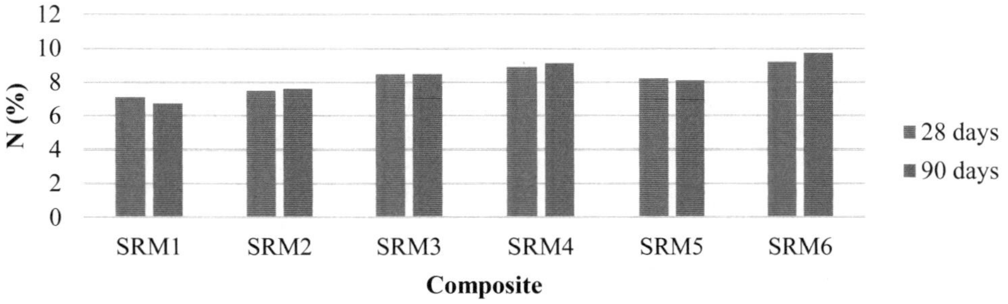

Figure 1. The total absorption water capacity of composites after 28 and 90 days of curing.

The absorptivity results of the prepared sediment-based composites with addition of fly ash and NaOH as an activator also showed that even the presence of NaOH in the mixture and also the form of NaOH application influences the resulting total absorption water capacity. Pouring the mixture with 5M NaOH slightly increased the absorptivity of the hardened composites compared to the mixtures prepared with the granulated NaOH added into the grinding process.

4. Conclusions

Presented results of testing the strength parameters of composites based on partial cement replacement by activated reservoir sediments have shown the possibility of sediment reuse in building materials. The use of higher masses of cement substitutes by fine-grained sediment (40%) from water reservoirs is furthermore environmentally and economically advantageous. The use of fly ash and sediment together as a partial cement substitution is advantageous in the case of requirement to reuse large volumes of sediment and fly ash at the same time. Addition of NaOH as an activator of sediment into composites in the concentration of 5M acts negatively to the compressive strength development and thus sodium hydroxide isn't effective pozzolanic activator for sediments.

Sediment-based composites as binder in concrete achieve the absorptivity values above 7. Addition of fly ash and sodium hydroxide increase the absorptivity. Also the form of NaOH application influences the resulting total absorption water capacity. Pouring the mixture with 5M NaOH has increased the absorptivity of the hardened composites compared to the mixtures prepared with the granulated NaOH added into the grinding process.

Acknowledgments

This research has been supported by the Slovak Grant Agency for Science (Grant No. 1/0563/15).

References

[1] Wohl E 2011 A World of Rivers: Environmental Change on Ten of the World's Greatest Rivers (Chicago, IL, USA: University of Chicago Press)
[2] Junakova N and Balintova M 2012 *Chem. Eng. Trans.* **26** 591-596
[3] Junakova N and Balintova M 2011 Model for predicting the nutrient content in reservoir bottom sediments *11th International Multidisciplinary Scientific Geoconference and EXPO - Modern Management of Mine Producing, Geology and Environmental Protection, SGEM 2011* 3 475-482
[4] Slezingr M 2016 Lowering the water level at the dam reservoir Brno *Public Recreation and Landscape Protection - With Nature Hand in Hand, Conference Proceedings* (Brno: Mendel university in Brno) pp 102-104
[5] Sedimentation Committee 2009 Sedimentation and Sustainable Use of Reservoirs and River Systems: http://www.icold-cigb.org/userfiles/files/CIRCULAR/CL1793Annex.pdf (online)
[6] Deibel I, Lampe C, Ulbricht J P, Cnudde T and van Dessel G. 2007 Beneficial Use *Sustainable Management of Sediment Resources: Sediment and Dredged Material Treatment* eds P Bortone, L Palumbo (Amsterdam, The Netherlands: Elsevier) pp 119–132

[7] Junakova N and Balintova M 2014 *Chem Eng Trans.* **39** 637-642
[8] Langland M and Cronin T 2003 A Summary Report of Sediment Processes in Chesapeake Bay and Watershed Available at: http://pa.water.usgs.gov/reports/wrir03-4123.pdf (accessed on 15 March 2017)
[9] PIANC-World Association for Waterborne Transport Infrastructure 2009 Dredged material as a resource. Options and constraints (Brussels, Belgium: PIANC)
[10] Hamer K, Hadeler A, Muschalla T, Schroter J and Timmer G 2003 *J Soils Sediments* **4** 284-291
[11] Hung M F, Hwang C L 2007 *Waste Manag. Res.* **25** 449–456
[12] Chen H J, Yang M D, Tang C W, Wang S Y 2012 *Constr Build Mater* **28** 387-394
[13] Samara M, Lafhaj Z and Chapiseau Ch 2009 *J Hazard Mater* **163** 701-710
[14] Romero M, Andrés A, Alonso R, Viguri J and Rincón J M 2009 Valorisation of contaminated marine sediments to produce ceramic construction materials. *Spanish National Conference on Advances in Materials Recycling and Eco – Energy* pp 45-48
[15] Torres P, Manjate R S, Fernandes H R, Olhero S M, Ferreira J M F 2009 *Ind Ceram* **1** 5-12
[16] Harborrock. Company summary. Available at: http://www.harborrock.com/HarborRockSummary.pdf (accessed on 15 March 2017).
[17] Mensinger M C 2008 Cement-Lock® Technology for Decontaminating Dredged Estuarine Sediments *Topical Report on Beneficial Use of Ecomelt from Passaic River Sediment at Montclair State University* (Essex County, NJ, USA: Montclair State University)
[18] Junakova N and Junak J and Balintova M 2015 *Clean Technol. Environ Policy* **17** 1161-1169
[19] Junakova N and Junak J 2017 Sustainable use of reservoir sediment through partial application in building material *Sustainability* **9** 1-13
[20] Granizo M L, Blanc-Varela M T, Puertas F and Palomo A 1997 Alkaline activation of metakaolin: Influence of synthesis parameters *Proceedings of the 10th International Congress on the Chemistry of Cement* (Gothemburg, Göteborg, Sweden)
[21] Palomo A, Grutzeck M W, Blanco M T 1999 *Cem. Concr. Res.* **29** 1323–1329
[22] STN EN 206 2017 Concrete. *Specification, Performance, Production and Conformity* (Slovakia, Bratislava: Slovak Office of Standards, Metrology and Testing)
[23] STN EN 12390-3 2010 Testing hardened concrete. Part 3: *Compressive strength of test specimens* (Slovakia, Bratislava: Slovak Office of Standards, Metrology and Testing)
[24] STN 73 1316 1989 *Determination of moisture content, absorptivity and capillarity of concrete* (Slovakia, Bratislava: Slovak Office of Standards, Metrology and Testing)
[25] Beeghly J and Schrock M 2009 *International Journal of Soil, Sediment and Water* **3** 1-21
[26] Abdullah A, Taufiq-Yap Y H, Jaafar M S, Alhozaimy A, Al-Negheimish A and Noorzaei J 2012 *Sci. Res. Essays* **7** 719–729
[27] Martinez-Ramirez S and Palomo A 2001 *Cem. Conc. Res.* **31** 1581-1585

Thermal resistance of selected thermal reflective insulations

F Kabelka, D Skricek and S Stastnik

Faculty of Civil Engineering, Brno University of Technology, Veveri 331/95, 60200-Brno, Czech

Abstract. The increasingly stringent requirements of thermal engineering standards lead to ever greater thicknesses of thermal insulating layers in the structures, which makes the insulation of the manufacturer to improve material parameters. Thermally reflective insulation currently occupies only a small volume of products in the building industry, but their use in buildings where a large temperature gradient is created for thermal insulation offers many advantages. The pressure on the thermal insulation efficiency is increasing, so that their thickness is as small as possible.

1. Introduction

Reflective thermal insulation layers are in literature attributed, in contrast to conventional thermal insulation layers, to a significant thermal insulation effect.

Reflective layers have been used in Europe since 2000 in France and Benelux where they were made of aluminium foil sewn together with hollow fibbers and various fabrics. At present, considerable attention is paid to the development of thermal insulation of this kind as it has great potential for use in the building industry, although it is obvious due to their layered layout and the materials used that they will not be as widely used as conventional thermal materials.

In our country, they have been proven to be a vapour-proof thermal insulation after 1990. The fabrics previously used were damp and therefore replaced by a blister foil or a polyethylene foam which was coated with a polished aluminium foil on one or both sides of the blister foil.

The advantage of these insulating materials is that they combine three advantages:

- the relatively high thermal resistance of the insulation layer,
- high reflectivity of the heat of aluminium foil,
- high diffusion resistance for water vapour spread.

The low water vapour permeability of the heat reflective counting stack provides non-porous layers, and the reflective films provide protection against electro smog, which is related to the electrical conductivity of the aluminium and the reflective films provide protection against electro smog, which is related to the electrical conductivity of the aluminium.

2. The current state of thermo-reflective layers on the market

Thermally reflective layers are used in the building industry especially for the insulation of walls, ceilings, roofs, floors, both in residential houses and light buildings, sports and warehouses. On the construction market today, we meet single-layer or multilayer heat-reflective formation.

This group includes vapour-tight single-layer thermal insulation, the thermal insulation of which forms a bubble foil with aluminium foil. In this group we find on the market an AB type - a bubble foil made of polyethylene with a polished aluminium foil or ABA type on one side - a polyethylene film laid on both sides with a polished aluminium foil (figure 1).

Figure 1. Layout diagram of single-layer insulation.

Single-layer heat reflective coatings are used to heat roofing, walls of wooden constructions and, finally, heavy walls with ventilated insulation [1]. Their thermal resistance R ranges from 0.3 m^2K/W for type AB to just above 0.5 m^2K/W for ABA. The value range is so wide because the prescribed measurement procedures provide worse results (ca. 0.3 m^2K/W) than in situ measurements or calculations (0.5 m^2K/W or better).

Another group consists of heat reflective foils, the thermal insulation of which is made of aluminium polished foil with fused polyethylene foil such as type AP3 - polyethylene foam of 3 mm thickness with a single-sided polished aluminium foil or AP5 - polyethylene foam of 5 mm thickness with a polished aluminium foil. The use of heat-reflective insulation is suitable for use in roof constructions, wooden walls, underfloor heating systems, heating radiators, etc.

The diffusion resistance factor for water vapour is given for simple reflective insulation $\mu = 3.46.10^5$, for ABA with two Al-foils then $\mu = 3.94.10^5$.

Multi-layered thermal reflection layers in which layers with melted foils or layers without melted foil are laid on top of each other. The thermal resistance R of the base stack AB or ABA is $R \in$ (0.3-0.6) m^2K/W, which corresponds to the equivalent value of the thermal conductivity coefficient of the homogeneous material layer $\lambda_{ekv} \in$ (0,001-0.005) W/(m.K). The same coefficient of thermal conductivity can also be considered for a stacked sandwich, so it can be very remarkable values, which can exceed six times the coefficient of thermal conductivity of conventional building insulation.

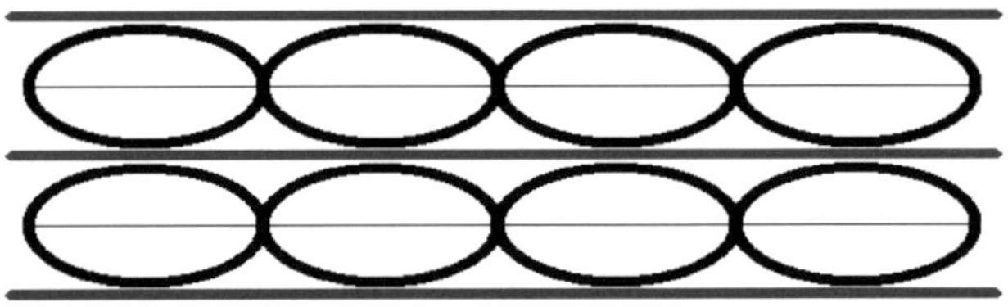

Figure 2. Layout diagram of multilayer insulation arrangement.

In this group a sandwich arrangement of polished aluminium, foam and blister foils of polyethylene, for example, with 5 layers (figure 2) or sandwich stack of polished aluminium, foam and blister foils of 13-layer polyethylene can be found. Thermally reflective foils are used as an additional or main insulation for thermal insulation of roof structures, walls of wooden buildings, passive and low energy houses, thermal insulation of sports and agricultural halls. These multilayer heat reflective films have the advantage of low thickness, high heat resistance, low bulk density and simple application.

All of the above-mentioned heat reflective films reflect up to 95% of the radiant heat component, thus protecting the object even in the summer against sunlight from the exterior.

The recently of these reflective insulation consists in the use of polished aluminium foils welded to sandwich with bubble or foam foil.

3. Reflective capability of thermal insulations

Building heat insulation is up to 98% filled with air and at room temperature heat is dissipated in about one-third of the radiation, at higher temperatures, for example under heated roof, it is up to 40%. As a result, in the range of practical temperatures from -30 °C to 60 °C the efficiency of classic thermal insulation changes by up to 30 percent. To do this, we have to add that for conventional thermal insulation, the lambda coefficient increases with temperature. More precisely, when the temperature rises from 15 °C to 60 °C, it will deteriorate by 15%.

With single-layer or multilayer reflective insulations provided with polished aluminium foil, we can almost eliminate the effect of radiant heat transfer mechanisms in isolation and achieve heat conductivity coefficients of isolation at a level close to the non-reflective thermal conductivity of the stationary air, approximately 0.027 W/(mK) [2]. The efficiency of these reflective foils is particularly high when performing roof structures.

4. Methods of heat transmission in thermal insulating materials

Heat dissipation is a typical physical phenomenon that always occurs from higher to lower temperature locations, it is important to have a temperature gradient between two places to spread the heat [3].

The basic ways of spreading heat include:

- conduction - according to Fourier's law,
- convection - according to Newton's law,
- radiation - in the sense of Stefan Boltzmann's law.

Most importantly, this type of reflection is used to characterize the ability of the material to reflect infrared radiation. It depends on the wavelength of the incident radiation, the angle, the type of material, its structure and its surface. For conventional reflective films, it reaches 0.8 to 0.9.

Emissivity indicates the ratio of the amount of light emitted from the material (grey surface) to the quantity that emits an absolutely black body at the same temperature. The intensity of the radiation depends only on the temperature and reaches the maximum value. According to Planck's radiation law, the spectral intensity of the black-body radiation is given. The emissivity value ε depends on the direction of radiation, the material type, the wavelength, the chemical composition of the material, the structure, and the state of the surface of the material.

Emissivity ε occurs in all directions according to the equation (1) [4]:

$$\varepsilon = \frac{M}{M_b} = \frac{C\cdot\left(\frac{T}{100}\right)^4}{C_b\cdot\left(\frac{T}{100}\right)^4} \leq 1 \tag{1}$$

M radiant intensity [W/m^2], M_b irradiance intensity of black emitter [W/m^2], T thermodynamic temperature [K], C grey radiant [W/(m^2 K^4)], C_b black radiant = 5.67 W/(m^2 K^4).

The absolutely black body reaches the emissivity values ε = 1 and emissivity values ε = 0 for an absolutely white body.

5. Measurement of thermal insulation properties of heat reflective layers

Thermal reflection layers can't be used to determine the thermal insulation properties. Only the Hot Box measurement method has been proposed for these insulation. For these insulation, the value of the thermal resistance R is neither of the decisive factor, not the coefficient of thermal conductivity. The Hot Box method describes a standard issued in 2015 by EN 16012 + A1, Thermal insulation for buildings - Reflective insulation products - Determination of the declared thermal performance [5]. This standard specifies, beyond the Hot Box method, three other options for obtaining a thermal resistance value. Therefore, the most accurate experimental measurement is to determine the thermal resistance of these layers.

The standard describes 4 different thermal resistance measurement methods:

- Method A: The method of a protected heating plate that meets the requirements of ISO 8301, EN 1946-2, EN 12664 and EN 12667.
- Method B: The method of a heat flux meter meeting the requirements of ISO 8301, EN 1946-3, EN 12664 and EN 12667.
- Method C: The Hot Box method, which meets the requirements of EN ISO 8990.
- Method D: Measurement of emissivity and calculations.

For this reason, an XPS expanded polystyrene foam measuring box was designed and assembled to measure the heat resistance R of the samples.

Samples need to be conditioned before testing. When samples delivered in a compressed form, they leave for some time to reach their functional status.

5.1. Hot Box Measurement method

A protected warm case is a measuring device that serves to measure the heat flow passing through the thermal insulation pattern at the given temperature difference, regardless of heat dissipation.

The case diagram is shown in figure 3. The cabinet is made of foamed polystyrene, the outer dimensions of the case are 1200 x 1200 x 1000 mm. In the compensating, measuring and cold parts are placed precision temperature sensors type Pt1000. In the cool part, the low temperature is ensured by the cool air from the cooling unit.

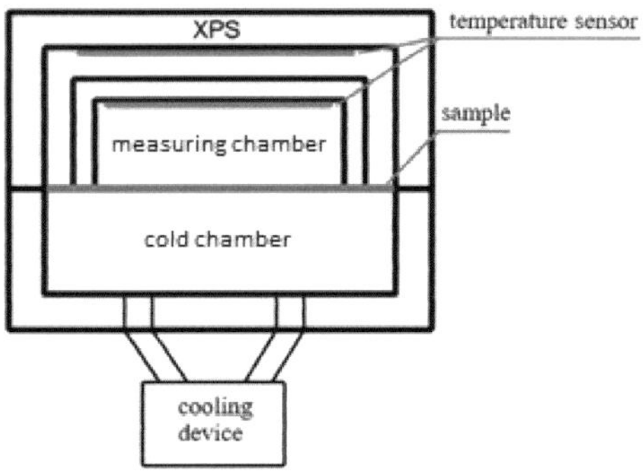

Figure 3. Scheme of the Hot Box measurement device.

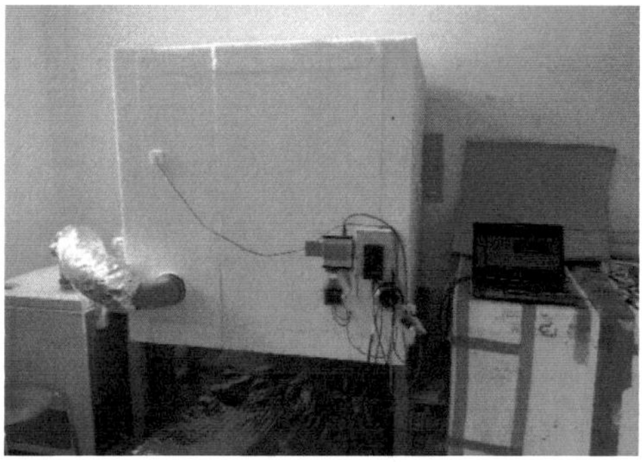

Figure 4. Photo of a measuring device with control program of a computer.

A 1000x1000 mm reflective insulation sample is attached to the wood frame to avoid heat leakage. Before measuring, it is necessary to determine the temperatures in the hot and cold parts, in which the thermal resistance of individual samples of thermo-reflective insulation will be measured. The temperature difference on the sample is usually 10 K, as well as the temperatures that can occur in the summer and also the temperatures simulating the winter season.

The measurement was carried out in a vertical position of insulation with double-sided air cavities of 50 mm thickness.

5.2. Measured samples of thermo-reflective foil insulation

All of the samples used for reflective foil insulation were from a LDPE bubble film with bubbles of 30 mm diameter and 10 mm height. The samples also contained a classic aluminium foil or a golden-coated foil. These are samples of the classic arrangement, ABA (Al foil - bubble foil - Al foil), but also multilayer insulation.

For the following marking of the insulation layer, the abbreviations of the individual layers will be used:

- A - aluminium foil and BB - "big bubble" bubble foil.

Samples used:

- Sample 1 - ABBA - Classic layout (2x Al, 1x BB).
- Sample 2 - ABBABBA - multilayer layout (3x Al, 2x BB).
- Sample 3 - ABBABBABBA - multilayer layout (4x Al, 3x BB).
- Sample 4 - ABBABBABABABBA - multilayer layout (5x Al, 4x BB).
- Sample 5 - ABBABBABBABBABBA - multilayer layout (6x Al, 5x BB).
- Sample 6 - ABBA - Classic layout, use of a gold-look surface.

After the measured values were entered into the calculation formula, the heat resistance values were plotted against the heat dissipation at a given temperature difference. It should be taken into consideration that these thermal resistance values define only the thermal insulating ability of the thermo-reflective insulation, which does not include unpowered airspaces that occur when the insulation is applied between the wooden grate, which will increase the value of the thermal resistance. The R_T thermal resistances given in table 1 were found at the highest set temperature gradient of 17 K and are shown.

Table 1. Overview of measured thermal R_T values by Hot Box method for temperature gradient θ_{EX} = -2 °C and θ_{IN} = +15 °C.

Reflective thermal insulation sample	R_T [m^2·K/W]
3 x Al + 4 x BB not melted on surface (Al + BB + BB + Al + BB + BB + Al)	1.555 ± 0.018
3 x Al + 4 x BB melted on surface (AlBB + BBAl + BB + BBAl)	2.424 ± 0.028
4 x Al + 6 x BB (Al + BB + BB + Al + BB + BB + Al + BB + BB + Al)	2.804 ± 0.033
5 x Al + 8 x BB (Al + BB + BB + Al + BB + BB + Al + BB + BB + Al + BB + BB + Al)	2.448 ± 0.029
Metallic Al foil ABBA (Al + BB + Al)	0.760 ± 0.009
3 x Al + 2 x BB (AlBB + Al + BBAl)	1.661 ± 0.020
2 x Al + 1 x BB (gold-look surface) (AlBB + Al/Au)	1.003 ± 0.012

Note: The uncertainty values given are calculated at 95% confidence level.

5.3. Discussion of the results

From the measured values it was found that almost all aluminum foils of silver appearance have very good reflectivity. If the aluminum foil is plastic foiled, the reflectivity is lower, in some cases by tens percent. The gold-look coated aluminum foil has approximately half the worse reflection value on the opposite side of the silver look. The reflectance is also dependent on the thickness of the aluminum foil, the thicker film labelled TL exhibiting lower reflectivity values. It is therefore best to use classical aluminum foils of silver appearance of a smaller thickness for the layer of reflective foil insulation. The highest radiant emitting ability exhibits PE foil BB, thus exhibiting behavior is close to the physically black surface.

6. Conclusion

Experimental measurements of reflex thermal insulations proved the great advantages of these reflective materials and their use in construction. The most important component of heat transfer is radiation. It is because of this phenomenon that the emissivity of reflective surfaces is very important, which should be as low as possible so that the heat loss rate is as low as possible. It follows from the above that the higher the temperature gradient in the building structures, the more thermal insulation properties of the thermal reflective insulation are achieved. If the temperature reaches -15 °C in the exterior, the thermal resistance R_T increases significantly.

Reflective insulation can be applied separately in the building without the use of any other insulators or combined with conventional insulators such as mineral wool. In any case, it is advisable to keep an air gap on both sides of the insulation which contributes little to its favourable thermal insulation properties.

Furthermore, the design, construction and calibration of a new measuring device for the thermal resistance measurement by the Hot Box method were realized. This measurement method serves to determine the thermal resistance R only for reflective bubble insulation, allows the measurement of the samples in various ways of incorporation into the structure and at various widths adjacent to the air cavity. The thermal resistance of the test specimens is highly dependent on the layout of the stack and the type of sub-materials used.

Experimental measurements of reflective foils have shown that they have great potential for building use, but are a new, not thoroughly researched material. Three important physical properties are combined in one building material: thermal insulation, vapour-proof properties and reflexive properties. The advantage of these materials is total recyclability, reducing the consumption of primary raw materials and reducing the insulation thickness required by current requirements for building structures.

References

[1] Remes M 2010 *Multilayer Reflective Insulation* http://www.asb-portal.cz/stavebnictvi/materialy-a-vyrobky/tepelne-izolace/vicevrstve-reflexni-izolace (online)
[2] Hejhalek J 2009 *Multilayer thermal insulation with reflective foils* http://www.stavebnictvi3000.cz/pdf/2969.pdf (online)
[3] Tenpierik M J and Hasselaar E 2013 *Journal of Energy and Buildings* **56** 233–243
[4] Halahyja M 1986 *Building thermal engineering, acoustics and lighting* (Bratislava: ALFA) (in Slovak)
[5] EN 16012: 2011 *Thermal insulation for buildings – Reflective insulation products – Determination of the declared thermal performance*

IOP Conf. Series: Materials Science and Engineering **385** (2018) 012025 doi:10.1088/1757-899X/385/1/012025

Properties of concrete intended for further testing measured by the Impact-Echo and the ultrasonic pulse method

O Karel, R Dvorak, I Rozsypalova and P Schmid

Faculty of Civil Engineering, Brno University of Technology, Veveri 331/95, 602 00 Brno, Czech Republic

Abstract. The aim of the paper is non-destructive measurement of differently degraded specimens by high temperature intended for further testing of joints of stainless steel helical reinforcement glued into the groove and differently degraded concrete. Measurement intended for determination of possibilities of estimation of future properties of named joints is performed by the Impact-Echo method and by the ultrasonic pulse velocity method on specimens of dimensions 400 × 100 × 100 mm made of concrete of the C20/25 strength class degraded by different elevated temperature. Five sets of specimens were manufactured - four sets of specimens were heated in the furnace at temperatures of 400 °C, 600 °C, 800 °C and 1000 °C and one set was kept intact as reference. Specimens will be afterwards additionally strengthened at the tensile side of specimens and broken by four-point flexural strength test. The non-destructive measurement aims to evaluate the residual physical-mechanical properties of plain concrete in terms of resonance frequency of test specimen, and sound velocity in tested specimen before and after the temperature degradation. This assessment will serve as material information basis for interpretation of the expected behaviour of used helical reinforcement for a retrofitting process and the thermally damaged concrete reaction to such intervention.

1. Introduction

Due to economical and serviceability demands on structures either in housing or industry buildings, a building structures damaged by fire is more often rehabilitated by retrofitting methods. There are many possible approaches and tools to strengthen, retrofit and repair the damaged elements such as partial replacement of structure elements, concrete or steel jacketing and retrofitting of structural members by fiber reinforced polymers (FRP) or stainless steel helical reinforcement [1].

Presented paper focus on the inspection process of thermally degraded concrete before the application of retrofitting tool – in this case a stainless steel helical reinforcement glued into groove in tensile area of test specimen. The inspection process in relation to retrofitting by this reinforcement is not described yet.

This kind of additional reinforcement is made of high strength (800 – 1300 MPa depending on type and manufacturer) stainless steel which is twisted due to better cohesion with mortar used to bond reinforcement with basic material. Its cross-section also ensures that even if it is able to carry a large load the worker is still able to bend it by hands to desired shape – this advantage is very useful for structure retrofitting [2].

Concrete can withstand up to 400 °C without notable changes in macrostructure and physical-mechanical properties. When silicate aggregate with a cement binder are burned, the greatest weight loss occurs due to the dehydration of physically bonded water and chemically bonded water from the individual components of the binder [3].

In this process of thermal degradation, a cement stone tends to shrink and opposingly the aggregate tends to expand in volume. In range 600 °C to 800 °C the main composite C-S-H gels starts to decompose at around 560 °C and it decomposes into ß-C2S at around 600 °C to 700 °C. The most visible reduction of compressive strength is in range 800 °C to 1000 °C when C-S-H decomposes and decarbonation of carbonates continues [4, 5].

For the purpose to test the joints of stainless steel helical reinforcement bonded to grooves with the basic material concrete beams of dimensions 400 × 100 × 100 mm were manufactured and then non-destructively measured. Concrete of the C20/25 strength class was heated in furnace to temperatures of 400 °C (set B), 600 °C (set C), 800 °C (set D) and 1000 °C (set E). Reference specimens (set A) that were not heated were laid aside under the laboratory conditions. For temperature - loaded specimens, various levels of degradation are expected, which will subsequently affect the tested joints. The intended objective of these measurements is the future comparison of the results of non-destructive measurements in relation to the real behavior of the joints.

For assessing the change in physical and mechanical properties before and after heating, measurements were made using the acoustic non-destructive method Impact-Echo [6] and the ultrasonic pulse method [7]. Similar assessment was conducted in previous studies [3], [8] which proves the IE method to be a suitable tool for assessing the thermally degraded concrete. In the vast assessment of Wisconsin highway bridge in 2003, where numerous test methods where used on thermally degraded concrete structures, Ghorbanpoor concluded [9] that acoustic non-destructive testing is one of the most suitable test tools for thermally degraded concrete structures.

2. Testing Methodology

The Impact-Echo method is based on excitement of the low-frequency vibration of the test specimen by a mechanical impact and subsequent measuring of excited vibrations [10]. The recorded signal is then converted by Fast Fourier Transform [11] into a spectrum that can be easily analyzed (figure 1, figure 2 and figure 3). The mechanical waves propagate in the form of longitudinal, transverse and torsional waves, and more easily pass through a solid homogeneous material without the presence of cavities. When impacting on the interface of two dissimilar materials, the reflection, shattering, absorption or transition of the waves to another type may occur, depending on the ratio of the acoustic impedance of these two materials. In the case of the mechanical wave in the material strikes a defect or a crack, the mechanical waves partially diminish, resulting in a change in the frequency spectrum [12]. Using this method, it is possible to locate defects within subtle structures, profile thickness or to determine the physical and mechanical state of the material that has succumbed to any degradation factor.

The testing specimens were made of concrete of expected strength class C20/25 and specific mix design of used concrete is shown in table 1. For the experiment was designed usual concrete with no special requirements.

Table 1. Mix proportions.

Material	Weight (kg/m^3)
Cement	376
Aggregate 0-4	1058
Aggregate 8-16	706
Water	226
w/c	0.6

Measurement by Impact-Echo was conducted on test specimens placed on rubber pads. As exciter of mechanical stress waves a spherical hammer of total mass 25.5 g was used. For recording of vibrations, a piezoelectric sensor MIDI 446s12 was used and for processing the signal was used a digital oscilloscope Handyscope HS3 with sampling speed set to 0.2 MHz.

The ultrasonic pulse method is based on the principle of measuring the shortest ultrasound signal passing through the specimen. The presence of cavities, cracks and defects forces the signal to travel more complicated and thus longer. From the knowledge of the velocity of the propagation of the mechanical waves in the material, the dimensions of the tested test body and the bulk density, the dynamic modulus of elasticity E_{cu} [7] can be calculated.

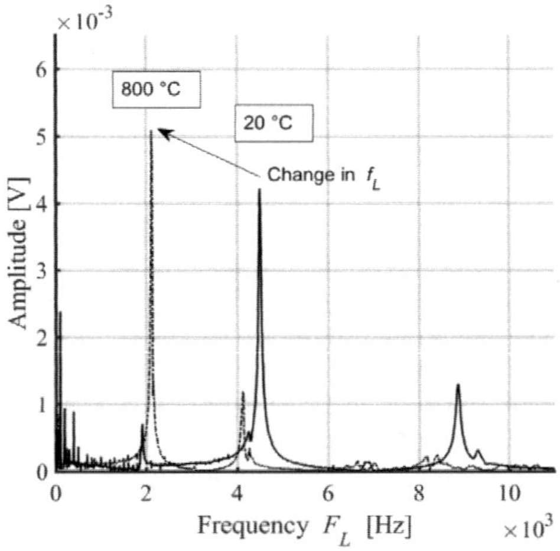

Figure 1. Change of first own longitudinal wave frequency for DP1 specimen before and after heating to 800 °C.

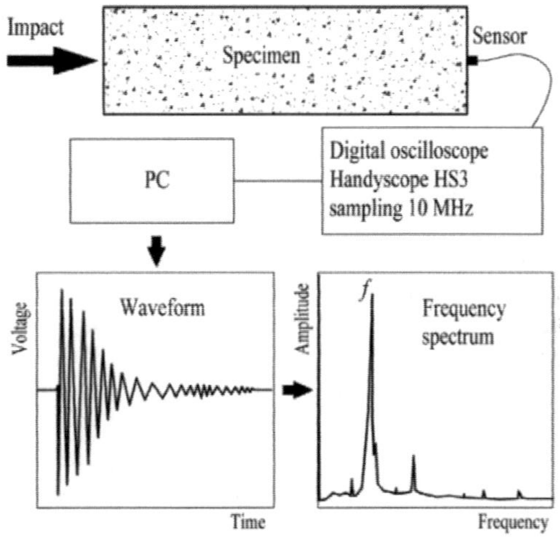

Figure 2. Impact-Echo test setup [3].

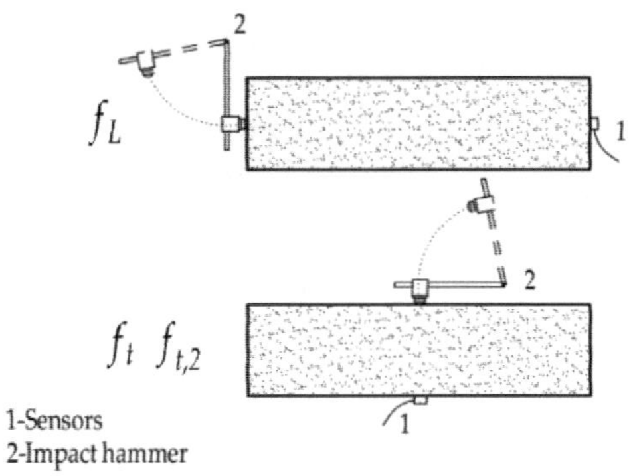

Figure 3. Test position of test specimens in longitudinal way f_L of measurement and transverse way f_t and $f_{t,2}$ [3].

3. Results and Discussion

Results of bulk density loss in relative scale for set E burned to 1000 °C are given in figure 4. The largest drop in the bulk density of 4.5 % occurred between the reference set A and set B burned at 400 °C can be seen. This temperature was higher than a critical water temperature of 374 °C, after which water can no longer be present in the form of physically bonded water. The lowest decrease of 1 % occurred between 800 °C and 1000 °C, which was mainly due to calcite decomposition ($CaCO_3 \rightarrow CaO + CO_2$) at about 930 - 960 °C.

The change in the bulk density is also associated with the change in mechanical properties of the material. As shown in figure 5, the dynamic modulus of elasticity reaches 38 GPa for the unheated set A, while the set B reaches 23 GPa. Like the bulk density, there was the largest drop between set A and B. The minimal modulus of elasticity was reached by set E, which showed the value of the dynamic modulus of elasticity of 0.5 GPa. The surface of these burnt test specimens was covered by cracks with widths ranged from 0.1 to 2 mm. The cover layer degraded to the greatest extent and where parts of the aggregate and the cement matrix lost its cohesion and fell apart.

The measurement of the test specimens by the resonance method of Impact-Echo also showed a change in the first dominant frequency of longitudinal and transverse waves (see figure 6). The resonance frequency of the longitudinal waves for the unloaded set A was 4.55 kHz and for the set E degraded to 1000 °C the resonant frequency was 0.75 kHz. Due to the chosen aging, hydration and recipe, the test specimens were tested by Impact-Echo in a worse condition than would be in the case of optimal aging according to CSN EN 12390 Part 2: Testing hardened concrete - Part 2: Making and curing specimens for strength tests. This reference state is therefore slightly degraded from the start compared to properly aged concrete [13]. It is important to note that the method of aging was chosen respectively to further testing of the application of stainless steel helical reinforcement in temperature-degraded concrete and was therefore not intended to produce perfectly "healthy" test specimens.

The highest drop of 28 % in the resonance frequency for longitudinal waves occurred between 800 and 1000 °C. Although the ultrasonic pulse velocity method showed the highest loss in dynamic modulus of elasticity between 20 °C and 400 °C, Impact-Echo measurement indicates the highest loss of physical-mechanical properties between 800 and 1000 °C. A second sharp drop in the resonance frequency of longitudinal waves by 21 % occurred between the reference set and the temperature of 400 °C. A similar behaviour is shown by the measurement of transverse waves.

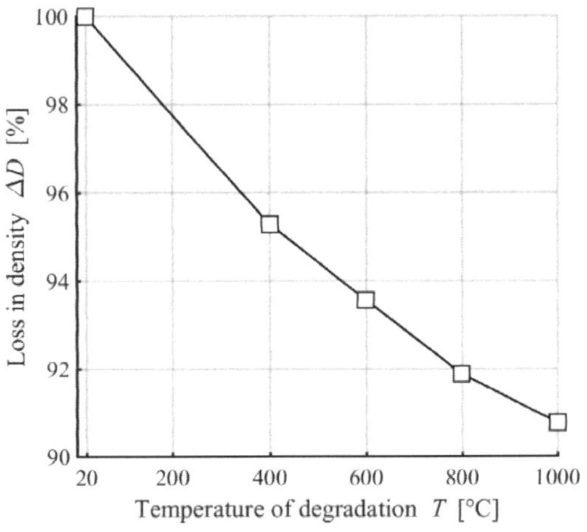

Figure 4. Relative bulk density change. The average bulk density before burning was 2350 kgm^{-3}.

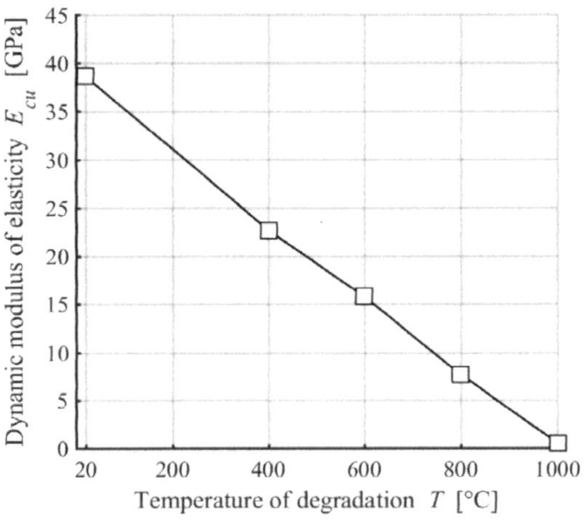

Figure 5. Change of the dynamic modulus of elasticity E_{cu} for each temperature set.

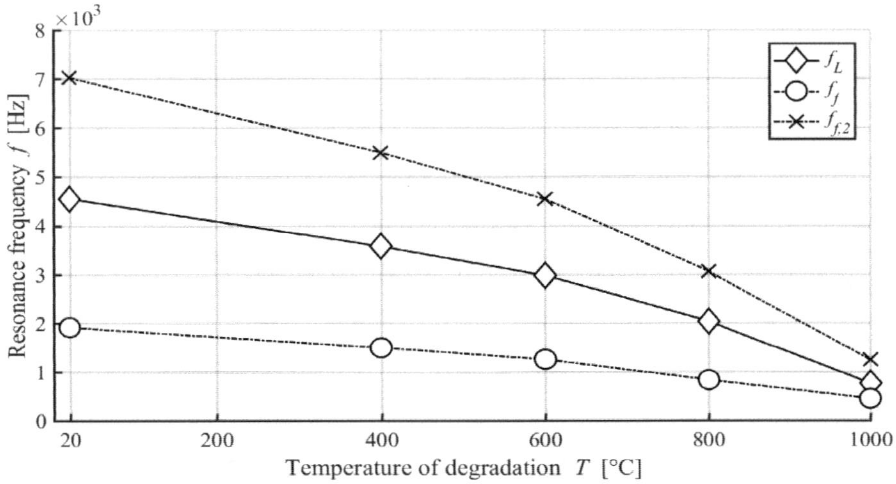

Figure 6. The first own frequencies of the longitudinal waves f_L and the first own frequencies f_f and the second own frequency $f_{f,2}$ of the transverse waves for each temperature set.

4. Conclusion

The sets of test specimens (a total of 42 specimens in six sets) were subjected to ultrasonic pulse and Impact-Echo methods measurement after exposure to various high temperatures. For these sets, the bulk density was also determined before and after burning. For all measured values it can be stated that they show changes that may indicate the subsequent behavior of the base material which will be surrounding the intended joint. The tested specimens will be strengthened on the tensile side with a stainless steel helical reinforcement placed into the groove according to the manual [2] and then tested by a four-point bending test similar to description in the standard for testing of the cohesion of steel reinforcing autoclaved aerated concrete [14]. The results from this measurement will be assigned to the quality of the connection failure in the differently degraded specimen sets.

Acknowledgement

This publication was created under the project FAST-J-18-5145 supported by Faculty of Civil Engineering of Brno University of Technology.

References

[1] Bazant Z and Klusacek L 2015 *Statics In Reconstruction Of Objects* **6** (Brno: Cerm publishing) (in Czech)
[2] Kubanek J and Schmid P 2006 *Manual And Methodology Of Designing And Application Of Retrofitted Helical Reinforcement Systems For Reinforcement Of Building Structures In The Conditions Of Construction* (Brno: Munipress) (in Czech)
[3] Dvorak R 2017 *Defektoskopie 2017* **47** 39-47 (in Czech)
[4] Simonova H, Rozsypalova I, Rovnanikova P, Danek P and Kersner Z 2018 *Sol. St. Phen.* **272** 47-52
[5] Bazant Z and Kaplan M 1996 *Concrete At High Temperatures: Material Properties And Mathematical Models* (UK: Longman)
[6] CSN 73 1372: 1993 *Non-destructive testing of concrete: Testing of concrete by resonance method*
[7] CSN 73 1371: 1983 *Non-destructive testing of concrete: Method of ultrasonic pulse testing of concrete*
[8] Hager I 2013 *Bull. Pol. Ac.: Tech.* **61** 1-10

[9] Ghorbanpoor A and Benish N 2003 *Final Report Non-destructive testing of wisconsin highway bridge: Wisconsin highway research program*

[10] Sansalone M and Streett W B 1997 *Impact-Echo: Non-Destructive Evaluation Of Concrete And Masonry* (Ithaca, N Y: Bullbrier Press)

[11] Bracewell R 1999 *The Fourier Transform & Its Applications* (USA: McGraw-Hill Science/Engineering/Math)

[12] Malhotra V M and Carino N J 2004 *Handbook On Nondestructive Testing Of Concrete* (Boca Raton, Fla.: CRC Press)

[13] Dvorak R, Hodulakova M and Topolar L 2017 *Proc. Int. Conf on Experimental stress analysis (Novy Smokovec)* (Kosice: TUKE) pp 426-434

[14] CSN EN 12269-2: 2010 *Determination of the bond behaviour between reinforcing steel and autoclaved aerated concrete by the beam test Part 2: Long term test*

IOP Conf. Series: Materials Science and Engineering **385** (2018) 012026 doi:10.1088/1757-899X/385/1/012026

The properties of water from a concrete plant to be used in cement composites

L Klus[1,2], V Vaclavik[1,2], T Dvorsky[1], J Svoboda[1,2] and R Papesch[1]

[1] Institute of Environmental Engineering, Faculty of Mining and Geology, VSB-Technical University of Ostrava, Czech Republic,
[2] Institute of Clean Technologies for Mining and Utilization of Raw Materials for Energy Use Faculty of Mining and Geology, VSB-Technical University of Ostrava, Czech Republic

Abstract: This article presents the requirements for concrete mixing water and evaluates the possibilities of using recycled waste sludge water generated in a concrete plant from rinsing water of the concrete plant agitating centre and agitation trucks. The production of concrete mix requires large amount of water and for this reason, it is necessary to use this resource in an economic way. For further use of these waters, it is vital to know their detailed properties in order to maximize their potential in the design of concrete formulations for the purpose of replacing pure mixing water with recycled sludge water so as to preserve or improve the physical and mechanical properties of fresh concrete mix or hardened concrete. The results of the wastewater tests from the concrete plant confirm compliance with the limit concentrations according to CSN EN 1008. The attention should be paid to the limit of concentration of sulphates.

1. Introduction

In order to maximize the potential utilization of mixing and recycled water, it is necessary to establish the chemical properties and their reactivity with other substances to facilitate their more efficient use in the design of concrete formulations. Nowadays, many research projects show the use of recycled concrete water, for example, to save fine particles such as fly ashes [1]. The problem of most of the waste and recycled waters in concrete plants is their content of acid substances and sulphate concentration. These waters must be chemically treated or diluted by adding pure mixing water for the optimum concrete mix [2-3]. Even the preparation of the so-called "gray water" (water containing a larger proportion of dissolved cement), which is added during concrete placing, can help to achieve higher initial strengths, i.e. acceleration of the setting and hardening of the concrete mix. This procedure does not have absolute validity, because the higher proportion of fine particles requires more water for proper hydration, as evidenced by many tests whose results are described in [4]. By modifying the mixing water, the exceptional physico-mechanical properties of the resulting concrete mix can be obtained. By applying the knowledge of chemical interactions, the desired properties of the concrete composites can be achieved [5-7]. This advantage is currently used in the design of new concrete formulations in civil engineering, since it is required by modern civil engineering so as to reach the required construction quality [8]. The aim of the presented research results is the possibility of using waste water from the concrete plant as the mixing water in the production of cement composites.

2. Materials and methods

This experimental study has examined water from natural sources (small water reservoirs) and recycled water (rinsing water from agitation trucks) for the use in the production of concrete. Based on the requirements of the European standards CSN EN 1008 [9], CSN EN 196-1 [10], EN 1015-3 [11],

EN 196-3 + A1 [12], the properties of these waters have been examined by means of laboratory methods such as: detailed chemical analysis and mineralogical analysis of dissolved particles.

2.1 Mixing water

Mixing water serves as the basic activator of chemical reactions in the production of concrete, causing the formation of new crystalline structures in concrete. The properties of mixing water can influence the character of the concrete mix, whether it will be used, for example, for concrete with reinforcement, without reinforcement or for aggressive environment. Water and its components change the resulting character of the concrete mix, influence the process of crystallization of minerals and, according to the present components (sulphates, chlorides, alkalis, nitrates and others), form the binding elements and create strong bonds. These bonds influence the final physico-mechanical properties of the hardened concrete.

2.2 Water sampling

Sampling of pure and recycled sludge waters took place in concrete plants in the Moravian-Silesian Region and their exclusive water resources. The used mixing water sources are almost identical, and the additives used in concrete are identical in the concrete plants where the samples of recycled sludge water were taken. The results of the sample tests were compared to find out their properties.

2.3 Testing methodology of water samples

The samples of pure mixing water and recycled waste water from the concrete plant were tested in order to determine the properties of the mixing water according to CSN EN 1008 [9] and the requirements of the concrete mix production technology.

2.3.1 Determination of the pH value, temperature and conductivity

The pH values of the samples of sludge water and pure mixing water were measured potentiometrically by means of an electrochemical analytical method based on the measurement of the equilibrium cell voltage, which is composed of indicative and reference electrode. The temperature and conductivity determination was performed using Merci pH/Cond 3401 instrument during each sampling.

2.4 Determination of the mineralogical composition of the sludge

The determination of the mineralogical composition of the sludge sample from the concrete plant waste water was carried out using Bruker Advance D8 powder diffractometer equipped with Lynx Eye linear semiconductor detector and SOL-XE energy dispersion detector.

2.5 *Determination of the chloride concentration*

A methodology in accordance with CSN ISO 9297 [13] was used to determine the concentration of chlorides in the samples of sludge water and pure mixing water.

2.6 *Determination of the sulphate concentration*

A methodology in accordance with CSN 75 7477 [14] was used to determine the concentration of sulphates in the samples of sludge water and pure mixing water.

2.7 *Determination of the nitrate concentration*

A methodology in accordance with CSN ISO 7890-3 [15] was used to determine the concentration of nitrates in the samples of sludge water and pure mixing water.

2.8 *Alkali analysis*

The tested samples were filtered to obtain filtrate which was used to determine its chemical composition by means of atomic absorption spectrophotometry (AAS), VARIAN AA 280FS instrument. AAS measurement takes advantage of the analytical absorption property of the radiation of free atoms of the observed element.

2.9 *Disolving substances (RL) and Insolable substances (NL) analysis*
The determination of all suspended solids was carried out according to the methodical guidance of CSN EN 872 [16], where the given sample volume is filtered through a filter and subsequently dried in order to determine the amount of Insolable substances (NL) in g/l. The determination of the quantity of all dissolved substances was carried out according to the methodical guidance of CSN 75 7346 [17].

2.10 *Determination of carbohydrates, humus substances and Fe*
The determination of carbohydrates was performed using a GlucoLab s 25 fast test. A 30 ml sample of humus substances was taken and treated with 3% sodium hydroxide according to CSN EN ISO 1008 standard [9] in order to determine the amount of the humus substances. The iron concentration was determined using an Eisen test set and colorimetric, ferospectral 1.14759 test method.

3. Results and discussion

Table 1 shows the average values of the test results of pure mixing water and recycled waste water from the concrete plant carried out in 2017 year (Stage I) and 2018 year (Stage II), as well.

Table 1. Results of pure and recycled water analysis.

Tested properties	Pure mixing water from a concrete plant		Waste water from a concrete plant		Limit concentration according to CSN EN 1008
	1st stage	2nd stage	1st stage	2nd stage	
pH	6.77	9.10	12.56	12.40	> 4
Temperature [°C]	18.78	11.5	20.20	14.2	-
Conductivity [μS/cm]	13.05	12.05	337.12	292.00	-
Humus substances	acceptable	acceptable	acceptable	acceptable	Palerthan yellow-brown
Chlorides [mg/L]	14.20	53.00	251.70	407.70	< 600< 2000< 4500
Sulphates [mg/L]	40.30	95.0	1702.10	2900.00	< 2000
Nitrates [mg/L]	47.00	90.00	180.00	234.00	< 500
$CHSK_{Cr}$ [mg/L]	5.980	5.236	60.637	60.351	-
Na [mg/L]	13.20	15.10	32.57	35.6	< 1000
Pb [mg/L]	0	0	0.1105	0.1864	< 100
Zn [mg/L]	0.058	0.061	0.022	0.028	< 100
Ca [mg/L]	32.68	24.90	1518.00	1610.00	-
Glucose = sucrose [mg/L]	< 100 + < 100	< 100 + < 100	< 100 + < 100	< 100 + < 100	< 100 + < 100
Fe [mg/L]	< 0.10	< 0.10	(immeasurable)	(immeasurabl)	-
BSK_5 [mg/L]	2.930	3.235	20.328	20.825	-
K [mg/L]	8.83	9.81	68.20	77.10	< 100
Mg [mg/L]	3.85	4.67	0.267	0.356	-
NL [mg/L]	0.004	0.004	0.075	0.082	-
RL [mg/L]	0.008	0.009	0.482	0.504	< 1000

Table 1 shows that, when the measured values of the monitored analytes in the 2nd and 1st stages of testing of the properties of pure and recycled waste water from the concrete plant are compared, the values of sulphates were considerably higher by 1 197.8 mg/l (41.3%), of chlorides by 156 mg/l (38%)

and of nitrates by 54 mg/l (23.1%). For clear specification the limit concentration of chlorides in reinforced concrete is 600 mg/L and for plain concrete is concentration to 4500 mg/L of chlorides. This can be explained by the fact that cast anhydrite screeds were prepared in the concrete plant prior to the sampling performed for testing the properties of pure and waste water, which confirms a significant increase of sulphates in the tested waste water samples. The results showed in table 1 clearly indicate that when using recycled waste water from the concrete plant, it is always necessary to take into account the different properties that could affect the properties of fresh concrete mix and hardened concrete. The other test values did not show significant difference when comparing the values from the 2nd and 1st stage of the property testing.

A mineralogical analysis presents the results of the represented minerals in the dry matter of the samples measured in the 1st and 2nd stage, resulting from the filtration of the individual samples of recycled sludge water. The highest amounts in the first stage included quartz (SiO_2), bassanite ($CaSO_4 \cdot 0.5H_2O$) and limestone ($CaCO_3$). Three new minerals were found in the second stage, namely thaumasite ($Ca_3Si(OH)_6(CO_3)(SO_4) \cdot 12H_2O$, gypsum ($CaSO_4 \cdot 2H_2O$) and oligoclase An16 (Ca, Na) (Al, $Si)_4O_8$. An overview of the amount of the individual minerals is presented in figures 1 and 2.

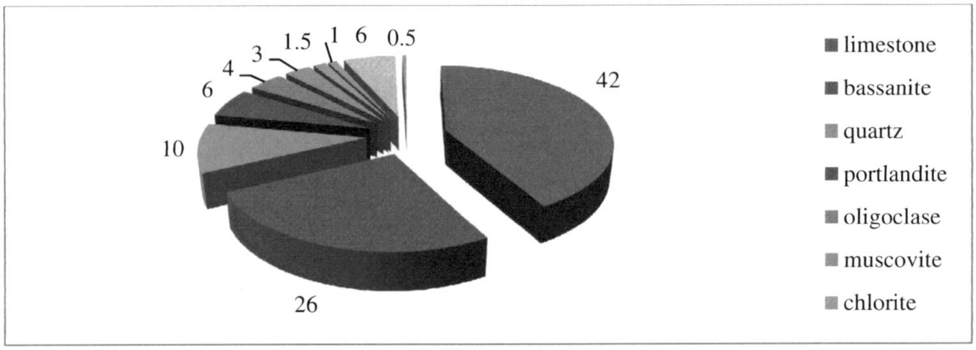

Figure 1. Mineralogical composition of waste water sludge from a concrete plant (1st stage) [%].

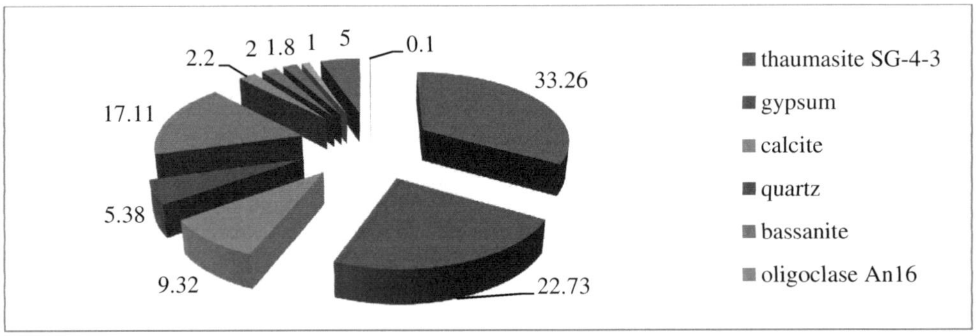

Figure 2. Mineralogical composition of waste water sludge from a concrete plant (2nd stage) [%].

4. Conclusion

This article presents the results of a research dealing with the testing of the properties of pure mixing water in the concrete plant and waste recycled water generated in the concrete plant. The aim of the subsequent research is to replace the mixing water with recycled waste water from the concrete plant in the amount of 25, 50, 75 and 100 in the production of cement composites. Prior to the start of the production of cement test specimens, it is necessary to know the properties of mixing water and to compare whether they comply with the standard values according to CSN EN 1008. During the research, the samples of pure mixing water and waste water from the concrete plant were taken in two stages. The 1st stage presents the average values from 2017 year and the 2nd stages the values from 2018 year. The results of the tests of the required properties according to CSN EN 1008 are presented in table 1. It is obvious that the different values of the examined analytes have to be taken into account, when testing the individual water samples, and they can affect the properties of fresh mix used for the production of cement composites. The properties of recycled waste water are directly proportional to the type of concrete produced in the concrete plant. Further research will be focused on the testing of the properties of fresh mix for the production of specimens of cement composites and the physico-mechanical properties of the produced test specimens based on waste recycled water from the concrete plant from the 2nd stage of research.

Acknowledgement
This article was written in connection with project Institute of clean technologies for mining and utilization of raw materials for energy use - Sustainability program. Identification code: LO1406. Project is supported by the National Programme for Sustainability I (2013-2020) financed by the state budget of the Czech Republic.
This research work has been supported by the student grant competition project SGS SP2018/12 Research on the use of sewage water from the concrete plant.

References
[1] Chatveera B and Lertwattanaruk P 2009 *J. Environ. Manage.* **90** 1901–1908
[2] Chatveera B, Lertwattanaruk P and Makul N 2006 *Cem. Concr. Comp.* **28** 441–450
[3] Tsimas S and Zervaki M 2011 *Manag Environ Qual Int J* **22** 7–17
[4] Ghrair A M, Al-Mashaqbeh O A, Sarireh M K, Al-Kouz N, Farfoura M and Megdal S B 2016 *Ain Shams Eng. Journal* 1–7
[5] Estokova A, Smolakova M, Luptakova A and Strigac J 2017 *IOP Conf. Ser.: Mater. Sci. Eng.* **251** 1–6
[6] Ondova M, Stevulova N, Palascakova L and Estokova A 2014 *Pollack Periodica* **9** 105–115
[7] Ondova M, Stevulova N, Palascakova L and Estokova A 2012 *Procedia Engineering* vol 42 *20th International Congress of Chemical and Process Engineering* (CHISA 2012) pp 1863–1872
[8] Tkach E V, Semenov V S, Tkach S A and Rozovskaya T A 2015 *Procedia Engineering* **111** *XXIV R-S-P seminar, Theoretical Foundation of Civil Engineering* (24RSP) (TFoCE 2015) pp 763–769
[9] CSN EN 1008: 2003 *Mixing water for concrete Specification for sampling, testing and assessing the suitability of water, including water recovered from processes in the concrete industry, as mixing water for concrete*
[10] CSN EN 196-1: 2016 *Methods of testing cement* Part 1: Determination of strength
[11] CSN EN 1015-3: 2008 *Methods of test for mortar for masonry* Part 3: Determination of consistence of fresh mortar (by flow table)
[12] CSN EN 196-3 + A1: 2008 *Methods of testing cement* Part 3: Determination of setting times and soundness
[13] CSN ISO 9297: 1996 *Water quality* Determination of chloride. Silver nitrate titration with chromate indicator (Mohr's method)
[14] CSN 75 7477: 2010 *Water quality* Determination of dissolved sulphate. Lead nitrate titrimetric method
[15] CSN ISO 7890-3: 1995 *Water quality* Determination of nitrate Part 3: Spectrometric method using sulfosalicylic acid
[16] CSN EN 872: 2005 *Water quality* Determination of suspended solids. Method by filtration through glass fibre filters
[17] CSN 75 7346: 2002 *Water quality* Determination of dissolved substances

New type of multifunctional silicate composite

K Kolar, P Konvalinka, J Litos, M Kocova and Z Rihova

Experimental Centre, Faculty of Civil Engineering CTU in Prague, Thakurova 7, 166 29 Prague, Czech Republic

Abstract. There are introduced basic information about the composition and the properties of the precast cement-based mixture in the paper. Studied multi-functional composite was developed for the easy and rapid reconstruction of the concrete structures. This composite exhibited rapid evolution of utility properties and volume stability during the process of hardening

1. Introduction

Currently is known prefabricated mixtures for preparation concrete used in the field of civil engineering and building structures. Concrete can be transported already done - ready for use or can be prepared directly on the site by adding the mixing water to the prefabricated mixture, optionally by addition of other substances.

Figure 1. Production of dry prefabricated mixture.

The basic components of dry prefabricated mixtures (figure 1) are binder matrix ensuring the cohesion of the setting concrete, aggregates and regulation systems ensuring setting and hardening concrete.

Current approach to composite materials is using different sources of aggregates. Those can be granular or fibrous character with different origin - natural or industrial. Binder matrix is based on the most commonly used silicate cement. This silicate base cement can be pure Portland cement, or blended cements containing, for technical, economic or environmental reasons, different admixtures. Used admixtures can be latently hydraulically active or inert, but it must be preferably such granulometry, which complements the granulometry of used binder, to obtain a continuous granulometry of the entire system of solid components in the final preparation of the freshly prepared mixture with water.

The use of commercially available silicate cements, declared as pure Portland cements, blended cements and standardized cements e.g. in accordance with CSN EN 197-1 or other normative regulations, brings to the formulation of special cement composites a number of problems arising from basic standard requirements for physicochemical parameters of commonly produced cement binders. This is primarily about the reliable control of hydration and setting mechanisms of used type of silicate cement, referred to as time development of the so-called setting and hardening processes, the achievement of acceptable workability of the freshly mixed mixture, simple processing and storage, bulk stability of the curing composite, achievement of predicted parameters and long-term stability under various conditions of use.

The available literature describes a number of possibilities for controlling the properties for various types of silicate binders, to modify or supplement them to produce a low or high-value cementitious composite. There are various options for modification so-called ratios of setting and hardening process with a variety of retarding or accelerating additives, the use of plasticizers or superplasticizers and stabilizers of various compositions, the mixing method for the necessary homogenization of the fresh mixture with the required amount of mixing water, or with other requirements for achieving the declared parameters such as the time consuming to preparation of composite, storage and compaction of freshly prepared composite with mixed water.

Mineral and pozzolanic components from natural or artificial sources such as micronized limestone, micro-silica, ground blast-furnace slag, fly ash, various aluminosilicates and others are used as super-fine admixtures. Their basic physicochemical parameters must be respected in order to avoid undesired effects on the control of setting and hardening processes and also to maintain a sufficient time for workability by the used flowability additives and to achieve the volume stability of solid structure of the composite.

Other fine aggregates above 0.1 mm are mainly sand fractions of different purity (high content of pure SiO_2), but can be also another dense or porous minerals (corundum, carborundum, fireclay, basalt, etc.). The use requires respecting their physical properties (absorption, etc.), which leads to the need for careful formulation of considered composite.

This paper presents the results of some preferred variants of a dry prefabricated multifunctional composite, which were mixed on slow-speed mixers with different mixing vessel volumes.

2. Experimental program

The individual components of multifunctional mixture are usually available at commercial distributor. The binder components are products of Czech cement companies from Čížkovice, Lochkov, Prachovice, Mokrá and Hranice. The additives in the powdered form, which are needed for this composite, are available on the domestic market by SIKA, BASF, STACHEMA, CHRYSO, RADKA and others [9].

The prefabricated mixture was mixed for 1.5 minutes. Mixing water was only 0,1 weight of all dry components.

Table 1. Properties of fresh mixture.

	Designation of mixture		
	1	2	3
Flow consistency without tapping [cm]	28	28	28
Start of setting [min.]	20	130	330

The setting ratios were tested by cementation methodology, placeability by the flow methodology according to mortar standards without the use of tapping. Further were tested also flexural and compressive strength (see table 2) and volume changes (see figure 4).

After summarizing the results of the extensive experimental program are there presented the values of three selected mixtures with different processing times ranging from a few tens of minutes to several hours see table 2. More over mixture 2 was made with different fiber content and for each of them fracture energy was tested (see table 3).

Table 2. Mechanical properties of study mixtures.

	Designation of mixture					
	1		2		3	
	Strength					
	flexural [MPa]	compressive [MPa]	flexural [MPa]	compressive [MPa]	flexural [MPa]	compressive [MPa]
1 hr.	0,9	5,8	-	-	-	-
2 hr.	1,4	8,0	-	-	-	-
5 hr.	3,7	22,1	0,3	1,6	-	-
10 hr.	-	-	4,5	26,1	0,1	0,8
15 hr.	-	-	-	-	1,6	9,5
1 day	8,1	47,4	8,7	50,2	7,6	45,5
7 day	13,2	79,4	13,1	78,3	11,2	65,3
28 day	19,7	106,1	16,2	95,6	19,2	95,5

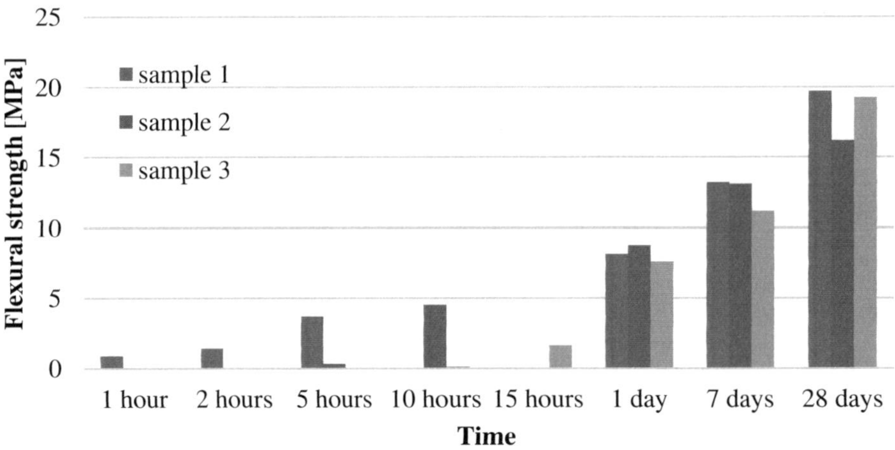

Figure 2. Development of flexural strength.

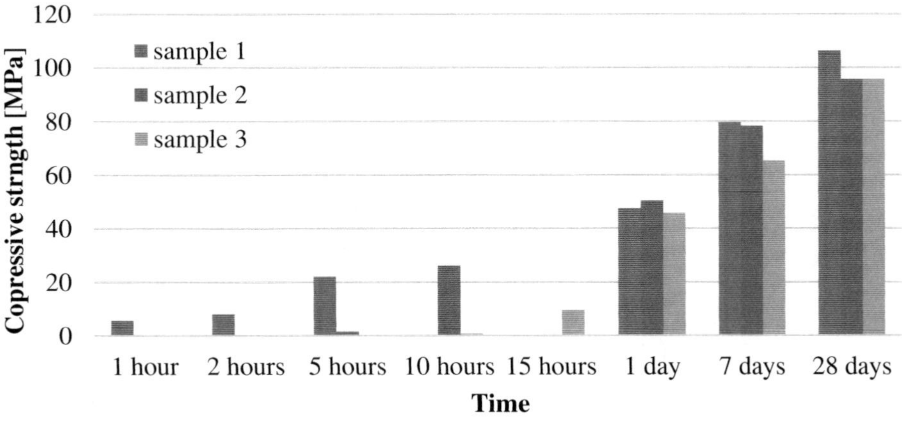

Figure 3. Development of compressive strength.

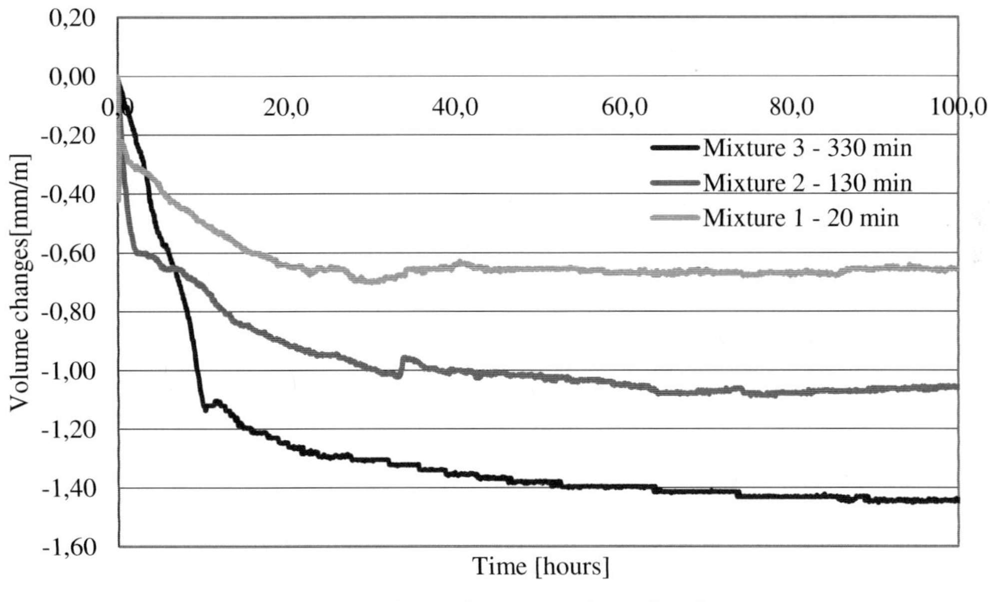

Figure 4. Volume changes vs. time of curing.

Table 3. Effective fracture energy values for mixture 2.

Fiber content	Effective fracture energy
% vol.	J/m^2
0	80
0,5	3 000
1	11 000
2	17 000
3	22 000

After 168 hours (7 days) of curing in water basin were samples subjected to freeze/thaw cycling and residual properties were studied. The course of flexural strength and compressive strength declare no damage during the testing. Moreover, gradual increase of studied properties was observed, what was probably caused by the gradual hydration. Destructive testing does not correspond with non-destructive methodology very well. Used pulse method is actually sensitive on the moisture content, which could noticeably misrepresent the results, that is why it is used only as an indicative procedure [5-6, 8].

Table 4. Index of frost resistance after 100 freeze/thaw cycles for mixture 2 [-].

Compressive strength	Flexural strength	Modulus of elasticity
1.23	1.63	1.00

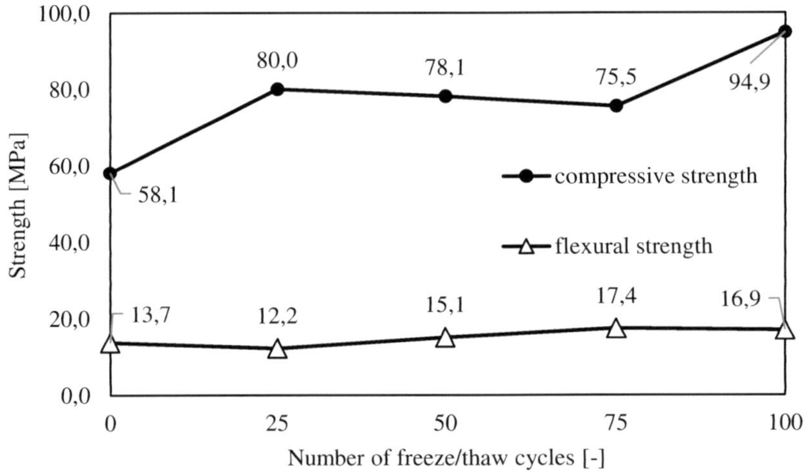

Figure 5. Evolution of dynamic modulus of elasticity during freeze/thaw cycling.

Frost resistance is usually expressed by the index of frost resistance calculated as a ratio of final and initial value of the observed property [4, 7]. Reached values of index of frost resistance for compressive strength, flexural strength and modulus of elasticity are shown in table 4 and figure 5. Obtained values well declare high resistance to action of frost of developed composite and its ability to hardening in such severe condition. Because binder system hydration was not demonstrably interrupted due to freeze/thaw cycling.

3. Conclusion

Dry mixture is made up of a new and very advantageous combination of commonly available ingredients. When combining their sometimes contradictory parameters, lead to a beneficial effect. The mixture is easily elaborate and needs a low dose of mixing water. Multifunctional composite prepared from dry prefabricate mixture is easily usable in practice with the wide range of applications, with a controlled hydration process and simply preparation technology. The composite has a sufficient workability time, a rapid increase in initial mechanical properties and reaches reliable stability after final hardening. Durability testing was presented through the freeze/thaw cycling what means one of the crucial parameter of the restoration of concrete structures. High performance concrete mixtures often exhibited low frost resistance due to extreme high density which does not allow transfer of the creating ice, since they have very low water adsorption.

Acknowledgement

This contribution was supported by CTU in Prague under the project No. SGS18/057/OHK1/1T/11.

References

[1] Kolar K, Litos J, Bazantova Z and Reiterman P 2016 *Appl. Mech. Mater.* **827** 328 - 331
[2] CSN EN 197-1: 2005 *Methods of testing cement – Part 1: Determination of strength*
[3] Valcuande M, Benito F, Parra C and Minano I 2015 *Constr. Build. Mater.* **76** 1 - 9
[4] Hora M and Reiterman P 2016 *Rev. Rom. Mater.* **46** 405- 413
[5] CSN EN 12504-4: 2005 *Testing concrete-Part 4: Determination of ultrasonic pulse velocity.*
[6] Rehman S K U, Ibrahim Z, Memon S A and Jamel M 2016 *Constr. Build. Mater.* **107** 58 -86
[7] Lazniewska - Piekarczyk B 2013 *Constr. Build. Mater.* **40** 659 - 671
[8] Brozovsky J, Bodnarova L, Hela R, Drochytka R and Hela V 2013 *Appl. Mech. Mater.* **284-287** 1315-1319
[9] Kolar K, Bazantova Z and Konvalinka P Czech patent No. CZ 306663

IOP Conf. Series: Materials Science and Engineering **385** (2018) 012028 doi:10.1088/1757-899X/385/1/012028

Experimental assessment of steel fibre reinforced concretes with different concentrations of fibres

T Komarkova[1], J Lanik[1], L Topolar[1], J Stoller[2] and P Stonis[2]

[1] Faculty of Civil Engineering, BUT in Brno, Veveri 95, Brno, Czech Republic
[2] Department of Engineer Technology, Faculty of Military Technology, University of Defence, Sumavska 4, Brno, Czech Republic

Abstract At present, steel fibre reinforced concrete can be classified as a composite material commonly used for load-bearing building structures. Adding steel fibers to fresh mixture improves its mechanical properties in comparison to plain concrete. The problem is, however, the production technology, which has not been standardized by regulations so far, and thus, proposals for the amount of steel fibers added to the mixture are still based on the long-term experience and performed tests. The paper describes an experiment focused on the determination and assessment of mechanical properties of steel fibre reinforced concretes with high concentrations of fibres (40 kg/m^3, 80 kg/m^3 and 120 kg/m^3). To better understand the behavior of test specimens during the three-point bending test, the acoustic emission method was used, which is able to capture active defects. The results show that the increasing amount of fibers has the influence on values of $F_{Rk,0.5}$ and $F_{Rk,res,1}$ (e.g. for concentration 40 kg/m^3 is $F_{Rk,0.5}$ = 13.27 kN and for 120 kg/m^3 is $F_{Rk,0.5}$ = 21.18 kN).

1. Introduction

Steel fibre reinforced concrete (SFRC) is a composite material with an added component of steel fibres guaranteeing an improvement of mainly the tensile load capacity and an increase in ductility. As a component, the fibres should be dispersed uniformly in the given volume, similarly, for example, to the dispersion of grains of the coarse fraction of aggregate in plain concrete. A uniform dispersion of steel fibers leads to a positive effect of their spatial action in the composite structure and to its overall stiffening. Steel fibers thus cannot be regarded as reinforcement in the traditional sense of the word but only as a solidifying element of the relatively fragile structure of plain concrete [1]. The use of various types of SFRCs is increasingly frequent in the building practice. The times when this concrete was only used for slab constructions of large dimensions, such as floors of production plants, etc., are long gone. In the course of the last few years, SFRC has been used for the production of prefabricated products, architectural panels, structures in seismic areas, tunnel lining, hydraulic structures and many others [2, 3]. SFRC is very often used also for military defensive structures, in which the selected concentration of steel fibres is many times higher than in civil structures.

The greatest problem of this building material is its production technology. Each composite material thus presents a complication for technologists considering the fact that the resulting characteristics depend partly on the properties of individual components, and partly, primarily, on their proportions. The more components a composite material contains, the higher attention must be paid to its production and selection of the composite's individual parts [4]. The efficiency of dispersed fibers does not depend on cohesion only, what is also very important is the choice of their correct amount, concentration, and mainly their uniform distribution in the material's structure. Under homogeneous steel fibre reinforced concrete, we understand a composite material in which the added steel fibres are uniformly distributed in the given volume and no direction of fibres significantly predominates. Some studies show that the final distribution of fibres depends on the process of pouring [5, 6]. For this reason, much attention is given to the technological procedure of manufacturing SFRC itself.

The criteria according to which the homogeneity of SFRC is assessed, have not been fixed yet. Under homogeneous hardened SFRC concrete, we understand a material with an acceptable dispersion of the results of its tensile strengths obtained in the prescribed tests, in which a potential inhomogeneity can manifest itself significantly [7]. To guarantee a homogeneity of SFRC concrete means to subordinate its production to the used fibres and their volume stage of stiffening.

2. Methods

Tensile strength in bending is in fact the most important laboratory test because tensile strength as such is considerably more important for SFRCs than for plain concrete. The test is performed with a controlled deformation and the specimen is loaded with one loading roller figure 1. The beam is loaded at a uniform rate in such a manner that the deflection in the middle of the beam under loading grows proportionally to time, i.e. at a uniform loading rate. During each test of individual test beams, the progress of loading and response $F_{Ri} - \delta_{ti}$ is recorded. For each diagram record obtained, it is necessary to determine force $F_{Rk,\,0.5}$ with an agreed deflection of $\delta_{t,0.5} = 0.5$ mm and force $F_{Rk,res,1}$ with an agreed deflection limit of $\delta_{tl} = 3.5$ mm in the middle of the span.

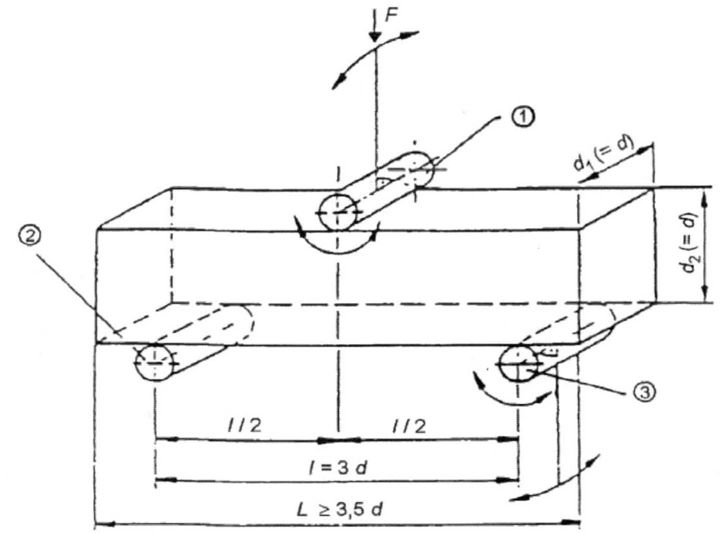

Figure 1. Schematic diagram of testing the SFRC specimen [8].

To better understand the behavior of test specimens during the three-point bending test, the acoustic emission method was used. The acoustic emission method, as an unusual non-destructive method, can often be used to detect a failure at a very early stage of damage long before the structure completely fails [9,10]. The paper also presents experiments focused on analyzing acoustic emission (AE) signals captured during commonly used three-point bending tests of specimens of SFRC without centre notch [11,12].

3. Experiment

The aim of the experiment was to compare the results of the tensile strength in bending tests on test specimens with dimensions of 100 × 100 × 400 mm from three different SFRC formulas. The test used was the so-called three-point bending test, in which the specimen is placed on two supports at a distance of 300 mm apart and loaded with an isolated load in the middle of the span according to CSN EN 12390-5 [8]. The formulas of SFRCs differed only in the concentration of steel fibres, and from each formula, 3 test specimens were prepared. As has already been mentioned, SFRC is also used in military defensive structures [13, 14]. Considering the use of the proposed formulas for building structures of this type, we chose concentrations of 40 kg/m^3, 80 kg/m^3 and 120 kg/m^3. The composition of individual mixtures and properties of steel fibers used are given in table 1 and table 2.

Table 1. Composition of the mixtures.

SFRC mixture	Cement 42.5 R [kg/m^3]	Water [kg/m^3]	Water ratio [-]	Aggregates 0 – 4 [kg/m^3]	4 – 8 [kg/m^3]	8 - 16 [kg/m^3]	Plasticizer [kg/m^3]	Steel fibers [kg/m^3]
T 1	390	160	0.41	900	250	670	4	40
T 2	390	160	0.41	900	250	670	4	80
T 3	390	160	0.41	900	250	670	4	120

Table 2. Properties of the steel fibers.

Fibers	Length [mm]	Diameter [mm]	Tensile strength [N/mm^2]	Young´s modulus [N/mm^2]
DRAMIX RC-65/35-BN	35	0.55	1.345	210.000

The acoustic emission measurements were performed by means of an acoustic emission measuring system XEDO made by DAKEL (Czech Republic). This system comprised five channels. The guard sensor eliminated mechanical and electrical noise. In this study, four acoustic emission sensors of the MIDI type (made by DAKEL) were used, all having the same frequency range, and they were attached to the surface by beeswax [15].

4. Results and discussion

The following figures show the processed average diagrams of the dependence of the force on deformation $F_{Ri} - \delta_{ti}$. Also marked are the average values of forces $F_{Rk,\,0.5}$ with an agreed deflection of $\delta_{t,0.5} = 0.5$ mm, and force $F_{Rk,res,1}$ with an agreed deflection limit of $\delta_{t1} = 3.5$ mm. It is apparent from the diagrams that increasing the concentration by 40 kg/m^3 causes both an increase in the tensile strength in bending, and an increase in the load capacity after the occurrence of a crack. Figure 1 shows the average diagram $F_{Ri} - \delta_{ti}t$ for a concentration of 40 kg/m^3 – set of specimens T1. This diagram is typical of common structural SFRC concrete, in which the occurrence of a crack is followed by a significant decrease in strength, but thanks to the ductility and activation of steel fibres, the load capacity of the element decreases slowly. The maximum force was $F_{max} = 16.7$ kN and it was achieved at a deflection of $\delta_t = 0.2$ mm.

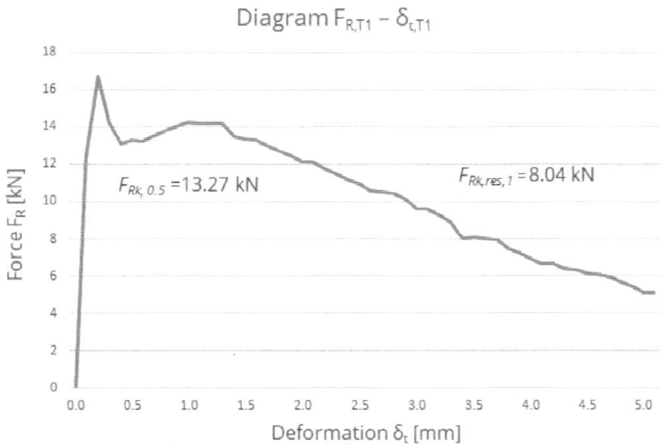

Figure 2. Average diagram of resistance for T1.

From the results of testing the T2 set of specimens (80 kg/m^3) and from their average diagram F_{Ri} – δ_{ti}t it is possible to conclude that the overall load capacity was increased, and force $F_{Rk,0.5}$ at a prescribed deflection of $\delta_{t,0.5}$ = 0.5 mm is at the same time equal to the maximum achieved force of F_{max} under loading. The very progress of the average diagram suggests that during the tests, steel fibres performed their function and were not pulled out.

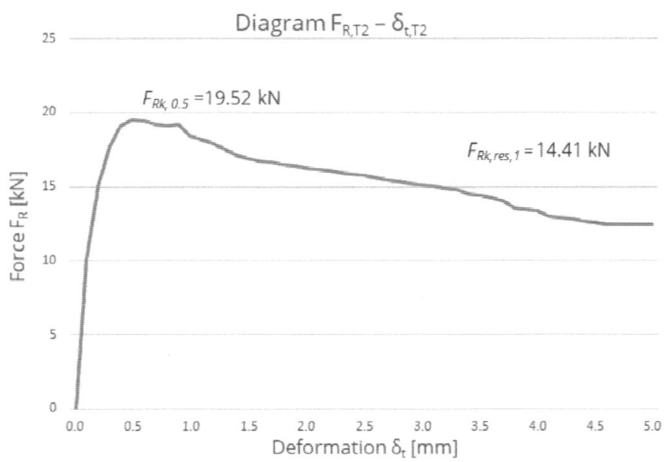

Figure 3. Average diagram of resistance for T2.

In the T3 set of specimens (120 kg/m^3), the achieved maximum force was F_{max} = 22.5 kN at a deflection of δ_t = 1.1 mm. For the specimens of the given dimensions and with this concentration, it seemed to be problematic to guarantee their homogeneity. Nevertheless, the progress of the average diagram in figure 4 shows that homogenous SFRC specimens were manufactured even with such a high concentration. Overall, it is possible to say that at a concentration of 120 kg/m^3, the SFRC concretes have a high load capacity and above all, ductility.

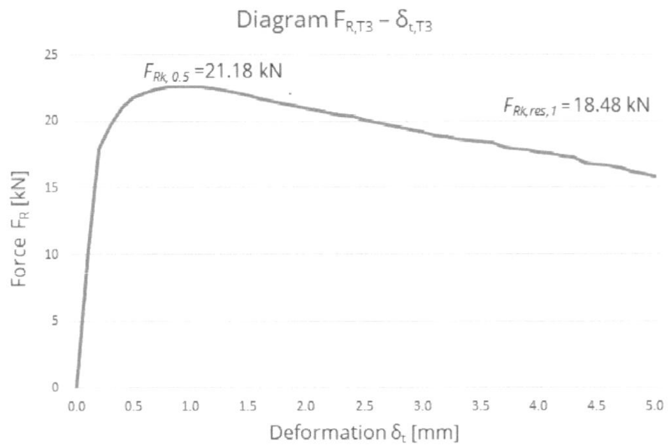

Figure 4. Average diagram of resistance for T3.

Acoustic emission signals which were received during the three-point bending testing of the specimens in the course of the entire measurement were analyzed. The attention was focused on three basic parameters: the number of events, and the amplitude and duration of the AE signals table 3.

Table 3. Mean values of the selected parameters obtained from AE measurements (coefficients of variation in %).

	Number of AE events /–	Amplitude of AE signals /*mV*	Duration of AE signals /*μs*
T1	10218 (1.17)	1647 (0.05)	1684 (0.03)
T2	11780 (1.16)	1601 (0.06)	1544 (0.04)
T3	14353 (1.18)	1585 (0.05)	1529 (0.05)

When the AE sensor captures a signal over a certain level, an AE event is recorded. The occurrence of a large number of cracks generates a rather large number of events during measurement. The highest number of events was shown by the T3 specimens with the greatest amount of steel fibres. As the amount of steel fibres decreases, the number of AE events becomes slightly lower. The amplitude is the greatest measured voltage in a waveform. This is an important parameter in the AE inspection because it determines the detectability of the signal. Signals with amplitudes below the operator-defined minimum threshold are not recorded. A higher amplitude indicates the occurrence of a larger and more significant crack. The amplitude values varied between 1585 and 1647 mV after 65 dB amplification. With the increased amount of steel fibres, the amplitude slightly decreased. The duration of an AE signal is the time difference between the crossing of the first and the last threshold. The duration of AE signals achieved for specimens with a higher amount of steel fibres was lower compared to the specimens with a lower amount of steel fibres. Nevertheless, the decrease in the duration of AE signals with increasing amounts of steel fibres was not so significant.

5. Conclusion

The paper presented the results of performed experiments which confirmed dependencies of the resulting mechanical properties of SFRC on the technological process of its production. With unusually high concentrations of steel fibres (40 kg/m^3, 80 kg/m^3 and 120 kg/m^3), which were used in this experiment, it was possible to manufacture homogenous SFRCs with very high values of load capacity at prescribed deformations. The results can be summarized as (related to 40 kg/m^3):

- the value of $F_{Rk,0.5}$ increased about 47 % (80 kg/m^3) and 60 % (120 kg/m^3),

- the value of $F_{Rk,res,1}$ enlarged about 79 % (80 kg/m^3) and the mixture with 120 kg/m^3 of fibres even about 130 %.

The resulting values of parameters of AE signals captured during the loading tests correspond very well with the results of the strength tests, thanks to which this method seems to be suitable for a better understanding of the behaviour of test specimens under loading with controlled deformations.

Acknowledgement
This paper has been worked out under the project No. LO1408 "AdMaS UP - Advanced Materials, Structures and Technologies", supported by Ministry of Education, Youth and Sports under the „National Sustainability Programme I".

References

[1] Gettu R, Gardner D R, Saldivar H and Barragan B E 2005 *Mater. Struct.* **1** 31-37
[2] Ferrara L and Meda A 2007 *Mater. Struct.* **4** 411-420
[3] Brandt A 2008 *Compos. Struc.* **86** 3-9
[4] Komarkova T, Lanik J and Dvorak P 2017 *Key Eng. Mat.* **755** 75-81
[5] Lataste J F, Behloul M and Breysse D 2008 *NDT&E Int.* **41** 8 638-647
[6] Stahli P, Custer R and Van Mier J 2008 *Mater. Struct.* **41** 189-196
[7] Blaszczynski T and Przybylska-Falek M 2015 *Procedia Engineer.* **122** 282-289
[8] CSN EN 12390-5: 2001 *Testing hardened concrete - Part 5: Flexural strength of test specimens*
[9] Fan J, Chen J, Jiang D, Chemenda A, Chen J and Ambre J 2017 *Int. J. Fatigue* **94** 140-144
[10] Topolar L, Pazdera L, Bilek V and Dedeckova L 2012 *Proc. Int. Conf. on Experimental Stress Analysis (Tabor)* vol 1 (Prague: Czech Society for Mechanics) pp 477-484
[11] Rovnanik P, Simonova H, Topolar L, Bayer P, Schmid P and Kersner Z 2016 *Constr. Build. Mater.* **119** 223-229
[12] Behnia A, Chai H K and Shiotani T 2014 *Constr. Build. Mater.* **65** 282-302
[13] Sahmenko G, Krasnikovs A, Lukasenoks A and Eiduks M 2015 *Proc. Int. Conf. Scientific and Practical Conference (Rezekne)* vol 1 (Rezekne: Rezekne Higher Education Institution) pp 193-199
[14] Holschemacher K, Mueller T and Ribakov Y 2010 *Mater. Design* **31** 2604-2615
[15] Topolar L, Pazdera L and Cikrle P 2014 *Proc. Int. Conf. on Experimental Stress Analysis (Litomerice)* vol 1 (Prague: Czech Society for Mechanics) pp 267-272

IOP Conf. Series: Materials Science and Engineering **385** (2018) 012029 doi:10.1088/1757-899X/385/1/012029

Experimental testing of the electromagnetic coil – determining the range of electromagnetic coil

T Komarkova, J Lanik, O Anton and V Hermankova

Department of Building Testing, Faculty of Civil Engineering, BUT in Brno, Veveri 95, Brno, Czech Republic

Abstract The principle of steel fibre reinforced concrete structures is a uniform distribution of steel fibers in the given volume (structure or structural element), which, however, is not always possible to guarantee in the production. This is closely related to a problem of performing diagnosis of the existing steel fibre reinforced concrete structures. Despite the fact that this material is not unknown, in the standards regulating quality control of the uniformity of fibre distribution, the assessment of distribution uniformity is still based on a destructive testing method. The paper describes a partial experiment which consisted in testing a proposed measuring device and in determining the depth of electromagnetic field range in the test specimens.

1. Introduction

In order to improve certain physical and mechanical properties of fresh concrete, additives and admixtures are added to it [1]. Steel fibres can also be regarded as a certain type of additive, giving rise to a building material called steel fibre reinforced concrete. Thanks to the specific properties obtained precisely by adding steel fibres of various sizes and shapes, steel fibre reinforced concrete can be used for achieving other required parameters of concrete elements with a view to their use in the structure [2]. Fibres improve resistance of concrete against mechanical stress, thermal stress and, above all, they improve the tensile resistance of concrete, even after the occurrence of cracks, when its ductility comes into play. These improvements of physical and mechanical properties of steel fibre reinforced concrete are highly welcome, especially because concrete is characteristic of brittle failure and the subsequent loss of its load-bearing capacity, which occurs immediately as a result of the failure [3-4]. Composite materials require much attention not only during designing the technological process of their production, but also during the production alone and during their subsequent testing. This is given by the fact that their properties are determined not only by the properties of individual components, but also by the proportions of their amounts per a unit of volume. Also important is the resulting homogeneity of fibre concrete, on which the improvement of its properties depends. Physical and mechanical properties of steel fibre reinforced concrete can change depending on the distribution and orientation of fibres [5-6].

It is problematic to perform a diagnosis of steel fibre reinforced concrete structures. It is not yet possible to verify compliance with the technological procedure for the production and execution of construction works by means of non-destructive methods, which are highly desired by the investors. One of their major advantages is the repeatability of tests performed and mainly their non-destructive result. Although the destructive testing methods performed on the extracted drill cores offer a much more precise interpretation of the measured values, many research institutions in Europe focus on the development of new NDT methods for steel fibre reinforced concrete testing.

Methods on the principle of which it is possible to assess the concentration of steel fibres in both fresh and hardened steel fibre reinforced concrete, already exist, however, assessing the orientation of fibres is considerably more difficult. Experiments show that it would be possible to check concentration and distribution of fibres on the basis of the ferromagnetic properties of steel fibers in steel fibre reinforced concrete non-destructively [7]. Diagnosis of steel fibre reinforced concrete structures is common, and the construction practice would highly welcome the possibility to use a measuring device proposed for checking the homogeneity of steel fibre reinforced concrete. The fact that design regulations should also include a method of non-destructive testing of the homogeneity of steel fibre reinforced concrete is mentioned also in [8]. Of course, development of every new method and methodology is time consuming and the final results not always confirm their applicability. On the basis of the requirements of construction practice, the authors participated in the development of a new non-destructive method of testing – proposing a new measuring device and methodology of testing steel fibre reinforced concrete. This device works on the principle of measuring physical quantities describing the electromagnetic field arisen by connecting an electromagnetic coil to the electric circuit. The principle of the measurement consists in measuring the inductance, during which the electromagnetic coil with a ferrite core is attached to the surface of concrete and gives rise to an alternating electromagnetic field. Presence of steel fibers increases the magnetic permeability of the composite material compared to a mixture without fibres, which implies an increase in the measured induction during the motion of the electromagnetic coil. Rotating the coil about its axis figure 1 and changing its position in the grid along the surface of the tested specimen thus allows to determine, from the measured values, the prevailing orientation of fibres [9].

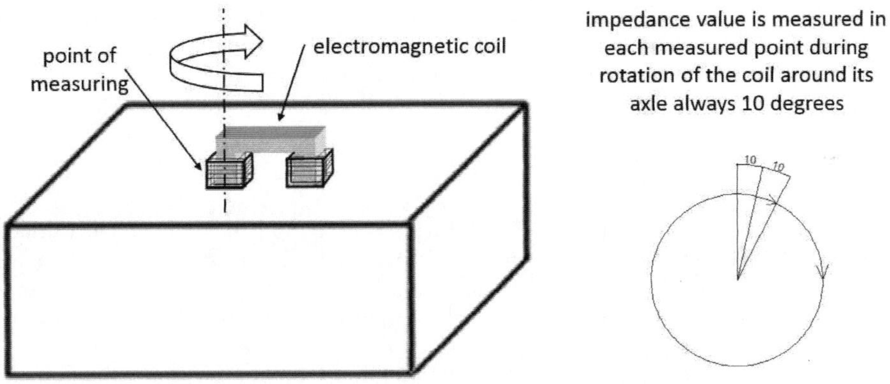

Figure 1. Methodology of testing the specimen.

Methods that are currently used are based on the evaluation of the electrical impedance of the tested specimen where the parameter monitored is the electrical impedance module. The prerequisite for the measurement proper is a dependence between the value of impedance and the concentration of fibres, which was demonstrated in the laboratory tests. The performed experiments showed that it is possible to monitor the concentration and distribution of fibres non-destructively using the electrical and ferromagnetic properties of steel fibres as a composite filler in the steel fibre reinforced concrete. The use of a suitable methodology of measurement and of the frequency band near the resonance of the sensor-specimen configuration allows for a more precise derivation and evaluation of the composition of the monitored composite component. The non-destructive methods used up to now utilize the properties of the electromagnetic field for a monitoring of fibre distribution in the tested composite. The properties of composite steel fibers are evaluated on the basis of the principle of electrical impedance in the area of the tested specimen. Details of the fields and other aspects are illustrated e.g. in figure 2.

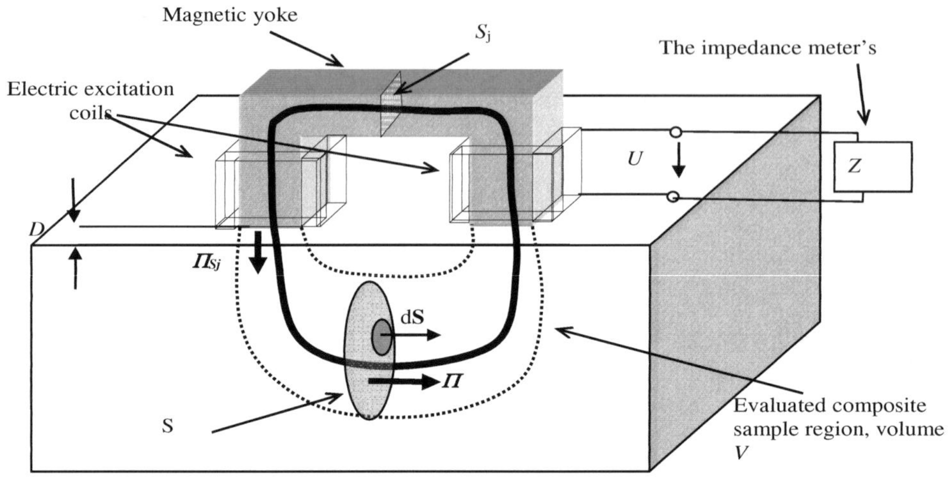

Figure 2. A scheme of the impedance-based NDT method: the evaluation of the tested sample parameters.

The hitherto experiments lead to determining the most suitable setting of the selected parameters, which is described in detail in. The first partial experiment determined the range of the electromagnetic field in the tested environment, and thus the maximum depth, in which it is possible to determine the steel fibre concentration and distribution uniformity from the data measured. The article describes results of the first experiments performed according to the newly proposed methodology and with the newly developed measuring device with the aim to determine the maximum range depth for the given setting [10].

2. Experiment

The experiment was performed using the methodology described in [9], where an electromagnetic coil with a ferrite core is attached to the surface of the test specimen. An electromagnetic field arises in the given volume, and by measuring impedance and power dissipation it is possible to assess especially the uniformity of steel fiber distribution in the controlled area and also in the given depth. When testing the new device, the initial effort was aimed at proving the range of the arising electromagnetic field and thus at confirming the correctness of the setting. 5 different resonance frequencies were selected assuming the differentiated sensitivity of the measurement and mainly the maximum depth of the electromagnetic field range [9-10]. Testing the range of the electromagnetic coil was performed using polystyrene slabs between which slabs with steel fibres were put at regular distances, always after 1 cm, see Fig 3. At the same time, it was ensured that no magnetically conductive material was found around the measuring device during measurements. This guaranteed that the electromagnetic coil reacts only to steel fibres placed on the polystyrene specimen.

Figure 3. a) Proposed measuring device - electromagnetic coil situated underneath the device (left)
b) Calibration of the electromagnetic field range (right).

3. Results and discussion

The measurement was performed for all the 5 settings of resonance frequencies (327 kHz, 145 kHz, 50,12 kHz, 21,9 kHz, 3,45 kHz), and the results achieved confirmed the assumption of varying ranges with the given settings. At the same time, uniform distribution of fibres in the selected area was confirmed, which can be observed in the chart of evaluated values in Fig. 4. The achieved results of testing the new measuring device are more than satisfactory. The following figure shows charts indicating the range of electromagnetic coil for the resonance frequency setting of 327 kHz. The measured values, which are compared, are power dissipation, impedance – angle and phase. The graphic representation indicates that at a steel fibre plate distance of 5 cm from the electromagnetic coil, none of the measured values changes, i.e. the maximum range of the electromagnetic field with the given setting is precisely 5 cm. The assessment was performed for all 5 resonance frequencies.

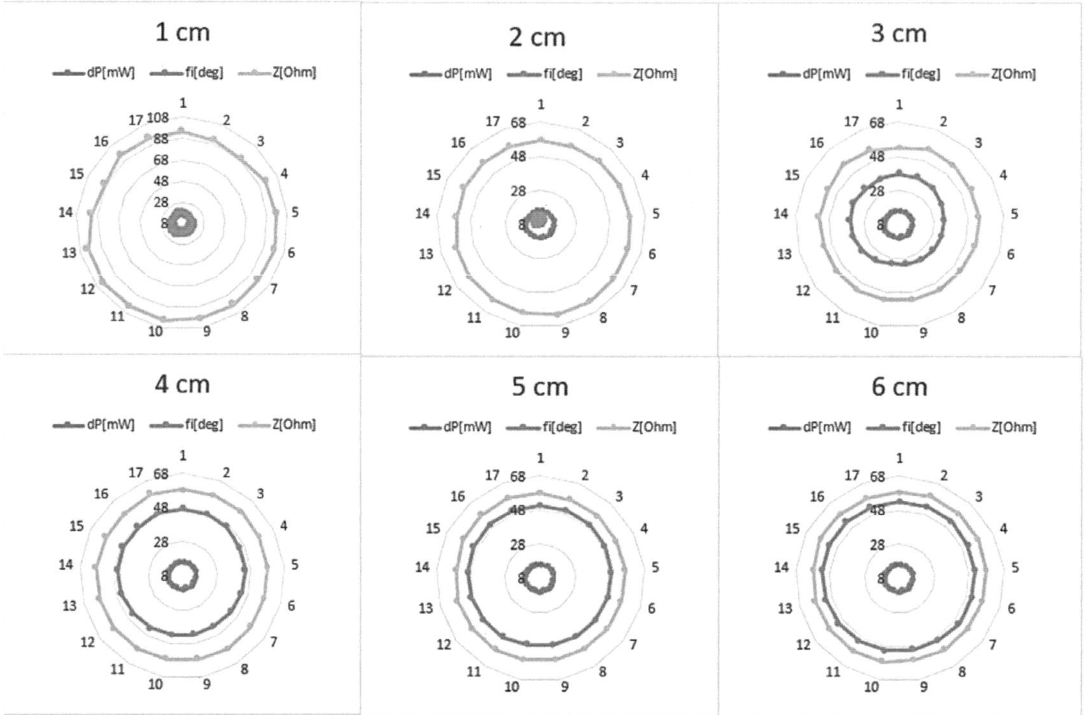

Figure 4. Graphical evaluation of the measured values – comparison of changes in the measured values depending on the distance of steel fibres from the surface of the electromagnetic coil.

4. Conclusion

By performing the experimental measurements in the laboratory on test specimens, it was possible to derive and prove the basic principles of the NDT method and methodology. The method is based on the physical principles of evaluation of the electrical impedance components of the magnetic circuit linked to the tested composite specimen, and the preliminary test results suggest its successful inclusion among the applicable NDT diagnostic methods used outside the laboratory. The potential of the proposed device is indisputable. The proposed methodology can be applied for a non-destructive diagnosis of a composite material either with a ferromagnetic addition, or with an electrically conductive addition with irregular shape of the composite component. The results of the first test experiments demonstrated the accuracy of the initial assumptions, and simultaneously, on their basis, it is possible to extend the experiment and determine validations of the methodology. The measurement results obtained from the first experiments show that with an appropriately selected setting, it is possible to perform measurements of steel fibre reinforced concrete to a set depth – for a frequency of 327kHz, to a depth of 5 cm. The hitherto positive measurement results proved a great potential of the proposed methodology, and after performing further laboratory tests, the proposed measuring device could be used in the diagnosis of steel fibre reinforced concrete structures.

Acknowledgement

This paper has been worked out under the project No. LO1408 "AdMaS UP - Advanced Materials, Structures and Technologies", supported by Ministry of Education, Youth and Sports under the „National Sustainability Programme I".

References

[1] Li Z 2011 *Advanced Concrete Technology* (Canada: John Wiley and Sons) p 506
[2] Vodicka J, Vesely V and Kratky J 2010 *Beton* (in Czech) **10**
[3] Vairagade V S and Kene K S 2012 *IJSTR* **1** 4
[4] Stoller J and Zezulova E 2017 *Key Eng. Mater.* **755** 374-381
[5] Lataste J F, Behloul M and Breysse D 2008 *NDT and E Int.* **41** 638-647
[6] Stahli P and G M van Mier J 2007 *Eng Fract Mech* **74** 223-242
[7] Zezulova E and Komarkova T 2017 *Key Eng. Mater.* **755** 153-158
[8] Shah A A and Ribakov Y 2011 *Mater. Des.* **32** 4122-4151
[9] Komarkova T 2016 *Key Eng. Mater.* **714** 179-185
[10] Komarkova T, Fiala P, Steinbauer M and Roubal Z 2018 *Meas. Sci. Rev.* **18** 35-40

IOP Conf. Series: Materials Science and Engineering **385** (2018) 012030 doi:10.1088/1757-899X/385/1/012030

Review of practical applications of alkali-activated systems in construction

J Konarik and J Bohacova

Department of Building Materials and Diagnostics of Structures, Faculty of Civil Engineering, VSB-Technical University of Ostrava, Ludvika Podeste 1875/17, 708 33 Ostrava - Poruba, Czech republic

Abstract. This paper reviews the possibilities of using alkali-activated systems in building industry. The history of research of alkali-activated systems and practical applications of these binders in building industry are summarized. This article shows where the evolution of these systems should be directed.

1. Introduction

Alkali-activated materials represent a group of alternative inorganic binders, where binder component consists of aluminosilicates and alkaline activators [1]. Wider interest in research and practical use of these systems in building industry is registered in the last 15 years.

Opportunity to use in production of these composites based on waste materials from various industrial sectors and also the excellent properties of the final product make this group of binders, suitable ecological alternative to Portland cements.

This paper summarizes the basic directions of the possible application of alkali-activated materials, primarily in the field of construction.

2. Materials used for alkaline activation

For alkaline activation, as binder are used latent hydraulic substances or pozzolans. Hydraulicity of these binders takes shape when a suitable activator is added with indispensable quantity of water. Binders are based on natural or technogenic origin, the most commonly used, are blast furnace slag, fly ash, metakaolin and also natural pozzolans [2].

The function of activator is performed by hydroxides, silicates, or carbonates of alkali metals. The final structure and properties of alkali-activated systems depend primarily on used raw materials, the ratio of particular components, type of aggregate, method of preparation and storage conditions. The diversity of the final products is increased in effort to prepare various binders of specific properties by using mixture of many materials as a raw material [1,2,3].

The best results of strength and durability were reached by alkali activated systems based on blast furnace slag. Composites based on alkali-activation of metakaolin reach low strengths, but the final materials are highly external influences resistant, especially resistant to frost, aggressive substances and high temperatures. Fly ash-based geopolymers reach lower durability, in comparison to metakaolin-based geopolymers, but generally achieve higher strength [1].

3. History of alkali activation research and application

Probably the first use of alkali-activated systems belongs to antiquity, is discussed its particular use in construction of the pyramids in Egypt and the ziggurats of Mesopotamia. There are also speculations about the use of alkali-activated systems by civilizations in Central and South America and many centuries later in relationship with the buildings of ancient Rome. Based on information about these advanced cultures, economic situation, geographical location and according to the results of chemical analyzes of preserved structures we suggest that people were able to produce these materials [6].

Current research has been started in the first half of the 20th century. One of the first reported use of alkali-activated materials in the industry was application of the mixture based on kaolin and sodium carbonate at 150 °C in a ceramic factory Olsen in 1934 [7]. In the 1930s research of suitability sodium hydroxide and potassium hydroxide for use with blast furnace slag as possible ingredients of Portland cements was conducted. During the research of the systems Purdon found, that the addition of alkali generates new, fast-hardening binder [8, 9].

In the 1950s, professor Gluchovskij led research of alkali activated slag in Ukraine, during the 1960s and the 1970s he contributed significantly to the identification of solidified products. The research revealed that the components react together to form zeolites. Concrete based on alkali activated slag called the "gruntocements" and described it in the book "Gruntosilikaty" published in 1959. In 1969, Gluchovskij acquired the first patent for the preparation of these materials [10]. In the follow to his research a first building was built in the 1960s in Ukraine. The building was built of prefabricated blocks prepared according to his recipe. Clays were substituted by blast furnace slag, prepared concrete was called slag-alkali, and it was used in the construction of sewers and other, by external influences highly stressed structures [7].

"Trief Cements" (alkali-activated slagcements) were, due to the generation of smaller hydration heat used in the 1950s mainly in the massive structures construction [9].

The nowadays first practical use of alkali-activated system for construction is known from the 60s of the last century. In Ukraine, the building was built of blocks made of recipe Professor Gluchovskij [1].

In 1953 alkali-activated systems were researched by the U.S. Army, the alkaline activation was carried out by mixing of 1.5 % NaCl, 1.5 % NaOH and 97 % ground slag [9].

From the 1960s sewer systems, roads and breakwaters were built of alkali-slag concretes in Russia, Poland, Finland, USA, Canada and later also in Spain, Germany and other countries [7].

In the 1970s alkali-activated systems were also studied by French scientists Besson Cailler and Henin, in 1969 the result of their work was the synthesis of various kaolinitic materials in concentrated sodium hydroxide solution at a temperature of 100 °C [10].

Then building products were fire-resistant wood chipboard designed the 1972 by Legrand. These were formed by a wood core encased with alkali-activated material. [10].

In 1972, a team of scientists Davidovics and Latapie prepared mixture for the production of water-resistant ceramic tiles. These were produced at temperatures lower than 450 ° C [11].

In the years 1977 - 1982 L.T.G.S. technology were created. It is a procedure where the kaolinite ceramic paste achieved in an alkaline environment high strength and water resistance [10].

In the 1970s team of Davidovits and Legrand developed technology based on geosynthesis. Joseph Davidovits, claiming to be the discoverer of geopolymerization, he used for alkali activated materials based on kaolin as the first term "geopolymer" in 1978 [11].

In 1983, the U.S. developed high strength geopolymer cement PYRAMENT™. It was made of Portland cement, fly ash, metakaolin, ground slag, and potassium carbonate. This cement was used to airport runways repair. PYRAMENT™ achieved high initial strength of about 20 MPa after 4 hours of hardening, after 28 days were measured strength in the range of 70 to 100 MPa [10].

In 1983, Foss developed F-cement in Finland, low-porosity cement designed for industrial applications with a rapid increase in strength. It was based on alkaline activation of slag or fly ash. F-cement contains as a superplasticizer a sulphonated polyelectrolyte, which in this type of concrete allowed to achieve optimal consistency even with minimal water coefficient [12].

In 1984, Langton and Roy examined specimens of mortar and plaster from ancient Greece, because of their high resistance. According to their findings, they divided the composite matrices into five groups, which are divided according to the content occurrence of individual hydraulic aluminosilicate and pozzolanese mastics. In the framework of this research it was found, that the mineralogy of the matrix, the particle size, the porosity, the dosage and the type of aggregate, as well as the method of deposition and the surrounding conditions, therefore the same factors that affect life of materials based on Portland cement, have also a crucial influence on the extremely long lifetime of these new type materials [13].

In 1986, Krivenko described the mechanism that controls the process of forming the structure of the alkali-activated mass, dividing the links into two main categories. In the first case, the alkali component is a monovalent cation of Na^+ or K^+, and in the latter case, the divalent Ca^{2+} cation is in the alkaline component. In general, these binder systems can be divided into the following two groups according to the theory:

$Me_2O - Me_2O_3 - SiO_2 - H_2O$

$Me_2O - MeO - Me_2O_3 - SiO_2 - H_2O$ [14,15]

In 1986, Malek and colleagues proposed slag-concrete for the disposal of radioactive salt (generated by crystallization of waste sludge) from the Savannah River area of South Carolina by placing this material in artificial ditches. The material capture matrix must have the required properties, in particular the minimum permeability of toxic substances to the environment. Therefore, a mixture of ash with granulated slag was proposed, variants of this mixture were tested using Portland cement and the variant without addition, and the data were compared with other formulations (based on a different type of ash, without slag, etc.) [16,17].

In 1989, in the Russian city of Lipetsk a twenty-floor house was built of alkali-slag concrete, without Portland cement [28].

Since the 1990s, the number of research studies dealing with alkali-activated systems has grown considerably. As a representative of this era, in addition to the ongoing experiments of the aforementioned scientists (Gluchovsky, Davidovits, Krivenko, Roy), it is possible to mention, for example, the results of Professor Brandštetr´s team, whose research focuses on the preparation of composites using waste materials, blast furnace slag, and the use of ash in non-cement mixtures. Furthermore, the team of prof. Paloma, which deals with alkali activation of fly ash, including observation and description of the resulting microstructure, as well as the Provis - Deventer team, dealing with, among others, the industrial use of alkaline activated materials [15].

From the point of view of practical use in recent years, the following applications of alkali-activated materials can be mentioned.

In Australia, in Brisbane, the first house was constructed for Queensland University, in which a geopolymer concrete was used. The four-storey building is built of prefabricated buildings. From the alkaline activated blast furnace slag and ash composite, 33 prefabricated ceiling panels were used [18].

The largest project for the use of alkali-activated material is the construction of the Brisbane West Wellcamp Airport in 2014, with the implementation of 40,000 cubic meters of geopolymer concrete. The material was used during construction for its high strength, low shrinkage and good workability. Geopolymer concrete was applied in a layer of 435 mm thickness [19].

Another example of the application of alkaline activated systems in building practice is E-Crete ™. This is concrete prepared on the basis of fly ashes from coal combustion in power plants and blast furnace slag. E-Crete ™ is a product of the Australian company Zeobond Group, which offers it in the form of both the concrete and prefabricated parts [22].

In Australia, "geopolymer concrete" is also produced, called Earth Friendly Concrete (EFC). It is a composite based on a combination of blast furnace slag and fly ash, wherein alkaline activation takes place at elevated temperatures using alkali in solid or liquid form. Various types of prefabricated components are produced from the given material, for example their use in tunneling and sewerage segments, using their high resistance to aggressive environments [21].

4. Alkali activation research in Czech Republic

In the Czech Republic research of alkali activated materials for construction first began professor Brandštetr in the 1960s in Brno University of Technology [4,5,7].

From the 1970s alkali activation was researched by the Department of Glass and Ceramics, Institute of Chemical Technology in Prague, later in cooperation with the Faculty of Civil Engineering, CTU in Prague [22].

In 1979 a team of CTU and University of Chemistry and Technology Prague formulated the principles of preparation no-gypsum Portland cement, called "BS cement". This is made of Portland cement, or slag, alkaline salt and anioactive surfactant. Final product is rapidly hardening cement. It hardens at temperatures as low as -30 °C and it is resistant to high temperatures up to 1200 °C. The maximum measured compressive strengths reach values up to 245 MPa. BS cement was in 1989 experimentally applied to industrially produced building materials, concretes reached strengths of 100 MPa [22].

In the 90th years, the Department of Glass and Ceramics at University of Chemistry and Technology Prague researched geopolymers based on metallurgical slags and fly ashes. The final product contained a mixture of fly ash and slag reached compressive strengths up to about 170 MPa. In 2002 a cement-free geopolymer concrete based on fly ash - POPbeton® was prepared. This material achieved compressive strengths up to 60 MPa, it is resistant to salts and high temperatures. The main advantage of this product is the use of waste material [22].

Research of Institute of Rock Structure and Mechanics ASCR, v.v.i. Prague is focused on potential use of geopolymers in disposal of radioactive and hazardous waste and for the fireproof, sound and thermal insulating materials preparation [24].

The Institute of Chemical Technology in Prague deals with the technology of alkali-activated mortars and concretes preparation in dependence on the final use of the product. The research is also focused on finding the optimal composition of the activator, the determination of micromechanical properties using nanointendance, a detailed description of the mechanisms of alkaline activation, achieving the required properties for each type of mixture, fixation of heavy metals in the polymer matrix, determination of the extracts of fly ash geopolymer, chemical resistance in strongly acidic and alkaline mediums, geopolymer adhesion to concrete reinforcement [22].

Institute of Industrial Ceramics FMME VSB - TUO studies the use of secondary metallurgy slags in application of alkali-activated systems [2,3].

Since 2008 alkali-activated materials has been researched also by Faculty of Civil Engineering, VSB-TU Ostrava. For the possible use of these systems in the construction industry were tested alkali activated blast furnace slags with fillers based on various types of construction and mining waste. Research is also focused on the preparation of thermal insulation alkali-activated building materials, compressed geopolymer mixtures and development of protective layers [20,23,25-27].

In cooperation with building companies (Prefa Troubelice, a.s., ŽPSV, a.s., etc.), practical applications are being carried out in the construction industry, the possibilities of using these systems in elements of transport structures, fine-grained mixtures, facade elements and prefabricated elements are tested.

An example of practical application of alkali-activated systems is alkali activated binder Baucis, a raw material based on kaolinite. This binder is produced by Ceske lupkove zavody. It is produced by the controlled heat treatment of kaolin and shale. After mixing the binder with water and alkali silicates, material chemically and structurally similar to natural rock is prepared. Firing temperature of this material is 750 °C, it hardens rapidly, 40-50 % of the final strength is achieved after 1 day and 90 % strength is reached in 7-14 days. The binder is suitable for fire-resistant products, and may be exposed to temperatures up to 1200 °C [29].

With the support of the Czech Development Agency are in the short-run lines produced various prototypes of products of alkaline activation, imitating various materials, such as bricks, stone, wood and metals [10].

Another example of the alkali-activated system used in practice is the binder Geopol®. This technology of SAND TEAM Ltd. is prepared for the production of molds and cores for ferrous and non-ferrous metals casting. The advantage of the technology is the possibility of the used materials recycling, the minimum mechanical abrasion, good workability and good binder properties at a very low dosage of binder [30]. The company is applying cast molds based on alkali activation especially in the U.S. where growing pressures on environmental protection [31].

5. Perspective

Alkali-activated materials have considerable potential for use not only in the building industry. Their widespread use in building practice prevents in particular the need for considerable technological discipline during the alkaline activation process, the failure of which ultimately leads to significant fluctuations in the values of the monitored parameters.

The wider application of alkali-activated systems is obstructed also by the absence of standards for testing, the use of liquid activators in the manufacture and the conservative attitudes. Builders are afraid of the use of non-traditional binding systems. Due to the requirements of sustainable development and conservation of mineral resources, however, we can expect in a relatively short period, a boom in the market of products based on alkali-activated systems.

6. Conclusion

This paper was prepared to summarize the historical and current scientific results and practical applications in the field of alkali-activated systems. Alkali-activated materials form a group of inorganic binders. The properties of these materials are verified in many experiments and represent a promising set of alternative masses. At present, they are used rather marginally in construction practice, but in the future they can be expected to expand into many sectors of the construction industry.

Acknowledgement

Contribution was realized with the financial support of Ministry of Industry and Trade - Partnerství znalostního transferu FAST, VŠB-TUO - Prefa Troubelice a.s., the identification number CZ.01.1.02/0.0/0.0/15_013/0004867.

References

[1] Skvara F 2007 *Alkali-Activated Materials-Geopolymers* http://old.vscht.cz/sil/pojiva/geo_2007.pdf (online) (in Czech)
[2] Vlcek J 2008 *Material Use of Slag from Iron and Steel Metallurgy by Alkaline Activation* (Ostrava: VŠB-TUO) p 99
[3] Tomkova V 2008 *The Influence of Phase Composition and Microstructure on the Functional Properties of Geopolymer Systems from Technogenic Pozzolans* (Ostrava: VŠB - TUO) p 28
[4] Brandstetr J and Krivenko P V 1999 *Silika* **10** 26-28 (in Czech)
[5] Brandstetr J, Meluzin O, Kohut J and Zavřel S 1986 *Stavivo* **3** 182-188 (in Czech)
[6] Davidovits J 2006 *They Built The Pyramids* (France: Institut Geopolymére)
[7] Brandstetr J, Kolousek D, Vorel J, Opravil T and Bayer P 2005 *Silika* **7** 208 – 211 (in Czech)
[8] Douglas E and Malhotra V M 1988 *Proc. Int. Conf. and Symp. on Ferrous and Non-ferrous Slags in Concrete, Energy, Mines and Resource* (Canada: Canada Centre for Mineral and Energy Technology) p 36
[9] Roy D M and Langton C A 1983 *Characterization of Cement-Based Ancient Building Materials in Support of Repository Seal Materials Studies* (Columbus: Bat-telle Memorial Institute)
[10] Bartonova P 2009 *Preparation and Properties of Pure Geopolymers* (Brno: BUT) p 58 (in Czech)
[11] Li C, Sun H and Li L 2010 *Cement Concrete Res.* **40** 1341-1349
[12] Malek R I A, Roy D M, Barnes M W and Langton C A 1985 Report *Slag Cement – Low-level*

Waste Froms at the Savannah River Plant (Pennsylvania State University, University Park, Materials Research Laboratorie Du Pont de Nemours (EI) and Co, Aiken, SC (USA) Savannah River Laboratory)

[13] Unitet States Nuclear Regulatory Commission Report *Savannah River Site Disposal Facility for Waste Incidental to Reprocessing* https://www.nrc.gov/waste/incidental-waste/wir-process/wir-locations/wir-srs.html (online)

[14] Hassell 2013 *University of Queensland Global Change Institute Retrieved from* http://www.hassellstudio.com/en/cms-projects/detail/the-university-of-queensland-global-change-institute/ (online)

[15] Technical Paper on Geopolymer Aircraft Pavement: https://www.geopolymer.org/library/technical-papers/technical-paper-on-geopolymer-aircraft-pavement/ (online)

[16] Zeobond group *E-crete*: http://www.zeobond.com/products-e-crete.html (online)

[17] Glasby T, Day J, Kemp M 2014 *Proc. Int. Conf. on Arabian Tunnelling Conference held (Dubai*)

[18] Davidovits J 2008 *Geopolymer*: *Chemistry* (Saint-Quentin: Institute Géopolymére) 587

[19] Davidovits J 1991 *J Therm Anal* **37** 1633 - 1656

[20] Vaclavík V 2014 *Adv Mat Res* **77** 204-214

[21] Czech Development Agency 2005 *Current state of Research in Geopolymers* (Prague: Czech Development Agency) 285 (in Czech)

[22] POP beton http://www.geopolymery.eu/ (online)

[23] Gola L, Vaclavik V, Valicek J, Harnicarova M, Kusnerová M and Dvorsky T 2015 *Mechanical and Materials Engineering of Modern Structure and Component Design* (Basel: Springer) p 155-165

[24] Institute of Rock Structure and Mechanics of AS *CR Institute of Rock Structure and Mechanics of AS CR* http://www.irsm.cas.cz/ (online)

[25] Koutnik P, Bortnovsky O, Antos P and Roubicek P 2010 *Proc. Int. Conf. on Remediation and reconstruction of buildings* (Prague) (Brno: VUT) p 130 – 135 (in Czech)

[26] Vavro M, Bujdos D, Pticen F and Vlcek J 2009 *Proc. Int. Conf. on Remediation and reconstruction of buildings* (Prague) (Prague: TAURIS) p 71-77 (in Czech)

[27] Bohacova J, Vavro M and Stanek S 2011 *Proc. Int. Conf. on Development and Research of Properties of Thermal-insulating Alkaline Activated Systems* (Ostrava) vol 2(Ostrava: Printo s.r.o.) p 65 – 74

[28] Vavro M, Bohacova J, Bujdos D, Tomková J, Vlček J 2011 *Proc. Int. Conf. on Alkali-Activated Materials as Primary Protection of Concrete* (Kosice) (Kosice: TU Košice) pp 244 – 248

[29] Brandštetr J, Havlica J 2000 *Materialy pro stavbu* vol 6 pp 48-50 (in Czech)

[30] ČLUZ *Geopolymers:* http://www.cluz.cz/ geopolymer obecne-informace (online)

[31] SandTeam: http://www.sandteam.cz/ technologie-geopol/ (in Czech)

Sensitive analysis of calculation of corrosion rate according to standard approach

P Kotes, M Strieska and M Brodnan

Department of Structures and Bridges, Civil Engineering Faculty, University of Zilina, Univerzitna 8215/1, 010 26 Zilina, Slovakia

Abstract. There is a lot of factors, which may affect the reliability of the construction, as durability, serviceability and safety of construction or its parts elements. Corrosion damage either reinforcement is one of the most important factor in reinforced concrete structures or structures made from structural steel. This corrosion damage can be determined by actual standard EN ISO 9223. This standard describe the equations, called dose-response functions, for calculation of the corrosion rate r_{corr} for standard materials like carbon steel, zinc and so on. These equations are determined from the corrosion rate of samples under outdoor atmospheric corrosion in cooperating with measuring of climatic data. Input parameters on calculation of the dose-response functions are sulphur dioxide, temperature, relative humidity and chloride ions. This article is focused on the calculation of the carbon steel corrosion rate according the dose-response function and its sensitive analysis for various concentration of the input parameters. This variability of the input parameters may represents the different conditions, in which the constructions may appears. In practice, this information may provide, which of the input parameter influence the most the calculated corrosion rate and if it is needed to more concentrate its measuring station network. The location of the outdoor stations with the specimens or choice of accelerated corrosion test may be also provided.

1. Introduction

All materials are affected by the surrounding environment [1-2]. It means that the climatic condition and air pollution like temperature, relative humidity, sulphur dioxide, chloride ions etc. may increase or decrease speed of the corrosion rate r_{corr} [3]. Many researches have been done to assess the correlation between corrosion damage (corrosion rate) and aggressiveness of the environment. Nowadays, the standard EN ISO 9223 [4] describe the designation of the yearly corrosion rate for the first year of exposure. The corrosion rate r_{corr} can be designated either on standard specimens under real environment or calculated according to equations so called as dose-response functions. Calculation of these functions is based on the climatic information obtained from the public's hydrometeorogical institutes. The dose-response function for the carbon steel is following:

$$r_{corr} = 1.77 \cdot \left[SO_2\right]^{0.52} \cdot \exp(0.020 \cdot Rh + f(t)) + 0.102 \cdot \left[Cl^-\right]^{0.62} \cdot \exp(0.033 \cdot Rh + 0.040 \cdot T) \quad (1)$$

where r_{corr} is the corrosion rate (μm/year), f(T) = 0.150·(T – 10) when T≤10°C; otherwise – 0.054·(T – 10), SO_2 is the sulphur dioxide (μg/m³ or mg/(m²·day), where SO_2 in units mg/(m²·day) is equal to 0.8·SO_2 in unit μg/m³), T is the temperature (°C), Rh is the relative humidity (%), Cl^- is the chloride deposition rate (mg/(m²·day)). The standard EN ISO 9223 [4] describe the interval of these input parameters, see table 1.

Table 1. Input parameters used in the dose-response functions, including description, symbol, interval, 30 - 70% of these input parameters, mean values and unit.

Description and symbol	Interval	30 – 70 percentage of input parameters	Mean value	Unit
Temperature T	-17.1 – 28.7	-11.97 – 20.09	4.06	°C
Relative humidity Rh	34 – 93	50.70 – 65.10	57.90	%
Deposition of SO_2	0.7 – 150.4	1.14 – 131.60	66.37	$mg/(m^2{\cdot}day)$
Deposition of SO_2	0.6 – 120.3	0.91 – 105.28	56,01	$\mu g.m^{-3}$
Deposition of Cl^-	0.4 – 760.5	0.52 - 532.34	266.44	$mg/(m^2{\cdot}day)$

2. Sensitive analysis of dose-response equation for carbon steel

The sensitive analysis may be carry out either when the fixed input parameter is represented by its mean value or the most frequent values, while only one parameter is changed. In this study, for better representation how is the sensitivity of the dose-response function, the range of values of the input parameters were chosen in the range of their interval according the standard EN ISO 9223 [4] and not in the range of maximum or minimum values, which were monitored by the Slovak Republic by Slovak hydrometeorogical institute (SHMU) [5]. For this reason, not the frequent but the mean values of input parameters were chosen as the fixed parameters. In addition, various combinations of these fixed input parameter were calculated and its results, as well as the discussion and conclusions were made.

Table 2 shows the changes of input parameters in the ten steps, as well as the calculation of the corrosion rate according equation (1).

Table 2. Changes of input parameters in the sensitive analysis and the calculated corrosion rate r_{corr} of carbon steel according the equation (1).

Percentage of input parameter	Temperature		Relative humidity		Deposition of SO_2		Deposition of Cl^-	
	Changing parameter	r_{corr}	Changing parameter	r_{corr}	Changing parameter	r_{corr}	Changing parameter	r_{corr}
(%)	(°C)	(μm/year)	(%)	(μm/year)	(mg/(m². .day))	(μm/year)	(mg/(m². .day))	(μm/year)
0	-11.97	15.27	50.70	36.19	1.14	28.07	0.52	18.78
10	-8.76	18.15	52.14	37.64	14.18	34.04	53.70	27.82
20	-5.56	21.92	53.58	39.16	27.23	37.35	106.89	32.92
30	-2.35	26.99	55.02	40.74	40.28	39.94	160.07	37.10
40	0.85	34.03	56.46	42.39	53.32	42.15	213.25	40.77
50	4.06	44.11	57.90	44.11	66.37	44.11	266.44	44.11
60	7.27	58.91	59.34	45.90	79.42	45.89	319.62	47.20
70	10.47	76.77	60.78	47.77	92.46	47.54	372.80	50.10
80	13.68	74.46	62.22	49.72	105.51	49.08	425.98	52.84
90	16.88	73.86	63.66	51.75	118.55	50.53	479.17	55.46
100	20.09	74.90	65.10	53.87	131.60	51.90	532.35	57.97

The sensitive analysis of the dose-response function for carbon steel was calculated on the basis of the following considerations:

- it was suitable to avoid the extreme values of input parameters, for this reason the table 1 describe us chosen interval between 30 % and 70 % of the input parameters,
- all input parameters was fixed in the mean value (see table 1), while only one parameter was changed,

- this parameter was changed in ten steps from its minimum value (represent 0%) to its maximum value (represent 100%), see table 2,
- the procedure described above, where only one parameter was changing while another parameters were fixed in the mean value, was made for all input parameters like temperature, relative humidity, deposition rate of 3ulphur dioxide and chloride, as can be seen in figure 1,
- for better demonstration, the intervals according the standard is described, in the table and graphs, not in the range from 30 to 70% of its value but in the range from 0 to a 100%.

The result from the table 2 are drawn in figure 1. The horizontal axis represent ten steps, the step zero, in which the changing parameter has value of 30% (0% on the horizontal axis) described in the standard to step ten where the changed value has 70% (100% on the horizontal axis) of its interval described in the standard. The vertical axis represent corrosion rate calculated according the equation (1).

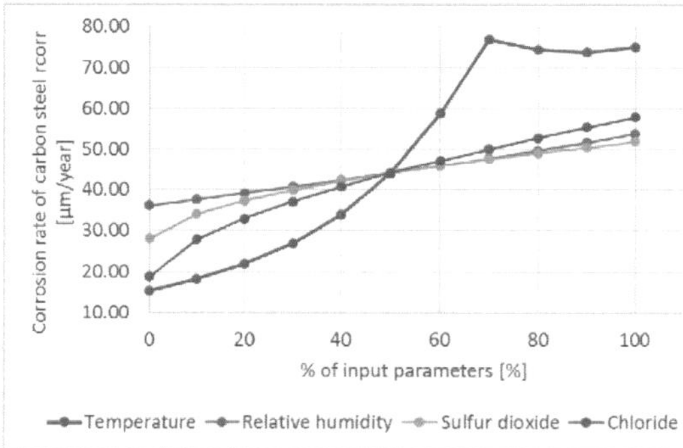

Figure 1. Input parameters are fixed at the mean values.

The sensitive analysis shows that the temperature the most affects the calculated corrosion rate of carbon steel. The corrosion rate is in the range from 15.27 to 76.77 μm/year (which represent increase by 61.50 μm/year). The chloride deposition rate is the second input parameter, which the most affect the calculated corrosion rate r_{corr}. It is in the range from 18.78 to 57.98 μm/year (increase by 39.20 μm/year).

In the real environment may occur various combination of values of the input parameters. It means that in the areas of high or low yearly concentration of temperature, relative humidity, sulphur dioxide or chloride deposition may occur the situation, where the fixed values are not represent by mean value but its high or low value. For this reason, these values are taken into account, in the second step of this research. These areas may be in the case of chloride deposition rate near the see or traffic infrastructure network where de-icing salt is applied [6-8]. The sulphur dioxide concentration is dominated in industrialized countries where is the higher level of atmospheric pollution [9]. The temperature and relative humidity may be dominated in the mountains or in the areas of different height above sea level.

Following figures show the situation, where all the parameters are fixed on the thirty (figure 2) or seventy percentages (figure 3) of its intervals according to standard and only one, calculated parameter, is changed. On these figures can be seen that the corrosion rate r_{corr} may be interfaced by changing of fixed input parameters. Figure 2 shows the situation when all fixed parameters are fixed on its thirty percentage and it is show that violet-chloride ions line is dominated and not red-temperature line as it was before. This leads to the conclusion that the calculated corrosion rate r_{corr} depends on the values of input fixed parameters (their mean, minimum, maximum and so on).

For this reason the graphs in figures 4-11 were made, where only one value was fixed on its 30 or 70 percentage.

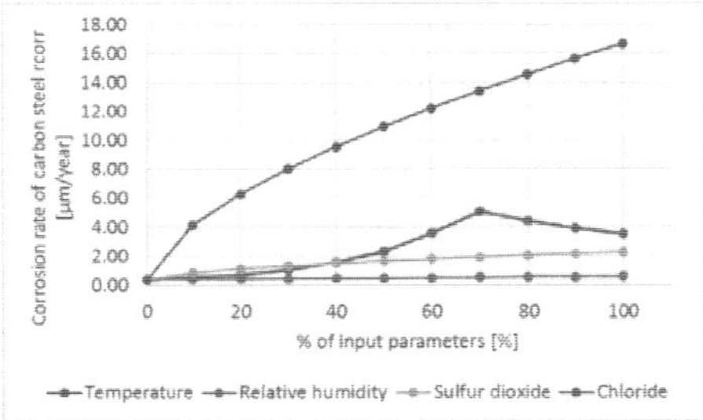

Figure 2. Input parameters are fixed at 30% of its value.

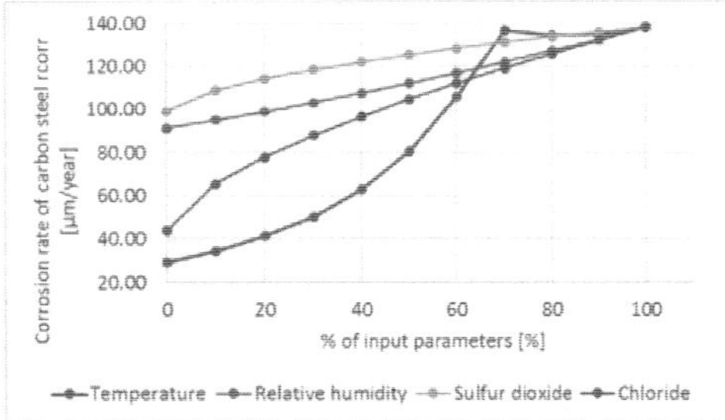

Figure 3. Input parameters are fixed at 70% of its value.

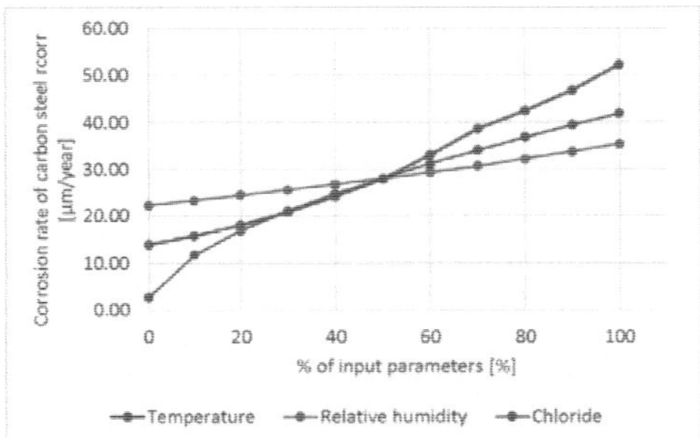

Figure 4. SO_2 is fixed on its 30% value.

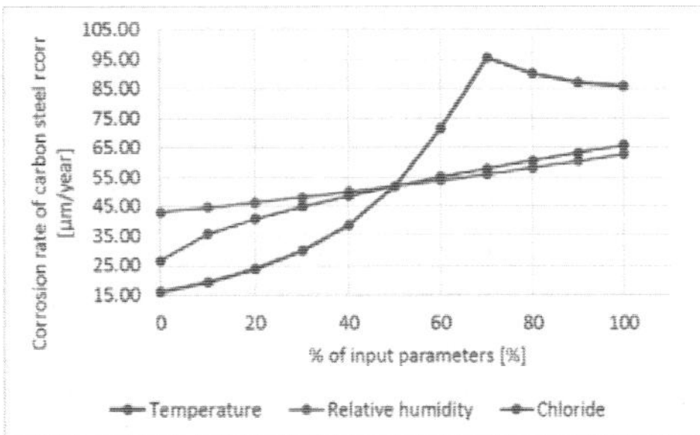

Figure 5. SO_2 is fixed on its 70% value.

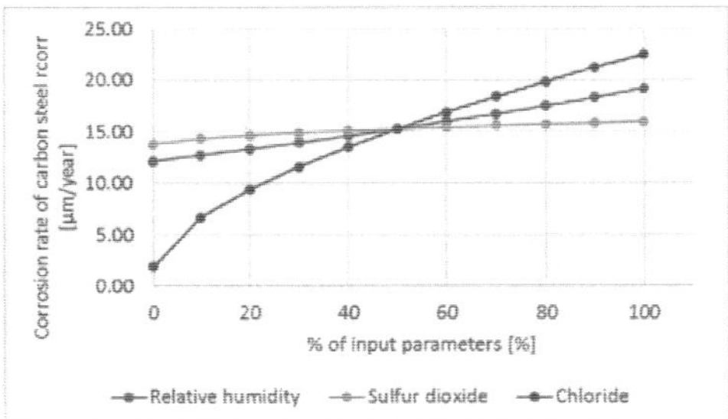

Figure 6. Temperature is fixed on its 30% value.

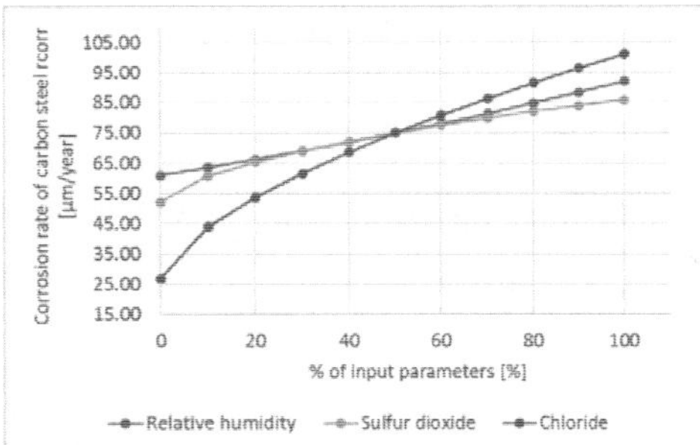

Figure 7. Temperature is fixed on its 70% value.

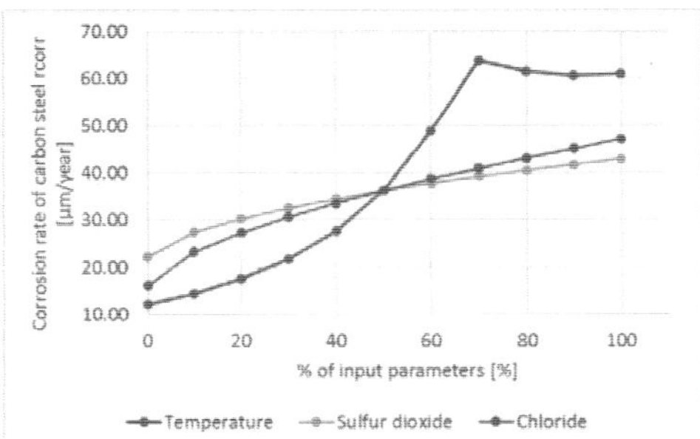

Figure 8. Relative humidity is fixed on its 30% value.

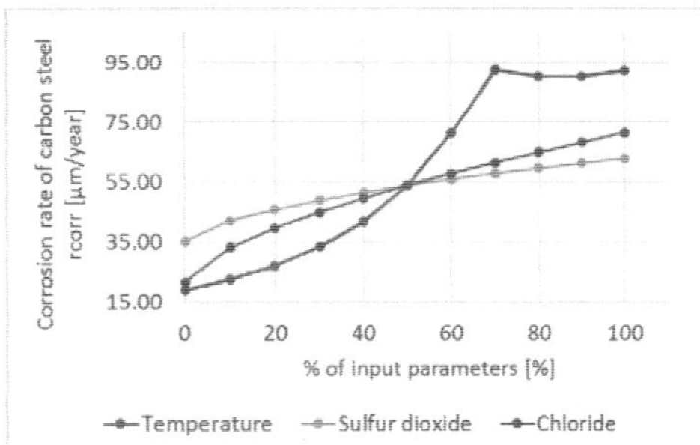

Figure 9. Relative humidity is fixed on its 70% value.

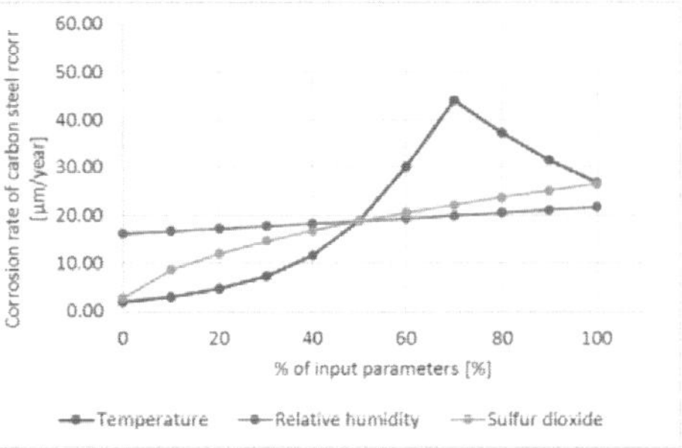

Figure 10. Chloride ions are fixed on its 30% value.

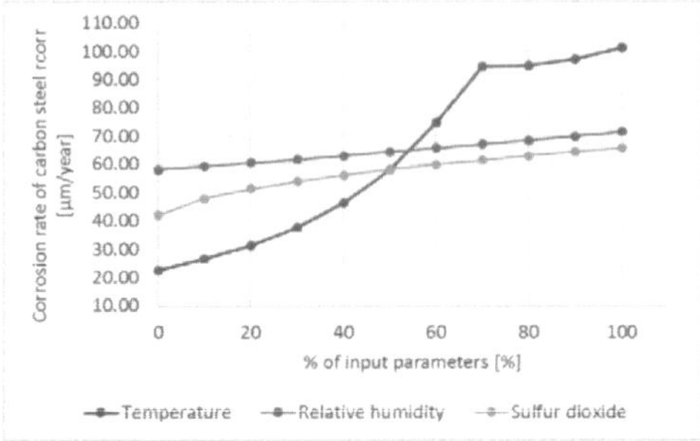

Figure 11. Chloride ions are fixed on its 70% value.

Figure 4 shows that the corrosion rate r_{corr} is in range from 2.74 µm/year to 41.94 µm/year (increases by 39.20 µm/year) for violet-chloride ions line. This corrosion rate (violet-chloride ions line) is calculated according to equation (1), when the input parameters like temperature and relative humidity are on their average values (T = 4.06 °C and Rh = 57,90 %), while the sulphur dioxide SO_2 value is on its 30% recommendation in the standard (SO_2 = 1.14 mg/(m^2·day)). The chloride deposition rate is changing parameter in calculation and is in range from 30 % to 70% recommendation in the standard (Cl^- = 0.52 – 532.34 mg/(m^2·day)). This extend (from 30% to 70%) is represented from zero to a hundred percentage, see horizontal axis of figure 4. Another two lines, in figure 4, were made in the same way, whereas the temperature (red-temperature line) or relative humidity (blue-relative humidity line) were changing parameters. Another graphs in figures 5-11 were made by the same procedure.

On these graphs, range of the corrosion rate r_{corr} was observed. In the case of violet-chloride ions line, it can be seen that the calculated corrosion rate r_{corr} is influenced by fixed SO_2 concentration on the 30 % of its value, figure 4 or by temperature concentration fixed on its 70 % of its value, figure 7. In these two cases, the calculated corrosion rate is in the range of 39.20 µm/year (in the case of 30 % of SO_2 concentration) and 74.42 µm/year (in the case of 70 % of temperature concentration). The results from figures 5-11 are summarized in the table below.

Table 3. Sensitive analysis of the dose-response function for the carbon steel.

	SO_2 - fixed on its		T - fixed on its		Rh - fixed on its		Cl^- - fixed on its	
	30%	70%	30%	70%	30%	70%	30%	70%
	Range of the corrosion rate r_{corr} (µm/year)							
Sulphur dioxide line	-	-	2.15	33.69	20.64	27.53	23.83	23.83
Temperature line	38.41	79.32	-	-	51.72	73.27	42.03	78.65
Relative humidity line	13,05	19.93	7.01	31.01	-	-	5.53	13.48
Chloride ions line	39.20	39.20	20.64	74.42	30.91	49.71	-	-
Figure	4	5	6	7	8	9	10	11

3. Conclusion

The article was focused on the sensitive analysis of the dose-response function for carbon steel. The standard EN ISO 9223 [4] recommends the intervals for input parameters as sulphur dioxide, relative humidity, temperature and chloride ions. The range from 30 to 70 percentage of these recommended intervals were chosen for the analysis of the function. Generally, it is recommended as a fixed input parameter, for sensitive analysis, choose either the mean or the most frequent value. For this reason, when it is not known the most frequent values, the various combinations of fixed parameters were selected. It has to be mentioned, that the dose-response function was derived from the corrosion rate of the specimens under real atmospheric environment and it is the reason, why it is possible to make same particular conclusions on the base of the corrosion rate according this function.

The individual results show that the corrosion rate r_{corr} is mostly depending on the changing of the temperature and then on the changing concentration of the chloride ions. In the case, when all fixed input parameters are fixed in their thirty percentage, the change of chloride ions value the most attached the calculated corrosion rate. In the last step of this research only one parameter was fixed on its 30 or 70 percentage and the chloride ions curves were observed. The highest corrosion range, for the chloride ions line were either, when the temperature was on its 70% whilst another two fixed parameters were on their mean value (relative humidity and sulphur dioxide) or when the sulphur dioxide was fixed on its 30% (temperature and relative humidity was fixed in their mean value). This leads to the conclusion that the decreasing of SO_2 and increasing of temperature (such as in the case of global climate change) may causes the higher sensitivity of the corrosion rate r_{corr} equation on the chloride ions concentration.

Based on the dose-response function, the maps of corrosion rate can be created. For creation of these maps, the number and frequency of measurement stations are important. The temperature is measured in about 120 measurement stations in Slovakia by Slovak hydrometeorogical institute, which may be sufficient for creating these maps [5]. But the chloride deposition rate is not measured and so it is important to create the measurement stations mainly near the traffic infrastructure network, where the chloride ions are dominated due to de-icing salt.

Creation of the new corrosion stations with the specimens of non-protected (structural steel) and protected carbon steel by concrete (reinforced concrete) should by created in the areas where the temperature or chloride ions are dominated (near the traffic infrastructure network and/or in the area, where the temperature is higher like western part of Slovakia).

It may by mentioned that another point of the research should be the sensitive analysis in the interval of Slovakia's input parameters measured by Slovakia by Slovak hydrometeorogical institute.

Acknowledgements

This research is supported by the Slovak Research and Development Agency under contract No. APVV-14-0772, and by Research Project No. 1/0413/18 of Slovak Grant Agency and also by the project DS-2016-0039 in frame of bilateral cooperation.

References

[1] Krejsa M, Brozovsky J, Mikolasek D, Lehner P and Parenica P 2017 *Proc. Conf. on Dynamics of Civil Engineering and Transport Structures and Wind Engineering* (Zilina) vol 107 (France: EDP Sciences) pp 1-8

[2] Bujnak J, Hlinka R, Odrobinak J and Vican J 2012 *Procedia Engineer.* **40** 56–61

[3] Borko K, Pastorek F, Fintova S and Hadzima B 2016 *Komunikacie* **18** 99-102 (in Slovak)

[4] EN ISO 9223: 2012 *Corrosion of metals and alloys - Corrosivity of atmospheres - classification, determination and estimation*

[5] Strieska M and Kotes P 2017 *Proc. Inter. Conf. on Civil engineering and architecture Young scientist* (Strbske Pleso) (Kosice: TUKE)

[6] Krivy V, Kubzova M, Kreislova K and Urban V 2017 *Metals* **7** 336

[7] Tibdlad J *Atmospheric corrosion of metals in 1020-2039 and 2070-2099* (Elsevier)

[8] Ghosh, P, Konecny P, Lehner P and Tikalsky P J *Comput. Concrete* **19** 305-313

[9] Adikary A A M T, Munasinghe R G N De S and Jayatileke S 2014 *Engineer: Journal of the Institutuion of Engineers* **47** 75-83

IOP Conf. Series: Materials Science and Engineering **385** (2018) 012032 doi:10.1088/1757-899X/385/1/012032

Early-age behaviour of cement-based self-leveling flooring compounds

B Kucharczykova and D Kocab

Faculty of Civil Engineering, Brno University of Technology, Veveri 331/95, 602 00 Brno, Czech Republic

Abstract. The article deals with the cement-based flooring compounds, especially with the behaviour during the early solidification process. A touch-less laser sensor and a cone-shaped mould were used for the measurement of the plastic settlement/shrinkage. The process of solidification and the strength development were determined using the ultrasonic pulse velocity measurement. Both test devices were equipped with the thermocouple to measure the real-time internal temperature evolution during solidification of the flooring compounds. The measurements presented in this paper proved to be useful for the clarification of the technological problems which arose during the manufacturing of the test specimens.

1. Introduction

The self-leveling flooring compounds are usually dry ready-mixed composite materials used mainly in indoor applications for constructing new or renovating existing floors. The demands posed to today's flooring compounds are quite high, which forces the manufacturers and implementing companies to continuously develop and improve the compounds' composition and laying procedure during its implementation. The basic requirements posed to these products are usually rapid setting time, which has to be adequate for finishing the application, early strength development, good adhesion to the existing subfloor-layer, low shrinkage deformations and high resistance to abrasion [1].

One of the factors, which takes a notable role in the characteristics of the final flooring compound, is the quality and characteristics of the raw materials themselves [2]. Another factor is the composition design which consists in the appropriate combination of the raw materials to obtain the product of high quality [3]. The sustainable development of building materials emphasizes also the eco-friendly approach, which motivates to the usage of industrial by-products as a component of the final product [4].

The most frequently discussed issue of the flooring compounds is still the shrinkage and high risk of cracking after pouring/laying the fresh mixture. Consequently, the enhanced effort of the producers is still aimed at the development and testing of appropriate binders with a low potential to shrinkage [5-9]. Note that the curing conditions often do not comply with the requirements stated in the technical sheets of the flooring products, which can be the main source of cracking in the finished floors.

The authors of the paper focused on the behavior of the cement-based flooring compounds during its early solidification process, especially on the measurement of the plastic settlement/shrinkage and on the strength development expressed by the development of ultrasonic pulse velocity.

2. Experiment

Three different self-leveling cement-based flooring compounds (FC), identified as FC 1, FC 2 and FC 3, were selected for the experiment. The main motivation to perform the measurements was the technological problems which arose during manufacturing and demoulding process of the test specimens, which were intended for the shrinkage and strengths' parameters measurement.

One of the problem was the excessive plastic settlement of the FC 1 resulting in a decrease in the top surface of the test specimens from the edges towards the center. Subtle test specimens showed uneven cavities in the top surface (see figure 1). Another problem was the demoulding of the test specimens made of FC 3 compound. The specimens were not hardened enough at the required time for demoulding (after 24 hours) which led to the spalling of the material and to the rupture of the subtle test specimens (see figure 2). The last problem was observed in the case of FC 2 which showed high sensitivity to the changes in the ambient relative humidity which led to cracking of the test specimens.

Figure 1. FC 1: Test specimen with dimensions of 40×40×160 mm (left) and 10×40×160 mm.

Figure 2. FC 2: Test specimen with dimensions of 40×40×160 mm (left) and 10×40×160 mm.

2.1. Materials
The composition of the tested products cannot be listed in the article because of the pilot tests of the floor compounds, which are currently under development. Generally speaking, the FC 1 is primarily intended for the thin-layer floors (approx. 10mm), while the FC 2 and FC 3 can be applicated in layers up to 40 mm. All tested materials are described by the strength characteristics, shrinkage values and mass losses values determined on the prismatic test specimens with dimensions of 40×40×160 mm at the age of 3, 7 and 28 days (see figure 3 and 4) [10].

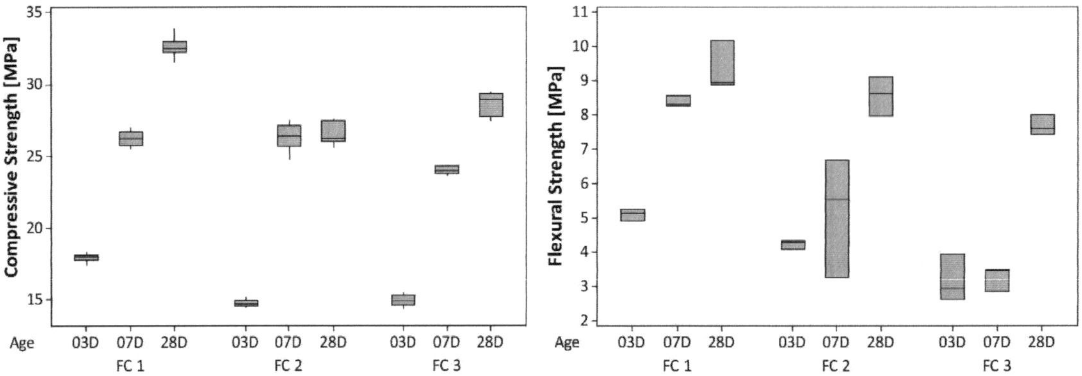

Figure 3. Compressive strength (left) and flexural strength of tested floor compounds.

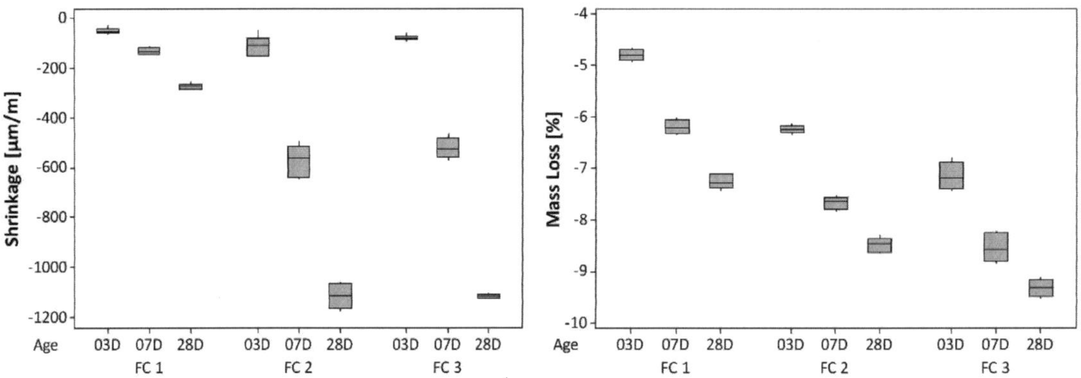

Figure 4. Shrinkage (left) and mass loss of tested floor compounds.

2.2. Test specimens and test methods

All mixtures were prepared from the dry-mixture mixed with the prescribed amount of water. The Hobart mixer was used for preparation of the fresh mixtures, whereas the mixing and manufacturing process followed the instructions stated in the technical sheet of each product. The Shrinkage-cone and Vikasonic apparatus were used for the determination of the plastic settlement and solidification process of selected flooring compounds.

The Shrinkage-cone was used especially for measurement of the plastic settlement. It is a test device equipped with a touchless laser sensor which enables to measure the changes in height (Δh) of the test specimen poured into the cone-shaped mould with height of 125 mm and volume of 682 cm^3 (diameter approx. 144 mm). The evolution of the internal temperature during the maturation process was continuously measured by the embedded thermocouple. The measurement in the Shrinkage-cone started immediately after filling the mould with the fresh mixture. The data were continuously recorded into the data-logger. In the experiment, the PTFE foil was placed into the mould before its filling to avoid the friction between the mould and poured material. The upper surface of the test specimens was not protected from drying during the whole time of measurement.

The Vikasonic apparatus is an ultrasonic (US) device which consists of measuring cell equipped with two transducers and a Vicat ring, and a control unit which enables to setup the measurement and also serves as a data storage device. The measurement principle is based on the measurement of the transit time of ultrasonic wave which pass through the solidifying material from the transmitter to the receiver. Based on the measured US transit time and the distance between transducers, the US pulse velocity can be calculated. The Vikasonic apparatus is also equipped with the thermocouple to record the internal temperature evolution during the solidification process of the compounds. Both

IOP Conf. Series: Materials Science and Engineering **385** (2018) 012032 doi:10.1088/1757-899X/385/1/012032

measurement devices are shown in figure 5. During the measurement, all of test equipment were placed in a climate chamber with an ambient temperature of (22 ± 2) °C and relative humidity RH = (50 ± 5) %. More details about Shrinkage-cone and Vikasonic see in [11].

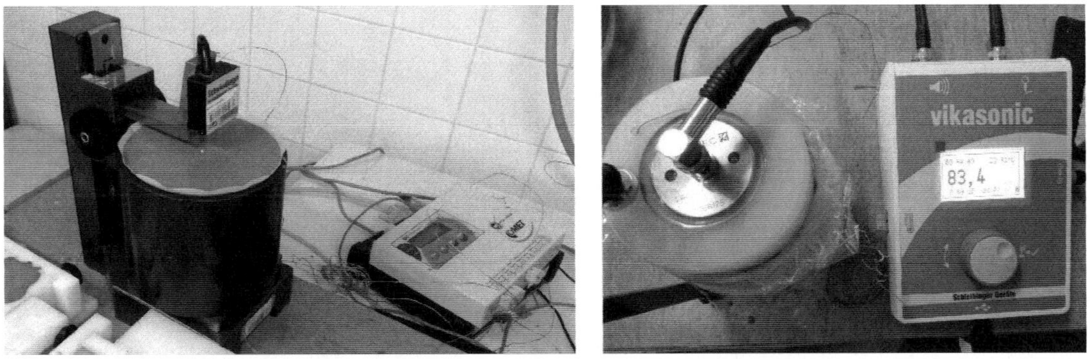

Figure 5. Shrinkage-cone (left); Vikasonic apparatus.

3. Results and discussion

The results of performed measurements are shown in figures 6 – 8. The outputs from the Shrinkage-cone are shown in the left column. The data obtained from the measurement performed by Vikasonic apparatus were processed and they are shown in the right column. The results show differences in the solidification process of particular flooring compounds.

In the case of FC 1, the initial plastic settlement is three to four times higher than in the case of FC 2 and FC 3. There is also a visible delay in the evolution of the internal temperature of the test specimen. The steep growth of the temperature occurred approx. 7 hours after the start of the measurement. In this time, the growth of plastic settlement/shrinkage had stopped and the vertical deformation stabilized (see figure 6 left). The above mentioned findings clarified the technological problems during the manufacturing of the test specimens using FC 1. It can be suggested that the process, magnitude and final impact of the plastic settlement on the quality of upper surface depends on the shape and area of this surface (see figure 1). The results obtained from the measurement of the US transit time for FC 1 correspond well to the measurements performed by Shrinkage-cone. The delay in the internal temperature evolution, which led to the delay in the setting time, was also detected. This delay is very well expressed by the progress of the US pulse velocity. The steep growth of the US pulse velocity occurred within the interval of temperature growth. Shortly after the temperature reached its peak, the US pulse velocity had stabilized (see figure 6 right). This process indicates rapid strength development.

The slight differences between the magnitude and evolution of the internal temperature recorded in the cone-shaped mould and Vicat ring-mould are supposedly caused by the different volume of test specimens and technical parameters of the moulds.

In the case of FC 2 and FC 3, two temperature peaks were detected during early solidification. The first temperature peak occurred shortly after the start of the measurement (approx. 2 hours for FC 2; 1 hour for FC 3), the second temperature peak was detected approx. 20 hours for FC 2 and 24 hours for FC 3 after the start of temperature measurement (see figures 6 and 7). The magnitude of the internal temperature of FC 2 and FC 3 mixtures was in both cases smaller than the one recorded for FC 1. Concerning the plastic settlement, the initial magnitude was the same for both FC 2 and FC 3. The subsequent process of the vertical deformation was quite different. While in the case of FC 3 the deformation stabilized shortly after the initial settlement, for FC 2 the expansion, which stabilized approximately at the time of second temperature peak, was detected.

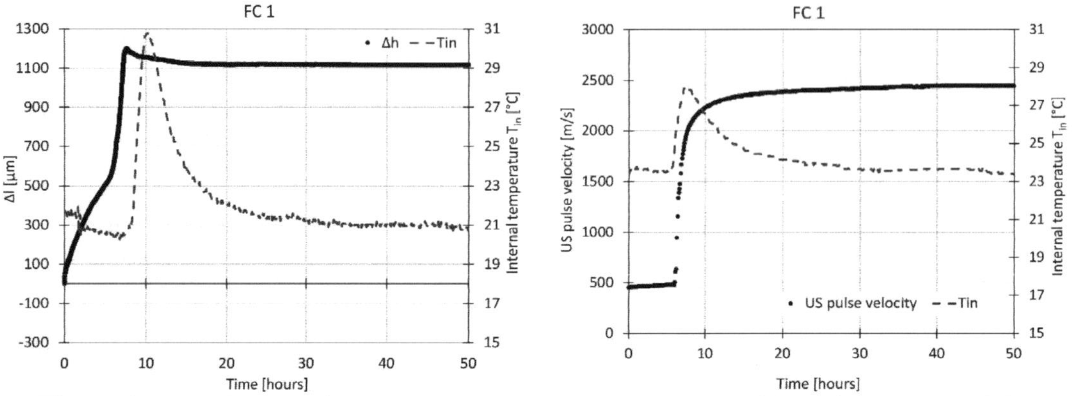

Figure 6. FC 1: Plastic settlement and internal temperature measured in Shrinkage-cone (left); US pulse velocity and internal temperature measured in Vikasonic.

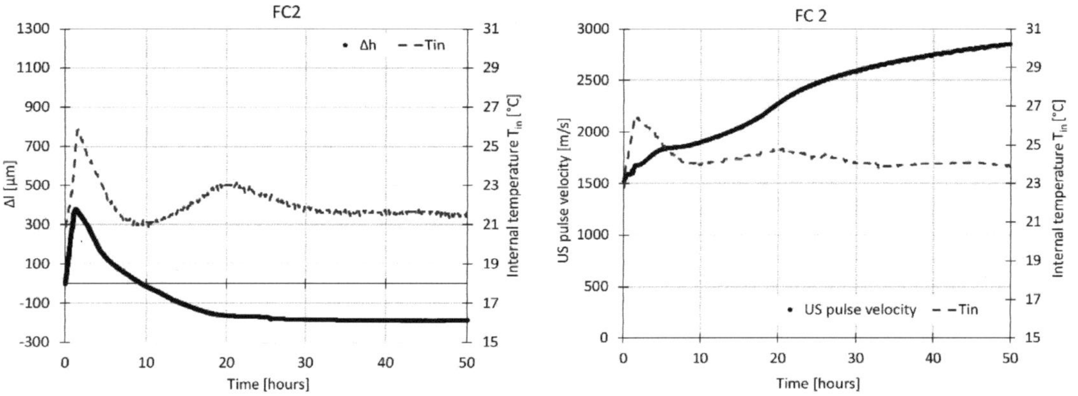

Figure 7. FC 2: Plastic settlement and internal temperature measured in Shrinkage-cone (left); US pulse velocity and internal temperature measured in Vikasonic.

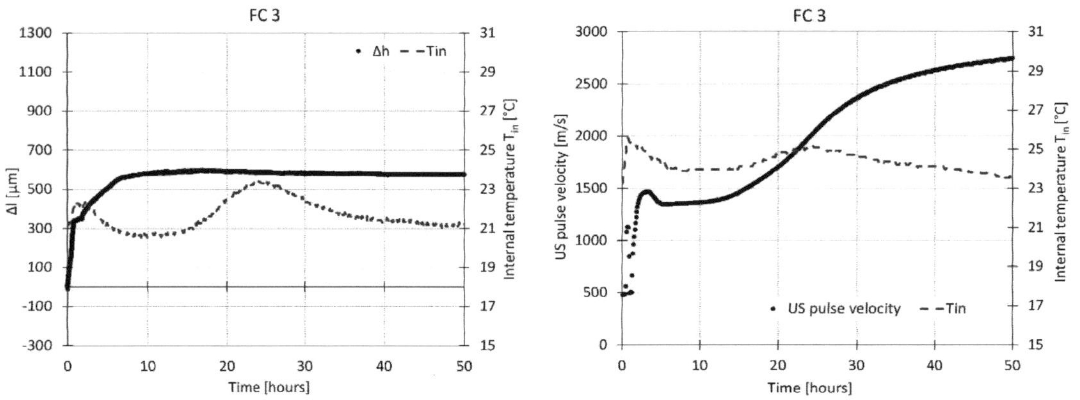

Figure 8. FC 3: Plastic settlement and internal temperature measured in Shrinkage-cone (left); US pulse velocity and internal temperature measured in Vikasonic.

The technological problem with demoulding process of FC 3 is clarified by the results of the US pulse velocity and internal temperature evolution (see figure 8 right). Based on the process of the US pulse velocity and internal temperature, it can be suggested that the strength development of FC 3 was slower in comparison to the FC 1 and FC 2. The second temperature peak in FC 3 occurred exactly at the time of required demoulding process (see figure 7).

4. Conclusions

The article focuses on the measurement of early age materials' characteristics which take place during the solidification process. The plastic settlement/shrinkage of the flooring compounds was measured by the Shrinkage-cone. The strength development was estimated based on the US pulse velocity development calculated from the US transit time measured by the Vikasonic apparatus. Both test devices were equipped with the thermocouple to measure the real-time internal temperature evolution during solidification of the flooring compounds. The measurements presented in this paper proved to be useful for the clarification of the technological problems which arose during the manufacturing of the test specimens. The Shrinkage-cone measurements proved excessive plastic settlement/shrinkage of the FC 1 mixture which led to the settlement and cavitation of the upper surface of the test specimens. The internal temperature evolution and process of the US pulse velocity determined by the ultrasonic measurements provided information about the strength development of tested products. Based on the results it can be stated, that the FC 3 mixture had not enough strength at the time of demoulding of the test specimens.

Acknowledgement
This work was supported by the Czech Science Foundation [grant No. GA17-14302S].

References

[1] Papayianni I, Anastasiou E, Avramidis P and Chatziagorastou P 2016 *Proc. Int. Conf. on Concrete Repair: Concrete Solution* vol 1 (Thessaloniki, Greece: CRC Press) pp 281-286
[2] Zhao Z G, Shi Z W and Liu S 2012 *Applied Mechanics and Materials* vol 117-119 (Schwitzerland: TTP) pp 1092-1094
[3] Canbaz M, Topcu I B and Atesin O 2016 *Constr. Build. Mater.* **116** 321-325
[4] Barluenga G and Hernandez-Olivares F 2010 *Constr. Build. Mater.* **24** 1601-1607
[5] Katsiadramis N J, Sotiropoulou A B and Pandermarakis Z G 2010 *EPJ Web of Conferences* **6** no 23002
[6] Dohnalek J 2016 *TZB-info* (Prague: Topinfo s.r.o.) pp 1-8 (in Czech)
[7] Kaddo M and Murgul V 2017 *MATEC Web of Conferences* **106** no 03024
[8] Hirano Y, Komatsu R and Ikeda K 2006 *Trans. Mater. Res. Soc. Jpn.* **31** 325-328
[9] Li N, Xu L, Wang R, Li L and Wang P 2018 *Cem. Concr. Res.* **108** 103-115
[10] Kocab D and Kucharczykova B 2018 *15th conference Technologie* (Jihlava: CBS) pp 1-7 (in Czech)
[11] *Schleibinger Testing Systems* http://www.schleibinger.com (online)

IOP Conf. Series: Materials Science and Engineering **385** (2018) 012033 doi:10.1088/1757-899X/385/1/012033

The use of environmentally acceptable plastic materials based on cement and polymer composites with out of consideration to their physical properties

M Ledererova, Z Stefunkova and V Gregorova

Department of Material Engineering, Faculty of Civil Engineering, Slovak University of Technology, Radlinského 11, 810 05 Bratislava, Slovakia

Abstract. The aim of the research is to find possibilities for the re-use of environmentally eligible plastic waste as one of the options as a filler in light concretes. The plastic waste was obtained by recycling the plastics from the packaging of electrical cables, Ethylene-vinyl acetate and waste from expanded polypropylene in various proportions. All plastics are included among the medium-sized recyclable plastics according to the degree of their recycling. The experimental part of this work is devoted to verify the impact of the dose of fillers on the observed physical properties (bulk density, heat-technical parameters, sound absorption) of light concrete.

1. Introductions

Currently, the recycling in Slovakia is only about 1.2 % of municipal waste [1]. Great attention is paid to waste on the use of recycled plastic [2-5], for example, Recycled polystyrene, polyurethane foam, PET bottles, waste from electrical and electronic devices, PVC pipes etc. One of the ways to use recycled materials of plastics is to include them in light concrete, the selection of the appropriate combination of filler and binder will form building material not only for the next construction, but also for remodeling, rehabilitation, new materials etc.

Plastics, with their weight, form less than 8% of the waste, but the volume is much bigger, this is particularly important in landfills, where volume is more important than weight. Plastics are mostly non-reproducible oil, whose stocks are estimated at 40 years. Only by their transportation, 3.5 million tons is annually released into the surface-water . The biological decomposition of plastics in the nature or in landfills is long-term, it takes several decades. In case of combustion of certain kinds of plastics, harmful substances escape into the air. SR recycles 16% of the plastics [1].

Plastics are used as some of the most common packaging, and thus are among the most common type of waste that is produced nowadays. Plastics were supposed to be a replacement for natural materials, natural resin, or glass fiber. Now they are considered to be a special group of materials with unique properties. Today, they are used as high-molecular-weight organic compounds. Initially these were mainly natural materials, such as cellulose, proteins, rubber, and resin. Plastics are produced by the process of the so-called polymerization, in the small short molecules together in a long variety of linked chains. Therefore, there are several hundred different polymers with very different properties [6].

2. Materials and methods

For experimental testing for the research of the recycled waste such as polypropylene (L1) ethylene vinyl acetate (EVA) (L3), and recycled material from cables (L5) was used as shown in figure1. Materials and components of mixtures were mixed in following proportions: fillers 100 % PP (L1) binder Conipur 360, fillers 50 % PP + 50% EVA (L2) binder Conipur 360, fillers 100% EVA (L3), binder Conipur 360, fillers 50% EVA + 50% PVC (L4), binder Conipur 360 and fillers 100 % PVC (L5) binder Conipur 360.

L1

L3

L5

Figure 1. Plastic waste fractions:
L1 of 4/8 mm L3 of 4/8 mm L5 of 0/4 mm.

Polypropylene is very light and strong graining. Polypropylene is a thermoplastic polymer from the group of polyolefin used in many sectors of the economy. Polypropylene features with very good chemical and mechanical resistance. For research were used the waste from the recycled polypropylene (L1). Ethylene-vinyl acetate (L3) is the elastomeric polymer with high elasticity, known as the ethylene-vinyl acetate (EVA) [7]. The recycled material of the cables is a pulp recycled cable PVC where the cable core is formed, in particular polyvinyl chloride (L5) [8].

The polypropylene and EVA were crushed and the fraction 4/8 mm screened cable for the fraction 0/4 mm, wherein the bulk density of polypropylene 17.52 kg.m^{-3}, EVA 104.41 kg.m^{-3} and cable 359.33 kg.m^{-3}. LB were produced with the use of a filler, or as a combination of two materials according to the percentage share: 100%, 50:50 % (table 1).

As binder was used undiluted one-component polyurethane adhesive Conipur 360. This adhesive is used in the implementation of the polyurethane surfaces. It has a firming effect, it is resistant to abrasion and very easily workable. Dose of glue in all mixtures was also a constant of 0.8 l per one mixture. Considering different surface and material was - moistened 2% water content, which regulates the surface tension.

Table 1. Composition of the individual mixtures.

Test samples	Binder (l) Conipur 360	Graining (l) PP	EVA	Cable
L1	0,8	8	0	0
L2	0,8	4	4	0
L3	0,8	0	8	0
L4	0,8	0	4	4
L5	0,8	0	0	8

Samples in the shape of 100×10×100 mm were made from homogeneous mixtures. These samples were treated 48 hours in the laboratory environment. Subsequently, they were removed from the mold and stored in laboratory environment (20 °C, humidity= 50%). Subsequently, the physical properties (bulk density, heat –technical parameter and thermal conductivity) were measured. The sound absorption was tested only on the EVA sample. Measurements of bulk density and heat of technical parameters were carried out in the laboratory conditions of the STU.

Sound absorption

Measurements of sound absorption were performed by the acoustic laboratory at KU Leuven (The Katholieke Universiteit Leuven, Belgium). Due to the complicated measurement and sample size (according to STN EN ISO 354 [9]. the sample's size is 10m^2) the sound absorption was verified only on small EVA samples (L3) in the impedance tube. These measurements were made only in KU Lueven. For this purpose, the impedance tube was measured on samples L3 (figure 2). Knowledge of the sound absorption of building materials, opens up the possibility of their applications also in terms of dampening the sound as in the interior (generally in diffusion audio field), as well as in the outdoors (free sound field) [10-11]. The experiment was focused on the measurement of the sound absorption factor. For this purpose, a set of L3 (EVA) samples for measurement in the impedance tube was modified. For the verification of sound absorption was used ethyl-vinyl acetate (EVA), in combination with the binder Conipur 360 (polyurethane glue). It was a sample with thickness of 6.5 cm. Diameter of cylinder of the sample was 4 cm. The measurement and the used measuring assembly operates on the principle described in the standard ISO 10534-2 [12-13].

Subsequently, the first results were measured in the acoustics laboratory. Figure 2 shows the measurement of sample L3.

a) b) c)

Figure 2. The measurement of the sample EVA (L3):
a) L3 b) The sample is inserted into the piston metering assembly c) impedance tube.

3. Results

Physical and thermophysical properties are shown in (figure 3,4).

While the bulk density of samples from PP (L5) and EVA (L3) ranged from 245 to 230 kg.m^{-3}, the cable (PVC) increased by an increasing proportion of PVC due to a higher bulk of the input material reaching values up to 700 kg.m^{-3} (figure 3).

With the increasing proportion of PP (L1, L2), the thermal conductivity coefficient of the test samples decreased, and vice versa with increasing proportion of cable fillers, the coefficient of thermal conductivity of the test samples increased (L4, L5) (figure 4).

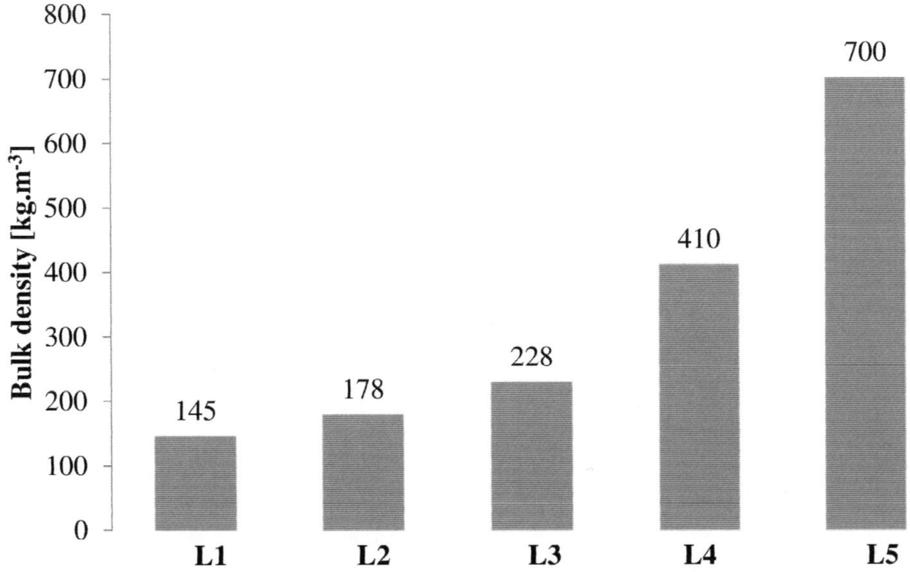

Figure 3. Comparison of the Bulk density of mixtures.

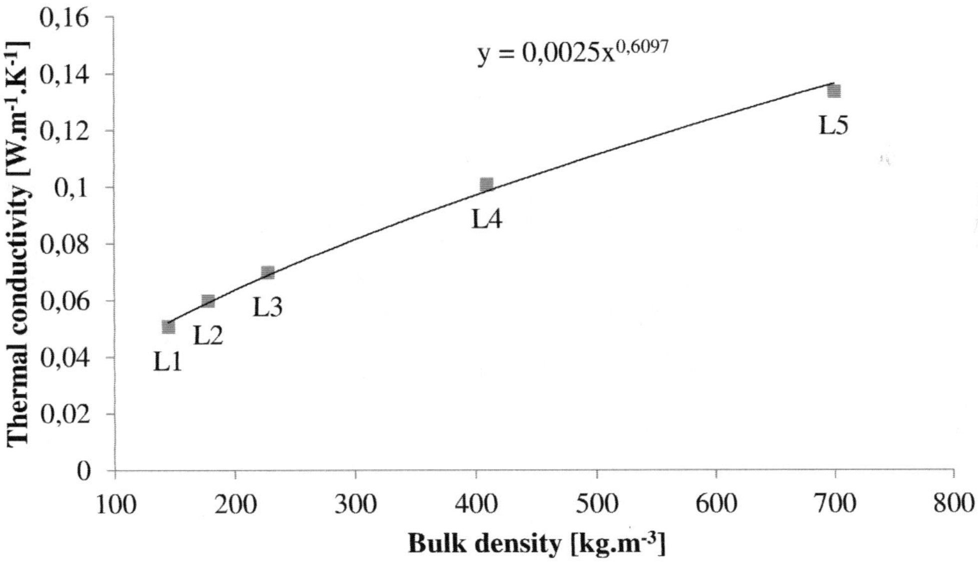

Figure 4. Relationship between thermal conductivity and bulk density of mixtures.

Figure 5 shows the results of the factor of sound absorption α (-) and factor of reflection of the acoustic pressure in the perpendicular impact of the sound of the r (-) measured in an acoustic laboratory at the KU Leuven (Belgium). Results obtained by this measurement (verified impedance tube with high precision, stability and all repeatability of this measurement) are in this work taken as basics. From results is clear, that already in the comparison of three samples, of the same composite, there have been deviations in the measured values up to 10%. It may predestinate the diversity of the measured sample, but also the way of installing the sample into the measuring report (samples were bigger than the diameter of the measuring piston, and so were manually modified).

The results of the KU Leuven gave lower values of absorption than assumed. The transformation (gear) method offers greater stability. Its advantage is also the duration of the measurement when the measurement is made within 10 minutes. Measured values are relevant for the frequency range 1250 - 1500Hz.

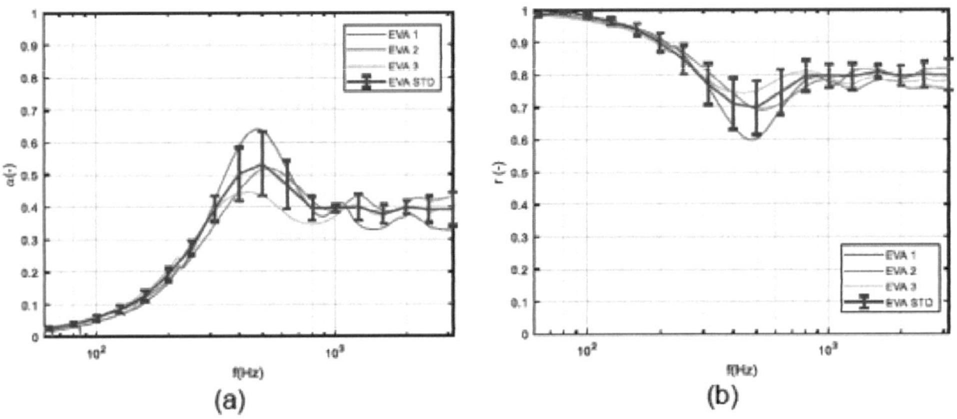

Figure 5. The resulting spectrum for the perpendicular impact of the sound waves with their standard deviation (L3):
a) Range of agent sound absorption; b) The absolute value of the factor of reflection.

4. Conclusion

On the basis of the results we can say that samples of plastic materials (L1, L2, L3) have favorable thermal properties with the binder. It can be stated that the use of recycled PP, EVA and PVC has a perspective in the production of lightweight concrete. Use of these materials can be envisaged in the construction industry. Standardized methods (Transformation Transition Method) have been considered in the Experiment of Determining Sound Absorption Factor (L3). The method does not require large samples and the assembly of the measurement set for basic sound absorption verification is financially less demanding. In summary, recycled plastic waste has a future in construction as well as in combination with various other plastic waste using a suitable binder. In view of their reuse in practice (eg with floor insulation), it would be advisable to further monitor and verify acoustic parameters.

Acknowledgements

This article was created with the support of the Ministry of Education, Science, Research and Sport of the Slovak Republic within the Research and Development Operational Program for the project "University Science Park of STU Bratislava", ITMS 26240220084, co-funded by the European Regional Development Fund and by the Slovak Research and Development Agency under the contract No. APVV-15-0681.

References

[1] Gregorova V 2015 *Proc. Int. Conf. on for PH.D Students and Yong Researchers (Vilnius)* pp 2484–2489

[2] Chandra S and Berntsson L 2002 *Lightweight Aggregate Concrete. Science, Technology, and Applicationsications.* (Norwich: William Andrew)

[3] Bouvard D, Chaix J M, Dendievel R, Fazekas A and Letang M *Cement and Concrete Res.* **37** 1666-1673

[4] Chen B and LIU J 2007 *Constr. Build. Mater.* **21** 7–11
[5] Pacheco – Torgal F, Ding Y and Jalali S 2012 *Constr. Build. Mater.* **30** 714–724
[6] Janickova B 2012 Wastes and waste management. SZSaVOS Chrudim (in Czech)
[7] *Ethylene-vinyl acetate*: https://en.wikipedia.org/wiki/Ethylene-vinyl_acetate (online)
[8] Sicakova A and Sad J 2009 *Build. Mater.* **5** 36-40
[9] STN EN ISO 354: 2004-01 *Acoustic. Measurement of sound absorption in a reverberat room.*
[10] SNTL1965: https://aukro.cz/tranzistorove-menice-sntl-1965-6928958662 (online)
[11] Hopkins C Sound insulation. Routledge.2012
[12] ISO 10534-1:1996 *Determination of sound absorption coefficient and impedance in impedance tubes - Part 1: Method using standing wave ratio, 1996*
[13] ISO 10534-2:1998 Acoustics - *Determination of sound absorption coefficient and impedance in impedance tubes - Part 2: Transfer-function method, 1998*

IOP Conf. Series: Materials Science and Engineering **385** (2018) 012034 doi:10.1088/1757-899X/385/1/012034

The influence of the addition of gypsum on some selected properties of lime-metakaolin mortars

J Majerova and R Drochytka

Institute of technology of building materials and components, Faculty of Civil Engineering, Brno University of Technology, Veveri 331/95, 602 00 Brno, Czech Republic

Abstract. The renovation and replacement of historical plasters is still a very current topic. Today's trend is for the production of dry mortar mixtures that can be used for the renovation of historic buildings. The article therefore focuses on the preparation of gypsum-modified lime-metakaolin mortar. An experiment consisted of two stages. In the first stage of the experiment, the basic technological properties of fresh and hardened mortars were examined (bulk density of fresh and hardener mortar, compressive and flexural strength). The second stage aimed at monitoring the hydration process in different environments. Environments should simulate real factors affecting the plaster (humidity, low and higher temperatures). To determinate of the phase composition, the X-ray diffraction analysis was used. The results showed that gypsum in the lime-metakaolin plaster caused an increase in compressive and flexural strength. The presence of gypsum also conduced to the formation of ettringite and this has caused volumetric changes.

1. Introduction

Almost all existing building materials (such as mortar, concrete and plaster mixtures) are designed using Portland cement. However, Portland cement is a modern binder and has been widely used only for the last hundred years, so it is not appropriate to use in the reconstruction of historical buildings. When reconstructing historic buildings, the materials used should be a combination of materials that best substitutes the original binders used [1].

The first lime paints and plasters date back to 7500 BC and come from the area of today's Jordan. In Syria, gypsum was used as a wall finishing from as early as 7000 BC. These plasters were often decorated with paintings. In some cases, pozzolanic admixtures have been added to the lime plasters because they speed up the hardening of fresh plaster and increase its strength. The beneficial effect of the addition of ash from volcanic areas has also been described by the Roman writer Marcus Vitruvius Pollio in his work Ten Books on Architecture [2-4].

In our region (Czech and Slovak Republic), limestone was for a long time the most commonly used material; this is due to the prevalent geological conditions in the area, especially the strong karst areas [5]. The first evidence of the use of lime comes from the remains of the furnaces used for lime burning that have been found in several Stone Age archaeological locations. It is known that lime was also used in the areas of Great Moravia and Bohemia, as a means of painting the walls. The lime-burning process was poorly supervised and even low-quality limestone was burned. This resulted in the presence of hydraulic oxides (Al_2O_3, Fe_2O_3, SiO_2), which formed the natural hydraulic lime. These limes are characterised by good plasticity and higher strengths than found in present-day lime. The plaster was added to the mortar mostly to accelerate the hardening process or to achieve higher strengths.

2. Materials

2.1. Metakaolin (·$Al_2O_32SiO_2$)

Metakaolin is one of the most reactive pozzolans produced by the calcination of kaolin, kaolinitic clays, shales and other suitable raw materials containing the mineral kaolinite ($Al_2O_3{\cdot}2SiO_2{\cdot}2H_2O$). The raw materials are calcinated at temperatures of 600 – 900°C. At optimum activation temperatures between 725°C and 850°C, the dehydroxylation of the kaolinite occurs, forming a highly reactive amorphous structure of pentacoordinated atoms Al(V). At the same time, the temperatures are still low enough to form the crystalline phase. At temperatures above 900°C, the formation of mullite ($3Al_2O_3{\cdot}2SiO_2$) and cristobalite (SiO_2) occurs and causes a decrease of pozzolanic activity. The burning is followed by grinding to increase the specific surface of metakaolin [6]. Metakaolin reacts with $Ca(OH)_2$ to form the C-S-H and C-A-H phases of C_4AH_{13}, C_2ASH and C_3AH_6 [7-11].

The metakaolin used here, commercially sold as Mefisto K05, is one of the products by the company České Lupkové Závody a.s.

2.2. Calcium hydrate ($Ca(OH)_2$)

Calcium hydrate CL 90-S was used in this experiment. It is a white, slaked lime which is used to produce mortars, plasters and stucco. According to the standard EN 459 – 1, the CaO + MgO content must be at least 90%. The apparent density of the lime hydrate is 425 $kg{\cdot}m^{-3}$.

2.3. Gypsum ($CaSO_4{\cdot}1/2\ H_2O$)

In addition to the above, normally hardening building gypsum was used.

2.4. Filler

Natural non-sorted aggregate from Bratčice with the fraction 0 – 4 mm was used as the filler. In the experiment, a grain fraction of 0 – 2 mm was used, so the sand was sifted through a 2 mm-mesh standard sieve and only the sieved part was used.

3. Methods and data

The study was carried out in two stages. The first stage was based on the modification of reference mortar mixture that was developed in the previous research conducted at the Institute of technology of building materials and components FCE VUT in Brno. This reference mixture was modified by the addition of plaster regarding its influence on basic technological properties of the material. In the second stage, the hydration process of the binder stored in various conditions was monitored.

3.1. Stage I

Mix design of raw material powder and amount of mixing water. In the first stage, the plaster was added to the standard mixture to substitute 5 or 10% of the dry mortar mix (including fine aggregate). The amount of mixing water was carefully determined so that all samples exhibited a constant consistency (175 ± 10 mm) corresponding to determination of consistence of fresh mortar (by flow table) according to CSN EN 1015 – 3. The value is characterised by the water coefficient w (ratio of the amount of water to the dry mortar mix). The mix composition of the historic mortars is presented in table 1.

The bulk density of fresh mortar. The fresh mortar bulk density test was carried out according to CSN EN 1015 – 6.

The bulk density of hardened mortar, compressive strength and flexural strength. The fresh mortar was poured into moulds to form prismatic specimens of 40 × 40 × 160 mm. After 24 hours, the samples were demoulded and stored in the laboratory conditions. The bulk density, compressive strength and flexural strength of hardened mortars were analysed after 7 and 28 days.

Table 1. Mix design for historic mortars.

	S0	S5	S10
Fine aggregate 0 – 2 mm Bratčice (%w)	82	77.9	73.8
Metakaolin Mefisto K05 (%w)	10	9.5	9
Lime hydrate (%w)	8	7.6	7.2
Gypsum (%w)	0	5	10
Water ratio (water/dry mortar mix)	0.26	0.24	0.23

3.2. Stage II

The entry raw materials that form the mortar (lime hydrate, metakaolin and gypsum) were mixed according to proposed recipes from the Stage I. The materials were homogenised and mixed with water to form a paste of standard consistency (according to Vicat). After 24 hours we made test microsamples of all mixtures and stored them in various moisturising environments. These environments simulated the effects of atmospheric exposure on the plaster (humidity, low and higher temperatures, etc.):

- L/20: Exposure in the laboratory environment,
- C/5: Exposure in water saturated with CO_2 at 5°C,
- C/40: Exposure in water saturated with CO_2 at 40°C,
- V/5: Storage in water at 5°C,
- V/40: Storage in water at 40°C.

From each of the environments described above, sampling was performed after one, three, seven, 14, 21, and 28 days. To prevent any further hydration, the samples were washed several times in isopropyl alcohol, then dried and milled. The phase composition of all samples was determined using X-ray diffraction analysis.

4. Results and discussion

4.1. Stage I

It is clear from the results that, in order to ensure a uniform mortar consistency, it is necessary to use the largest quantity of mixing water in the reference mixture S0 (i.e., the mixture without the addition of gypsum). Furthermore, it can be assumed that the higher value of the water ratio in the S0 sample (table 1) results in lower values in the bulk density of the fresh mortar and in lower strength values (table 2).

When the gypsum content is increased, bulk density decreases after seven days. This is as a result of the more intensive drying process used to remove the mixing water. After 28 days, the bulk densities were virtually equal.

The compressive strengths and flexural tensile strengths of samples containing gypsum were already quite high after only seven days of hydration. After 28 of hydration, the strength increase in these samples is not so significant. The opposite phenomenon can be observed in the reference sample S0 (which is without gypsum content). The high initial strengths of samples containing gypsum can be attributed to the hydration of calcium suplhate hemihydrate to gypsum and, apparently, to a moderate increase in sample densities caused by the formation of primary ettringite. It can be assumed that, in the presence of sufficient water, the strength of the reference sample S0 will grow significantly in the long term. This will be caused by the pozzolanic reactions between metakaolin and lime hydrate during the formation of C-S-H products.

Table 2. The technological properties of fresh and hardened mortar.

	S0	S5	S10
Bulk density of fresh mortar (kg·m^{-3})	1,970	1,990	2,010
Bulk density of hardened mortar after 7 days (kg·m^{-3})	1,760	1,710	1,670
Bulk density of hardened mortar after 28 days (kg·m^{-3})	1,650	1,670	1,660
Compressive strength after 7 days (N·mm^{-2})	0.6	6.7	8.6
Compressive strength after 28 days (N·mm^{-2})	2.4	8.3	8.8
Flexural strength after 7 days (N·mm^{-2})	0.3	1.5	1.6
Flexural strength after 28 days (N·mm^{-2})	0.7	1.7	1.9

4.2. Stage II

X-ray diffraction analysis has proven the presence of residues of inert mineral, specifically kaolinite, quartz, and portlandite from lime hydrate. From the new hydration products, the following minerals were identified: gehlenite octahydrate (C_2AH_8), C-A-H phase-type C_2AH_8, carbonate complex ($C_3A{\cdot}CaCO_3{\cdot}12H_2O$), gypsum ($CaSO_4{\cdot}2H_2O$), ettringite ($C_3A{\cdot}3CaSO_4{\cdot}32H_2O$), monosulfate ($C_3A{\cdot}CaSO_4{\cdot}12H_2O$) and calcium carbonate ($CaCO_3$) in the form of vaterite, aragonite and calcite.

In the case of samples with the addition of gypsum (S5, S10), the same minerals formed. Thus, from now on only the results from mixtures without gypsum (S0) and the mixture with the 5% gypsum content (S5) will be reported.

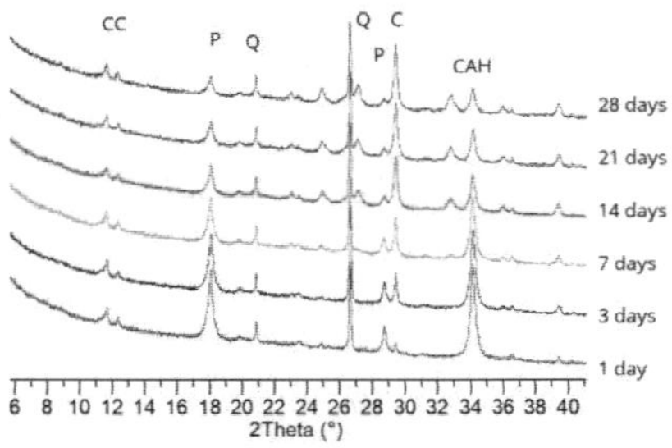

Figure 1. L/20 sample S0 (note: CC – carbonate complex; P – portlandite; Q – quartz; CAH – C-A-H phase; C – calcite).

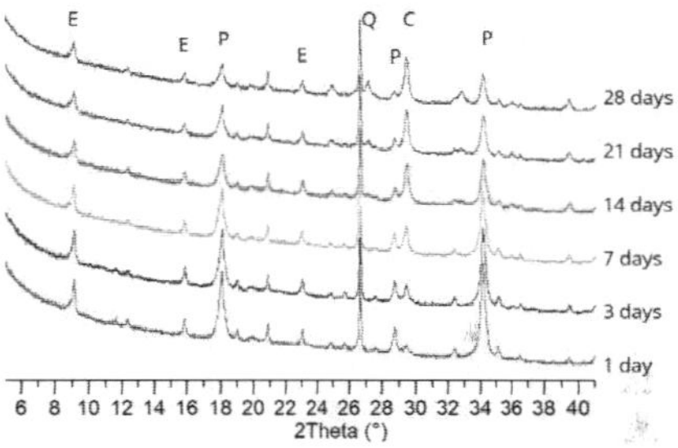

Figure 2. L/20 sample S5 (note: C – calcite; E – ettringite; P – portlandite; Q – quartz).

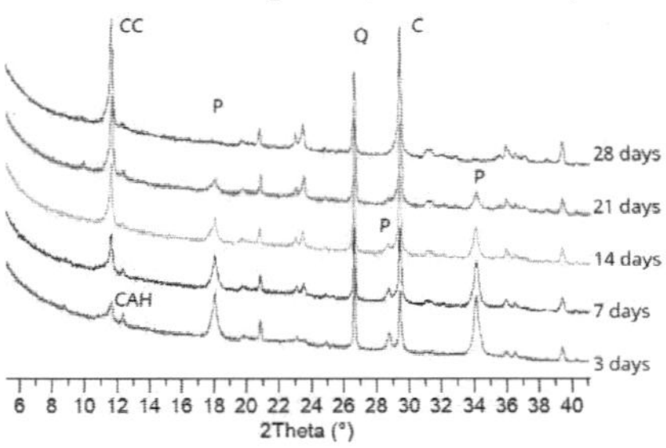

Figure 3. C/5 sample S0 (note: CAH – C-A-H phase; C – calcite; CC – carbonate complex; Q – quartz; P – portlandite).

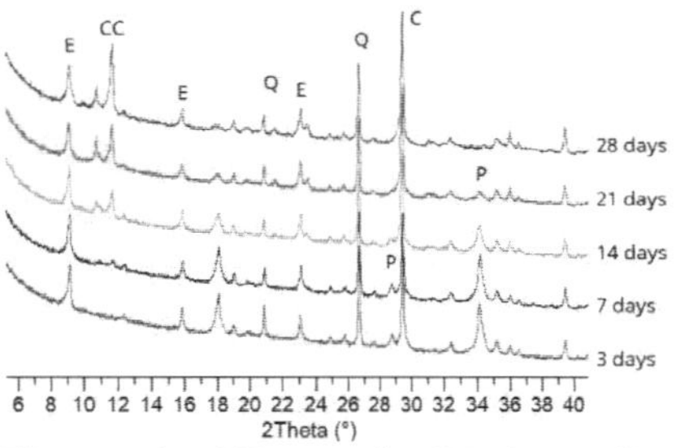

Figure 4. C/5 sample S5 (note: C – calcite; CC – carbonate complex; E – ettringite; P – portlandite; Q – quartz).

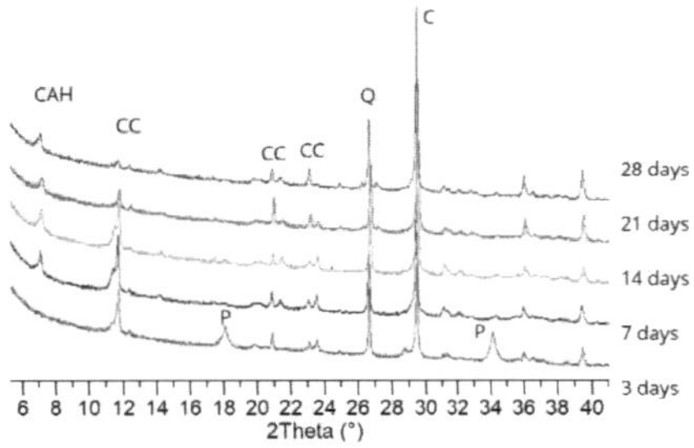

Figure 5. C/40 sample S0 (note: CAH – C-A-H phase; CC – carbonate complex; C – calcite; P – portlandite; Q – quartz).

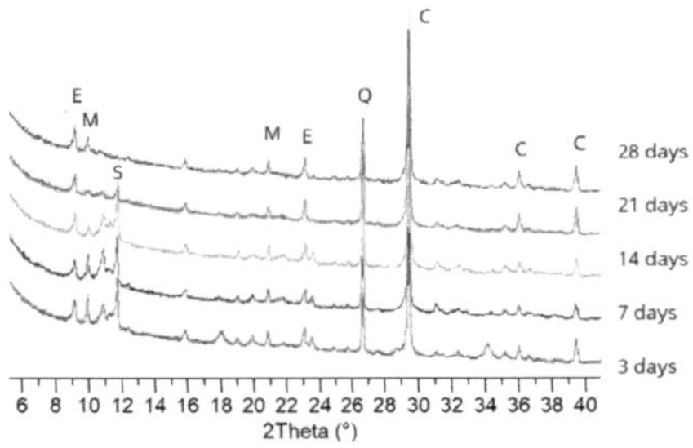

Figure 6. C/40 sample S5 (note: C – calcite; E – ettringite; M – monosulfate; Q – quartz; S – gypsum).

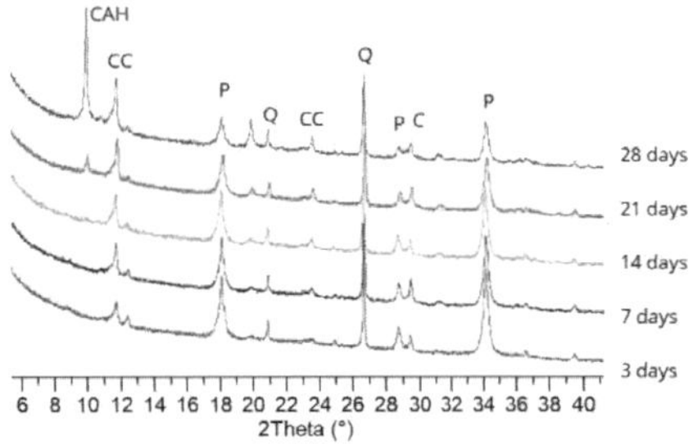

Figure 7. V/5 sample S0 (note.: CAH – C-A-H phase; CC – carbonate complex, C – calcite P – portlandite, Q – quartz).

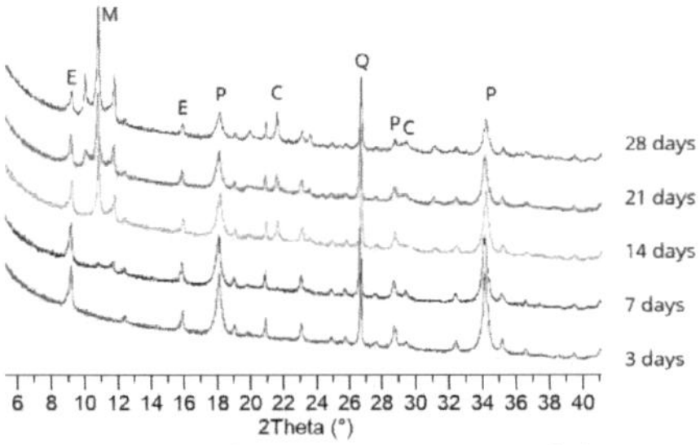

Figure 8. V/5 sample S5 (note: C – calcite; E – ettringite; M – monosulfate; P – portlandite;Q – quartz).

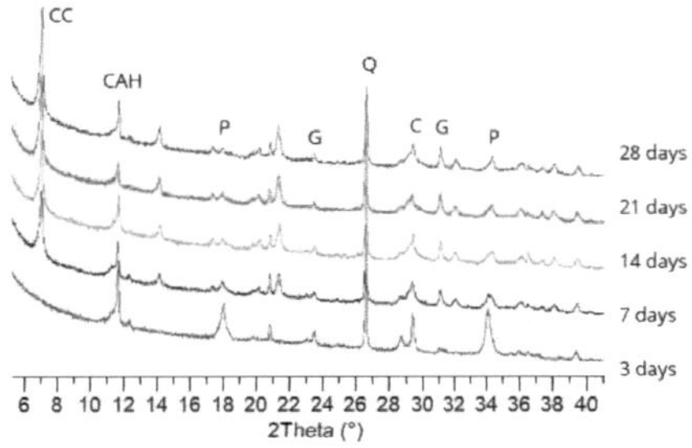

Figure 9. V/40 sample S0 (note: CAH – C-A-H phase; CC – carbonate complex; G – gehlenite octahydrate; P – portlandite;; Q – quartz).

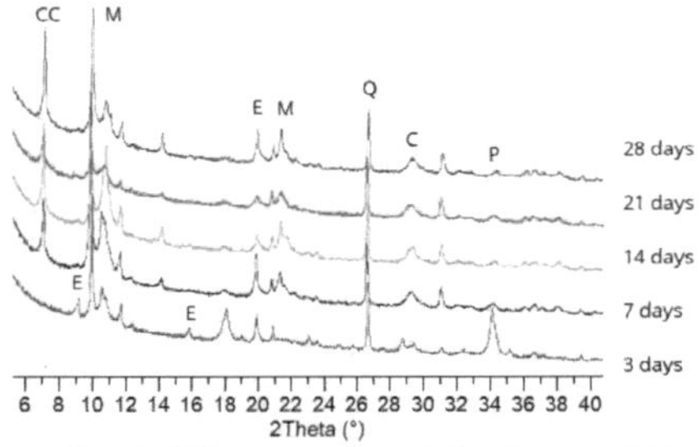

Figure 10. V/40 sample S5 (note: CC – carbonate complex; C – calcite; E – ettringite; M – monosulfate; P – portlandite; Q – quartz).

By following the phase composition in various hydration environments, it has been discovered that, in pure lime-metakaolin mortars, new hydration products from the C-A-H phase are formed (figures 1, 3, 5, 7, 9); these then gradually transform into an AFm phase of the carbonate complex type. This transformation takes place also partly at lower temperatures. Higher temperatures help create gehlenit octahydrate in a water environment (figure 9). When replacing a part of the lime-metakaolin mortar with gypsum, the ettringite is present from the beginning figures 2, 4, 6, 8, 10); this is unstable and transforms into monosulfate at 40 °C (figure 6 and figure 10). In the long run, this process may have a negative effect on the volume constancy and, consequently, on the technological properties of the hardened mortars

5. Conclusion

The aim of the experiment was to research the influence of plaster on some properties of lime-metakaolin mortar intended for the preparation of historic plasters. It was discovered that replacing 5% of the original reference mixture by gypsum positively affects both the compressive and the flexural strength. The compressive strength after 7 days was ten times higher than the compressive strength of the reference sample. The flexural strength after 7 days was five times higher than the flexural strength of the reference sample. This is principally caused by the hydration of calcium sulphate hemihydrate to gypsum and the formation of primary ettringite. However, the presence of sulphates and water in the plaster leads to the formation of ettringite, which can transform to monosulfate at temperatures of 40 °C and above. This transition is characterised by volumetric changes and can, in extreme cases, lead to a defects within the plaster.

6. Acknowledgment

This paper has been worked out with financial support from the budget under the project FAST-S-18-5406, "Study of advanced silicate composites structure with emphasis on the characterization of secondary raw materials influence" and under the project No. LO1408 "AdMaS UP - Advanced Materials, Structures and Technologies", supported by Ministry of Education, Youth and Sports under the "National Sustainability Programme I".

References

[1] Pavlik V and Uzakova M 2016 *Constr. Build. Mater.* **102** 14-25
[2] Hosek J and Losos L 2007 *Historical omissions: researches, redevelopment, typology* (Praha: Grada Publishing) (in Czech)
[3] Rovnanikova P 2002 *Omissions: chemical and technological properties* (Praha: STOP) (in Czech)
[4] Pavlik V and Uzakova M 2016 *Constr. Build. Mater.* **102** 14 - 25
[5] Rovnanikova P 2016 Omissions of historic buildings: complexes, analyzes, restoration (in Czech)
[6] Grilo J, Santos S A, Faria P, Gameiro A,Veiga R and Velosa A 2014 *Constr. Build. Mater.* **51** 287-294
[7] Bakolas A, Moropoulou A and Aggelakopoulou E 2011 *Appl. Clay Sci.* **53** 15-19
[8] Cerny R, Tudlitat V, Drchalova J and Rovnanikova P 2006 *Constr. Build. Mater.* **20** 849-857
[9] Andrejkovicova S, Velosa A L, Ferrat E and Rocha F 2014 *Constr. Build. Mater.* **65** 132-139
[10] Vimmrova A, Keppert M, Michalko O and Cerny R 2014 *Cem. Concr. Compos.* **52** 91-96
[11] Melichar T and Bydzovsky J 2013 *Appl. Mech. Mater.* **405-408** 2806-2809

Effect of wood chips properties on mineralogical composition and microstructure of silicate matrix based composites

T Melichar, J Bydzovsky and A Dufka

Faculty of Civil Engineering, Brno University of Technology, Veveri 331/95, 602 00 Brno, Czech Republic

Abstract. The research presented in this article was focused on proving the effect of wood chips parameters on the hardening process of silicate matrix in cement-bonded particleboard. Namely content of humidity in the chips (in range 30 to 90%) was significant. This amount of humidity was chosen with respect to real conditions of cement-bonded particleboards production. Influence of humidity could be reflected in the content of sugar (cellulose etc.) in wooden chips. The sugars affect the hardening of matrix negatively. Thus, phase composition and microstructure of matrix of cement-bonded particle boards were analysed during 2 to 28 days. Investigation of physico-chemical parameters and microstructure was also supplemented by verification of the basic material characteristics of cement-bonded particleboards (strength and modulus of elasticity in bending, transverse tensile strength perpendicular to the plane of the board, etc.). Slightly negative effect of increasing chip humidity content on the mineralogical composition and microstructure including physical parameters was shown in scope of the experiments. Therefore attention during cement-bonded particleboards production has to be paid even to moisture of the chips.

1. Introduction

The production process of cement-bonded particleboards consists of several phases. First, considering the given humidity, the basic raw materials are weighed – chips, cement, water, additives. The additives are mainly used for closing of board surface part structures. Another reason is to prevent release of sugars which are contained in the chips. During the boards production a mixture with humidity up to 45% is put in layers on steel plates by using machine equipment. The layering is done in three steps – first the bottom visual layer, then core and finally the visual one again. The visual layer contains more cement bonding material and smaller chips resulting in more compact structure. Thickness of visual layer moves within several mm. Metal plates with mixture are stacked one over another. Then the stack is pressed in a hydraulic press. By pressing the original height is reduced to 1/3. The stack in this pressed condition is placed in a steaming chamber for several hours. The temperature here does not exceed 55°C. Further approximate 7 days the boards mature freely in the production plant with subsequent drying in tunnel drying plant where the temperature does not exceed 90°C. In the last production phase the boards are processed to required formats, surface quality with eventual application of surface treatment. The mixture for production of cement - bonded particleboards is characterized with a certain optimum humidity as it was stated above. The chip humidity is closely related to this. Considering the wood processing, transport and storage condition (also the weather) the chips are delivered to the production plant with variable humidity. In order to preserve the production process continuity it is not possible to dry the chips to a constant weight or humidity respectively. The chips are usually delivered with absolute humidity content within 30 to

90%. In case of lower humidity content just adding of water will do. However with humidity content close to the upper limit, not only the amount of mixed water batch must be reduced but also the amount of water for dissolving of mineralization mixtures should be reduced. Theoretically the hydration process could be negatively influenced due to insufficient mineralization of chips. It can also be assumed that during pressing water out of the chips over its surface parts the already mineralized layer could be damaged - „drifted out" of the mineralizing component from these surface parts. Presence of sugars in cement matrix changes reaction of tricalciumaluminate (C_3A), which is the fastest reacting component of cement. Extraction components from wood may be also absorbed on the core of calcium hydroxide, therefore they slow down hydration of tricalciumsilicate (C_3S) [3]. According to [4-5] inhibiting components (hemicellulose etc.) act so that an impermeable film is created around non-hydrated grains of cement and for this reason, formation of calciumhydrosilicates (CSH) and portlandite (CH) are subsequently slowed down. Also the pressing the water out of the chips could result in drifting out ("pushing away" respectively) of the cement from the chip surface. This effect could participate on imperfect coverage of chips with cement paste which is connected namely with radical decrease of strengths. Objective of the research presented in the article is proving the effect of wood chips parameters on the hardening process of silicate matrix in cement-bonded particleboard.

2. Methodology of testing

Three formulas were tested in total. Differences were only in water distribution in the given mixture. Particularly the chips with humidity 30%, 60% and 90% were used (formulas - mixtures – W30, W60 and W90). For particular formulas the water amount was set in a way its content was the same in the resulting mixture for all the formulas. Therefore the mixing water amount was gradually reduced in dependence on growing chip humidity. In case of the last formula containing chips with 90% humidity the mixing water was completely eliminated and in addition also a part of the water necessary for dissolution of mineralizing components. The prepared mixture was layered in forms and subsequently pressed with thrust approximately 25 $N.mm^{-2}$. The mixture pressed in this way was moved into the steaming chamber. The specimens were removed from this ambient and taken out of forms after approximately 10 hours. In respect of the production of boards the very phase of mixture pressing can be considered as problematic one. With the mixtures containing chips with higher humidity content a water displacement may occur. At water leak also a certain amount of mineralizing additives or cement eventually could be drifted out. With the humidity contained in chips compared to the mixture where the "free" mixing water is included this risk is much higher. The reason is the fact that the water travels from chips (therefore over their surface) which contains water solution including mineralizing components and it is wrapped with cement paste. This would then negatively result in the hydration process course and therefore on final properties of cement-bonded particleboards. The hydration process was monitored at age of 2, 3, 8, 14 and 28 days. The mineralogical composition of the matrix was analyzed – identification and quantification of typical hydration phases. XRD and DTA were used for evaluation of crystalline phases. Microstructure was examined using SEM. Namely the board matrix subject of interest during microstructure investigation. At age of 8, 14 and 28 days also basic physical and mechanical parameters – bulk density, strength and modulus of elasticity in bending, transversal tensile strength vertically to board plane were analyzed according to CSN EN 310, 319 and 323.

3. Results and discussion

3.1. Physical and mechanical characteristics of tested boards

The physical and mechanical parameters were monitored at age of 8 days (after second heating cycle), 14 days (expedition of the boards in real production conditions) and 28 days (normative requirements). As it is clear from test results all the formulas are balanced with respect of the monitored properties (see table 1). The boards made from chips with 30% humidity are evaluated as the best ones. Further it is evident that with growing age the differences between particular formulas decrease. Also the comparison of strength parameters with modulus of elasticity is interesting. Here we can see differences in course of particular quantities. Bending strengths are characterized with growing tendency of dependence on growing input humidity of chips. On the contrary the highest tensile strengths vertically to board plane had the formula with chips of 60% humidity. Upon the course of physical and mechanical properties a slight influence of chip humidity can be assumed.

Table 1. Parameters of mixtures containing chips with different amount of humidity, age up to 28 days.

Formula – age (days)	Bulk density (kg.m^{-3})	Bending strength (N.mm^{-2})	Modulus of elasticity in bending (N.mm^{-2})	Transverse tensile strength (N.mm^{-2})
W30-08	1340	11.2	5326	0.54
W60-08	1360	10.6	5384	0.52
W90-08	1330	9.9	4992	0.51
W30-14	1310	12.2	5531	0.57
W60-14	1320	12.0	5491	0.55
W90-14	1310	11.3	5283	0.53
W30-28	1270	13.1	6359	0.63
W60-28	1260	12.9	6385	0.65
W90-28	1240	12.4	6302	0.59

3.2. Mineralogical composition and microstructure

Base phase components were identified in the silicate matrix by XRD. The emphasis was namely placed on clinker minerals and hydration products. Basically the identical mineralogical components with only imperceptible differences were identified in all analyzed samples. However the courses of particular diffraction lines are far more interesting. Two facts can be partially assessed from these courses. In the first line the peak intensity of identified components give information on their amount in the analyzed material. Further the intensity value of particular minerals characterizes their morphology - i.e. how is the given material developed. In the figures 1-4 there are selected diffraction schemes. Complete development of diffraction lines is shown in figure 1 (chips saturated by 30% water) and figure 2 (chips saturated by 90% water). For illustration representative samples analyzed at age of 2 and 28 days were selected (see figure 3 and 4). It can be seen very well how portlandite and calcite develop what is accompanied by decrease of diffraction maximums of clinker materials. The differences are also notable between the formulas. In figure 3 a quite significant difference can be seen at $2\theta = 18°$, which is a peak characteristic for portlandite (decreasing peak intensity with growing input chip humidity and lower amount of mixing water).

Table 2. DTA results – content of CaO from portlandite and calcite, age from 2 up to 28 days.

Age (days)	CaO – W30 *(%)*	CaO – W60 *(%)*	CaO – W90 *(%)*
2	9.31	8.54	8.59
3	14.64	13.85	12.71
8	15.94	16.18	14.79
14	17.76	17.12	16.85
28	19.21	18.63	18.05

For hydration grade quantification the content of CaO falling on $Ca(OH)_2$ (portlandite) and $CaCO_3$ (calcite) was selected. This value can be considered as relatively objective indicator of hydration reaction course. However the weight loss related to exothermic delay at temperature from 200°C to

IOP Conf. Series: Materials Science and Engineering **385** (2018) 012035 doi:10.1088/1757-899X/385/1/012035

500°C had to be excluded as first. Within this interval a thermal decomposition of mineralized chips occurred. Regarding objectivity the values were related only to the matrix during calculations. Results of setting of CaO content falling on portlandite and calcite are stated in the following table (see table 2). The resulting values of CaO content circa correspond to bending strength (see table 1).

Amount of CaO (falling on monitored components – $Ca(OH)_2$ and $CaCO_3$) clearly points to a slightly negative influence of chip humidity as filler for production of cement-bonded particleboards. Only with samples at age of 8 days the CaO content was increased with formula W60. Further it is clear that differences are stronger between the formulas W60 and W90. The influence of water amount reduction on account of necessary dose for mixing of mineralizing additives can be considered as well.

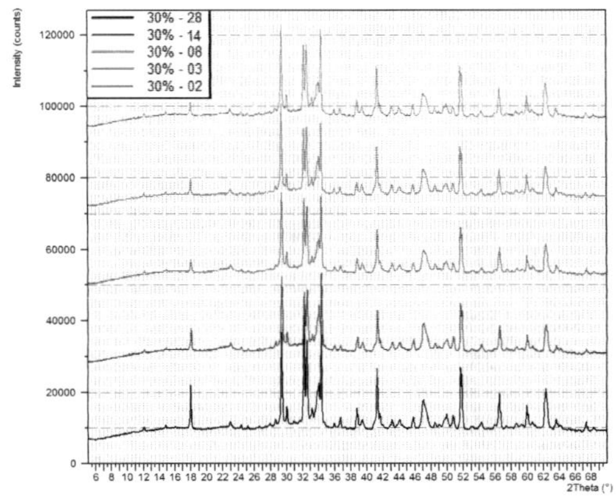

Figure 1. XRD patterns of W30 from 2 (top brown curve) to 28 days (bottom black curve).

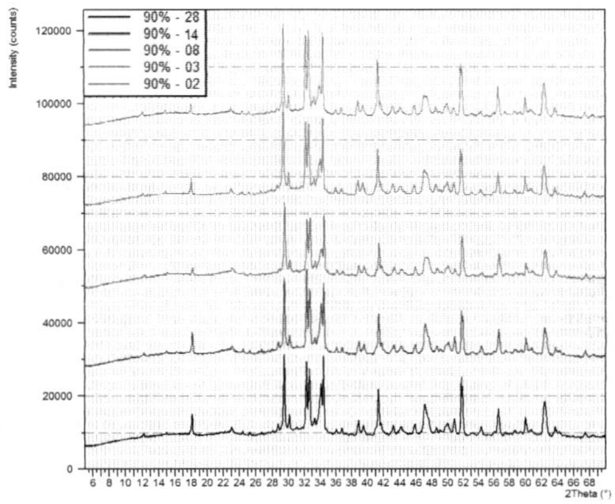

Figure 2. XRD patterns of W90 from 2 (top brown curve) to 28 days (bottom black curve).

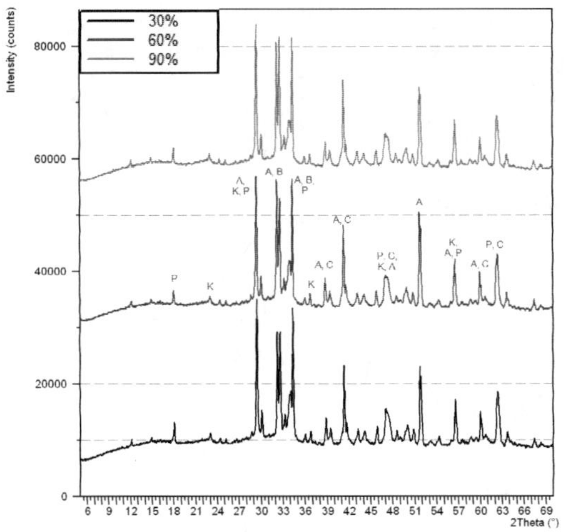

Figure 3. XRD patterns of mixtures W30-02 (bottom curve), W60-02 (middle curve) and W90-02 (top curve) (P-portlandite, K-calcite, A-alite, B-belite, C-celite).

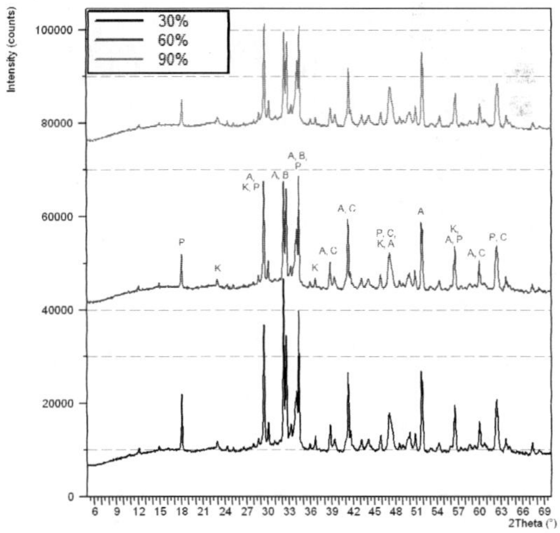

Figure 4. XRD patterns of mixtures W30-28 (bottom curve), W60-28 (middle curve) and W90-28 (top curve) (P-portlandite, K-calcite, A-alite, B-belite, C-celite).

Figure 5. Microstructure of (a) W30-02; (b) W60-02; (c) W90-02 – mag. 15,000×.

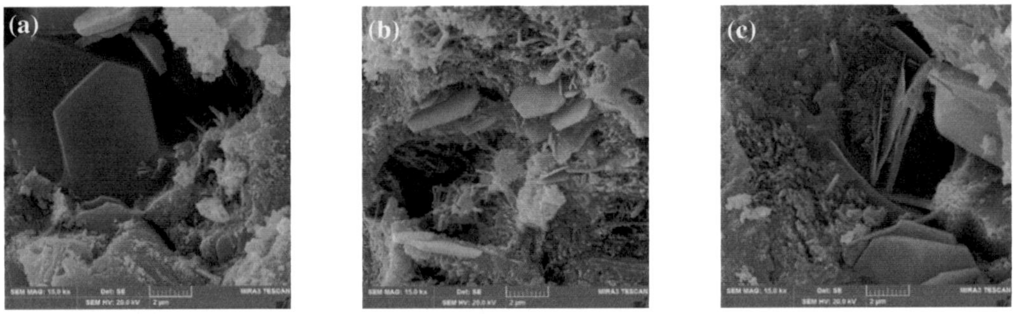

Figure 6. Microstructure of (a) W30-08; (b) W60-08; (c) W90-08 – mag. 15,000×.

Figure 7. Microstructure of (a) W30-14; (b) W60-14; (c) W90-14 – mag. 15,000×.

Figure 8. Microstructure of (a) W30-28; (b) W60-28; (c) W90-28 – mag. 15,000×.

The figures 5-8 show the differences in microstructure of particular formulas at age of 2, 8, 14 and 28 days. Developed crystals are perceptible mainly from 8-day age. Figure 6 represent clear difference in progress of portlandite crystals development whereas significant change of this mineralogical component proportion is evident in case of W30. General, in the case of the mixture W30 (i.e. 30% water saturated chips) it can be stated that portlandite crystals were identified slightly often within microstructure. Finding of clinker minerals was very difficult. Developed CSH phases in the gel form could be identified in all mixtures, however in boards W30 could be evaluated slightly better.

4. Conclusions

Moderately negative influence of chip humidity on hydration course and final properties of cement-bonded particleboards was proved by the performed experiments and their detailed assessment. This fact was confirmed either by physical, mechanical tests and physico-chemical analyses as well. The differences are more noticeable in case of application of chips with 90% humidity (W90), when the mixing water had to be completely eliminated and also partially the water necessary for mineralization of chips. It is probable that mineralization layer or coating of chips was upset and thus sugars contained in chips probably released (extraction of sugars from wood by water washing is one of the possible ways). This resulted in slowing down of the hydration process. This fact can be seen well with particular diffraction lines and further with quantified CaO amount falling on portlandite and calcite. In spite of the above mentioned they are only minor scale changes. In addition it can be stated that requirements in regard of board properties are met (the strengths and modulus of elasticity meet the minimum standard criteria) even with the formula W90. Considering also the microstructural analysis outputs it can be stated that probably no significant drifting out of cement paste wrapping chips occurred. With respect of the found knowledge it is necessary to monitor the water content in chips with aim to use chips with the lowest possible humidity. In case of chips with higher humidity it is necessary to suitably regulate the mixing water amount or amount of the water necessary for dissolving of mineralizing additives. The reason is to preserve the constant humidity of the fresh mixture of the cement-bonded particleboards and therefore the provision of required parameters (strengths, modulus of elasticity, bulk density etc.).

Ackowledgements

This paper has been worked out with financial support from the budget under the project FAST-S-18-5406, "Study of advanced silicate composites structure with emphasis on the characterization of secondary raw materials influence" and under the project No. LO1408 "AdMaS UP - Advanced Materials, Structures and Technologies", supported by Ministry of Education, Youth and Sports under the "National Sustainability Programme I".

References

[1] Maail R, Umemura K, Aizawa H and Kawai S 2011 *J. Wood. Sci.* **57** 302-307
[2] Wang L, Chen S S, Tsang D C W, Poon C S and Dai J-G 2017 *Journal of CO_2 Utilization* **18** 107-116
[3] Bentz D P, Coveney P V, Garboczi E J, Kleyn M F and Stutzman P E 1994 *Modell. Simul. Mater. Sci. Eng.* **2** 783-808
[4] Fan M Z, Ndikontar M K, Zhou X M and Ngamveng J N 2012 *Cement Concrete Compos.* **36** 135–140
[5] Quiroga A, Marzocchi V and Rintoul I 2016 *Composites Part B* **84** 25–32

IOP Conf. Series: Materials Science and Engineering **385** (2018) 012036 doi:10.1088/1757-899X/385/1/012036

Techniques of soil modification for re-use in construction

M Michalcikova and R Drochytka

Faculty of Civil Engineering, University of Technology, Veveri 95, 602 00 Brno, Czech Republic

Abstract. Nowadays effort is made to study what effects may impact the environment and to address the issue of waste management. The majority of all waste produced is construction waste and demolition debris (approx. 65%). This category includes soil excavated during the construction of e.g. foundations or utility infrastructure. In most cases, this soil finds no use at the original construction site and is landfilled. The challenge is to find a useful purpose for this soil. In order to re-use soils in construction, they must be modified. Currently, there are several techniques of soil modification, most notably compaction and stabilisation. Stabilisation is usually achieved by the addition of binders (lime, cement). A lesser known technology of soil modification is fluidification by means of fluidifiers, plasticisers, or other admixtures. In the Czech Republic, this technology is currently not being used. Within the following contribution is described the testing of two types of soil in which was verify the possibility of liquefaction by using binders and additives. As optimal was appeared the use of 1.5% lime, 4% cement and 1% plasticiser SVC-1062. The binders soil effectively stabilized and the plasticiser positively influenced the rheological properties of the mixture.

1. Introduction

A large amount of soil can be excavated during construction. Because of their lacking properties, these soils rarely find use at the construction site and are landfilled. The currently effective Act no. 185/2001 Coll. on Waste and the Amendment of Some Other Acts describes this issue and defines the usage of the soils [1]. The problematic qualities of soils are their softness, susceptibility to liquefaction and a high content of moisture and organic material. This prevents their re-use in construction and is the reason why they are disposed of at landfills, which is neither eco-friendly nor cost-effective. That is why people seek ways of modifying and re-using them.

There are several soil-modification techniques currently in existence. Methods of soil modification can generally be divided into two categories - mechanical stabilisation (e.g. vibration, compaction, barriers, etc.) and chemical stabilisation achieved by binders or other admixtures. Research presented in this paper focused on chemical stabilization [2-4].

A lesser known method of modifying soils is fluidification. It is a new technique, which is currently not very well known [5]. In the Czech Republic, it is seeing no use and abroad, it is likewise described and implemented only rarely. Mention of using fluidified soils is made e.g in papers published in Germany, Switzerland, France, Italy, the United Kingdom, USA, and Japan. Soil fluidification according to the researchers [5] involves the use of materials such as stabilisers (mostly binders), fluidifiers, or other alternative materials. There is a lack of knowledge concerning the type of admixtures used in this particular technology. Similarly, standards do not define specific requirements for soil fluidification, which makes it difficult to verify their properties. These are the reasons why this paper addresses the topic of soil fluidification.

2. Requirements for work with soil

Why is excavated soil considered waste? Currently effective legislation, i.e. Act no. 185/2001 Coll. on Waste and the Amendment of Some Other Acts [1], considers most excavated soils to be waste. This is due to the properties of the soils and the fact that they rarely find use at the construction site.
In addition, it is geotechnically not possible to obtain a soil that would meet construction requirements without needing any modification [2]. This sparks the effort to seek an optimal technology or methods that would improve the parameters of soils so that they may be re-used during construction. When choosing a technique, one must be aware of the purpose the soil in question is to fulfil.

2.1. Soil modification

As stated above, in order for it to be possible to re-use a soil, it must be modified. Methods of soil modification can be divided into two broad categories. The first is mechanical stabilisation (achieved by changing the physical properties of the soil particles e.g. by vibration compaction, barrier compaction, etc.) and chemical stabilisation (accomplished by a chemical reaction between a stabiliser/binder and minerals present in the soil) [2-3]. Another possible modification can be, for instance, the removal of an unsuitable fraction, removal of organic material, or size reduction of the clay parts of the soil. A lesser known technique is soil fluidification. The research described in this paper focuses primarily on chemical stabilisation.

2.1.1. Stabilisation. Stabilisation is a process meant to improve the properties of soils. The correct choice of stabiliser can improve e.g. its workability, compactability, strength, moisture content, etc. Stabilisation also results in a reduction in moisture content, improving the load bearing capacity (CBR_{sat}), increasing the modulus of deformation ($E_{def,2}$), reducing the index of plasticity (I_p), and decreasing frost susceptibility (β) [6-7].

The simplest contemporary processes of soil stabilisation are compaction and dewatering. Compaction is performed to improve the load-bearing capacity of soils, preventing consolidation, improving stability, etc. The purpose of dewatering is according to Rogers [3] to improve workability, because a high moisture content causes the soil particles to stick together, making the material more difficult to handle.

The best known type of chemical modification is according to researchers [3-4] a form of stabilisation aimed at adjusting a soil's water content and more importantly improve its strength. The stabilisation process thus results in a better resistance to softening. A number of additives are used for soil stabilisation, most commonly binders (lime, cement). The addition of lime improves e.g. workability, while cement can improve the strength of the soil [2]. Soils can be stabilised by the addition of other admixtures as well, e.g. fly ash, bitumens, etc. The choice of additive depends on the properties of the soil and its intended purpose [2, 6-7]. This is why the type and amount of binder added to the soil must be chosen on the basis of laboratory test results.

A number of studies have focused on soil stabilisation by the use of lime and cement [8]. They concluded that it the choice of type and amount of stabiliser is subject to moisture content, index of plasticity (I_p), and clay particle content (f). According to [9], lime (3 – 9%) is optimal for soil stabilisation if $I_P > 10$ and clay particle content $f > 10\%$, (3 – 10%) cement if $I_p \leq 10$ and $f < 20\%$, or a combination of lime and cement (3 – 10%), if $10 < I_p < 20$ and $f < 10\%$.

2.1.2. Fluidification. Fluidification is a process that reduces viscosity so that the soil has the highest possible flowability at a maximum content of solid particles [10]. Soil fluidification (especially clay soils) consumes a large amount of water, which is why researchers are searching for ways how this amount could be reduced without disrupting the optimum consistency. This can be achieved by fluidifiers, plasticisers, or superplasticisers. According to ISO 14688-1 [11] the correct type and amount of chemical agents is chosen based on the specific type of soil being modified and on the basis of laboratory tests. Soils with a higher content of siliceous components which would require more binder (especially cement), can benefit from the addition of plasticisers or superplasticisers. On the other hand soils with a higher clay content are better suited for modification by fluidifiers, which are used in the ceramic industry.

3. Testing methodology and materials

In the Czech Republic especially, flowable fills are very rarely used. This is partly due to the fact that they and their properties are not defined in any standards. The determination of their properties was thus performed using foreign standards, most notably ASTM (American Society for Testing and Materials). These standards list some requirements for flowable fills, known therein as CLSM (Controlled Low Strength Material) and means of their testing.

3.1. Testing Methodology

Flowable fills are generally divided to excavatable and non-excavatable. It is expected that excavatable fills will have lower strength in case they need to be removed to allow access to civil networks (water pipes, sewage, etc.). According to Adaska [12] a flowable fill should optimally reach a 20-day compressive strength of 0.2 MPa to 1.38 MPa. In order to perform our tests (in the Czech Republic), it was necessary to optimise testing in accordance with valid standards. Using Czech and foreign standards, suitable tests for determining specific properties (e.g. consistency, setting and hardening time, compressive strength, volume changes, etc.) were selected. This paper only presents the consistency and compressive strength tests.

Flowability is a primary requirement for flowable fills, since it is necessary for the mix to fill even the least accessible areas around civil infrastructure. Flowability testing was based on CSN EN 12350-8 Testing fresh concrete – Part 8: Self-compacting concrete – Slump-flow test [13]. The tests focused on flow values belonging in category SF2 (660 - 750 mm). This flow value is appropriate in terms of maintaining optimal flowability while keeping the right water content.

Compressive strength is another important property. As mentioned above, this property should not be too high so as to allow later excavation if needed. The compressive strength tests were performed with 100 mm cube specimens according to CSN EN 12390-3 [14] at an age of 28 days.

3.2. Materials

The research was conducted with two types of soil (saCl - a clay soil and fgrCSa - sandy clay soil), binders (lime, cement), plasticisers (based on polycarboxylates), and water. For comparison, the study also used reference soils without any binders or admixtures (saCl (REF), fgrCSa (REF)).

Soil is the main component of a flowable fill. Its properties determine the resulting consistency and strength. Table 1 shows the properties of the soils. According to the above-described specifications and a determination of binder amounts, the mixtures were made with quicklime produced by Kotouč Štramberk, spol. s r.o. at an amount of 1.5% and Portland cement CEM II/B-M (S-LL) 32.5 R from Českomoravský cement, a. s. in Mokrá at an amount of 4%. This cement contains 21 – 35% of added limestone and blast-furnace slag. Both types of binder readily reacted with the soils used. Moisture content was reduced and workability improved by the addition of plasticisers SVC-1062, SVC-21, and SVC-4088 at an amount of 1%. Tests showed that this quantity was optimal.

Table 1. Soil properties.

Name		Clay soil	Sandy clay
Symbol		saCl	fgrCSa
Liquid limit	w_L [%]	38.0	31.0
Plastic limit	w_p [%]	21.6	18.5
Index of plasticity	I_p [%]	16.4	14.3
Moisture content	w [%]	19.7	4.2
Consistency index	I_c [-]	1.12	0.00
Uniformity coefficient	C_u [-]	25.3	21.92
Coefficient of permeability	k [$m \cdot s^{-1}$]	$2.064 \cdot 10^{-9}$	$1.599 \cdot 10^{-4}$

4. Results and discussion

Drawing on experience gained during previous research, a set of raw materials suitable for fluidification was selected. Once mixed together, these components make up a mixture of very specific properties. In physical terms, it is a rather complex environment which merges three phases – solid, liquid, and gaseous. The ratio of the components affects their joint behaviour as parts of the whole. If we wish to examine the behaviour of flowable fills under their own weight or under external load after application in civil infrastructure, we must know the properties of each component in terms of its influence on the whole. These properties are most commonly determined by laboratory or in-situ tests. It is assumed that the mixture's rheological properties (consistency, flow value, etc.) will be most affected by fluidifiers, plasticisers, or superplasticisers. Hardening and final strength of the mixture was then affected by stabilisers. Figures 1 and 2 show the results of consistency and compressive strength tests.

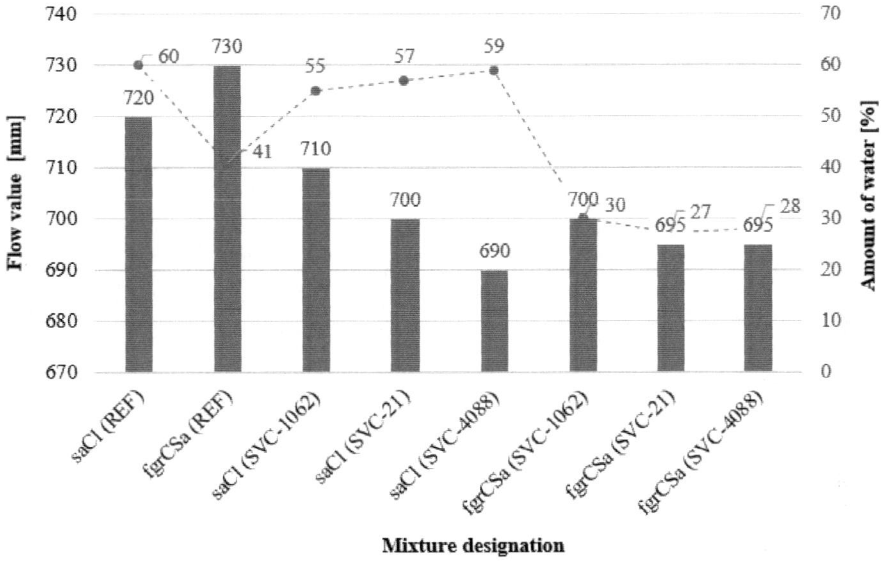

Figure 1. Flow value according to CSN EN 12350-8 and amount of water necessary for flowing.

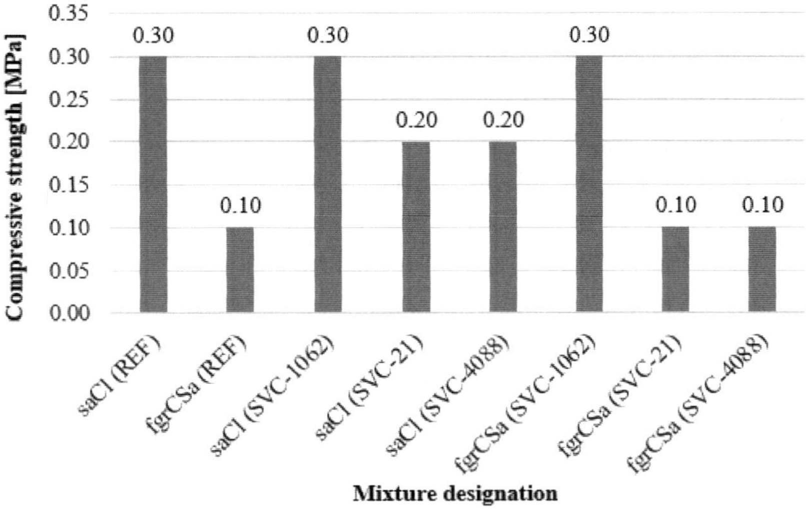

Figure 2. 28-day compressive strength according to CSN EN 12390-3.

Reference soils without any binders or admixtures are described as saCl (REF) – clay soil, fgrCSa (REF) - sandy clay soil. The mixtures containing 1,5% lime, 4% CEM II/B-M (S-LL) 32.5 R and 1% of plasticizer are described as saCl(SVC-1062), saCl(SVC-21), saCl(SVC-4088), fgrCSa(SVC-1062), fgrCSa (SVC-21), fgrCSa (SVC-4088).

The tests showed that even a small amount of lime (1.5%) resulted in a change in the properties of the soils, most notably a reduction in the moisture content of the clay soil (saCl). Moisture makes soils with a high clay particle content difficult to work with. However, the addition of lime made an improvement in these properties. The effect of lime manifested itself in hydration ($CaO + H_2O = Ca(OH)_2 + 15.5$ kcal), evaporation of some part of the water and a chemical reaction known as clay flocculation. In general, clay particles present in a soil tend to react with lime very slowly. These changes are visible in an improvement of geotechnical properties of the mixture (better compressive strength, saturated stability, frost resistance, etc.). After assessing the combined effect of lime and cement, it can be said that cement reacted better with the sandy clay (fgrCSa), which resulted in higher values of compressive strength.

In the fresh state, cement also proved to be an effective stabiliser. The stabilisation manifested itself mainly through a reduction in particle sedimentation during mixing and placement. The reaction between cement and the soil generated hydration products (soluble gels), which gradually crystallised and formed an interlocking matrix. Since cement CEM II/B-M (S-LL) 32.5 R also contained a small amount of free lime, the hydration and stabilisation were supported by its presence as well.

The flow test (figure 1) shows that the highest flow values were reached by mixtures that contained 1% of plasticiser SVC-1062 (saCl (SVC-1062) – 710 mm, fgrCSa (SVC-1062) – 700 mm). The flow value of both versions was within category SF2. This admixture had a greater effect in combination with the sandy soil. These mixtures also appear to require a smaller amount of water to produce a fluid mixture. The effect of the plasticisers, SVC-21 and SVC-4088, was not quite as visible. These results show that an optimal improvement in rheological properties was achieved by the addition of 1 % of plasticiser SVC-1062 to both soils.

The laboratory tests showed that the best way to improve compressive strength was to include cement. The measurements indicate that the compressive strength of soils can be modified by the amount of cement added. In excavatable clay soils where compressive strength ranges within 0.2 – 0.3 MPa the amount of 2% of cement (under normal conditions) should be sufficient. In sandy soils, due to their low load-bearing capacity, it is better to add 4% of cement, which, however, is costly. The progress of strength was also influenced by plasticisers, mainly SVC-1062. Mixtures which contained plasticiser SVC-1062 saw values of compressive strength up to 0.3 MPa (see figure 2). The use of plasticisers SVC-21 and SVC-4088 had a negligible effect on improving compressive strength.

5. Conclusion

The goal of the research described herein was to design and test soil modification by means of fluidification. It is a new technique, which is not yet broadly known. The soils were fluidified by the addition of suitable stabilisation materials and fluidifiers.

Laboratory tests showed that excavated soils contain some moisture, have varying grain size distribution, and may be contaminated by materials that prevent their re-use. This is why they need to be modified. The most suitable technique of modification appeared to be the addition of quicklime at an amount of 1.5% and 4% of Portland cement CEM II/B-M (S-LL) 32.5 R. The lime positively affected the moisture content of the soil by reducing it while improving workability. It readily reacted especially with the clay soil (saCl), forming hydration products that stabilised the mixture. Cement also proved to be an effective stabiliser, mainly in the sandy soil (fgr CSa).

In order for the mixtures to be used as flowable fills in civil infrastructure it is necessary that they are able to reach and fill all the required areas of the structure. This is the reason for the effort to find a mixture with a flow value of at least class SF2 (660 – 750 mm) and SF3 (760 – 850 mm). The flow value can be increased either by adding water or plasticisers/fluidifiers. The experiment showed that the use of plasticiser SVC-1062 at an amount of 1 % proved to be effective. This admixture had a positive effect on increasing the flow value as well as improving compressive strength. The results show that the correct choice of binders and plasticizers can positively affect the properties of the soils and can therefore be reused. The effect of other types of plasticisers/fluidifiers in combination with different soils will be subject of future research.

Acknowledgments
This paper has been worked out under the project of specific research r. n. FAST-J-18-5347 "Study of the usability of soils from excavation works and the possibility of their modification by flowing form".

References
[1] ACT no. 185/2001 2001 *Coll., on Waste and the Amendment of Some Other Acts*. Act of the Czech Republic (in Czech)
[2] Keller Inc. Keller 2018 *Improvement of Weak Soils by the Deep Soil Mixing Method* Keller Bronchure
[3] Rogers C D F, Glendinning S 1993 *Modification of clay soils using lime* (London: Thomas Telford) p 99-114
[4] EuroSoilStab 2002 *Development of Design and Construction Methods to Stabilize Soft Organic Soils: Design Guide for soft soil stabilization.* CT97-0351, European Commission, Industrial and Materials Technologies Programme, Bryssel
[5] Liquid soil – A Technical challenge – UNITRACC 2018 - Underground Infrastructure Training and Competence Center 2018 *Liquid Soil – a technical Challenge* (Germany: Knowledge Factory GmbH) http://www.unitracc.com/aktuelles/artikel/liquid-soil--a-technical-challenge-en (online)
[6] Kaun M and Lehovec F 2004 *Roads 20* (Prague: CVUT) p 238 (in Czech)
[7] Little D N 1995 *Handbook for stabilization of pavement subgrades and base courses with lime* (Iowa: Kendall Hunt Pub. Co.) p 219
[8] Al-Rawas A A and Goosen M F A 2006 *Expansive Soils- Recent Advances In Characterization And Treatment* (London: Taylor & Francis Group) p 540
[9] Bell F G 1987 *Ground Engineer's Reference Book* (London: Butterworths) p 1-17
[10] Sokolar R 2006 *Ceramics. Module BJ01-M01* (Brno: University of technology) p 176
[11] CSN EN ISO 14688-1: 2018 *Geotechnical investigation and testing - Identification and classification of soil Part 1: Identification and description*
[12] Adaska W S 1997 *Concrete International* **19** 41-43
[13] CSN EN 12350-8: 2010 *Testing fresh concrete Part 8: Self-compacting concrete – Slump-flow test (in Czech)*
[14] CSN EN 12390-3: 2009 *Testing hardened concrete Part 3: Compressive strength of test specimens (in Czech)*

IOP Conf. Series: Materials Science and Engineering **385** (2018) 012037 doi:10.1088/1757-899X/385/1/012037

Effect of polyethylene glycol on the rheological properties and heat of hydration of alkali activated slag pastes

O Mikhailova and P Rovnanik

Faculty of Civil Engineering, Brno University of Technology, Veveri 95, Brno, Czech Republic

Abstract. In this work the effect of two types of polyethylene glycol on the rheological properties of alkali activated slag pastes have been investigated. Polyethylene glycols were incorporated in the alkali activated slag pastes in quantities ranging between 0.5 and 5% by mass of slag. All rheological measurements were performed on all pastes with water-to-slag ratio of 0.4. Rheological parameters (shear stress and viscosity) of obtained fresh pastes were studied. Rheological behaviour of pastes modified with polyethylene glycols fits the Newtonian model. In this study also isothermal calorimerty test were performed with different amount of polyethylene glycols concerning time. Results showed that the heat flow generally decreases with increasing polymer dosage.

1. Introduction

Portland cement based materials are the most used construction materials worldwide. However, in recent years new systems have been developed to reduce the cement content due high energy consumption of Portland cement production and high greenhouse gas emissions. One of the alternative materials to Portland cement is alkali-activated slag (AAS), which has shown comparable performance while offering major reductions in CO_2 emissions compared to Portland cement [1-3].However, AAS provides some challenges regarding its rheological properties, and the workability is considerably lower at the same water to solids mass ratio compared to Portland cement [4].

The rheological behavior of Portland cement pastes has been studied in detail. In the case of alkali activated materials this research field is ongoing yet. Chemical admixtures, which have been proved effective for cement-based materials, are not suitable for alkali-activated systems [5]. Very few studies have been conducted on alkaline activated slag pastes rheology. Palacios et al. [6] observed that polycarboxylate-based, melamine formaldehyde derivative, naphthalene formaldehyde derivative, vinyl copolymer, and shrinkage-reducing agent do not significantly modify rheological parameters. In other studies [7], the rheological properties and hydration process of slag-blended cement pastes in the presence of polycarboxylate superplasticizers with different structure were studied. Polycarboxylate-based admixtures do not modify significantly the plastic viscosity of any of studied pastes.

A.S. El-Dieb [8] investigated the effect of using water-soluble polymers as polyethylene glycol (PEG) and poly acrylamide. Investigation showed a denser microstructure and a lower tendency for microcrack formation. Basil M Joseph [9] studied influence of PEG 400 on the Portland cement concrete and found that the optimum dosage of PEG 400 for maximum strengths (compressive, tensile and modulus of rupture) is 1%.

Another important characteristic for cementitious materials is hydration parameters. These parameters are affected by polymers admixtures. Calorimetric tests are usually used to capture these hydration parameters. Isothermal calorimetry is used to monitor heat flow changes during alkali-activation reaction. Group of authors [10] concluded that higher and earlier heat production rate due to hydration is found for higher amounts of polyethylene glycol added to the concrete.

Since AAS have an important disadvantage – significant shrinkage followed by formation of microcraks, the main aim of present study is to investigate the rheological behavior of AAS pastes in relation to the effect of polyethylene glycol's amount and how heat release changes during time while changing rheological properties The present study provides new information regarding the effect of polyethylene glycols and their dosage on the rheology and hydration process of AAS pastes by performing isothermal calorimetry.

2. Experimental

2.1. Materials

In this study Czech ground-granulated blast-furnace slag (Kotouč Štramberk, spol. s r.o.) was used as a primary raw material. The chemical composition of slag is given in table 1. The specific surface of the slag was 380 m^2/kg. The particle size, determined through laser granulometry was 19 μm (D50). The commercial solid sodium water glass SUSIL ($SiO_2/Na_2O = 2.0$) was used as alkaline activator.

Table 1. Chemical analysis of blast furnace slag, [%].

SiO_2	Al_2O_3	Fe_2O_3	CaO	MgO	Na_2O	K_2O	MnO	Cl	LOI
39.66	6.45	0.47	40.12	9.5	0.33	0.55	0.65	0.05	1.25

The effect of two types of commercial PEG (ROTIPURAN®, Ph.Eur), with different molecular weight was used at different concentration: PEG 6000 and PEG 20000 as modifier. PEG is a water soluble linear polymer [11] condensated of ethylene oxide and water with the general formula $H{-}(O{-}CH_2{-}CH_2)_n{-}OH$, where n is the average number of repeating oxyethylene groups [12] with n=5–182 for typical PEGs (average molecular mass 200-8000 g/mol). PEG 6000 is powdery polymer with molecular mass 6000 g/mol. PEG 20000 is solid polymer in the form of scales with molecular mass 20000 g/mol. Both types of PEG were added at dosage from 0.5 up to 10% by mass of raw material.

2.2. Rheological measurements

All rheology measurements were performed on all pastes with a water-to-slag ratio of 0.4. The reference (REF) AAS paste mix design represents mix of slag (300 g) and activator (60 g). Other tested (experimental) AAS pastes were modified by different types of PEG at dosage 0.5, 1 and 5% by mass of slag. The addition of 10 % wasn't used since the consistency of the mixture was stiff (at given w/s ratio) what is not sufficient for rheological measurements in used rheometer. The activator and polymer admixtures were dispersed and partially dissolved in water prior to mixing with slag.

Rheological measurements were performed at 25 °C on the stress-controlled rheometer Discovery HR-1. A gap thickness of 20 mm was selected for the paste. Rheological properties were measured using 2 program methods: Flow Sweep and Flow Ramp. For both program measurements, the pastes were initially mixed by hand in a plastic bowl with a spatula for approx. 2-3 minutes, depending on the polymer dosage (lower dosages required more time), and then they were inserted into the sample holder of a coaxial cylinder device mounted on the rheometer.

Flow Sweep procedure is based in measuring shear stress at changing shear rate. The rheological properties of AAS pastes were determined from curves at increasing and decreasing rates of shearing in the range 0.1−100 s^{-1}. The paste was pre-sheared for 60 s at 100 s^{-1}. After a rest period of 60 s, the rheological measurements were started. The testing routine consisted of a shear rate increase (from 0.1 to 100 s^{-1}) applied through 30 steps with 15 s of measuring time at each shear rate followed by a decrease of shear rate at the same conditions. The results were expressed as shear rate vs. shear stress.

With the Flow Ramp method shear stress at a constant shear rate in AAS paste was measured. This procedure started with pre-shearing at 100 s^{-1} for 1 min and was followed by ramping at the constant shear rate of 100 s^{-1} during 30 min. The results were expressed as shear stress vs. time.

2.3. *Isothermal calorimetry test*

TAM Air calorimeter (TA Instruments) was used to measure the generated heat in AAS pastes when different dosages of polyethylene glycol were used. For this purpose, several mixtures were prepared, containing slag (4 g), activator (0.8 g), water (2 g) and different types of PEG at dosages 0.5, 1, 5 and 10% by mass of slag. Mixing of each composition was performed in identical glass vials at 25 °C, and then placed into calorimeter. Heat flow was measured at bath temperature of 25° C for 235 hours for each sample.

3. Results and discussions

3.1. *Rheological properties*

Figure 1 shows the shear stress versus shear rate flow curves for AAS pastes with addition of PEG 6000 and PEG 20000.

a) b)

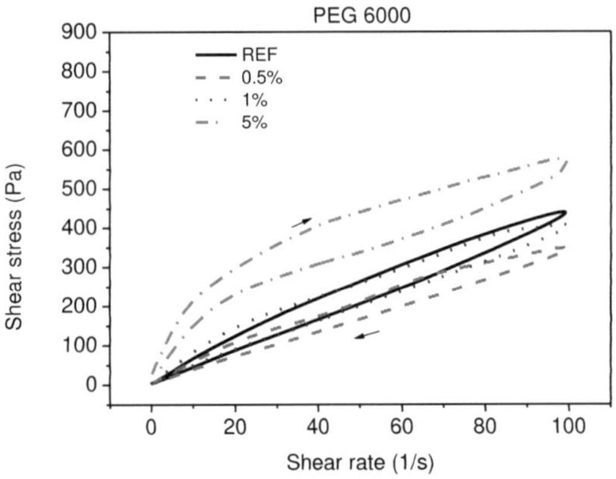

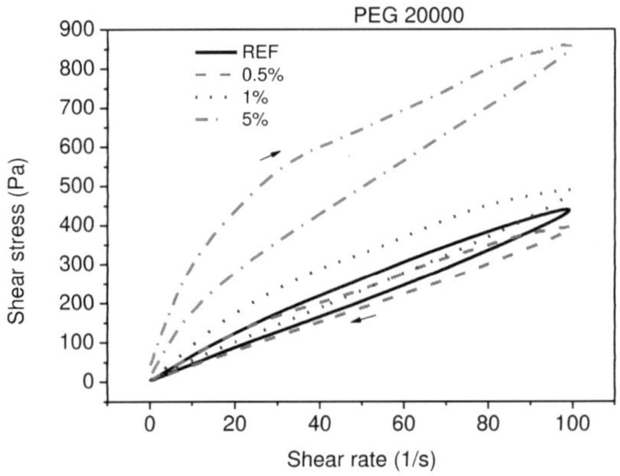

Figure 1. Flow curves of AAS pastes with PEG 6000 (a) and PEG 20000 (b) at different dosage.

The shear stress values found for pastes modified with PEG 6000 (figure 1a) at 0.5 and 1% dosages are lower than those observed for REF paste. The addition of 5% of PEG 6000 has higher value of shear stress. However, 0.5% of PEG 20000 reduces values of shear stress (figure 1b). Whereas the addition of 1 and 5% of PEG 20000 increases the shears stress values. 5 % of both types of PEG significantly change the viscosity of the pastes. Increasing the amount of PEG affected shear stress.

All AAS pastes containing PEG display Newtonian fluid properties, since yield stress was not indicated. When the relationship between the shear stress τ and the shear rate γ is a constant, the material constitutes a so-called Newtonian fluid [13]. Plastic viscosity was calculates using Newtonian model and presented in table 2. The viscosity of the activation solutions was determined by fitting the rheological data to the Newtonian flow model shown in equation (1):

$$\tau = \eta \cdot \gamma \qquad (1)$$

Table 2. Viscosity of AAS pastes, depending on type and dosage of PEG.

	Viscosity [Pa·s]		
Type and dosage of admixtures	0.5%	1%	5%
REF	4.23	4.23	4.23
PEG 6000	3.37	3.98	5.99
PEG 20000	3.80	4.69	9.10

Figure 2 and 3 shows the variations in shear stress at a constant shear rate in pastes with and without PEG 6000 and 20000. Shear stress is lower in REF pastes than in pastes with different amount of PEG 6000 (figure 2). In the AAS pastes the initial shear stress rose with concentration of PEG 6000. Furthermore, shear stress values for pastes with PEG 6000 decline slightly over time and by the end of the test progressively return to almost the initial values.

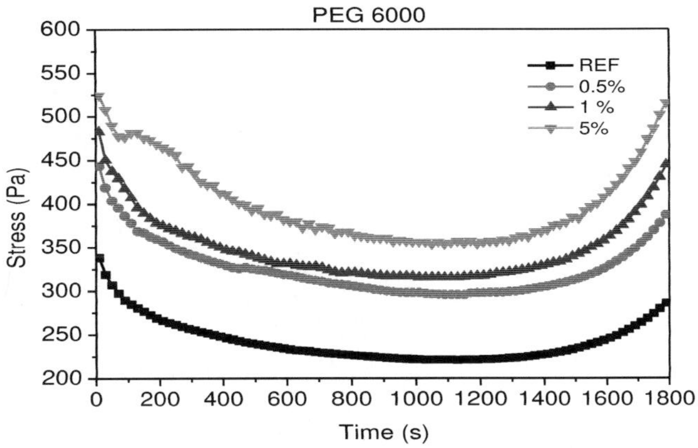

Figure 2. Shear stress versus time of AAS pastes containing PEG 6000.

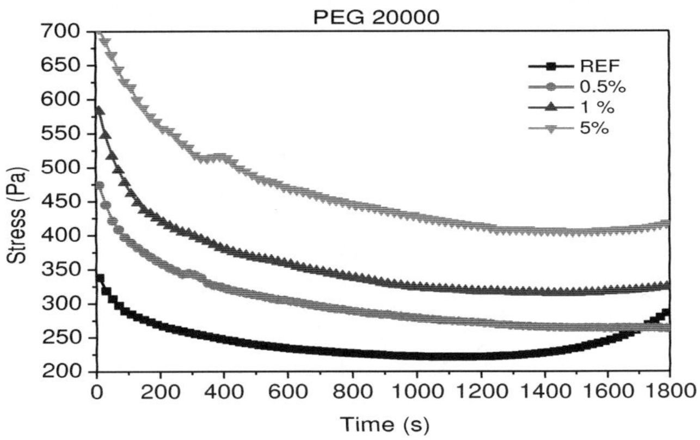

Figure 3. Shear stress versus time of AAS pastes containing PEG 20000.

PEG 20000 raised the initial shear stress values with increasing its concentration (figure 3). The addition of 5 % of PEG 20000 doubled the initial shear stress value in comparison with REF values (from 350 to 700 Pa). Also shear stress values for AAS pastes with addition of PEG 20000 decline over time to the end of the test. However, after 1200 s remained substantially constant for 1% and 5% contents.

3.2. Isothermal calorimetry

The isothermal test results of normalized heat flow of all different mixtures are illustrated in figures 4 and 5. The graphs are presented the initial heat peak for the first 3 hours (figure 4a and 5a) and peak which appeared in period from 20 to 120 h (figure 4b and 5b). The heat peak which was discovered after the first 30 min (figure 4a and 5a) is common to all alkali-activated systems as well as to the majority of other cements [14].

The addition of PEG 6000 reduced heat flow values in comparison with REF values (figure 4b). The addition from 1 to 10 % of PEG 6000 shifted the heat flow peak to the left compared to the REF peak. However, PEG 6000 at dosage 0.5% did not change heat flow peak and heat flow value significantly. The greatest effect had 10% addition of PEG 6000. First peak was higher (figure 4a), however the second peak (figure 4b) was significantly lower and was significantly shifted to the left. It shows that heat generation accelerates with the increasing concentration of PEG 6000.

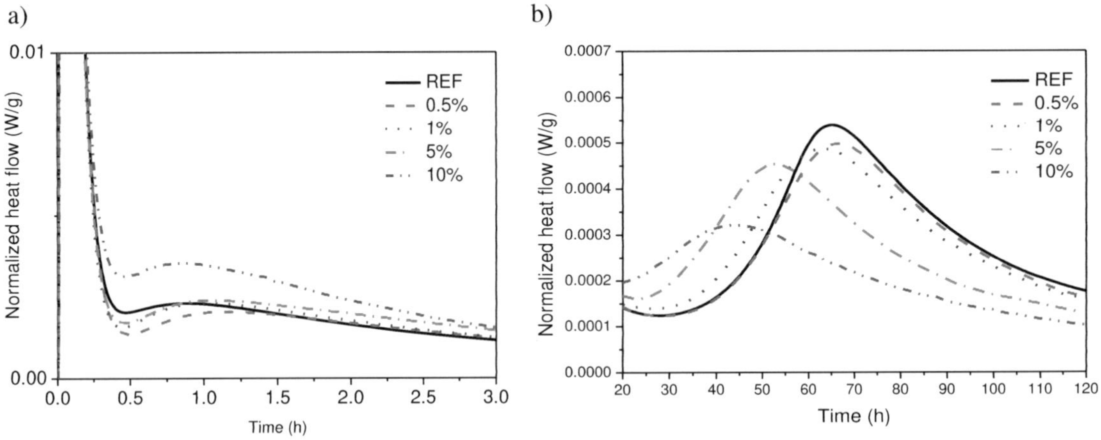

Figure 4. Calorimetric major peaks from 0 to 3 h (a), from 20 to 120 h (b) of sample with PEG 6000.

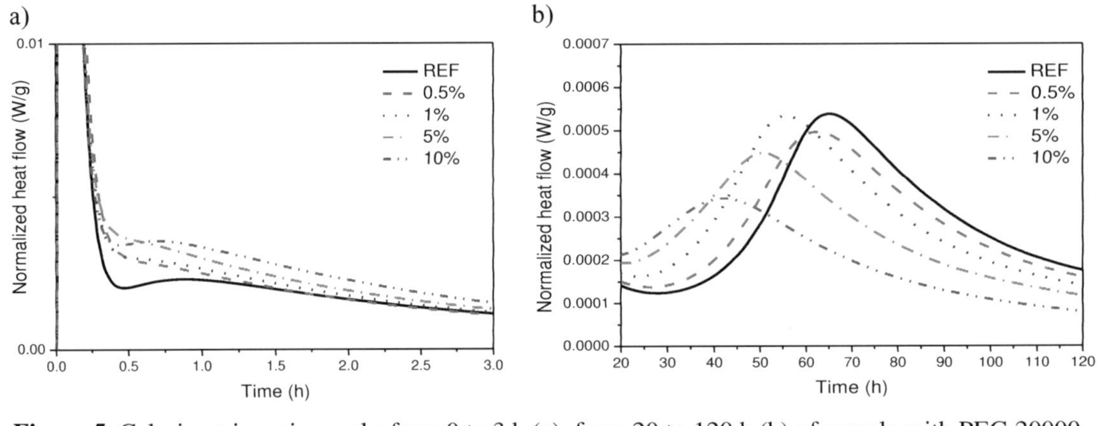

Figure 5. Calorimetric major peaks from 0 to 3 h (a), from 20 to 120 h (b) of sample with PEG 20000.

The addition of PEG 20000 at concentration 0.5, 5 and 10 % showed that heat flow decreased compare to REF sample (figure 5b). 1% of PEG did not affect to the heat flow values. Results showed that mixtures with addition of PEG 20000 mostly shifted peaks to the left side of REF mix; however the addition of 0.5% didn't shift significantly. Moreover, with the increasing the concentrations of PEG 20000 the peaks were shifted more to the left, which indicates accelerating effect on the hydration.

It can be concluded that alkali activated reaction in samples with addition of PEG 6000 and PEG 20000 starts earlier (second peak at figure 4b and 5b). At the same time intensity of reaction reduces to compare with REF sample, since the heights of heat peaks become lower and less sharp. This phenomenon can be explained, that PEG covers the slag grains what can lead to reduction the intensity of alkali activation. Due to the fact that the reaction of alkali activation take place on the slag surface and in the liquid regions [15] This reaction process will affect inter-particle forces what can result to slow formation of calcium-sodium aluminosilicate hydrate gel (the main product of alkali activation reaction) [16].

4. Conclusion

This paper presented procedures and analyses for elucidating rheological and hydration characteristics and the key conclusions are listed below:

- An analysis of the present findings showed that the polymers PEG 6000 and 20000 affected AAS pastes rheology depends on its concentration. The shear stress values were higher with increasing the concentration of both types of PEG. The rheological behaviour in AAS pastes fit a Newtonian model.
- An increase in both types of PEG dosage results in faster reaction in AAS pastes, with the calorimetric data showing main peak at earlier time than REF paste mix. However, the employment of PEG effects on reduction the intensity of reaction of alkali activation, since the height of main peak is higher and sharper of REF samples than peaks of modified samples by PEG.

Acknowledgements

This paper was elaborated with the financial support of research foundation of Brno University of Technology under the specific research project FAST-J-18-5568.

References

[1] Leonelli C and Romagnoli M 2015 *Handbook of Alkali-Activated Cements, Mortars and Concretes* ed Pacheco-Torgal F, Labrincha J A, Leonelli C, Palomo A and Chindaprasirt P (Cambridge: Elsevier) chapter 6 pp 133-170
[2] Bakhareva T, Sanjayana J G and Chengb Y B 2002 *Cem. and Con. Res.* **32** 211-216
[3] Shi C, Roy D and Krivenko V P 2003 *Alkali-Activated Cements and Concretes* (London: CRC Press) p 1-333
[4] Feys D, Cepuritis R, Jacobsen S, Lesage K, Secrieru E and Yahia A 2018 *Rilem Technical Letters* **2** 129–135
[5] Puertas F, Varga C and Alonso M M 2014 *Cem Concr Compos* **53** 279-288
[6] Palacios M, Marta, Banfill, Phillip and Puertas F 2008 *ACI Mater J* **105** 140-148
[7] Palacios M, Puertas F, Bowen P, Houst Y F 2009 *J Mater Sci* **44** 2714–2723
[8] Amr S, El-Dieb Tamer A, El-Maaddawy A and A M Mahmoud 2012 *Adv. in Cem. Res.* **24** 291-299
[9] Basil M J 2016 *Iosr-Jmce* 12-17
[10] Thiru Chelve S R, Sivakumar S, Kumar M Raj, Shanmugaraja G, Nallathambi Vishnu P M 2017 *Int. Jour. of Inn. Res. in Sc, Eng .and Tech.* **6** 3521-3524
[11] Hirata T, Ye J, Branicio P, Zheng J, Lange A, Plank J and Sullivan M 2017 *Scien. Rep.* **7** 1-10
[12] Brady J, Dürig T, Lee P I, Li J X 2017 *Pharm. Th. and Pr* 183-223
[13] Wallevik O H, Wallevik J E 2011 *Cem. and Con. Res.* **41** 1279–1288
[14] Qinwu X, Jiong H, Ruiza J M, Wang K and Ge Z 2010 *Thermochimica Acta* **499** 91-99
[15] Stein P, Poulesquen A, Diat O, Frizon 2012 *Langmuir* **28** 8502-8510
[16] Kashania A, Provis J L,Qiaoa G G, Jannie S J van D 2014 *Constr. Buld. Mater.* **65** 583-591

IOP Conf. Series: Materials Science and Engineering **385** (2018) 012038 doi:10.1088/1757-899X/385/1/012038

Raw materials for mullite grog manufacturing

L Nevrivova

Institute of Technology of Building Materials and Components, Faculty of Civil Engineering, Brno University of Technology, Veveri 95, 602 00 Brno, Czech Republic

Abstract. The starting raw materials for aluminosilicate grog manufacturing are fissile claystones and shales. These raw materials are available in the Czech Republic and grogs with up to 45 % aluminium oxide content are manufactured. However, refractory production cannot do without higher-Al_2O_3 -content grogs. This group includes grogs with alumina content of 45 % to 70 %, which constitute a transition between fire clay and mullite grogs. Because of the high market price of these grogs, however, there is an effort to seek ways to manufacture these in the Czech Republic. The primary condition for manufacturing is the correct choice of raw material, a suitable treatment facility, and a kiln capable of firing at temperatures above 1400 °C. This paper assesses some local raw materials for grog manufacturing and compares the results with a mullite grog of similar properties that is available on the European market.

1. Introduction
Refractory materials can be divided according to the dominant oxide content (silica, aluminosilica, alumina, magnesia, etc.), the manufacturing technology (shaped or unshaped), the temperature they can withstand, materials they contain (corundum, mullite, sillimanite, etc.), bulk density (dense, insulating), etc. The refractoriness of these aluminosilicate grogs depends mainly on Al_2O_3 content, which is also how they are classified; see table 1. The grog with a high Al_2O_3 content is used in the manufacturing of a great many shaped, as well as unshaped, refractories. The properties of the final product are determined by the chemical and mineralogical composition of the grog and its density. This paper discusses an aluminosilicate with Al_2O_3 content above 45 %. Currently, this grog is being imported to the Czech Republic and is used in the production of a number of refractory materials, such as blast furnace stack lining, torpedo ladles, backup lining in steel ladles, reheating furnaces in ferrous metallurgy, lime kilns, melting and holding furnaces, pot furnaces, anode baking furnaces in aluminium metallurgy, and many other areas where the use of these refractories is unavoidable [2].

This paper assesses several local raw materials suitable for grog manufacturing and compares the results with mullite grogs available on the European market.

2. Mullite Characterization
The only aluminosilica crystalline phase which is stable at high temperature is mullite, most commonly expressed as $3Al_2O_3.2SiO_2$. Mullite is a very stable compound with a high melting point. It also has very low thermal expansion, good strength, and high modulus of elasticity. This is why it is usually desirable that the mullite content be as high as possible. The mullite content of a material is determined primarily by the chemical composition of raw materials and firing temperature. Mullite begins to form typical needle crystals (figure 2) at temperatures above 1300 °C.

Sintered mullites are produced from different raw materials which are mixed in powdered form and then sintered at an optimum temperature after compaction. Oxides, hydroxides, salts and silicates are used as starting materials. Mullitization takes place by solid-solid or transient liquid phase reactions of the starting materials by aluminium, silicon and oxygen atom inter-diffusion [2]. These mullites tend to have "stoichiometric", i.e., 3/2 - composition ($3Al_2O_3 \cdot 2SiO_2$, ≈72 wt. % Al_2O_3).

Fused mullites are commercially produced by melting the starting materials above 2000 °C, casting the aluminium silicate melt into the ingot moulds and cooling it to room temperature. Raw materials with low impurity levels, e.g. Bayers alumina, quartz sand and fused silica are used. For lower quality mullite, bauxite and a mixture of alumina and kaolinite have also been used. These mullites tend to be Al_2O_3-rich with approximate 2/1-composition ($2Al_2O_3 \cdot SiO_2$, ≈ 78 wt. % Al_2O_3).

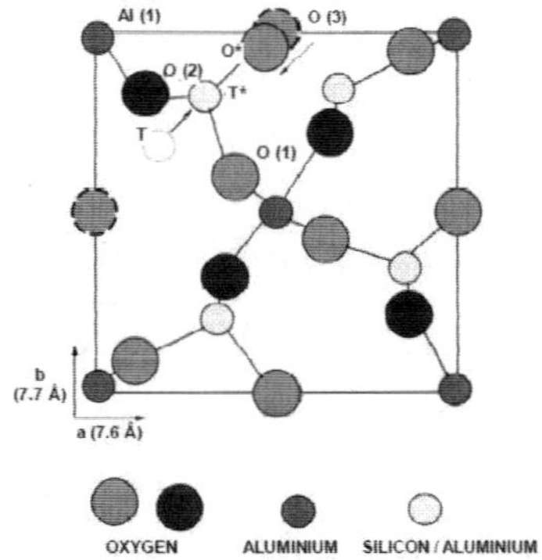

Figure 1. Atom positions in unit cell of mullite [4].

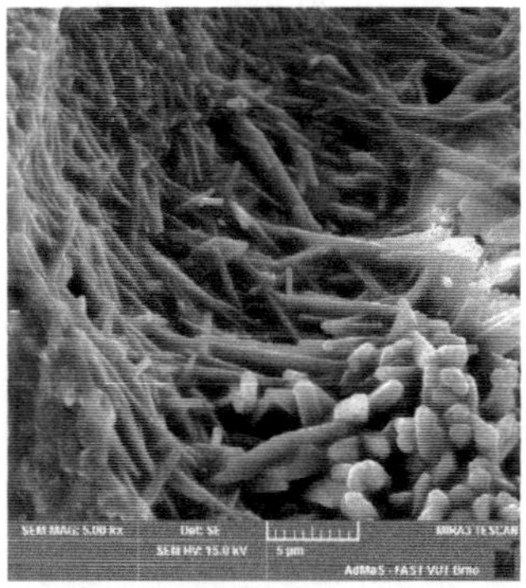

Figure 2. Scanning electron micrograph of mullite, firing temperature 1420 °C.

Chemical-mullites are produced by heat treatment of organic or inorganic precursors prepared by advanced processing techniques, such as sol-gel, hydrolysis, precipitation, spray pyrolysis, chemical vapour deposition, etc. The composition of such mullites strongly depends on the starting materials and the temperature treatment. Extremely Al_2O_3-rich compounds (>90 wt.% Al_2O_3) have been identified at synthesis temperatures <1000 °C [1].

Mullite has orthorhombic crystal structure with unit cell dimensions a = 0.7540 nm, b = 0.7680 nm and c = 0.2885 nm for the stoichiometric composition [3]. Mullite is intermediate in composition between alumina (Al_2O_3) and sillimanite (Al_2SiO_5). It can be seen from figure 1 that mullite consists of chains of AlO_6 octahedra at the edges and the centre of the unit cell. These chains are joined by Al_2SiO_4 tetrahedra [4].

3. Mullite grogs available on the European market

3.1. *Symulox*

Mullite identified as M72 contains 72 % Al_2O_3 and is typically extremely high in purity, which is its primary difference from natural aluminosilicates. Products made with clinkered mullite have a high chemical resistance and excellent refractory properties. Good strength combined with a low thermal expansion results in outstanding thermal shock resistance. The products can be applied at temperatures up to 1700 °C. The bulk density of Symulox is 2800 $kg \cdot m^{-3}$, water absorption is around 1 %, and open porosity is 1.5 %. It contains 90-95 % mullite, 5-10 % glass phase and 1% corundum [5]. Other products from this product line are sintered mullite Z72 or reactive aluminium oxide named M 672.

Table 1. Chemical and mineralogical composition of mullite grog Symulox M72.

Oxide	Content [%]
Al_2O_3	72.00
SiO_2	26.00
TiO_2	0.20
Fe_2O_3	0.30
CaO	0.05
Na_2O	0.20

Mineral	Content [%]
Mullite	90 - 95
Glass phase	5 - 10
α-corundum	1

3.2. Virginia Mullite
This type of mullite is produced by kyanite calcination at a temperature of 1450 °C. The material thus produced, known as Virginia Mullite, consists of approx. 80 % mullite, 11 % amorphous glass, 7 % quartz and less than 1 % cristobalite. It differs from other synthetic mullites in particle shape and material purity, which is higher than in mullites produced by clay mineral calcination [6].

Table 2. Chemical and mineralogical composition of Virginia Mullite grog.

Oxide	Content [%]
Al_2O_3	58.00
SiO_2	40.20
TiO_2	1.10
Fe_2O_3	0.50
CaO	0.04
Na_2O	0.10

Mineral	Content [%]
Mullite	79-85
Glass phase	8-12
Quartz	4-8
Cristobalite	<1

3.3. Vulcasil H
Vulcan Refractories produce a broad spectrum of materials containing 55-99.5 % of Al_2O_3. Their fused mullite is known for its high purity, resistance to thermal shock and chemical resistance. The HRP group of grogs contains 79 % Al_2O_3 and 21 % SiO_2. The bulk density is 2700 $kg{\cdot}m^{-3}$ and apparent porosity is around 18 % [7].

Table 3. Chemical composition of Vulcasil H.

Oxide	Content [%]	Oxide	Content [%]
Al_2O_3	78.0	Fe_2O_3	< 0.5
SiO_2	20.0	CaO	< 0.5
TiO_2	< 0.5	Na_2O	< 0.5

3.4. Henan Cunse Refractory - M47
The Chinese company Henan Cunse Refractory from Dengfeng offers a great many products and materials suitable for refractory roles. Grogs containing mullite are identified with an M followed by a number that specifies the average content of Al_2O_3. The company also offers grogs with a reduced Fe_2O_3 content; they are identified by the letters DM followed again by a number denoting the content of Al_2O_3.[8] Grog M47 has a bulk density of 2600 $kg.m^{-3}$, apparent porosity of 3 % and a refractoriness of 1780 °C.

Table 4. Chemical and mineralogical composition of a mullite grog from Henan Cunse Refractory.

ID	Al_2O_3 [%]	Fe_2O_3 [%]	K_2O+Na_2O [%]	Mullite content [%]
M47	45-49	<1	<0.3	55-65
M60	58-62	<1.3	<0.3	70-80
M70	68-72	<1.5	<0.3	85-90

DM45	43-47	<0.4	<0.3	50-60
DM60	58-62	<0.5	<0.3	70-80
DM70	68-72	<0.6	<0.3	85-90

4. Testing raw materials for suitability in grog production

Aluminosilicate grogs are typically produced from refractory clays, claystones, shales and kaolin.

4.1. Raw materials

This investigation tested passing fractions of W Super claystone, which had been heaped for half a year, 5 years and 10 years. Since claystone W Super contains a maximum of 44 % of Al_2O_3, a waste mullite produced by MOTIM Electrocorundum Ltd. was also tested as a potential source of Al_2O_3. Tables and diagrams refer to the waste mullite Motim by the letter M.

Table 5. Chemical and mineralogical composition of waste mullite Motim, claystone W Super and a test mixture containing 90 % of W Super and 10 % of Motim.

Motim	Content [%]	W Super	[%]	90 % W Super 10 % Motim	Content [%]
Al_2O_3	73.470	Loss on ignition	-13.51	Al_2O_3	46.78
SiO_2	22.710	SiO_2	52.92	SiO_2	49.67
TiO_2	0.014	Al_2O_3	43.17	TiO_2	1.63
Fe_2O_3	3.295	TiO_2	1.83	Fe_2O_3	1.02
CaO	0.008	Fe_2O_3	0.73	CaO	0.14
Na_2O	0.080	CaO	0.16	Na_2O	0.07
Mullite	93	MgO	0.13		
Glass	4	K_2O	0.63		
		Na_2O	0.07		
α-corundum	3				

The first set of specimens was made with a passing fraction of claystone W Super of varying age. The second set of specimens contained a passing fraction of W Super (90 %) and waste mullite Motim (10 %). The passing fractions of W Super were identified as Super 0.5, Super 5 and Super 10, with the number indicating heaping time in years.

The specimens were pressed using a hydraulic press at 20, 25 and 30 MPa. The mixture's water content was 8%. Once pressed, the specimens were dried in a laboratory dryer and fired in a laboratory superkanthal furnace at 1500 °C with a holding time of 4 hours.

4.2. Laboratory tests

The freshly pressed specimens were weighed and measured to determine their bulk density. After firing at 1500 °C, they were tested using the hydrostatic weighing method to measure their bulk density, apparent porosity, water absorption and apparent density. Selected specimens were tested for pore size distribution by high-pressure mercury intrusion porosimetry. The mineralogical composition was determined by XRD analysis. A scanning electron microscope was used to examine the internal structure of the claystone used in the specimen production. The values listed in the tables represent an average of three independent measurements.

Table 6. Properties of claystone W Super after firing at 1500 °C.

100 % W Super	Pressure	Water absorption	Bulk density	Apparent porosity	Apparent density	Overall shrinkage
	[MPa]	[%]	[$kg.m^{-3}$]	[%]	[$kg.m^{-3}$]	[%]
Super 0.5	20	2.0	2390	4.7	2507	-10.1
Super 0.5	25	3.2	2346	5.5	2535	-10.2
Super 0.5	30	2.5	2341	5.9	2489	-9.8

Super 5	20	3.6	2358	8.4	2575	-9.3
Super 5	25	2.3	2401	7.6	2543	-9.7
Super 5	30	2.6	2368	6.2	2524	-9.5
Super 10	20	3.7	2298	8.5	2510	-10.0
Super 10	25	4.3	2249	9.8	2493	-9.8
Super 10	30	4.2	2271	9.6	2512	-10.0

Table 7. Properties of the mixture of 90 % of W Super and 10 % of Motim after firing at 1500 °C.

90 % W Super 10 % Motim	Pressure	Water absorption	Bulk density	Apparent porosity	Apparent density	Overall shrinkage
	[MPa]	[%]	[kg.m^{-3}]	[%]	[kg.m^{-3}]	[%]
M-Super 0.5	20	2.3	2443	5.6	2588	-9.5
M-Super 0.5	25	2.6	2407	6.4	2571	-9.6
M Super 0.5	30	3.2	2390	7.6	2588	-9.3
M- Super 5	20	4.1	2368	9.7	2624	-8.1
M- Super 5	25	4.6	2365	10.9	2653	-8.2
M- Super 5	30	3.9	2428	9.4	2680	-8.1
M- Super 10	20	4.4	2336	10.3	2604	-8.9
M- Super 10	25	4.5	2359	10.7	2643	-8.8
M- Super 10	30	3.4	2378	8.1	2586	-8.9

4.3. Result discussion

Claystone W-Super was observed for the influence of heaping time and compaction pressure on the resulting material's mineral composition and pore structure. The apparent porosity of specimens made with claystone SW-Super 0.5 ranged from 4.5 to 5.5 %. The claystone which was heaped for 10 years reached values of apparent porosity between 8.5 and 9.8 %. As its heaping increases the porosity of the fired material (figure 4), the results show that it is best to use freshly mined claystone for dense grog manufacturing. The porosity of the fired grog is also affected by the pressure at which it was compacted; it is necessary to set an optimal pressure for each raw material. W-super 0.5 was the most suitable for manufacturing, with an optimal compaction pressure of 20 MPa. The firing temperature decisively influences the grog's density and mineralogical composition. 1500 °C is sufficient for achieving the desired mineralogical composition. The content of low-temperature quartz is 5-6 % and mullite content is 66 % for Super 5, and 70 % for M Super 5 sample.

W-Super contains the fissile mineral kaolinite. If a 30-mm passing fraction of claystone is heaped for several years, the kaolinite contained in it degrades, which manifests itself in fissures in its structure, thereby increasing the number of pores in the claystone; figure 3. Ten years of heaping causes kaolinite particle size to drop below 1 μm; the layers no longer have sharp edges and the structure is "aerated", figure 3 c). This "aeration" reduces density and increases porosity both after pressing and after firing.

a b c

Figure 3. Morphology of kaolinite a) half a year, b) 5 years, c) 10 year; magnified 50 000x.

Given its chemical and mineralogical composition, clay W-Super is suitable for high-alumina grog; however, it must be combined with a material with a high Al_2O_3 content. This study identified waste mullite Motim (aluminium oxide content of 73-75 %) to be a suitable source of Al_2O_3; table 1. The dosage of 10 % meant the total Al_2O_3 content in the mixture was 46.8 %. For details on the properties of the grog fired at 1500 °C see table 7. In some cases, the addition of waste mullite increased apparent porosity by 10 - 20 %.

The test results indicate the following:

- The application of the waste mullite Motim increases the apparent porosity of the fired product, figure 4, 5.
- The waste mullite causes the formation of 5-10 μm pores and increases the average pore size of the fired grog.
- The addition of Motim increases mullite content in the fired grog.

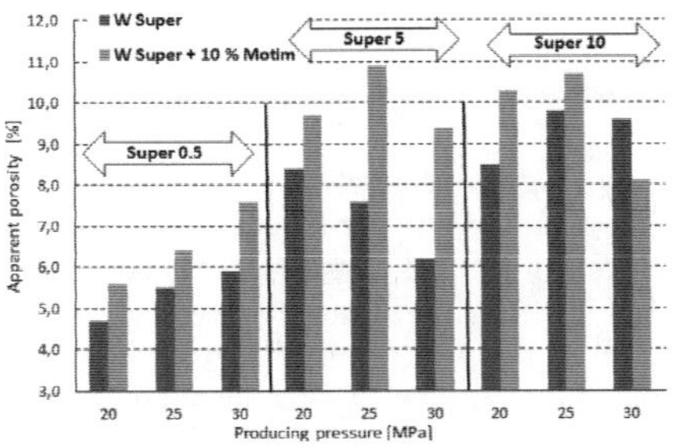

Figure 4. Influence of 10 % of waste mullite on apparent porosity of fired grog.

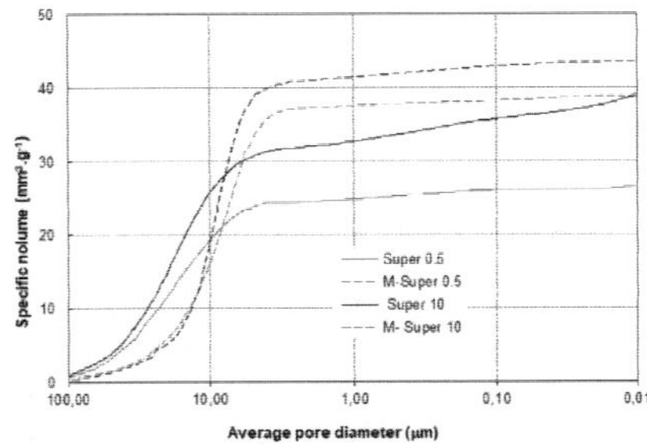

Figure 5. Comparison of pore size distribution.

5. Conclusion

The study demonstrates that the raw materials considered herein are well suited for mullite grog manufacturing. Grog W Super 0.5 with a 10 % addition of waste mullite was found to be the most suitable. The chemical composition (Al_2O_3 content = 46.7 %) meets the requirement for grogs with a high Al_2O_3 content. The lowest porosity achieved by any of the specimens was 5.6 %. This value approaches grog M47 with an apparent porosity of 3 %. The porosity can be further decreased by firing at a lower temperature, or by modifying the fresh mixture's composition. This will be subject to further research.

Acknowledgements

The paper was created within the project no. 2016TH02020321 "Production of Mullite Grog in a Shaft Kiln" sponsored by the Technical Agency of the Czech Republic.

References

[1] Fischer R X. and Schneider, H. 2005 Mulite *The mullite-type family of crystal structures* ed Schneider H and Komarneni S (Weinheim: Wiley-VCH) chapter 1 pp 1–46
[2] Kausik D, Somnath S, Himansu S Tripathi and Arup G 2014 *T. Indian. Ceram. Soc.* **73** 1-13
[3] Angel R J, Mcmullan R K and Prewitt C T 1991 *American Mineralogist* **76** 332-342
[4] Aksay I A, Dabbs D M and Sarikaya M 1991 *Applications Am. Ceram. Soc.* **74** 2343-2358
[5] Nabaltec: http://www.nabaltec.de/download/produkte/SYMULOX_M7272_Datasheet_EN.PDF (online)
[6] Kyanite mining corporation http://www.kyanite.com/assets/Kyanite_Mullite_Process_Sheet.pdf (online)
[7] Vulcan Refractories http://www.vulcanrefractories.com/wp-content/DataSheets/VulcasilH.pdf (online)
[8] Henan Cunse Refractory http://www.cncunse.com/basic-materials/mullite.html (online)

IOP Conf. Series: Materials Science and Engineering **385** (2018) 012039 doi:10.1088/1757-899X/385/1/012039

The influence of the raw materials mixture on the properties of forsterite ceramics

M Nguyen and R Sokolar

Brno University of Technology, Faculty of Civil Engineering, Institute of Technology of Building Materials and Components, Veveri 331/95, 602 00 Brno, Czech Republic

Abstract. Forsterite ceramics is a part of the refractory magnesium ceramics. The aim of this paper is to characterize the properties of forsterite ceramics in dependence on different compositions of the raw material mixture. Forsterite ceramics is not currently produced in the Czech Republic. Forsterite ceramics is used in industry for its high refractoriness and also for its coefficient of linear thermal expansion which is close to metals. The primary objective of this article is to synthesize forsterite ceramics via solid state reaction. Different raw materials were used for materials mixtures and properties of fired samples were investigated, considering the influence of the raw materials mixtures. Specifically the apparent porosity, water absorption, mineralogical composition, coefficient of linear thermal expansion and refractoriness were investigated.

1. Introduction

Forsterite is an end-member mineral of the olivine solid solution series which contains magnesium in its structure. Chemical formula of forsterite is $2MgO{\cdot}SiO_2$. In nature, forsterite can be found as a solid solution of olivine which has the chemical formula $(Mg^{2+}, Fe^{2+})_2[SiO_4]$ [1].

In the industry, forsterite is used as a refractory material due to its high refractoriness up to 1850°C and its refractoriness under load of 1600°C [2]. Another significant property of forsterite is its coefficient of linear thermal expansion, which is very similar to the metals coefficient with increasing temperature. This is the reason why is forsterite ceramics widely used in electrical engineering. In the civil engineering industry, forsterite can be used as a lining of the sintering zone of a rotary kiln for cement production or as a lining of metallurgical furnaces [3]. Table 1 contains physical and chemical properties of forsterite.

Table 1. Physical and chemical properties of forsterite.

chemical formula:	$2MgO \cdot SiO_2$	
chemical composition:	57.29 % MgO + 42.71 % SiO_2	
crystal system:	orthorhombic	
Mohs scale of mineral hardness:	7	
molar mass:	140.69 g/mol	
density:	3 270 $kg \cdot m^{-3}$	
melting point:	1 890°C	
refractoriness:	1 850°C	[2]
refractoriness under load:	1 600°C	[2]
coefficient of linear thermal expansion $\alpha_{30\text{-}100°C}$:	$8 \cdot 10^{-6} - 11 \cdot 10^{-6}\ K^{-1}$	[3]

2. Experimental

Powders of magnesite (purity 98 %, MgO: 44.2 wt%, d_{50} = 2 μm, Fichema, s. r. o.) and talc (purity 97 %, d_{50} = 4.7 μm, Fichema, s. r. o.) were mixed with a weight proportion corresponding to forsterite (table 2 – mixture A). Powders of talc and magnesium oxide (pharmaceutical purity 99.7 %, Fichema, s.r.o.) were mixed with a weight proportion corresponding to forsterite (table 2 – mixture B) and powders of magnesium oxide and micro-milled silica sand (high purity 99.6 %, d_{50} = 6 μm, surface area 8592 $m^2 \cdot kg^{-1}$, Sklopísek Střeleč, a. s.) were mixed with a weight proportion corresponding to forsterite (table 2 – mixture C). Chemical composition of used materials is presented in table 3.

Individual mixtures (MIX A, B, C) were mixed in a rotary homogenizer for 24 hours to ensure homogenous mixture. The prepared mixtures were uniaxially compacted at about 20 MPa into samples with dimensions 100×50×10 mm. The green bodies were dried in a laboratory drier at 110°C and then sintered at temperatures 1200°C and 1300°C with a heating rate of 4°C/min and soaking time of 2 hours at the maximum temperature in an electrical furnace in the air atmosphere. Pyrometric cones were also prepared from all mixtures to determine refractoriness.

Table 2. Raw material mixtures composition.

raw material	mixture A	mixture B	mixture C
magnesite	53.8 wt%	-	-
talc	36.2 wt%	56.4 wt%	-
magnesium oxide	-	33.6 wt%	54.3 wt%
silica sand	-	-	35.7 wt%
clay B1	10.0 wt%	10.0 wt%	10.0 wt%

Table 3. Chemical composition of used materials.

raw material	MgO [%]	SiO_2 [%]	CaO [%]	Fe_2O_3 [%]	Al_2O_3 [%]	K_2O+Na_2O [%]	TiO_2 [%]	LOI [%]	Total [%]
magnesite	44.2	0.0	0.4	0.04	0.0	0.0	0.0	52	96.64
talc	31.5	59.1	1.0	0.7	1.0	0.2	0.0	6.5	100.00
magnesium oxide	99.7	0.0	0.05	0.003	0.0	0.0	0.0	0.25	100.00
silica sand	0.05	99.6	0.02	0.03	0.2	0.1	0.0	0.0	100.00
clay B1	0.36	48.7	0.28	2.66	33.84	1.99	0.83	11.25	99.91

After firing the apparent porosity and water absorption were determined by the official standard EN 993-1:1995 [4] (vacuum water absorption). Pyrometric cones were also prepared according to standard EN 993-12 [5] to determine refractoriness. Coefficient of linear thermal expansion was determined according to standard EN 993-19: Determination of thermal expansion by a differential method [6].

The phases presented in the synthesized forsterite powders were identified by X-ray diffraction (PANALYTICAL Empyrean XRD) with Cu Kα as the radiation source, accelerating voltage 45 kV, beam current 40 mA in the range from 5° to 80° with a step scan of 0.01°. The obtained X-ray diffraction patterns were compared with the Crystallography Open Database (COD) [7].

3. Results and discussion

The most important properties of forsterite are high refractoriness and coefficient of linear thermal expansion which is close to metals. Coefficient of linear thermal expansion was determined from dilatometric curves (figure 1) by differential method. Higher firing shrinkage of mixture B (MIX B) is caused by talc, which is eutectic flux and accelerates sintering of the samples for temperatures above 1200°C.

Mean coefficient of linear thermal expansion of mixture B (MIX B) for temperature interval 20-1300°C is $9.81\cdot10^{-6}$ K^{-1} and for mixture C (MIX C) is $15.25\cdot10^{-6}$ K^{-1}. M. Bouhifd et al. [8] determined the coefficient of linear thermal expansion for forsterite in temperature interval 20-1300°C has linear growth and was found to be in the interval $(10\text{-}36)\cdot10^{-6}$ K^{-1}. Coefficient of tested samples is within the range given by the reference values.

Apparent porosity and water absorption values of tested samples are presented in table 4. Mean apparent porosity of mixture A (MIX A) was high due to loss on ignition of magnesite (magnesium carbonate), which decomposes to magnesium oxide and carbon dioxide. This decomposition is the cause of higher apparent porosity and water absorption in mixture A. Samples of mixture B (MIX B) fired at 1300°C were sintered with low values of apparent porosity and water absorption.

Table 4. Apparent porosity and water absorption of mixtures A, B and C at different firing temperatures.

mixture	firing temperature [°C]	apparent porosity [%]	water absorption [%]
MIX A	1200	53.7	36.4
	1300	48.9	29.6
MIX B	1200	36.1	18.1
	1300	11.6	5.4
MIX C	1200	42.7	23.8
	1300	39.1	20.7

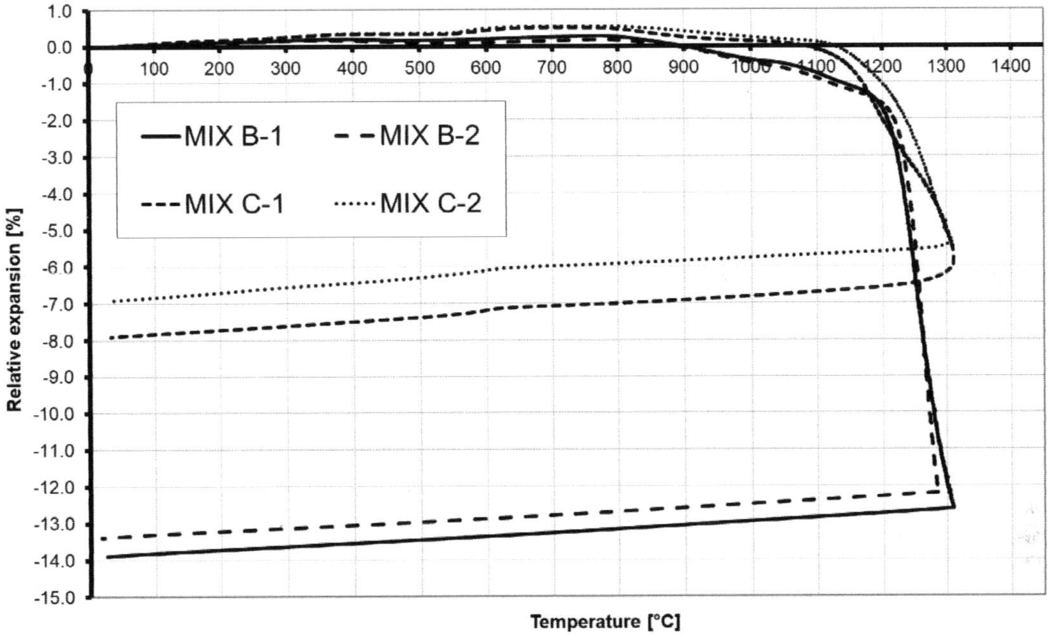

Figure 1. Thermal dilatometric analysis during firing (1300°C, heating rate 5°C/min).

Mineralogical composition of all mixtures is presented in table 5. The result of phase analysis by X-ray diffraction (XRD) of fired samples is shown in figure 2. Quantitative representation of minerals was carried out by Rietveld refinement (table 5).

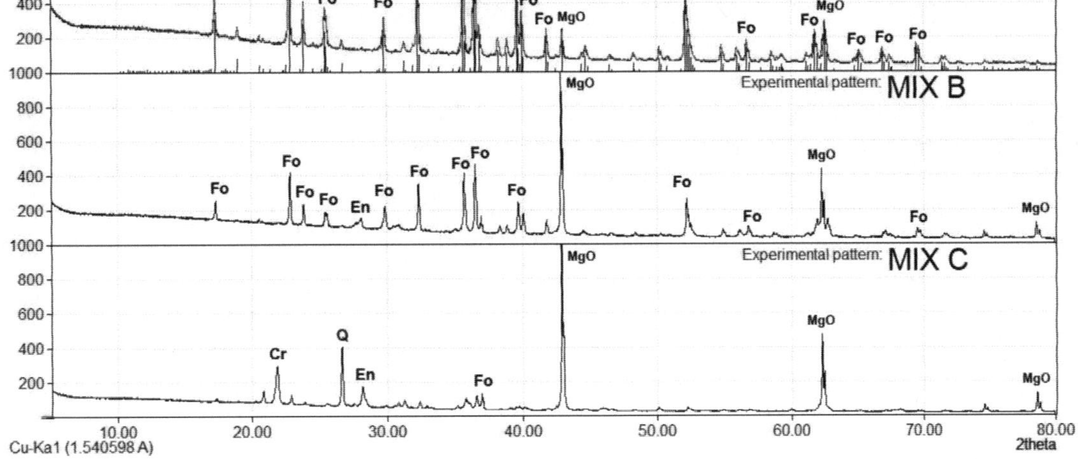

Figure 2. X-ray diffraction (XRD) of all mixtures (Fo = forsterite, En = enstatite, MgO = periclase, Q = quartz, Cr = cristobalite).

Samples from mixture A were almost pure crystalline forsterite. This occurred due to magnesite, which decomposes at temperatures above 500°C and released carbon dioxide resulted in higher reactivity to form forsterite. Intensive synthesis of forsterite begins at 1100°C if magnesite and talc are used as raw materials [3].

Samples from mixture B contained high amount of forsterite, but secondary phases (periclase MgO and enstatite $MgSiO_3$) were also present. This was probably due to merging of fine magnesium oxide particles, which resulted in micro-inhomogeneity of mixture B. With increasing temperature, the amount of forsterite in samples from mixture B was also increasing. This effect occurred due to flux ability of talc, which is present in the mixture B.

Samples of mixture C contained relatively low amount of forsterite with secondary phases of periclase (MgO), enstatite ($MgSiO_3$), quartz (SiO_2) and at temperature of 1300°C also cristobalite. With increasing temperature, the amount of forsterite and enstatite in samples from mixture C was also increasing. However, the temperature of 1300°C was not high enough for intense sintering of raw materials to occur.

Table 5. Results of X-ray diffraction and Rietveld refinement of samples fired at different temperature.

mixture	firing temperature [°C]	identified minerals	quantitative representation of minerals using the Rietveld refinement
MIX A	1200	forsterite (Fo), periclase (P)	95 % Fo + 5 % P
	1300	forsterite (Fo), periclase (P)	95 % Fo + 5 % P
MIX B	1200	forsterite (Fo), periclase (P), enstatite (En)	59 % Fo + 29 % P + 12 % En
	1300	forsterite (Fo), periclase (P), enstatite (En)	73 % Fo + 22 % P + 4 % En
MIX C	1200	forsterite (Fo), periclase (P), enstatite (En), quartz (Q)	14 % Fo + 42 % P + 18 % En + 26 % Q
	1300	forsterite (Fo), periclase (P), enstatite (En), quartz (Q), cristobalite (Cr)	22 % Fo + 44 % P + 20 % En + 7 % Q + 7 % Cr

The pyrometric cone equivalent test was carried out to determine the refractoriness of all mixtures. The effect of mineralogical composition on the refractoriness of all mixtures is shown in figure 2. Pyrometric cone from mixture B bent at 1490°C (figure 2 on the left) and pyrometric cone from mixture C bent at 1580°C. The test was terminated at 1620°C, because the pyrometric cone from mixture A fell down. This occurred probably due to creation of eutectic melt in the pad, which was formed from clay B1 and corundum.

Pyrometric cone from mixture A was not bent after the test, refractoriness of this mixture is therefore above 1620°C. Refractoriness of pure forsterite is approximately 1850°C [2]. Lower refractoriness of mixtures B and C was due to secondary phase of enstatite, whose melting point is 1557°C.

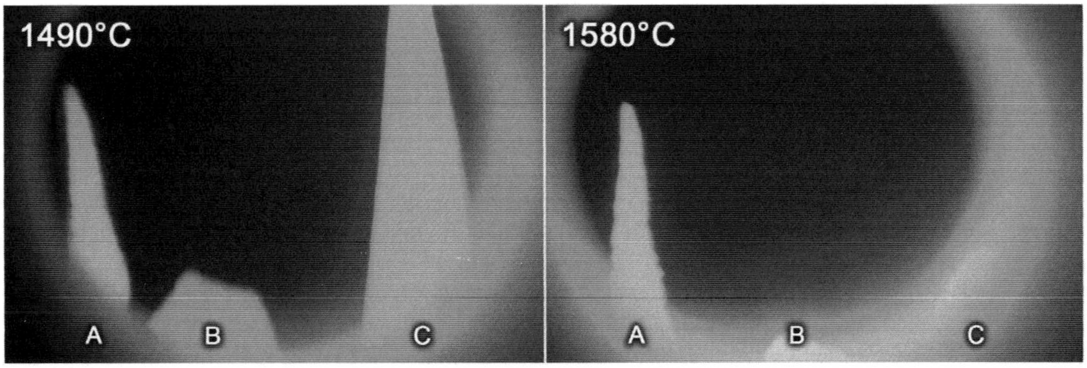

Figure 3. Pyrometric cone equivalent test: left – bending of cone from mixture B, right – bending of cone from mixture C.

4. Conclusion

In this article, forsterite was successfully synthesized via solid-state reaction from all mixtures. The influence of the raw material mixture was analyzed in terms of mineralogical composition, apparent porosity, water absorption, refractoriness and coefficient of linear thermal expansion.

Mixture A was composed of magnesite and talc, mixture B was composed of talc and magnesium oxide and mixture C was composed of magnesium oxide and silica sand. All mixtures contained 10 % of clay B1 which improves binding properties of the green body and sinterability during firing. Dry pressed samples were sintered at 1200°C and 1300°C. Mineralogical composition of individual mixtures varied. Samples from mixture A (magnesite and talc) were nearly pure forsterite with secondary phase of periclase, samples from mixture B (talc and magnesium oxide) contained high amounts of forsterite with secondary phases of periclase and enstatite and samples from mixture C (magnesium oxide and silica sand) contained minor amounts of forsterite with secondary phases of periclase, enstatite, quartz and cristobalite.

In terms of apparent porosity and water absorption, the raw materials and the sintering temperature plays and important role. Samples from mixture A had high apparent porosity and water absorption due to decomposition of magnesite (magnesium carbonate). Samples from mixture B sintered at 1300°C were densificated with low apparent porosity and water absorption due to talc in the mixture B which is eutectic flux and accelerates sintering.

Refractoriness of samples from mixture B is 1490°C, from mixture C is 1580°C and from mixture A is above 1620°C. Refractoriness of pure forsterite is above 1800°C. Lower refractoriness of samples from mixtures B and C was due to the presence of secondary phase enstatite. Melting point of enstatite is 1557°C. Mean coefficient of linear thermal expansion of mixtures for temperature interval 20-1300°C is between $(9.8\text{-}15.3)\cdot10^{-6}\ K^{-1}$, which is in the range of reference values for forsterite.

In conclusion, the results revealed that firing the samples from mixture A (magnesite, talc) are almost pure forsterite ceramics with high porosity and refractoriness. Samples from mixture B (talc, magnesium oxide) contain high amount of forsterite and are densified with low porosity and water absorption and samples from mixture C (magnesium oxide, silica sand) are porous and to achieve higher amounts of forsterite, sintering temperature needs to be above 1500°C.

It could be possible to utilize fly ash as a source of silicon dioxide (SiO_2). Fly ash is a secondary energy product from coal combustion in power plants. Therefore it is inexpensive and its recycling is important for the environment and sustainable development. Raw materials mixture from fly ash to form forsterite would consist of 53.7 wt% of fly ash and 46.3 wt% of magnesium oxide. Utilization of fly ash as a raw material could have potential economic benefit in the final price of forsterite refractory ceramics.

Acknowledgments
This article was supported under the Czech Science Foundation GAČR grant no. 18-02815S with project name: Elimination of Sulphur oxide emission during the firing of ceramic bodies based on fly ashes of class C.

References
[1] Kutzendorfer J and Tomsu F 2008 *Refractory Materials (*Prague: Czech Ceramic Society)
[2] Budnikov P P 1960 *Technology of Ceramics and Refractory Products* (Prague: State publishing of technical literature) (in Czech)
[3] Hanykyr V and Kutzendorfer J 2000 *Technology of Ceramics* (Hradec Kralove: Vega) (in Czech)
[4] EN 993-1:1995 *Methods of test for dense shaped refractory products Part 1: Determination of bulk density, apparent porosity and true porosity*
[5] EN 993-12:1997 *Methods of test for dense shaped refractory products Part 12: Determination of pyrometric cone equivalent (refractoriness)*
[6] EN 993-19:2004 *Methods of test for dense shaped refractory products Part 19: Determination of thermal expansion by a differential method*
[7] Crystallography Open Database: an open-access collection of crystal structures: http://www.crystallography.net/ (online)
[8] Bouhifd M, Andrault D, Fiquet G and Richet P 1996 *Geophysical research letters* **23** 1143–46

IOP Conf. Series: Materials Science and Engineering **385** (2018) 012040 doi:10.1088/1757-899X/385/1/012040

The effect of hydrophobization on the properties of mortar mixtures

V Novak and J Zach

Brno University of Technology, Faculty of Civil Engineering, Institute of Technology of Building Materials and Components, Veveri 331/95, 602 00 Brno, Czech Republic

Abstract. One of the possible methods to reduce degradation effects due to the influence of moisture on building structures and materials is the use of hydrophobic agents - hydrophobization. The article deals with the influence of the hydrophobic preparations use on the change of mortar mixtures properties, where not only the impact on capillary absorption, but also other key properties which can be influenced by the application of the hydrophobic agent, are monitored. By realizing a wide range of measurements, optimal concentrations were determined to achieve properties comparable with the rehabilitation materials without negative effect on other key properties of mortar mixtures, when is mainly a significant reduction of the capillary absorption factor of mortar mixtures only at tiny changing the thermal conductivity factor of the mortar mixtures and without proved changing the other monitored properties.

1. Introduction

One of the main factors that can significantly reduce the durability of the building is the influence effect of moisture [1]. In particular, this is a liquid form of moisture, which is primarily getting in construction due to a direct exposure of the construction to the moisture. In this case, the moisture in the construction spreads through the capillary system of materials - capillary absorption [2-3]. The most effective protection against this exposure of moisture is to prevent contact of moisture in the liquid form with the construction or the use of special materials for this purpose - waterproofing. In case of failure of these protection methods, either due to inappropriate design and use of the building, unprofessional application or mechanical damage of waterproofing, to reduce the impact of moisture degradation it is possible to use reducing capillary absorption of building materials. One of the possible measures is the hydrophobization of building materials and mortar mixtures from which the structures are made [4-5].

2. Tested materials

For the experimental part, four basic mortar mixtures were chosen - masonry mortars and plasters. This is a basic masonry mortar for all masonry works, rough coating plaster and masonry mortar applicable to masonry and coating plastering. A detailed description is given in table 1.

Table 1. Overview and description of the tested mortar mixtures [6].

	Description	Application
Baumit MM 50	Industrially manufactured dry mortar mixture for manual and machine processing.	Masonry mortar for all types of common masonry elements, for supporting walls, partitions and chimney bodies.
Baumit Manu 4	Industrially manufactured dry plaster mixture for manual processing in the exterior and interior.	Rough coating plaster for manual plastering, usable both in exterior and interior.
Baumit DuoDur	Industrially manufactured dry mixture suitable as masonry mortar as a universal plaster for manual processing.	Masonry mortar for all types of common masonry elements, for supporting walls and partitions. Coating plaster for manual plastering, usable both in exterior and interior.

The selected mixtures were mixed with water according to the manufacturer's recommendations and the individual consistencies were monitored by spill on a jogging table according to European standard
EN 1015-3 Testing methods for masonry mortars. Test specimens were prepared from individual mixtures, resp. from the pure reference mixture and subsequently with the addition of a hydrophobic agent. The hydrophobic agent was dosed to a total volume of mixture in an amount of 0.2 %, 0.6 % and 1.0 %. It was a silicone emulsion hydrophobic agent with a recommended dosage of 0.5 to 0.8 % of the finished mortar mixture volume. Silicone based hydrophobic agents work on principle of changing the physical-chemical properties of the material, in particular in a substantial increase of the wetting angle, which significantly reduces the absorption of moisture in the liquid form into the material.

3. Testing methodology

Test specimens were produced from individual mortar mixtures, namely 40 mm x 40 mm x 160 mm beam for the density in fresh and matured state and the capillary water absorption determining. Slabs 300 mm x 300 mm x 50 mm were also produced to determine the thermal conductivity factor and circular samples to determine the diffusion resistance factor.

3.1. Bulk density

The density in the fresh state was determined on 40 mm x 40 mm x 160 mm beams. By weighing the triple forms during production of the test specimens [7]. The density in the matured state was determined after 7 and 28 days of aging.

3.2. Adhesion

The effect of the hydrophobic agent on the adhesion both on the reference surface - concrete and ceramic masonry blocks was tested in the recipes. In case of adhesion to the concrete surface, the individual mixtures were applied in a uniform 10 mm thick layer to pre-moistened concrete slabs. In the case of adhesion to ceramic masonry fittings, the individual mixtures were also applied at a thickness of 10 mm in width of the ceramic wall fittings. In both cases, after 28 days of sample maturation, steel targets with a diameter of 50 mm by using two-component Sikadur glue were glued on the specimens. After curing, the targets were cut with the angle grinder in accordance with the relevant standard, always through the tested mixture to the subbase, and subsequently the adhesion by the DYNA Z 16 device was determined [8].

3.3. Capillary absorption
Determination of capillary absorption was performed on the beams made for this purpose. All of the longer surfaces must first be provided with a waterproofing coating so that water could then flow through only the lower square surface. After the waterproofing coat was matured, the beams were broken transversely into two halves. Each beam was weighed and with broken side down was put into a plastic container on a grid. The beams were watered up to a height of 5 to 10 mm from the bottom surface. At specified times, the weight gain of the test specimens was monitored as a result of water absorption. After the last weighing, each beam was longitudinally broken by a hammer and a steel chisel. On the fracture area of the sample was measured the height of the water rising [9].

3.4. Diffusion resistance
The principle of determining the diffusion resistance of mixtures consists of accurately monitoring the amount of moisture passing through by the material defined surface and the thickness of the material at fixed boundary environmental conditions. Prior to the start of testing, individual specimens were covered with silicone throughout their height to ensure a defined area. The specimens thus prepared were fixed by silicone on special testing vessels filled with dried silica gel. The vessels were placed in climate chamber with a defined environment of +23 °C and 80 % relative humidity. Approximately every 12 hours, the vessels with test specimens were weighed and the diffusion resistance factor was determined from the weight gain due to moisture passing through the silica gel [10].

3.5. Thermal conductivity
The thermal conductivity factor was determined on plates 300 mm x 300 mm x 50 mm by a stationary plate method on a Holometrix Lambda 2300 device based on a stationary plate method. Prior to the measurement, the flatness of the surface of each slab was checked and the slab was grounded if necessary. The determination of the thermal conductivity factor was performed in a steady state at an average temperature of +10 °C and a temperature gradient of 10 K. The measurement was always made on three test specimens (on each specimen was realized the determination of the thermal conductivity factor in total at 5 consecutive intervals) which were dried at a temperature of +75 °C to a constant weight. The final result presented is the arithmetic mean of these three measurements [11-12].

4. Influence of the hydrophobic agent on the properties of mortar mixtures
For mortar mixtures, the effect of the hydrophobic agent on the density was monitored both in the fresh state and at the maturation period after 7 and 28 days (table 2).

From referred values, it appears that by adding the hydrophobic agent to the mortar mixtures, in almost all of the specimens is density in the fresh state and during maturation reduced. There was no dependence between the size of the hydrophobic agent addition to the mixture and its density. In general, the addition of the hydrophobic agent to the mixtures does not have a significant effect on the density.

Table 2. Summary overview of densities of hydrophobic mortar mixtures.

	Addition of hydrophobic agent [%]	Baumit Manu 4	Baumit DuoDur	Baumit MM 50
Density fresh state ρ [kg/m^3]	0.0	1947	1950	2051
	0.2	1878	1953	1653
	0.6	1911	1924	1859
	1.0	1855	1979	1784
Density 7 days ρ [kg/m^3]	0.0	1721	1694	1854
	0.2	1631	1720	1748
	0.6	1656	1705	1643
	1.0	1638	1654	1594

	0.0	1749	1724	1824
Density 28 days	0.2	1575	1696	1711
ρ [kg/m³]	0,6	1642	1688	1592
	1.0	1620	1601	1581

After 28 days of maturation, the hydrophobic agent effect on the properties of the mortar mixtures was monitored (table 3). Capillary absorption, adhesion, diffusion properties and thermal insulating properties were monitored.

Table 3. Summary overview of adhesions of hydrophobic mortar mixtures.

	Addition of hydrophobic agent [%]	Baumit Manu 4	Baumit DuoDur	Baumit MM 50
	0.0	0.12	0.10	0.54
Adhesion strength on concrete	0.2	0.14	0.11	0.10
f_t [MPa]	0.6	0.11	0.21	0.54
	1.0	0.13	0.11	0.34
	0.0	0.03	0.05	0.03
Adhesion strength on brick	0.2	0.06	0.03	0.12
f_t [MPa]	0.6	0.07	0.07	0.05
	1.0	0.03	0.05	0.12

From the resulting values, no effect of the hydrophobic agent on adhesion of mixtures to concrete or ceramic surfaces is apparent.

Table 4. Summary overview of changes in properties of mortar mixtures.

	Addition of hydrophobic agent [%]	Capillary absorption C [kg/m²]	Thermal conductivity factor λ [W/(m.K)]	Diffusion resistance factor μ [-]
	0.0	18.25	0.4651	4.26
Baumit Manu 4	0.2	17.30	0.5251	4.06
	0.6	12.30	0.4313	4.49
	1.0	2.85	0.4060	4.75
	0.0	17.95	0.4668	3.81
Baumit DuoDur	0.2	16.05	0.5455	3.75
	0.6	14.00	0.4895	4.10
	1.0	2.55	0.4635	4.04
	0.0	14.45	0.6227	4.60
Baumit MM 50	0.2	14.20	0.6759	4.95
	0.6	5.75	0.6165	4.54
	1.0	1.50	0.5340	4.41

From the values of the capillary absorptions listed in table 4 the positive effect of the hydrophobic agent addition to the mortar mixtures clearly evident. There was an indirect proportionality between the increasing dose of the hydrophobic agent and the capillary absorption coefficients; therefore, the capillary absorption coefficients are reduced by increasing the hydrophobic agent dosage in all test mixtures.

With Manu 4, DuoDur and MM 50 mortar mixtures, the same effect of the addition of the hydrophobic agent on the thermal insulation properties is evident. Adding a small amount, namely 0.2 % will increase the thermal conductivity factor. On the contrary, with increasing amounts of added hydrophobic agent in plaster mixtures, the factor of thermal conductivity decreases, at a concentration of 1.0 % will be reached lower values than to plaster mixtures without the use of a hydrophobic agent.

No dependence of the diffusion resistance factor due to the addition of the hydrophobic agent to the mixture has been proved for the tested mortar mixtures. The addition of the hydrophobic agent has always had a different effect on the diffusion resistance factor.

5. Conclusion

The results of the previous chapter clearly confirm the positive effect of the hydrophobic agent on mortar mixtures. This is mainly a significant reduction of the capillary absorption factor of mortar mixtures, when similar properties of capillary absorption as achieved by rehabilitation mixtures were achieved by adding 1.0% hydrophobic agent to the Baumit Manu 4, Baumit DuoDur and Baumit MM 50 tested mortar mixtures. The addition of hydrophobic agent in an amount of 1.0 % also seems optimal in terms of the change in thermal insulation properties, when by the addition is slightly reduced the thermal conductivity factor of the mortar mixtures. The effect of the hydrophobic agent has not been proved for the other monitored properties, which is also positive, especially in terms of adhesion, which did not show a significant decrease with the addition of 1.0% hydrophobic agent.

In case it is desired to achieve capillary absorption values of mortar mixtures comparable to rehabilitation plaster values, without effect on changes in other properties of mortar mixtures, the optimal dosing of the hydrophobic agent was proved to be 1.0%.

Acknowledgements

This paper has been worked out under the project No. LO1408 "AdMaS UP - Advanced Materials, Structures and Technologies", supported by Ministry of Education, Youth and Sports under the „National Sustainability Programme I".

References

[1] Annila P J, Hellemaa M, Pakkala TA, Lahdensivu J, Suonketo J and Pentti M 2017 *Case Studies in Construction Materials* (Amsterdam: Elsevier) p 103-108
[2] Zhao J and Meissener F 2017 *Energy Procedia* (Amsterdam: Elsevier) p 261-266
[3] Hall Ch, Hoff W D, Viles H A and Eklund J A 2010 *Proceedings of The Royal Society A* (London: Royal Society Open Science) p 194-211
[4] CSN 73 0600: 2000 *Waterproofing of buildings - Basic provisions*
[5] CSN 73 0610: 2000 *Waterproofing of buildings - The rehabilitation of damp masonry and additional protection of buildings against ground moisture and against atmospheric water – Basic provision*
[6] Baumit https://www.baumit.cz/produkty (online)
[7] EN 1015-6: 1998 *Methods of test for mortar for masonry - Part 6: Determination of bulk density of fresh mortar*
[8] EN 1015-12: 2016 *Methods of test for mortar for masonry - Part 12: Determination of adhesive strength of hardened rendering and plastering mortars on substrates*
[9] EN 1015-18: 2002 *Methods of test for mortar for masonry - Part 18: Determination of water absorption coefficient due to capillarity action of hardened mortar*
[10] EN 1015-19: 1998 *Methods of test for mortar for masonry - Part 19: Determination of water vapour permeability of hardened rendering and plastering mortars*
[11] EN 12667: 2001 *Thermal performance of building materials and products - Determination of thermal resistance by means of guarded hot plate and heat flow meter methods - Products of high and medium thermal resistance*
[12] CSN 72 7012-1: 1994 *Determination of Steady State Thermal Conductivity of Materials. Hot Plate Methods*

IOP Conf. Series: Materials Science and Engineering **385** (2018) 012041 doi:10.1088/1757-899X/385/1/012041

Testing thermophysical and mechanical properties of cement-based composites with plastic and horse manure

G Pavlendova[1], **V Bohac**[2] **and P Sin**[1]

[1] Department of Physics, Faculty of Civil Engineering, Slovak University of Technology, Radlinskeho 11, Bratislava, SK 81005, Slovakia
[2] Institute of Physics, Slovak Academy of Sciences, Dubravska cesta 9, Bratislava, SK 84511, Slovakia

Abstract. Cement-based composites change their properties in dependence on added materials. Scientific papers display results on changes in various mechanical and thermophysical properties. In this paper, the effect of different amount of plastic in two shapes (granules and fibers) and horse manure on thermophysical and mechanical properties of cement-based composites are tested. To understand the structural changes 1-D process of water uptake and porosity measurements were done. The economic growth brings new possibilities for industry development but also increasing problem with waste disposal. Our tests were aimed at finding the optimal balance between plastic disposal in cement based materials and changes in some important properties. The best results of thermophysical properties were for plastic and natural fibers, where horse manure samples had even better mechanical properties than plastic ones.

1. Introduction

The 20th century witnessed substantial growth in the consumption of plastic, which is considered the most significant innovation of this century. In the 21st century, plastic belongs to the top 25 environmental concerns. Plastic recycling in concrete is often given as a good example of help with plastic disposal problems [1, 2].

The idea of plastic recycling or its disposal in cement based materials is based on the following facts. Cement-based composites are the most widely used materials in building construction. Available plastic polymers have a number of vital properties which could be beneficial for construction needs, e.g. corrosion resistance, good sound and heat insulation properties [3, 4]. Plastic is already used in concrete reinforcing as one kind of fibers (steel fiber, glass fiber, natural fiber, synthetic fiber) [5].

Several review articles, e.g. [6,7] have dealt with the contemporary state of knowledge in the properties of cement - based composites with some kind of plastic. The increase of concrete production means the increased amount of required mineral aggregates. Hence some researchers started with plastic waste substitution of mineral aggregates. The substituted amount was from 3-50% [7]. The measurements showed improvement in thermophysical properties, but decrease in mechanical properties. Plastic has different affinity to water and entraps air molecules, so the porosity changes [7]. Our approach was different. We did not replace the sand; plastic waste was added in two shapes: granules and fibers.

When testing fibers' addition, the plastic fibers were often compared to the natural ones. Natural fibers are biodegradable and it means shorter service life. New kinds of plastic fibers were produced and so the initial idea of recycling led to increased production of plastic. Therefore we also used horse manure, which fibers harden with time, and we compare it to plastic fibers. Our initial testing with plastic waste from recycled polyethylene (LDPE) Bralen Na7-25, Slovnaft, Bratislava, Slovakia, previously used for packaging (plastic bags, plastic films, containers including bottles, etc.) in the form of the granules with the size ∅4x4 mm showed the improvement of thermophysical properties. The results were better for the samples with 20 percent of waste granules and the same was true after the artificial ageing (to test changes in the course of time) [8]. The aim of this work is to compare influence of added materials on thermophysical properties.

2. Materials and Methods

Plastic waste in the form of the granules with the size ⊲4x4 mm was obtained from recycled polyethylene (LDPE) Bralen Na7-25, Slovnaft, Bratislava, Slovakia, previously used for packaging (plastic bags, plastic films, containers including bottles, etc.). Plastic fibers (PP) with the length 12-18 mm were made by Adfill Co., Hull, Great Britain. Horse manure was taken from the horse training center Carpathia, Bratislava, Slovakia. We used Cement Ladce (Portland-composite cement) CEMII / B-M (S-L) 32.5R, Bratislava, Slovakia and the finest fraction sand, Bratislava, Slovakia with the diameter <0.064 mm.

Samples with 10% (calculated from volumetric amount of cement and sand together) of plastic granules, plastic fibers and horse manure were tested on thermophysical and some mechanical properties. The densities of samples can be seen in table 1 and the composition of samples can be seen in table 2.

Table 1. Density of components.

Component	Cement (kgm^{-3})	Sand (kgm^{-3})	Water (kgm^{-3})	LDPE (kgm^{-3})	PP (kgm^{-3})	HM (kgm^{-3})
Density (kgm^{-3})	1292	1717	1000	560	338	153

Table 2. Composition of samples.

Sample	Cement (cm^3)	Sand (cm^3)	Water (cm^3)	Additive (cm^3)
Pure concrete	350	1050	350	000 (none)
C. with 10 % LDPE	350	1050	350	140 (granules)
C. with 10 % PP	350	1050	350	140 (fibers)
C. with 10 % HM	350	1050	350	140 (manure)

The samples were prepared according to the norm BS EN 12390-1:2012, dried for 30 days, and afterwards put into the oven with ventilation at the temperature 90°C for 7 days. The surfaces of all samples were polished to achieve smooth surfaces. LDPE = recycled polyethylene, PP = polypropylene, HM = horse manure, C. = concrete.

The samples with the base 100 x 100 mm and prescribed thickness (ranging from 14 to 15 mm) of the individual parts were measured for thermophysical properties at the temperatures: -30°C, -15°C, 0°C, 15°C and 30°C. The resulting values of the quantities were determined by the arithmetic average of ten measurements on four samples. The reproducibility for thermal conductivity measurement is 3% of reading + 0.001 W/m.K and for specific heat 3% of reading + 1 J/kg.K.

The principle of pulse transient method is in figure 1. The samples and additives are shown in figures 2-3.

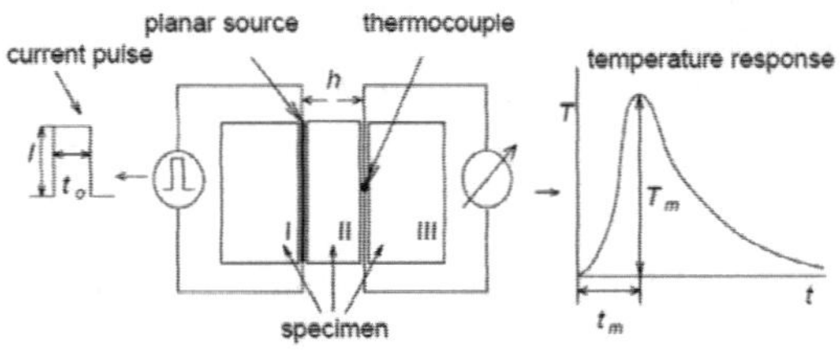

Figure 1. Principle of the pulse transient method where T = temperature and t = time.

Figure 2. Samples (left picture: upper left - pure concrete, upper right - concrete with spheres, lower left - concrete with fibers, lower right - concrete with horse manure).

Figure 3. Additives (left - spheres, middle - fibers, right - dry horse manure). Caliper is depicted for size comparisons of additives.

The thermophysical parameters - specific heat, thermal diffusivity, and thermal conductivity - are calculated according to the following relations [9]: specific heat $c = Q/\sqrt{2\pi e}\rho h T_m$, thermal diffusivity $a = h^2/2T_m$ and thermal conductivity, $\lambda = ac\rho$, where $Q = RI^2 t_0$, R is the electrical resistance of the heat source, t_0 is the width of the current pulse, ρ is the sample density, I is the heating current of planar heat source, h is the width of the sample, T_m is the temperature increase of the sample, as defined in figure 1 and e is Euler's number.

Expected uncertainty for each of the properties is 5% depending on pore size distribution. The diameter of thermocouple is about two orders of magnitude higher than the ones of material pores and so the distribution of pores under the thermocouple might be slightly different.

Thermophysical measurements were carried out before and after artificial ageing in climatic chamber BINDER MKF (E3) after 60 temperature and humidity cycles of 12 hours, or 30 days in total, table 3. We tested the longevity of achieved properties. The results are in figures 4-6.

Table 3. The regime of climatic chamber - number of cycles 60 times 12 hours = 30 days total.

Section	1	2	3	4
Time (hours)	3	3	3	3
Temperature (°C)	-10	20	60	20
Humidity (%)	0	70	70	80

Water sorptivity measurement using a free water intake experiment as described in Hall [10] was performed. The water absorption coefficient A, defined as the slope of the initial stage of the cumulative mass of water versus square root of time curve.

The effective pore volume and open porosity were measured and calculated. The whole volume of samples was put into the water and boiled for 50 minutes. After cooling at room temperature the samples were weighted. From the weight difference, before and after boiling, the open porosity was calculated [11]. The results are in the table 4.

The samples for flexural and compressive strength with dimensions 40 x 40 x 160 mm were tested according to the norm EN 196-1 „Cement testing“. Flexural strength was tested with Controls Model 65-L0019B apparatus at loading force 50 N/s. Compressive strength was tested using the Controls Model 1551246001 apparatus at loading force 2000 N/s. Measured quantities are in the table 5.

3. Results

Samples with 10% plastic granules, plastic fibers and horse manure were tested on thermophysical properties.

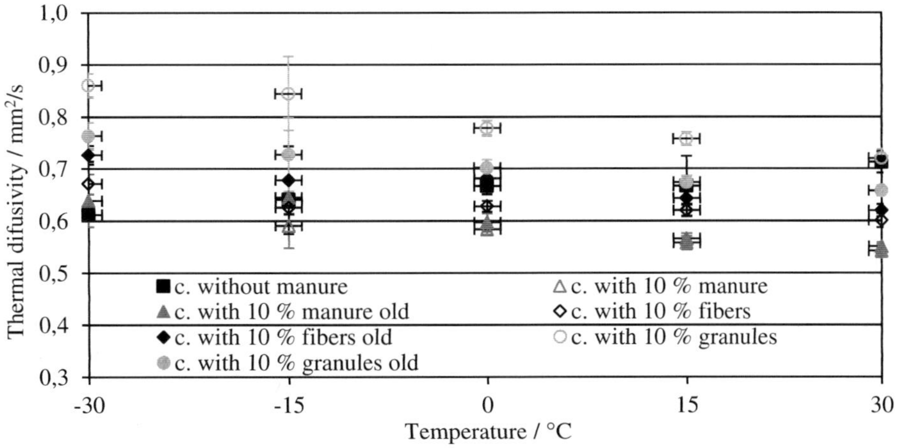

Figure 4. Thermal diffusivity values of concrete and concrete with 10% of different additives before and after ageing.

Thermal diffusivity in the temperature range -30 to 30 degrees of Celsius decreased by 12 to 17% for concrete based composites and increased by 17% for concrete. The decrease of thermal diffusivity means higher thermal inertia. The ageing process did not influence the temperature dependence of thermal diffusivity for concrete with additives (figure 4).

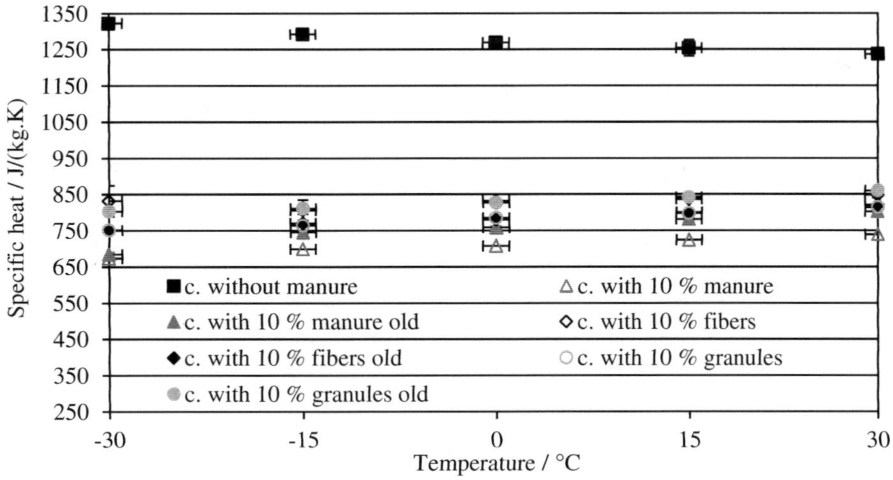

Figure 5. Specific heat values of concrete and concrete with 10% of different additives before and after ageing.

The difference between specific heat of concrete and concrete with additives was 60%. The smallest increase of specific heat value within temperature interval (-30, 30°C) was for concrete with 10% of plastic fibers (2%). Temperature changes of specific heat (after ageing) within the tested temperature range were by 5% (statistically not significant) except for concrete with horse manure (17%) (figure 5).

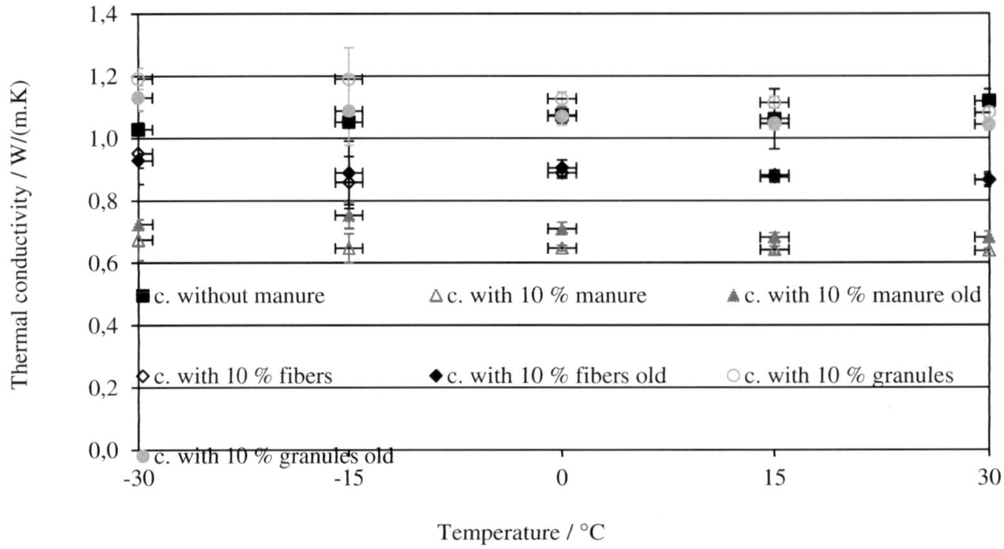

Figure 6. Thermal conductivity values of concrete and concrete with 10% of different additives before and after ageing.

Thermal conductivity values decreased for all samples with additives within the tested temperature range by 9-28%. Best observed improvement was for concrete with horse manure. Only for plastic granules thermal conductivity increased compared to pure concrete by 15% at -30°C. The thermal conductivity of materials depends upon many factors, including their structure, material mixture proportioning, size and type of added materials, density, porosity, the way of mixing, air content etc. Higher volume of plastic waste results in higher air contents, thereby decreasing the unit weight of the composite and its thermal conductivity. It might be due to the non-polar nature of granules and their tendency to entrap air in their rough surfaces. Also the tendency of plastic molecules to repel water enables air adherence to the granules. As for the best results for samples with horse manure they correlate with highest open porosity values of horse manure samples. The values of thermal conductivity after artificial ageing did not change significantly (6-8%) within the tested temperature interval (figure 6).

Table 4. Water absorption coefficient A and open porosity of non-aged concrete and concrete with 10% of different additives.

Property	Pure concrete	C. with 10 % LDPE	C. with 10 % PP	C. with 10 % HM
Density (kg/m^3)	2045	1845	1701	1586
Open porosity (-)	0.195	0.204	0.322	0.362
A (kg.m^2.s$^{-1/2}$)	0.075	0.061	0.127	0.193

Water absorption and porosity measurements were carried on to understand the material structure. The results are in table 4. Test so far hint to the following understanding. Plastic granules disrupt the capillaries, which leads to lower sorptivity value, but also smaller porosity. Fibers do not change the capillary structure so much and their sorptivity and porosity values are higher. As for 52% differences in sorptivity value between natural and plastic fibers we suppose the reason is "breathing" of natural fibers. They absorb and release water sooner. Mechanical properties' testing is important for the future usage of cement based materials. According to density values, we had lightweight concrete samples. Flexural and compressive strength of our samples are listed in table 5.

Table 5. Strength of the samples.

Sample	Flexural strength (MPa)	Compressive strength (MPa)
Pure concrete	4.9	21.1
C. with 10 % HM	3.5	13.3
C. with 10 % PET	2.4	8.2

4. Discussion

Putting the results together we can say that adding plastic waste and horse manure improved thermophysical properties before artificial ageing and after it the results were even better. The best improvement of thermophysical properties was for horse manure fibers then plastic fibers and the least improvement was for plastic granules. Horse manure fibers also proved to have better compressive and flexural strength values.

From the point of environmental concern let us do one simple calculation, which strongly depends on available data, which can vary from source to source, but it is just the basic idea. The annual production of concrete is 25 billion tons [12]. The annual production of plastic is around 350 million tons [13]. Assuming horse population of 58 million [14] and 17 kilograms of horse manure per day per horse [15], the annual production of horse manure is about 350 million tons. Assuming the whole plastic waste would be put into the concrete, its content should be 1.4 wt. %. Assuming the whole horse manure would be put into the concrete, its content should be 1.4 wt. %. Another 8.3 billion tons of plastic waste is already in the landfills [16]. If the plastic content could be increased to 4 wt. %, in about 12 years the plastic waste in the landfills could be eliminated. As the production of plastic grows about 3.3% annually, the elimination time would be prolonged to about 18 years. Assuming hard plastic density of 72 kg/m^3, and concrete density around 830 kg/m^3 [17] and horse manure density around 830 kg/m^3 [18], this translates to 16 vol. % of plastic in concrete to eliminate annual plastic waste production, respectively 46 vol. % of plastic in concrete to eliminate the plastic waste form landfills or 1.4 vol. % of horse manure to eliminate annual horse manure production. There are some drawbacks of these calculations. We did not exclude the high performance concrete production, where addition of plastic is not appropriate, secondly, there are different types of plastic (PP, HDPE, LDPE) with different material properties which determine the feasibility of their use [19].

5. Conclusion

Thermophysical properties of concrete with added horse manure fibers, plastic fibers and granules were measured. Best improvement was for horse manure fibers, where compressive and flexural strength values were also better.

Acknowledgments

Authors gratefully acknowledge the financial support of Slovak Research and Development Agency under contact No. APVV-15-0681.

References

[1] Meyer C 2009 *Cem. Concr. Comp.* vol **31** 601-605
[2] Siddique R, Khatib J and Kaur I 2008 *Waste Man.* **28** 1835-1852
[3] Iucolano B, Liguori B, Caputo D and Coffi R C 2013 *Mater. Design* **52** 916-922
[4] Pavlik Z and Zaleska M 2014 *TPH 2014 Conf. Proc. (Podkylava, SK)* pp 25-31
[5] Ochi T, Okubo S and Fukui K. 2007 *Cem. Concr. Com.* **29** 448-455
[6] Ozger O B 2013 *Mater. Design* **51** 989-997
[7] Sakia N and de Brito J 2012 *Constr. Build. Mat.* **34** 385-401
[8] Sin P, Pavlendova G, Lukovicová J and Kopcok M 2017 *AIP Conf. Proc.* **1866** 040371
[9] Kubicar L 1990 *Compr. Anal. Chem.* **12** 350
[10] Hall C 1977 *Build. Environm.* **12** 117-125

[11] STN EN 1936: 2007 *Determination of real density and apparent density, and of total and open porosity*
[12] 2009 *Cement Sustainability Initiative - Recycling concrete - Executive Summary*
[13] 2016 ISO - The Plastic Industry (Berlin)
[14] Rupak K 2015 *Global Horse Pop. with respect to Breeds and Risk Status* Master Thesis
[15] A Horse Keeper's Guide to A Horse Keeper's Guide to Manure Management
[16] UCLA - Carbon Neutrality Initiative: How do we break our initiative to plastic?
[17] Environment Protection Authority of Victoria: Waste Materials - Density Data.
[18] United States Department of Agriculture: Natural Resources Conservation Service.
[19] Yin S 2015 *Constr.Build. Mat.* **93** 180-188

Cementitious and pozzolanic behaviour of fly ash

V Pavlik

Department of Material Engineering, Faculty of Civil Engineering, Slovak University of Technology in Bratislava, Radlinského 11, 810 05, Bratislava, Slovak Republic

Abstract. Pozzolanic behaviour of fly ash was studied in mortars and pastes from lime-fly ash mixtures. Compressive strength increase of the mortar specimens that were cured in humid environment was a direct manifestation of a pozzolanic reaction of the fly ash with lime. Mineral composition of the reaction products was studied by X-ray diffraction analysis. Morphology of the samples was observed using SEM equipped with EDS detector. Analysis of a paste with the lime to fly ash ratio of 1:1 revealed that virtually all $Ca(OH)_2$ in the paste was consumed within 5 years. The crystalline products of the pozzolanic reaction included calcium hemicarboaluminate hydrate and monocarboaluminate hydrate and some ettringite.

1. Introduction

Siliceous fly ashes from high temperature coal combustion in power plant are materials that usually possess a satisfactory pozzolanic properties. Particles of the fly ashes with spherical shape are formed by the rapid cooling of molten ash particles and have an amorphous, glassy character. The proportion of the glass phase in the ash is usually 60-85%. Siliceous fly ashes contains high content of silica and alumina and also a smaller content of other compounds, e.g. Fe_2O_3. They can be used as inorganic reactive additions to concrete, as a component in some cements, in production or autoclaved aerated concrete, etc. When used, they must meet the required criteria. Their pozzolanic ability to react with calcium hydroxide in the presence of water at ambient temperature can be used also in lime-fly ash mixtures. The rate of the pozzolan reaction of ash with lime at normal temperatures is relatively slow. The reason is the relatively large size of the fly ash particles as well as their relatively small and "smooth" surface and hard and glassy nature. Contribution of fly ash to the strength in lime-ash or cement-ash materials develops only slowly. A limiting factor is moisture content in materials because reduction of the moisture content lowers the rate of the puzzolan reaction. After the moisture content drops below a certain threshold value, the fly ash in the mixture acts just as an inert filler. In the course of reaction between calcium hydroxide and fly ash different compounds can form, depending on the composition of ash and lime. The reaction product mainly include a nearly amorphous C-S-H or C-S-(A)-H phase [1]. Crystalline phases usually include mainly the AFm phases or some ettringite if sulphate anions are available in the system [1-3].

Metakaolin is much more reactive than the fly ash. Its main reaction products in the presence of small amounts of $CaCO_3$ in lime is C-S-H phase, calcium hemicarboaluminate hydrate $3CaO{\cdot}Al_2O_3{\cdot}0.5CaCO_3{\cdot}0.5Ca(OH)_2{\cdot}11.5H_2O$, monocarboaluminate $3CaO{\cdot}Al_2O_3{\cdot}CaCO_3{\cdot}11.5H_2O$ and also strätlingite $2CaO.Al_2O_3.2SiO_2.8H_2O$ that forms in the case when $Ca(OH)_2$ is consumed [1-7].

Our research was focused on the study of strength development and mineral composition changes in lime-fly ash mortars and pastes that were cured in moist conditions at relative humidity RH 100% in closed containers. These conditions enabled to monitor the development of strength in the samples as the result of the pozzolanic reaction.

2. Experimental

2.1. Materials

a) Natural standard silica sand with continuous grading according CEN EN 196-1.
b) Hydrated calcium lime powder designated CL90-S according to CEN EN 459-1 was produced by Calmit s.r.o. from the plants: (1) Tisovec, Slovakia and (2) Žirany, Slovakia.
c) Fly ash from high-temperature coal burning power plant Nováky (size fraction 0-0.25 mm). SEM images of particles in fly ash are given in figure 1 (a, b). Its composition is given in table 1.
d) Metakaolin produced by Keramost a.s. Czech Republic, designed as KM 60.
e) Plasticizing admixture (PA): A polycarboxylate based superplastificator was used. It was dosed in mass percentage of a lime-fly ash binder.
f) Tap water.

Table 1. Chemical composition of used materials: lime (L), Fly ash (FA) and metakaolin (MK).

Material	Chemical composition in wt. %											
	SiO_2	Al_2O_3	Fe_2O_3	CaO	MgO	TiO_2	MnO	K_2O	Na_2O	P_2O_5	SO_3	LOI*
Lime (1)	2.00	0.82	0.32	70.4	1.63	-	n	0.31	0.03	n	0.17	24.20
Lime (2)	0.35	0.13	0.08	73.3	0.60	-	n	0.16	< 0.2	< 0.01	0.55	24.60
Fly ash	53.50	17.8	11.7	8.0	2.13	0.56	0.20	1.60	0.90	0.40	0.98	1.98
MK	52.61	40.5	1.2	0.1	0.34	0.52	n	1.54	< 0.2	0.10	0.02	2.78

* LOI – Loss of ignition

Figure 1 (a, b). SEM of particles in fly ash: (a) - the typical spherical shape of the fly-ash particles; (b) diatomite – typical silicate shells of algae (Diatomaceae) from the coal.

2.2. Manufacture of test samples

Two types of samples were prepared:

a) Lime-fly ash pastes were prepared by mixing lime (1), fly ash and water. The ratio of lime and ash was 1: 1. Thin plates of about 15 mm thick were prepared from the paste that were stored in a closed plastic container in a humid environment at RH about 100%. Fragments of the sample were used for X-ray diffraction analysis and electron microscopy scanning (SEM).

b) Lime-fly ash mortars, lime-fly ash-metakaolin mortar and lime-fly ash paste were mixed in a laboratory mixer. The composition of the mixes is given in table 2. Cylindrical test specimens of diameter and length of 30 mm were prepared in cylindrical forms from the prepared mixes. Their consistency was measured by flow table test. A carboxylate based plasticizing admixture (PA) was used to prepare the MFA-2P-700 mortar. The produced mortar extremely fluid during the consistency test. All the samples were demoulded after 4 days and subsequently stored in a humid environment with RH 95-100%. The samples were used for compressive strength testing. Small fragments were used for electron scanning microscopy.

Table 2. Composition of a paste (P) and mortars (M) and their consistency determined by a flow table. Under the term binder we mean the common content of lime (L), fly ash (FA) and metakaolin (MK).

Sample	L (g)	FA (g)	MK (g)	Binder (g)	Sand (g)	Water (ml)	W/B (-)	PA (%)	Flow (mm)
MFA-1-700	350	350	-	700	1350	400	0.57	-	147
MFA-2P-700	350	350	-	700	1350	340	0.49	2	>180
MFA-50-350	175	175	-	350	1350	260	0.74	-	148
P-FA-3	300	300	-	600	-	320	0.53	-	-
M-MK-FA-50	175	125	50	350	1350	290	0.83	-	160
M-MK-FA-30	245	75	30	350	1350	290	0.83	-	148

2.3. Methods

Compressive strength of mortar and paste specimens was determined by using universal testing machine. Moist pastes were tested. Mineralogical composition of the pulverized lime-fly-ash pastes was determined by X-ray diffraction analysis on a Bruker D2 PHASER powder diffractometer using CuKa radiation (λ = 1.5418 Å) at step 0.02 2-theta and time per step 0.5 sec. Scanning electron microscope (SEM) Vega 3, Tescan was used for observation of samples. Acceleration voltage used was 30 kV. The SEM was equipped with EDS detector by Oxford Instruments.

2.4. Results

2.4.1 Results of X-ray diffraction analysis (XRD) of the lime-fly ash paste. The results X-ray diffraction analysis (XRD) are shown in figure 2 and figure 3 shows changes in the mineralogical composition of the lime-ash paste that occurred during its deposition at RV 100% over time. Samples were analysed in the wet state. The reason was the high sensitivity of calcium alumina hydrates to water loss during drying at elevated temperatures shown in figure 3.

For example, the calcium hemicarboaluminate hydrate $3CaO \cdot Al_2O_3 \cdot 0.5CaCO_3 \cdot 0.5Ca(OH)_2 \cdot 11.5H_2O$ (Hc) loses part of its crystal water already at drying at 40 °C. Due to these changes the intensity of its dominant basal line with d_{hkl} 8.2 Å decreases (figure 3). At drying temperature above 100 °C total disappearance of the peak was observed, but also disappearance of the dominant line of ettringite at about 9° 2 theta was observed.

XRD diffraction patterns in figure 2 revealed a decrease in portlandite content with time. The rate of reaction was relatively slow and the portlandite lines were observed even after 200 days of the paste curing. Later the lines of portlandite were only hardly observable and after 2000 days they were not present. The major crystalline products of the pozzolan reaction were calcium hemicarboaluminate hydrate (Hc) and monocarboaluminate hydrate $3CaO \cdot Al_2O_3 \cdot CaCO_3 \cdot 11H_2O$. These phases formed early as the result reaction of alumina with lime in the presence of carbonate ions ($CaCO_3$ in lime). The presence of carbonate ions eliminates the formation of C_4AH_{13}. Over time, the intensity of the Hc lines decreased due to the conversion of hemicarboaluminate to monocarboaluminate in the presence of $CaCO_3$. The results correspond to finding of [1-3]. The presence of amorphous C-S-H phases is virtually undetectable by diffraction analysis. (Their rise was mainly due to the increase in the strength of wet slurries and mortars).

Small amount of ettringite was also identified in the pastes throughout all curing period. Ettringite was probably formed as a result of the reaction of Al_2O_3 from the fly ash with a small amount of sulphate ions present in the mixture and calcium hydroxide. Ettringite presence in the paste was most probably stabilized by the presence of carbonate anions in the system. Presence of carbonate ions supports preferential formation of Hc and Mc phases instead of monosulphate and "stabilization" of ettringite, because the monocarboaluminate is a more stable phase than the monosulfate. However, a certain amount of hexagonal AFm phases containing some SO_4^{2-} ions cannot be excluded [1].

The presence of amorphous C-S-H phases was not identified by XRD analysis as these phase is virtually undetectable by diffraction analysis. However, its formation by the pozzolanic reaction probably mostly contributed to the compressive strength increase of lime-fly-ash mixtures.

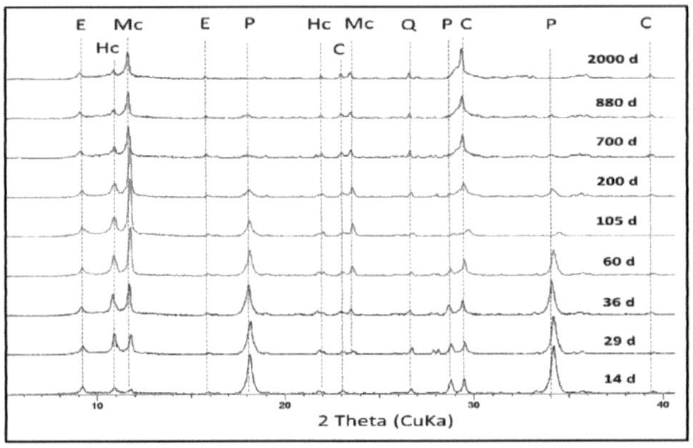

Figure 2. Changes in mineralogical composition of the lime-fly ash paste cured at RH 100% over time. Symbols used: (P) portlandite, (E) ettringite, (Hc) hemicarboaluminate, (Mc) monocarboaluminate and (C) calcite, (Q) quartz.

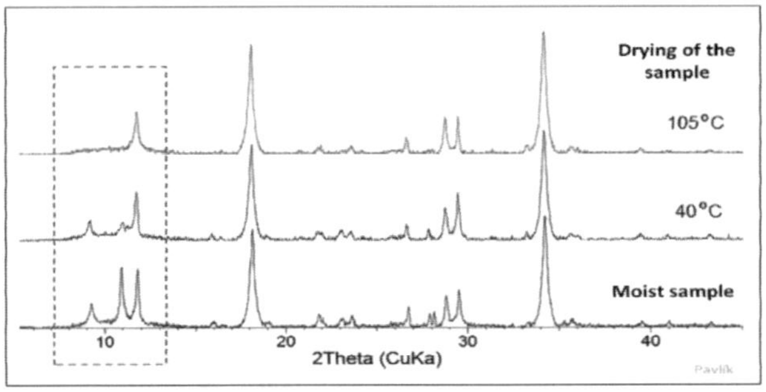

Figure 3. Effect of drying of the lime-fly ash paste cured 29 days on the basal XRD lines of aluminate phases.

2.4.2 Compressive strength results. The compressive strength of the lime-ash mortars slowly but gradually increased during the whole testing period (figure 4). The highest values of compressive strengths were achieved by MFA-2P-700 mortars with the lowest water/binder ratio (table 2) where plasticizing admixture was used. Compressive strength of this mortar stored in water for about 700 days was even slightly higher.

Compressive strength of mortars with lower binder content was lower. The lowest values of compressive strength were obtained in mortars where some part of the fly ash was replaced by metakaolin. These mixtures contained an increased content of Al_2O_3 at the expense of SiO_2 and therefore less C-S-H phase is formed. This might suggest that the crystalline AFm phases contribute less to the compressive strength development than does the C-S-H phase. However, the lower strength development at later ages in the mixtures could also be affected by a partial loss of moisture content.

All the results underline the bonding potential of the fly ash. This potential is only exploited when there is a sufficient amount of water available for the pozzolanic reaction. Decrease in the compressive strength of lime-fly ash paste at later times in figure 4 also can be explained as the result of partial drying of the samples during its storage.

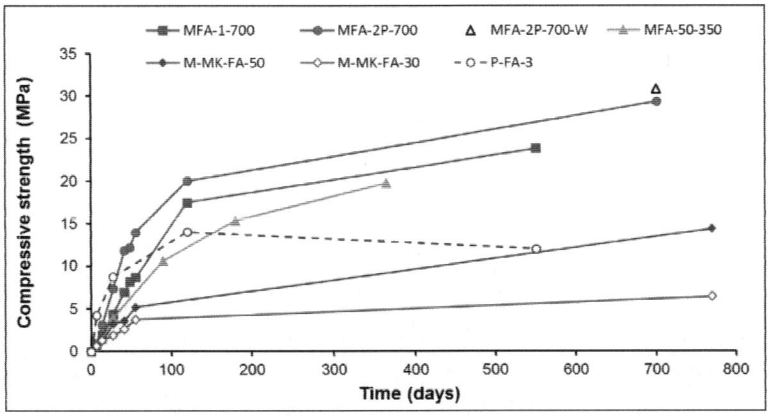

Figure 4. Compressive strength development of specimens cured in humid environment.

2.4.3 Scanning electron microscopy of samples. Appearance of the microstructure by SEM of FAM-2P-700 mortar and the X-ray microanalysis (EDS) of large hexagonal crystals that is visible in a cavity is shown in figure 5. The analysis of the hexagonal phase (spectrum 8) revealed that it probably corresponds an AFm phase containing sulphate ions, possibly monosulfate, or its solid solution with carbonate AFm.

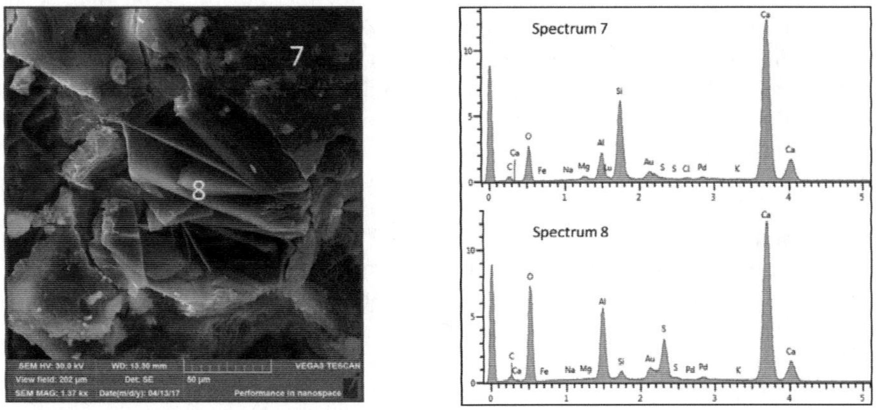

Figure 5. SEM image of lime-ash mortar FAM-2P-700 after 2 years of storage in water (left) with EDS spectra of the elemental analysis in selected points (right).

Figures 6 (a, b) show spherical fly ash particles in mortars that are covered with reaction products and a surrounding distinct edge zone. figure 6 (c) shows hexagonal AFm phases in a lime-fly ash mortar and figures (d-f) show products of pozzolanic reaction in vicinity of fly ash particles in mortars.

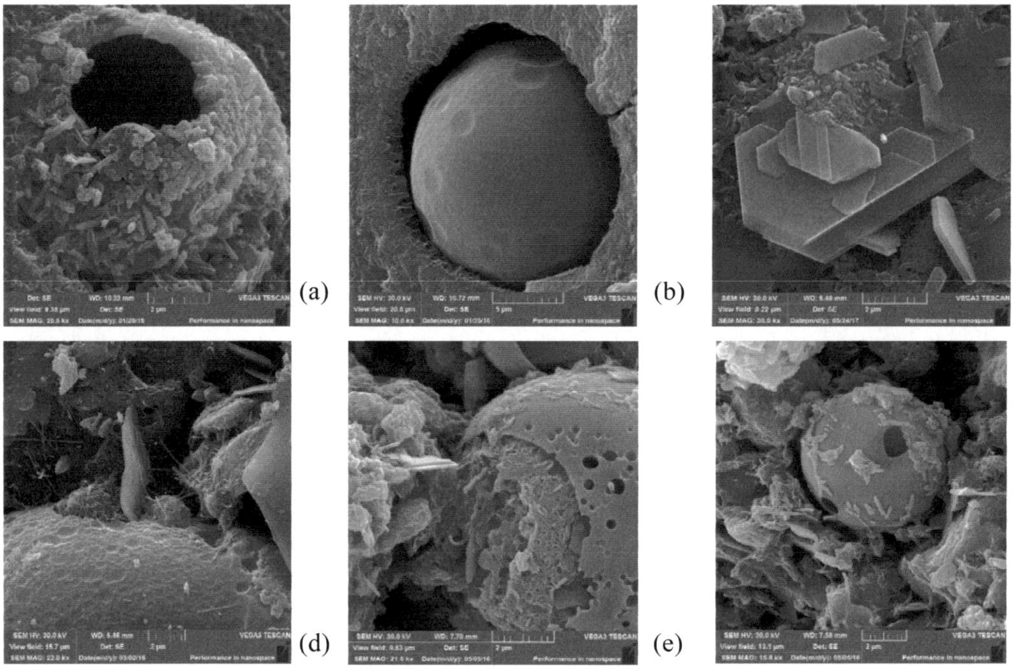

Figure 6. SEM image of mortars MFA-L-50 after 3 years (a, b) and 107 days (c); and M-MK-FA-50 after 8 days (d, e) and 30 days (f) of storage in humid conditions.

3. Conclusion

High-temperature siliceous fly ash has a great pozzolanic bonding potential. The increase in compressive strength of lime-fly ash composites in a humid environment enabled to estimate the contribution of the puzzolanic reaction to the compressive strength of the composites. At a 1:1 ratio of lime/ fly ash practically all calcium hydroxide reacted to produce the puzzolanic reaction products. The most important strength giving product of the pozzolanic reaction was C-S-H. The crystalline products were carboaluminate phases of AFm type. At higher content of sulphate ions in lime or fly ash some ettringite also formed. Its stability was likely to increase in the presence of $CaCO_3$ in lime.

Acknowledgements

This article was created with the support of the Ministry of Education, Science, Research and Sport of the Slovak Republic within the Research and Development Operational Programme for the project "University Science Park of STU Bratislava", ITMS 26240220084, co-funded by the European Regional Development Fund.

References

[1] Taylor H F W 1997 *Cement Chemistry* (London: Thomas Telford Publishing)
[2] Matschei T, Lothenbach B and Glasser F P 2007 *Cem. Concr. Res.* **37** 118–130
[3] Damidot D and Glasser F P 1995 *Cem. Concr. Res.* **25** 22-28
[4] Matschei T, Lothenbach B and Glasser F P 2007 *Cem. Concr. Res.* **37** 1379-1410
[5] Santos Silva A, Gameiro, A, Grilo J, Veiga R and Velosa A 2014 *Appl. Clay Sci.* **88-89** 49-55
[6] Cabrera J and Rojas M F 2001 *Cem. Concr. Res.* **31** 177-182
[7] Gameiro A L, Santos-Silva A, Veiga M R and Velosa A L 2012 *Materiali in technologije/Materials and technology* **2** 145-148

IOP Conf. Series: Materials Science and Engineering **385** (2018) 012043 doi:10.1088/1757-899X/385/1/012043

Internal curing of HPC with hydrogel

D Pytlik[1], V Bilek[2] and M Bambuchova[3]

[1,2,3] Faculty of Civil Engineering, VSB – Technical University of Ostrava, L. Podeste 1875/17, Ostrava – Poruba, Czech Republic

Abstract. High performance concrete (HPC) is concrete with a very high cement content and very low cement/water ratio and this leads to better properties of concrete. These concretes have a problem with traditional methods of curing because of very dense cement matrix. Water from outside cannot reach the interior of the concrete to achieve hydration and it may not have enough curing water for cement hydration. In this paper is discussed effect of internal curing of HPC. The objective of internal curing is to create water storage inside concrete. This should be achieved with hydrogel which was selected to examine influence on internal curing. Hydrogel has a high water-absorption capacity about 300% so it should create a sufficient supply of water in internal reservoir. The prediction is that the hydrogel will release water after the concrete get hardened and provide water to another cement hydration. This should lead to the concretes with better properties.

1. Introduction

Internal curing is basic method how to provide sufficient amount of water for HPC curing during hydration. Especially in case of High Performance concretes, it is not easy to supply ordinary curing water through the very dense cement matrix. The deficiency of internal curing water is caused by cement hydration which consumes water. The un-hydrated cement grains serve only as a filler and do not participate in hydration [1-3]. The mechanical properties of HPC do not show an increase, oppositely, microcracks occurrence can reduce some properties - flexural strength, fracture properties and also durability. There are several options to supply concrete with internal curing water [4-6].

The aim of the internal curing is to create an internal "reservoir" that will release the water after the concrete get hardened and help cement to hydration. It will then serve as curing water and will not change the real water/binder ratio [7]. In accordance with [8], the water/cement ratio should be more than 0.38 to avoid microcracking and achieve good cement hydration. This means that the main goal of internal curing is to have really low water/binder ratio when mixing to create very dense cement matrix and have enough water in "reservoir" for later to get better hydration of cement.

Self-curing is not common in practice. There are only the traditional methods of curing concrete from "outside". This can be done by using wet geotextiles and moistening concrete surface with water. These methods are highly functional, but in the case of high performance concrete the cement matrix is very dense and traditional curing does not guarantee a sufficient amount of curing water in the concrete.

The objective of this study is to investigate the effect of hydrogel on the flexural and compressive strength. In this paper mechanical properties of fresh and hardened concrete was discussed.

2. Experimental details and materials

The objective of this paper is to test the effect of internal curing on properties of HPC. The composition of HPC mortars is shown in table 1. This concrete has a very high dosage of cement and very low water/binder ratio. Fraction 0-4 mm of sand was used as an aggregate and Portland CEM 42.5 R was used as a binder. Limestone was used as a micro filler and metakaolin was used like a substitute part of binder to enhance the properties of fresh and also hardened concrete.

Table 1. Composition of concrete to 1m^3.

TYPE	Cement [kg/m^3]	Metakaolin [kg/m^3]	Limestone [kg/m^3]	Plasticizer [l/m^3]	Water [l/m^3]	Sand [kg/m^3]	hydrogel [kg/m^3]
C	650	90	227.5	10	206	1200	0
H 2.5%	650	90	227.5	10	206	1200	25
H 5.0%	650	90	227.5	10	206	1200	50
H 7.5%	650	90	227.5	10	206	1200	75
H 10%	650	90	227.5	10	206	1200	100

C – Control
H - with hydrogel

We can divide types of concrete in two group. One group of concrete is "control" without internal curing agent. The other concrete has agent in composition. The percentage of agent dosage was 2.5 %; 5 %; 7.5 % and 10 % volume of fresh mortar. All mortars were mixed in laboratory mixer with a volume of 1 liter.

The composition of internal curing agent was made from hydrogel, water and chemical admixture PCA, so hydrogel was pre-soaked with PCA and water. Hydrogel is commercial product especially for soil treatment. According to the technical data sheet hydrogel has absorption 300 % by weight. That means that 1 gram of hydrogel can absorb 300 grams of water it prevents drying of soil. Hydrogel is dry and granulated. The dry grain size is about 1 mm as it can be seen in figure 1. Chemical admixture PCA was added to hydrogel to enhance internal curing. PCA is commercial product of Chryso chemie. Acording to the data sheet it should help stabilized water in concrete. PCA should help stabilized water in hydrogel, because PCA can bind water like starch and release this water supply gradually for a long time [9]. The composition of internal curing agent was (2 % - PCA, 4 % - hydrogel and 96 % - water). The precondition was that water "trapped" in hydrogel does not affect the amount of mixing water so the internal curing agent creates internal "reservoir" of water. According to this the total amount of water should be higher than calculated from water/binder ratio.

Figure 1. Photo of dry (left) and soaked (right) grains.

3. Experimental procedures

All mortars were mixed in laboratory mixer in accordance to EN 196-1 [10] with a volume of 1 liter. The workability of mortars was measured by a truncated cone flow (low diameter 100 mm, upper diameter 70 mm, height 60 mm). The flow was measured on a jolting table immediately after remove from cone and after 15 jolts in two perpendicular directions.

After mixing the mortars were put into molds and compacted in accordance to EN 196-1 [10]. The prims were demolded at the age of 1 day and they were immediately wrapped in plastic foil and stored under laboratory conditions (temperature 23 °C and 52 % humidity), except prims for 24 hours strength. Prims were stored in plastic foil because the effect of the internal curing is more noticeable especially witch such a low water/binder ratio. Flexural and compressive strengths were measured on the prisms (40x40x160 mm) at the curing age of 1, 7 and 28 days (2 prisms at each age).

4. Results and discussion

4.1. Bulk density development

One of the evidence that internal curing agent really creates a reservoir with internal water is bulk density. Because if concrete has a pore filled by internal curing agent, with density lower than concrete, it has to be lighter than control concrete sample which has lower porosity. As shown in figure 2 shows that concrete with high dosage of internal curing agent has a lower bulk density than control concrete sample.

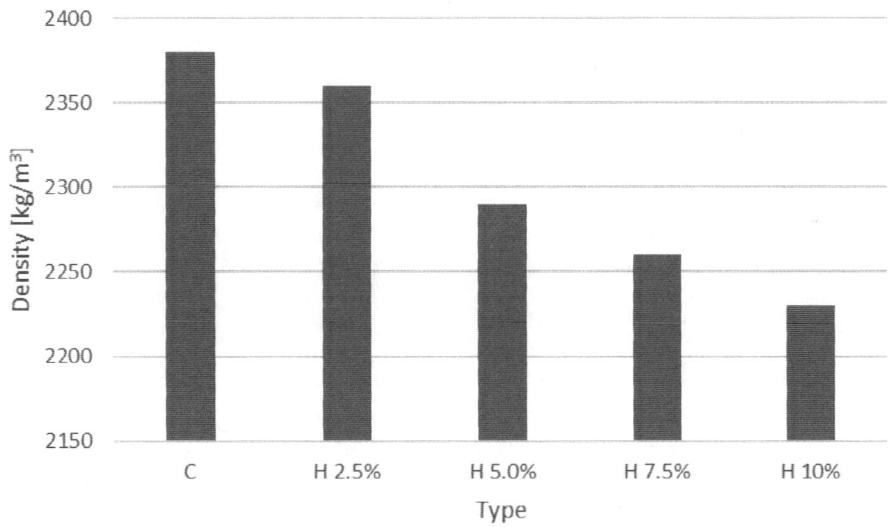

Figure 2. Density of different types concrete.

4.2. Workability of mortar

As it can be seen in figure 3 the workability of fresh mortar has changed with higher dosage of internal curing agent. Apparently, the hydrogel did not "lock" all water in itself, so some water was released while mixing. This has a negative influence on fresh concrete and hardened concrete too. The target workability was 250 mm. This test was mainly performed to determine if the water was released or not while fresh concrete was mixed. Maximum value (300 mm) of flow cannot be measured due to the device size. In further research it will be necessary to focus on how to better "lock" water in hydrogel.

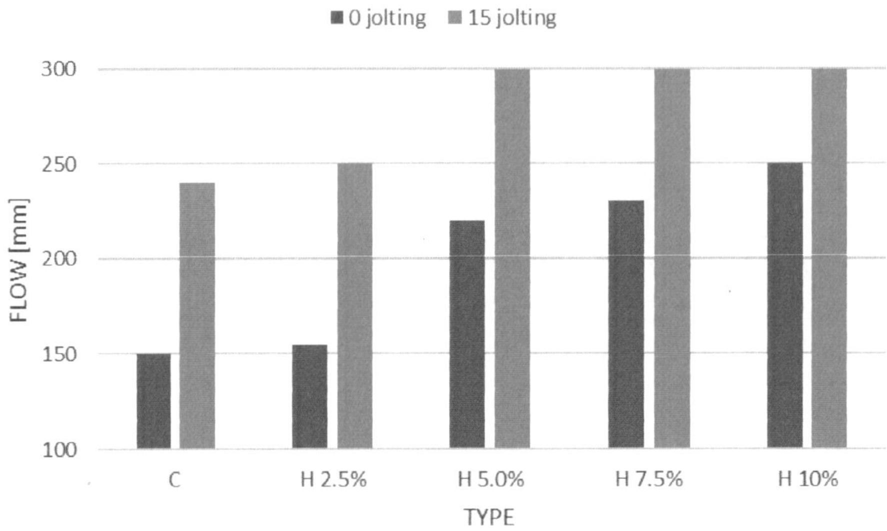

Figure 3. Workability of fresh mortar.

4.3. Strengths development

When internal curing agent is used, the flexural strength is lower than strength of control sample as it can be seen in figure 4. The interesting is difference between 7 and 28 days flexural strength. Controls sample has 28 days strength higher than 7 days. But the strength of sample with internal curing agent were similar in 7 days and 28 days. Hydrogel probably during hydration produce aggressive substances that react with the cement components. Aggressive substances produce either soluble compounds or insoluble compounds which do not have binding properties. Because of this in next research will be important to focus on chemical analysis of hydrogel and synergy of hydrogel with cement.

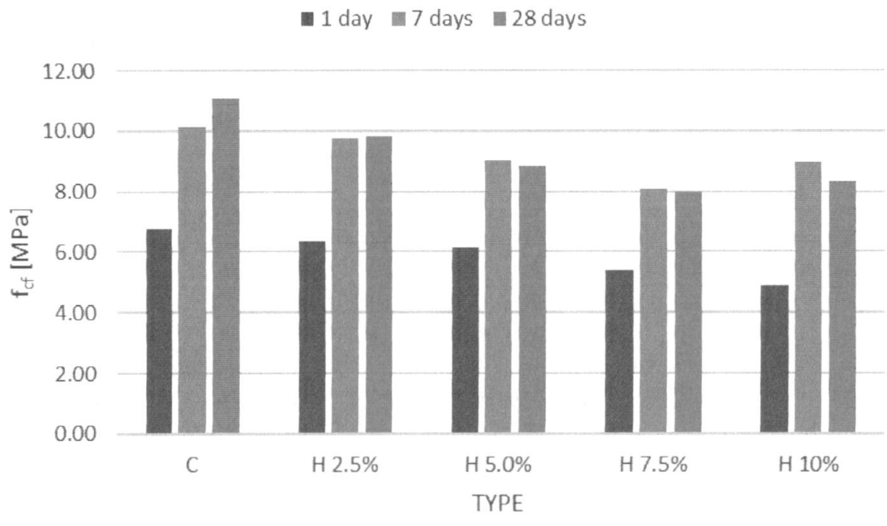

Figure 4. Flexural strength of concrete.

Figure 5 shows that agent is likely to have a negative effect on the water/binder ratio. The higher internal curing agent dosage, the lower the flexural strength as in case of flexural strength and also by changed water/binder ratio. It could be seen on ratio between compressive strength. The ratio between 1 day and 28 days compressive strength is around 50 percent higher for sample with 10 % dosage of internal curing agent than ration between strength of control sample and this is probably due to the changed water/binder ratio. This will be examined in further research.

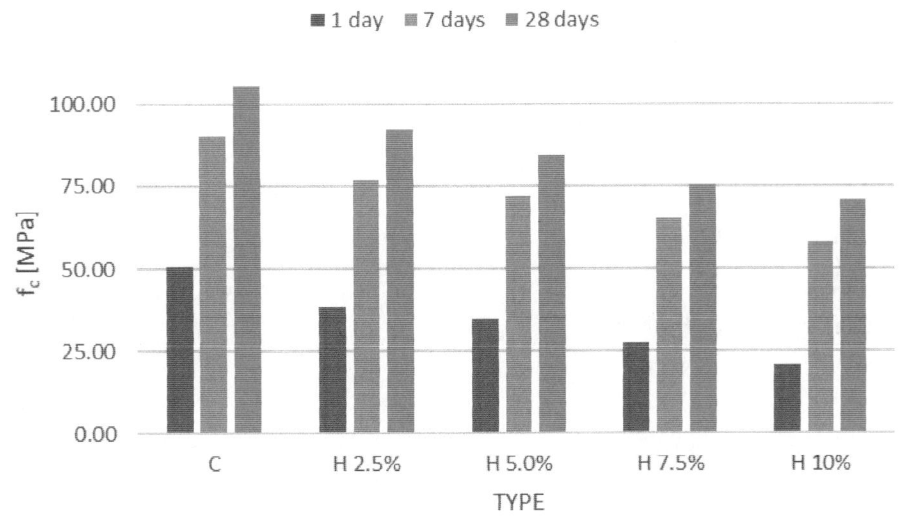

Figure 5. Compressive strength of concrete.

5. Summary

Internal curing was provided by internal curing agent which was in this case composition of hydrogel, PCA and water. The precondition was that hydrogel would not release water immediately after mixing the mixture, but after some time. This should suppress the negative effects (change of water/binder ratio, change of workability etc.) by adding water.

This internal curing agent do not seem very effective. But hydrogel itself seem to be very promising internal curing agent because of its cost and absorption ability. Because precondition was not fully fulfilled. Water in hydrogel were partly release during mixing especially in high dosage of hydrogel. In this case both flexural and compressive strength of sample with agent was lower than strength of control sample. Flexural strength of control sample was about 11 MPa. But the higher internal curing agent dosage was, the lower the flexural strength was. This happens probably due to chemical reaction between hydrogel and cement and changed water/binder ratio while mixing. Necessary will be chemical analysis of hydrogel and synergy with cement.

Released water while mixing change the workability of fresh mortar. When dosage of internal curing agent was higher than 5 % of volume the mortar was self-compacted.

Concrete with higher dosage of internal curing agent has lower density due to higher porosity of concrete.

In next research would be interesting combine with another way of internal curing. However, it is necessary to improve the methodology of the comparison of individual result and the effect of internal curing on other factor influencing mechanical and physical properties of concrete.

Acknowledgement

This work was supported by the project SGS SP 2018/182 granted by Faculty of Civil Engineering VŠB-TU Ostrava.

References

[1] Aitcin P C 2008 *Proc. Int. Conf. on NTCC (Brno)* (Brno: Brno University of Technology) pp 1-7
[2] Aitcin P C, Mindess S and Tagnit-Hamou A 2014 *Proc. Int. Conf. of NTCC (Brno)* (Brno: Brno University of Technology) pp 1-7
[3] Zhutovsky S 2012 *Mater. Struct.* **35** 97-101
[4] Suzuki M, Meddah M S and Sato R 2009 *Cem. Concr. Res.* **39** 373-381
[5] Weber S and Reinhardt H W 1997 *Advn.Cem.Bas.Mat.* **6** 59-68
[6] Bilek V, Mosler T, Kersner Z and Schmid P 2015 *Innovations and Developments In Concrete Materials And Construction* (Scotland: Thomas Telford) p 51-60
[7] Maduru S R CH, Ratna-Giri P S N, Kumar P R, Kumar G R and Raveena C 2016 *Contr. Build.Mater.* **107** 356-364
[8] Morin R, Haddad G and Aitcin P C 2002 *Concr. Int.* **24** 43-48
[9] Pytlik D, Bilek V 2017 *Sol. St. Phen.* **272** 88 - 93.
[10] CSN EN 196-1. 2005 *Methods for testing cement* - Part 1: Determination of strength. Prague: Czech Standards Institute

IOP Conf. Series: Materials Science and Engineering **385** (2018) 012044 doi:10.1088/1757-899X/385/1/012044

Impact of the grinding process on the granulometric properties of triclinic alite

S Ravaszova and K Dvorak

Faculty of Civil engineering, Brno University of Technology, Veveri 331/95, 602 00 Brno, Czech Republic

Abstract. Four major clinker minerals, alite, belite, tricalcium aluminate and tetracalcium aluminate ferrite are crucial for cement properties. This work deals with alite as the most important clinker minerals. Alite is affected by the milling process technology more than other clinker minerals. It contains micro-cracks, which are formed during fast cooling, and is therefore less resistant to grinding than other clinker minerals. The paper deals with the influence of length and the technology of the milling process in 3 types of laboratory mills on the particle size and distribution, agglomeration and the change of the grain shape of the synthetically prepared triclinic alite.

1. Introduction

The most important part of the Portland cement is clinker, which is produced by burning raw meal in a cement kiln at 1450°C. The final properties of the clinker and cement do not directly determine the chemical composition but the mineralogical composition. The mineral composition of the clinker is the result of used raw materials and production technology.

Up to now, more than 25 minerals are known in Portland clinker. It is necessary to distinguish pure minerals from technical ones that are usually altered in the clinker by the presence of other components or impurities in the solid solution. The most important technical phases occurring in the clinker are alite (C_3S), belite (C_2S), tricalcium aluminate (C_3A) and brownmillerite (C_4AF) (according to Törnebohm) and their polymorphs [1-3].

Alite (Ca_3SiO_5, C_3S) is the dominant phase of Portland clinker, the amount of which is generally above 60%. It is essentially tricalcium silicate (C_3S) with a certain amount of Al_2O_3, MgO, Fe_2O_3 and other solids in its structure. C_3S occurs at temperatures above 1250°C and in Portland clinker is in an unstable subcool condition, because its stability field lies between 1250 - 1900°C [4].

Alite is characterized by significant polymorphism. Up to date, seven modifications have been identified in the aliquot, 3 triclinic (T1, T2, T3), 3 monoclinic (M1, M2, M3) and 1 trigonal (R) dependent on temperature and impurity content. Under laboratory conditions there is a stable modification of T1. It is simplest to prepare because it is produced by cooling the pure C_3S compound [5-6].

This work is based on the preparation of the pure triclinic phase of alite, with the aim of monitoring the influence of milling in laboratory mills on its grindability.

The grinding time is one of the most important parameters affecting the milling process. With the growing time of grinding, the material gets so refined that agglomeration and aggregation may occur, which means that when the material is grinded, the individual particles are distributed but subsequently very compressed. There is an electrostatic charging of the particle surface with opposite charges.

2. Materials and methods

The Wesselsky-Jensen method was used to prepare pure triclinic alite. For solid phase synthesis, a ventricular high temperature kiln with superkanthal heating elements was used. Solid phase synthesis was carried out in two steps at 1600°C in accordance with the Wesselsky-Jensen method. Phase composition control was performed by XRD analysis on PANalatical Empyrean (Cu-cathode λ = 1.540598 for Ka1).

The triclinic alite was adjusted to an entry specific surface area of 400 m^2/kg. To achieve a specific surface area, the material was milled at 500 rpm in a Fritsch Pulverisette 6 planetary mill. Blaine specific surface area was measured using a PC-Blaine-Star automatic device with a measurement cell capacity of 7.95 cubic centimeters. The measurement was performed three times to eliminate errors, and the resultant value was the average of the three readings.

Alite subsequently underwent grinding in three types of laboratory mills with dry and wet technology. The milling was done using the PULVERISETTE 6 planetary mill, the RS 200 vibratory disk mill from RETSCH and the McCrone Micronising mill. Fritsch planetary ball mill Pulverisette 6 in a bowl with the volume of 0.5 dm^3 was used. The bowl was filled by 25 grinding balls with 20 mm in diameter. The comminution of the material takes place primarily through the high-energy impact of the grinding balls. The grinding bowl, containing the grinding balls and the material to be ground, rotates around its own axis on the main disk whilst rotating rapidly in the opposite direction. Centrifugal force causes the ground sample material and the grinding balls to separate from the inner wall of the grinding bowl. The grinding balls then cross the bowl at high speed and further grind the sample material by impact against the opposite wall. The Vibratory Disc Mill RS 200 comminutes by impact and friction. The grinding set is firmly attached to the vibration plate with a quick-action lever. The plate with the grinding set is subjected to circular horizontal vibrations. The centrifugal force acting on the grinding rings in the dish results in extreme pressure. Impact and frictional forces acting on the sample produced analytical fineness in 1-3 minutes. The circular vibrations are produced by a frequency controlled 1.5 kW 3-phase motor.

Isopropyl alcohol was used as an inert liquid for wet milling. McCrone Micronising Mill is designed only for wet milling. 5 periods of milling were set at 1, 5, 20, 60 and 120 minutes. The milling in the planetary mill was set at the optimum speed for the alite 300 rpm and in a vibratory disk mill at 1200 rpm. The grinding times in the McCrone Micronising Mill were the same as in the previous two types of mills.

The particle distribution and fineness were determined using the Malvern Mastersizer 2000 laser granulometer.

The monitoring of grain shape and topography and the agglomeration process for the various milling technologies were determined using a scanning electron microscope TESCAN MIRA3 XMU.

3. Results

The quality control of the mineralogical composition was performed after the 1st and 2nd firing steps by XRD analysis, the scans of which is shown in figure1.

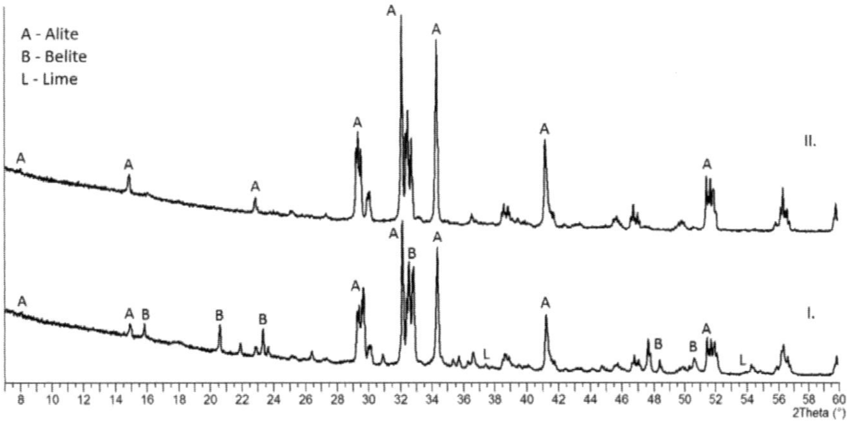

Figure 1. XRD patterns after first and second burning.

After the first firing, alite, belite and residual lime are identified in the diffraction record. After the second firing, this is a triclinic alite.

Comparison of the influence of the milling technology on the distribution and the particle fineness was done using laser granulometry. In the following graphs, the granulometry is compared at a milling time of 1 and 120 minutes for all three types of laboratory mills and milling technologies where the change in particle distribution and fineness is the most pronounced.

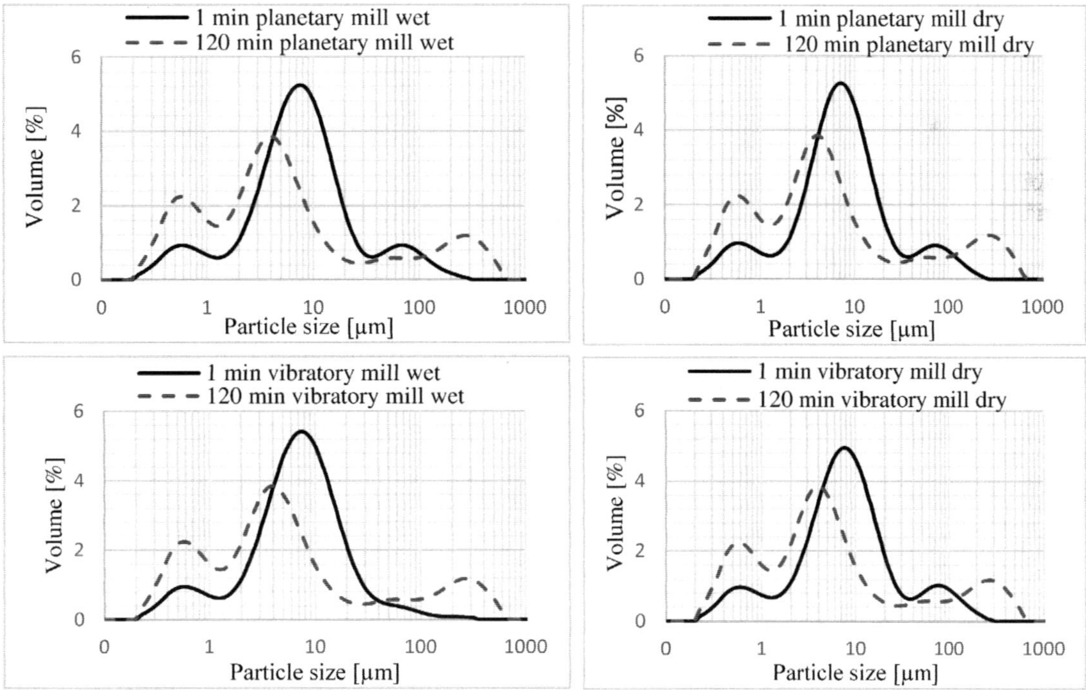

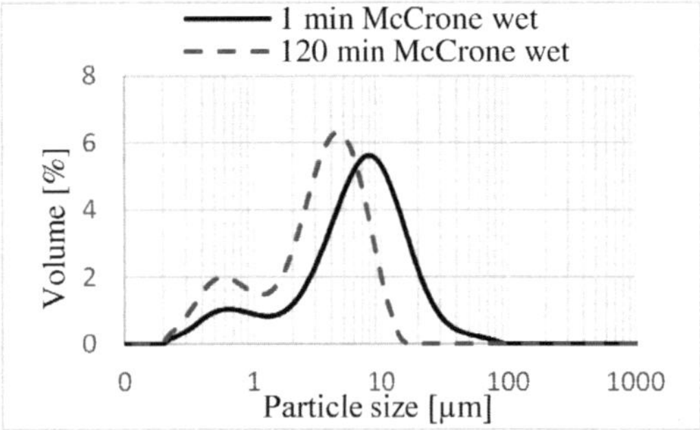

Figure 2. Particle distribution at milling time of 1 and 120 minutes for all types of mills and grinding technology.

Intensive milling in the vibratory and planetary mills results in the formation of large quantities of fine particles. These particles formed agglomerates which caused higher volume of particles above 100 μm. This phenomenon was more pronounced when grinding the material with dry technology.

Comparison of the particle distribution during grinding of material in all types of mills and milling technologies in 120 minutes is shown in figure 3.

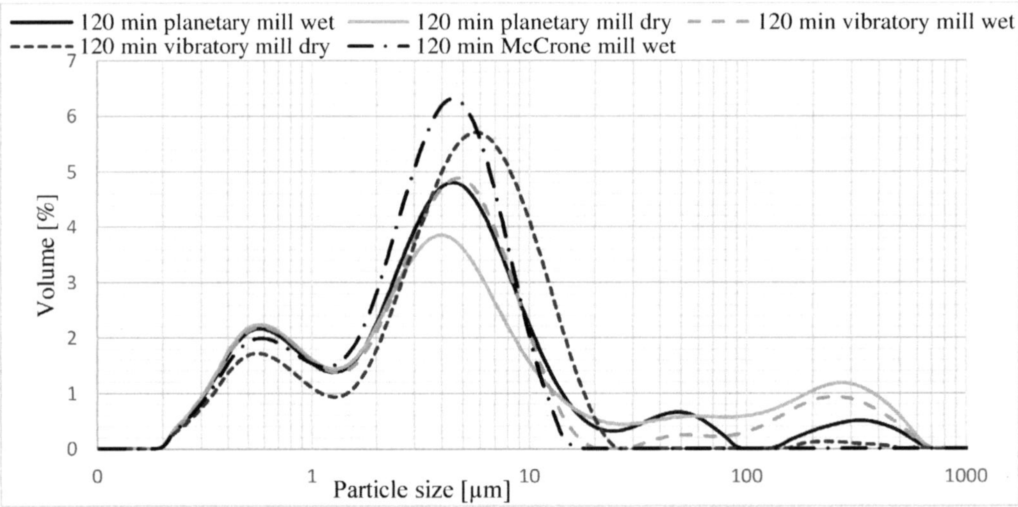

Figure 3. Particle distribution in 120 minutes.

The largest amount of very fine particles is produced by dry grinding in a planetary mill. The trend of the milling shows good grindability of the material, even the most pronounced increase in the volume of agglomerates which has been produced at the expense of fine particles up to 10 μm.

In wet grinding in a planetary mill, the amount of agglomerates is smaller than in the case of dry grinding, as it is to some extent suppressed by the liquid. In the area of very fine and fine particles, the wet milling trend in the vibratory mill is almost the same as in the wet milling in the planetary mill. At both mills, the formation of agglomerates, the amount of which in the case of the vibratory mill is greater.

With dry grinding in the vibratory mill, the smallest amount of very fine particles is observed, which is reflected by a relatively high amount of fine particles. We see only a slight increase in agglomerates, as there has been no significant increase in the amount of very fine particles that would tend to clump to the agglomerates in dry milling.

When grinding at McCrone mill, the material is well pulverized, increasing the amount of fine particles at the expense of the coarser portions without the agglomeration that is observed when grinding in other types of mills.

The influence of grinding on the grain shape was evaluated by electron microscopy. In figure 4 are images of material ground in each type of mills with dry and wet technology.

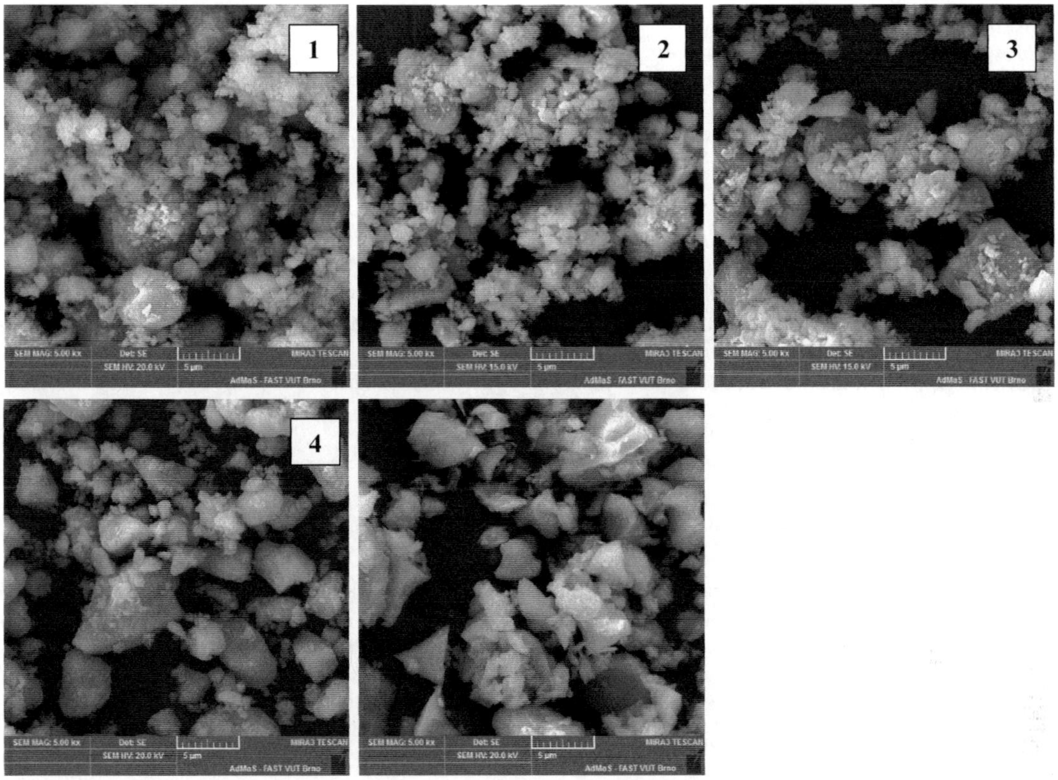

Figure 4. SEM images for dry milling planetary mill (1), wet milling planetary mill (2), wet milling vibratory mill (3), dry milling vibratory mill (4), wet milling McCrone mill (5).

In the image of dry grinding in a planetary mill, we see a considerable refinement of the material, a very large amount of very fine particles and massive agglomerates.

Wet grinding in a planetary mill, the agglomeration is less pronounced, the grain material is nicely rounded.

In the case of dry grinding in the vibratory mill, we do not see much of the material softening as in previous mills and milling technologies. Agglomeration is minimal.

In the image of dry grinding in McCrone Mill, we see that the grains are not isometric as seen in previous mills but are sharp-edged. Cleavage planes are visible on them. The amounts of fine particles are minimal agglomeration does not occur.

4. Discussion

The pure triclinic alite was prepared by the Wesselsky-Jensen method and subjected to various milling technologies in three types of laboratory mills.

Dry grinding in a planetary mill seems very effective in particle size reduction. The problem arises with a longer grinding time when a large amount of energy and very fine particles accumulate in the grinding capsule, which tend to stick together, and after 120 minutes of grinding clump into massive agglomerates, as confirmed by electron microscopy images.

Wet milling in a planetary mill, the agglomeration is suppressed to a certain extent in the liquid, but after 120 minutes of milling, agglomerates also occur, regardless of whether they are wet in the planetary or vibratory milled. The grinding efficiency in terms of particle size is in both cases substantially identical with the grinding parameters selected.

The material was milled at 300 rpm in the planetary mill and at the vibratory mill at 1200 rpm, indicating that the two grinding modes are comparable to one another.

The wet milling trend in the vibratory mill is therefore almost identical to the trend of wet milling in the planetary mill, with the exception that a larger amount of agglomerates is produced in the vibratory mill than in the case of a planetary mill. The reason for agglomeration in wet milling is the fact that when milling in a liquid and at high energy in a grinding capsule, the material is melted into very fine particles which are poured into themselves and after drying, tend to aggregate into agglomerates.

In the case of dry grinding in the vibratory mill, we observe the lowest amount of very fine particles, and thus a small amount of agglomerate which was confirmed again by electron microscopy. In this case, and for this material, the use of a vibratory mill in a dry medium appears to be the least appropriate.

McCrone-milled material does not tend to form agglomerates after 120 minutes. McCrone mill does not destroy the structure, but it mills a very small amount of material so it is not suitable for larger laboratory milling.

5. Conclusion

The greatest influence on the granulometric properties of triclinic alite has a dry milling in a planetary mill, which appears to be the most effective among all types of mills and technologies used. There is a very rapid refinement of the material. The problem occurs with long grinding times when very fine particles tend to clump to the agglomerates. In the case of selected grinding modes, specifically wet grinding, the efficiency and influence of the grinding is the same as for the vibratory mill in terms of its impact on the granulometry. It has been confirmed that McCrone mill is designed very well to prepare samples for XRD analysis.

Acknowledgements

This work was financially supported by project No. FAST-J-18-5432: "Preparation of alite polymorphs".

References

[1] Duda H W 1988 *Cement Data-book* (Wiesbaden: Berlin: Bauverl)
[2] Taylor H F W 1997 *Cement chemistry* (London: Thomas Telford Ltd.)
[3] Hewlett P 2004 *Lea's Chemistry of Cement and Concrete 4* (Oxford: Elsevier Butterworth-Heinemann)
[4] Bigare M, Guinier A, Mazieres C, Regourd M, Yannaquis N, Eysbl W, Hahn T H and Woermann E 1967 *J. Am. Ceram. Soc* **50** 609-619
[5] Dunstetter F, DE Noirfontaine M N and Courtial M 2006 *Cement. Conrete. Res.* **36** 39-53
[6] Dunstetter F, DE Noirfontaine M N and Courtial M, Gasecki G and Signes-Frehel M 2006 *Cement. Concre. Res.* **36** 54-54

IOP Conf. Series: Materials Science and Engineering **385** (2018) 012045 doi:10.1088/1757-899X/385/1/012045

Quality control of the concrete of the bridge reg. No. M117 in Pardubice

S Rehacek, D Citek and J Kolisko

Klokner Institute, Czech Technical University in Prague, Solinova 7, Prague 6, Czech Republic

Abstract. This paper is introducing results of diagnostic survey of bridge reg. No. M117 in Pardubice. Which included the quality control of the concrete used in the construction after more than 40 years of operation. Compressive strength of concrete, identification of content of chlorides in concrete and identification of resistance to CH.R.L. – method C are presented in this paper. This paper come into existence with cooperation with the state-funded organization Road Administration and Maintenance of the Pardubice Region and TOP CON SERVIS s.r.o. company.

1. Introduction

The bridge ev. No. M117 bridges over the railway line, street of kpt. Jarose (R36) and several local roads. The bridge has 20 sections and transfers the Kyjevska road.

The load-bearing structure consists, in the cross-section, of 9 pcs pre-stressed prefabricated beams KA 61, which were extended, furthermore, by 0.8 m. The length of the beam is, therefore, 14.4 m. The span is 13.85 m. During the repair of the bridge in the years 1992 – 1993, the exposed and ceiling surfaces of the load-bearing structure were provided with a unifying and protective coat.

Both supports are made of concrete with reinforced-concrete embedded groundsills. The middle supports are of reinforced concrete with cast concrete reinforced-concrete pre-stressed cap sills. During the repair of the bridge in the years 1992 – 1993, the bottom construction was treated with scouted sprayed concrete. The construction-technical investigation was focused only on the bridge fields in the surroundings of the railway line Prague – Ostrava and the highway R36. Shape and characteristics of the structure are in more detail shown below, on Figure 1.

Figure 1. Bridge No. M117 in Pardubice.

2. Concrete

2.1. Compressive strength of concrete – summary

The summary of the results of destructive tests of compressive strength of concrete and thereto corresponding strength classes or, as the case may be, classes of concrete, are identified in the following Tables 1 and 2.

Table 1. Summary of the results of the tests of the compressive strengths of concrete.

Diagnosed construction elements		Compressive strength of concrete (MPa)		Variation coefficient v*
		test average	characteristic	
Cap sills of pillars	destructively	57.3	50.3	3.4
Columns of pillars	destructively	44.1	38.1	4.4

* ČSN 73 2011 [1] specifies the limit value of the variation coefficient for the homogenous concrete C30/37 and higher **v** = 12 % (homogeneity from the standpoint of strengths).

Table 2. Strength class or, as the case may be, class of concrete on the basis of the performed tests.

Diagnosed structural elements		Class of concrete or, as the case may be / strength class of concrete	
		ČSN EN 1992	Requirement of documentation
Cap sills of pillars	destructively	C 45/55	-
Columns of pillars	destructively	C 35/45	-

The following conclusions can be stated on the basis of the performed tests of the compressive strength of concrete:

- On the basis of destructive tests of compressive strength of concrete within the limited scope, we recommend considering the following classes of concrete for the monitored reinforced-concrete structure of the bridge, according to ČSN EN 1992:
  - Columns of pillars: C 30/37
  - Cap sills of pillars: C 40/45

2.2. Identification of content of chlorides in concrete

The content of chloride ions above a certain limit level significantly increases the risk of the corrosion of the reinforcement. That is why a chemical analysis of the concrete was performed within the diagnostic works in order to identify the content of the chloride ions in concrete [4].

Sampling was evenly distributed along the structure of the bridge. Altogether, 20 samples of concrete were taken in 10 places (always two samples from a different depth in one place).

An informative recalculation was carried out for the following prerequisites and qualified estimates and conditions:

- Quantity of cement used for the production of 1 m^3 of concrete is 350 kg for the bottom constructions and 420 kg for the concretes of the beams.
- Volume weight of concrete of the beams was estimated on the level of 2350 kg/m^3, for the cap sills of pillars on the level of 2350 kg/m^3 and 2230 kg/m^3.

The limit content of Cl^- [% of weight] related to weight of cement is, according to ČSN EN 206 [2], for common concrete 1 % of volume weight; reinforced-concrete 0.4 % of weight; pre-stressed concrete 0.2 % of weight.

From the performed analysis of the content of the chlorides, the following can be stated:

- Average content of Cl^- [% of weight] identified by means of a laboratory analysis for the concrete of the cap sills is 0.33 % for the depth of sampling of 0-15 and 0.26 % for the depth of sampling of 15-30 mm. The limit content of Cl^- [% of weight] related to weight of the cement for the pre-strained concrete is, according to ČSN EN 206, 0.2 % of weight
- Average content of Cl^- [% of weight] identified by means of a laboratory analysis for the concrete of the columns of pillars is 0.19 % for the depth of sampling of 0-15 and 0.13 % for the depth of sampling of 15-30 mm. The limit content of Cl^- [% of weight] related to weight of the cement for the reinforced-concrete concrete is, according to ČSN EN 206, 0.4 % of weight
- Average content of Cl^- [% of weight] identified by means of a laboratory analysis for the concrete joints between KA beams is 0.53 % for the depth of sampling of 0-15 and 0.52 % for the depth of sampling of 15-30 mm. The limit content of Cl^- [% of weight] related to

weight of the cement for the reinforced-concrete is, according to ČSN EN 206, 0.4 % of weight

- Average content of Cl^- [% of weight] identified by means of a laboratory analysis for the concrete of the KA beams is 0.09 % for the depth of sampling of 0-15 and 0.05 % for the depth of sampling of 15-30 mm. The limit content of Cl^- [% of weight] related to weight of the cement is, according to ČSN EN 206, for the pre-strained concrete 0.2 % of weight
- In the concrete of the cap sills of pillars and joints between the KA beams, the content of chloride ions within the whole range of the depth of sampling (0 – 30 mm) is high and does not comply with the requirements of ČSN EN 206.
- In the concrete of the pillars and KA beams, the content of chloride ions within the whole range of the depth of the sampling (0 – 30 mm) is low and complies with the requirements ČSN EN 206.

2.3. Identification of resistance to CH.R.L. – method C

For this test, a bore with the diameter of 150 mm was taken within the investigation from the structure 4. The bores V1 and V7 from the cap sills of the pillars and the bores V4 and V9 from the columns of pillars.

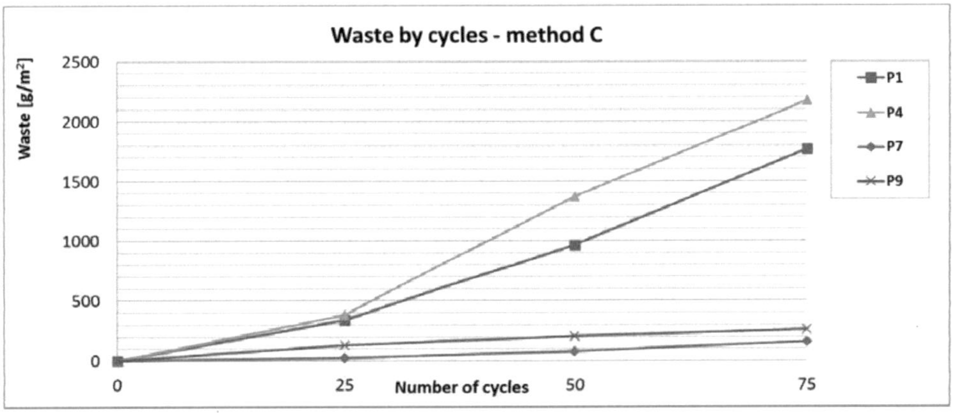

Figure 2. Graphic illustration of wastes from samples V1, V4, V7 and V9.

The following results from the performed measurement:

- Sample V1 (cap sill of the pillar) features the waste of 1769 g/m^2 after 75 cycles, sample V4 (column of the pillar) features the waste of 2176 g/m^2 after 75 cycles, sample V7 (cap sill of the pillar) features the waste of 156 g/m^2 after 75 cycles and sample V9 (column of the pillar) features the waste of 258 g/m^2 after 75 cycles.
- For the CH.R.L. test, the standard ČSN 73 1326 requires a core bore with Ø150 mm and for the reinforced concrete a core bore taken from the structure with Ø100 mm.
- Based on the results of the CH.R.L test, method C, according to ČSN 73 1326 [3], the concrete of the bore V7 and V9 has a high resistance against CH.R.L. and complies with the requirements for concretes of the grade of the environment XF required nowadays for this type of the structure (the highest grade of the impact of the environment XF4 according to TKP 18 (waste < 1000 g/m^2 after 75 cycles of method C).
- With respect to the samples V4, the resistance against the effect of CH.R.L. is incompliant already after 50 cycles and with respect to the sample V1 after 75 cycles; the breakdown occurs along the whole surface of the test specimen.

3. Summary

The construction-technical investigation was focused only in the bridge fields in the surroundings of the railway line Prague – Ostrava and the highway R36.

On the basis of the performed works, the following can be stated, therefore:

- On the basis of a limited number of the destructive tests of compressive strength of concrete, it is recommend considering the class of the concrete:
  - Columns of pillars: C 30/37
  - Cap sills of pillars: C 40/45
- In the concrete of the cap sills of pillars and joints between the KA beams, the content of chloride ions within the whole range of depth of the sampling (0 – 30 mm) is high and it does not comply with the requirements of CSN EN 206. In the concrete of the columns of the pillars and the KA beams, the content of chloride ions within the whole range of the depth of the sampling (0 – 30 mm) is low and complies with the requirements of CSN EN 206.
- On the basis of the results of the CH.R.L test, method C, according to CSN 73 1326, the concrete of the bore V7 and V9 has a high resistance to CH.R.L. and complies with the requirements for the concretes of the grade of the environment XF required nowadays for this type of the structure (the highest grade of the impact of the environment XF4 according to TKP 18 (waste < 1000 g/m^2 after 75 cycles, method C). With respect to the samples V4 (column of the pillar No. 14), the resistance against the effect of CH.R.L. is already incompliant after 50 cycles and with respect to the sample V1 (cap sill of the pillar No. 15) after 75 cycles; the breakdown occurs along the whole surface of the test specimen.
- By means of destructive probes into the Columns of pillars, the reinforcement of the type 10 400 (Rs40) was identified. On the reinforcement, there is an apparent surface corrosion of the reinforcement. On the binder bars, there is an apparent surface corrosion of the reinforcement, locally up to deep corrosion of the reinforcement related to peeling-off of the corrosion products in the layers and considerable weakening of the surface of the cross-section of approx. 20 - 30% of the surface of the cross-section, refer to the probe S3, pillar No. 14.
- By means of destructive probes into the prestressed cap sills of pillars, the pre-stressed reinforcement was revealed in the form of cables with individual diameters of the wires of approx. 4 mm. In the place of the probes into the sides of the cap sills (probes S1, S6 and S7), prestressed cables were found in the middle of the span. Individual cables featured signs of surface corrosion of the reinforcement; the protectors were usually not fully filled with the injection mixture.
- With the probe S8 into the face of the cap sill, 11 anchors of the prestressed reinforcement were revealed. Due to the symmetry of the element, it can be considered that the cap sill is reinforced probably by 18 prestressed cables.

According to the main bridge inspection of the bridge [6] of 05/2013, the situation of the load-bearing structure and the bottom of the construction is assessed as satisfactory (construction state IV according to CSN 73 6221 [5]). The usability is assessed as conditionally usable (state II according to CSN 73 6221).

With respect to the fact that the whole actual concrete surface of the structures is hidden, be that under the unifying coat (load-bearing structure) or sprayed concrete (bottom construction), it is very difficult to assess the actual structure as a whole on the basis of inspections.

For further use of the bridge structure in the following medium-term horizon (decade), the following must be performed unconditionally:

- A complete renewal of the hydro insulation of the load-bearing structure and, thereby, preventing from water entering the chambers of the KA beams, and also the cap sills and stems of the pillars. It must include a suitable modification of the dilatation areas above individual cap sills so as to avoid the follow-up damage to the insulation in such places due to the effect of dilatations of the individual fields.

- Elimination of cornice prefabricates and their replacement with others. The client of the diagnostics did not require the verification of the conditions of the attachment of the cornice prefabricates. However, as the fall of the prefabricate in autumn 2017 demonstrated, there has been a real risk of a strong corrosion violation of the anchoring of the cornice elements. However, a real inspection of each anchoring for that particular structural arrangement and without a drastic limitation of the operation in the surroundings of the bridge is not possible.
- Plan and, depending on the possibilities (e.g. at the moment of the replacement of the cornice elements or another uncovering of the bridge), perform detailed diagnostics of the cap sills. Depending on the result, propose intervals and methods of the control monitoring of the prestressed reinforcement of the cap sills of the pillars or, as the case may be, a maintenance intervention (strengthening, replacement). A localized diagnostic investigation of a small range revealed that the steel protectors of the prestressed reinforcement might not be fully filled with the injection material and on individual wires of the cables, surface corrosion of the reinforcement could already be seen. At this moment, the range of the finding is, however, only indicative and informative.

All aforesaid points are only necessary for a temporary fixation and preservation of the current conditions of the structure so as to avoid a significant deterioration of the conditions.

As it has been demonstrated by both stages of the diagnostic investigation, parts of the structures are strongly contaminated by chloride ions and they do not comply with the requirements of CSN EN 206.

The same applies in case of tests of resistance against frost and spreading salts. A considerable part of the samples does not comply with the requirements for concretes of the grade of the environment XF nowadays required for this type of the structure (the highest grade of the effect of the environment XF4 according to TKP 18 (waste < 1000 g/m^2 after 75 cycles, method C). Without the realization of the recommended measures, especially the renewal of the hydro insulation, the degradation processes will escalate in time and the conditions of the structure will deteriorate progressively.

Prestressed cap sills are a sensitive structure with respect to corrosion and possibly potential failures. In case of a replacement of the cornice prefabricates and/or at the moment of the renewal (replacement) of the hydro insulation, we recommend performing detailed diagnostics of all revealed and available parts of the cap sills (anchoring areas and adjacent parts, conditions of cable channels and prestressing). On the basis of the result, propose the following procedure of monitoring and possible steps of repairs (e.g. strengthening) of the cap sills.

Acknowledgements
This contribution is the result of the research supported by the grant projects GACR 105/12/G059.

References

[1] CSN 73 2011: 2012 *Non-destructive testing of concrete structures*
[2] CSN EN 206+A1: 2017 *Concrete - Specifications, properties, production and compliance*
[3] CSN 73 1326: 1985 *Identification of resistance of cement concrete surface to water and defrosting chemicals*
[4] CSN EN ISO 10304-1: 2009 *Water quality – Determination of dissolved anions by liquid chromatography of ions – Part 1: Determination of bromide, chloride, fluoride, nitrate, nitrite, phosphate and sulphate*
[5] CSN 73 6221: 1996 *Inspections of road bridges*
[6] HMP of bridge No. M117 in Pardubice, from 5. 5. 2013, Ing Pavel Doubravsky, electronic copy provided by the customer

IOP Conf. Series: Materials Science and Engineering **385** (2018) 012046 doi:10.1088/1757-899X/385/1/012046

Estimation of pozzolanic activity of artificial mineral additive using thermal analysis

P Reiterman

University Centre for Energy Efficient Buildings, Czech Technical University in Prague, Trinecka 1024, 273 43 Bustehrad, Czech Republic

Abstract. Application of various supplementary cementitious materials was adopted currently by number of research institutions worldwide. The main motivation is to find out predominantly environmental friendly technologies applicable in cement industry; however this technology is not new. Historically were various active mineral additives used in Ancient times and this technology was rediscovered at the end of 19th century with gradual development of cement and lime industry. This paper is focused on the long-term experimental study of pozzolanic reaction monitored in terms of thermogravimetry by the determination of the consumption of lime. Various types of ceramic powder were studied during two years and their activity was estimated using predictive method. Obtained results confirmed significant impact of the finesses on the reaction kinetics and also on the total hydration potential.

1. Introduction

Utilization of hydraulic lime or lime with addition of active mineral additive has long tradition in building industry. Development of this technology allowed dynamic of various human civilizations. Probably the most famous example offers Ancient Rome, where hydraulic lime was used for the most exposed structures of advanced Roman infrastructure. It was applied for the substructure of roads, bridges, and other iconic structures such as aqueducts or Pantheon, which is the most famous building made of predecessor of modern concrete. Romans used lime with addition of volcanic ash, which was used on the construction of the port in Caesarea [1-3]. However, as time went on, the natural pozzolan was replaced by artificial material in form of waste ceramic. Proofs of its utilization are confirmed by number of studies [4-5]. Due to limited finesses served applied ceramics predominantly as active filler. This technology was over time forgotten and replaced by other building procedures based on the utilization of clay, wood and stone. Further development of utilization of hydraulic lime started during 19th century in relation with the production of Portland cement [6].

Lime mortars with addition of pozzolanic additives were intensively used in Bohemia at the turn of the 19th century for the production of modern mosaic. This decorative element is typical architectural component of building structures from this period. They were used in interiors, however number of works was situated in exterior. In the exterior applications, it was obviously necessary to use mortar with suitable resistance to the environment. These mortars could be based on the cement or hydraulic lime; some of them were created from lime with addition of milled bricks, which ensured formation of hydraulic bond. These mortars primarily served for the fixation of tesseraes, which was most frequently represented by pieces of colored glass.

Actual state of binder system of these mortars is crucial information for the assessment of mosaics' properties and possibly for the procedure of their protection. The main aim of present paper is to estimate hydration process of lime mortar with various types of ceramic powder. This research partially follows previous research [7], where pozzolanic activity of ceramic powder using thermogravimetry was monitored. In this paper, predictive model for hydration monitoring was applied for the description of chemical processes with determined hydration capacity. The used mathematical model is based on Michaelis-Menten equation, which is commonly applied in chemical engineering [8-10].

2. Experimental program

Ceramic powder produced by Heluz Ltd. (Hevlín clay quarry) was used in performed experimental program. Subject material originates during hollow brick blocks production by grinding of load areas, what allows precious brick-laying; thickness of loading gap is approximately just about 1 mm. Chemical composition of produced ceramic in the studied factory is very stable in time (table 1). This composition perform annual average, however particular components are varying in order of tenth of percent.

Table 1. Chemical composition of studied ceramic powder [%].

SiO_2	Al_2O_3	Fe_2O_3	CaO	MgO	K_2O	Na_2O	TiO_2	SO_3
51.3	20.0	6.0	11.5	4.5	3.2	1.3	0.8	1.0

Illitic clay is used as raw material for the production in given factory. Mineralogical composition of studied material determining its pozzolanic activity is shown in table 2. Amorphous part of the ceramic powder indicates its pozzolanic activity. This pozzolanic material was studied by Keppert *et al.* [11], where ceramic powder with D_{max} 125 μm exhibited pozzolanic activity in terms of Chapelle test consumption of $Ca(OH)_2$ after 7 days 1066 mg/g. The finesses of ceramic powder has crucial impact on the reactivity, what was also published by Navrátilová et al. [12]. Different types of ceramic powders were also used in this program; particular materials were prepared by sieving with sieves 0.125, 0.063, 0.050, 0.030 and 0.015 μm, respectively. They were designated with increased finesses CP-1 to CP-5. Particle size distribution is shown on Figure 1.

Table 2. Mineralogical composition of studied ceramic powder [%] [7,11].

Amorphous components	45.3
Quartz	26.8
Albite	7.2
Microcline	6.3
Muscovite	3.1
Åkermanite	3.2
Hedenbergite	4.1
Gypsum	0.6
Hematite	1.9

Such prepared ceramic powders were mixed with slaked lime CL90 – S, which contains over 99% of CaO. Mixing ratio was 50:50 by mass, water to binder ratio was set to constant 0.75. Prepared fresh mixtures were placed into plastic containers with sealed lid to eliminate carbonation during hydration. The progress of pozzolanic reaction was monitored by using thermogravimetry and determination of the content of $Ca(OH)_2$ in time. Measurement was carried out using Schimadzu DTG60-H with heating rate 10K/min. Exact boundaries of the occurring portlandite decomposition were derived from derivative curves of TG curve. Single measurements were conducted after 7, 14, 21, 28, 60, 180, 360 and 720 days, respectively. Mass of samples subjected to thermal analysis was about 40 mg.

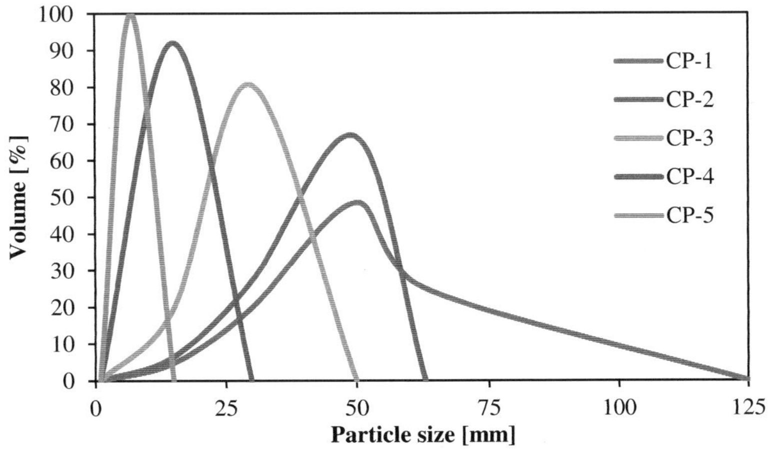

Figure 1. Particle size distribution of used ceramic powders.

Gradual consumption of $Ca(OH)_2$ was monitored in time and processed using Michaelis-Menten equation (1), which is often used for the prediction of entire reaction capacity. In knowledge of a few points of subject reaction, it is possible to estimate ultimate value (V_{max}). T means time (hours), v is partial value of studied property and K is rate constant. Illustrative example is drawn on Figure 2. It is obvious that present shape has good fit with progress of hydration of lime with pozzolanic addition.

$$v = V_{max} \frac{T}{K+T} \tag{1}$$

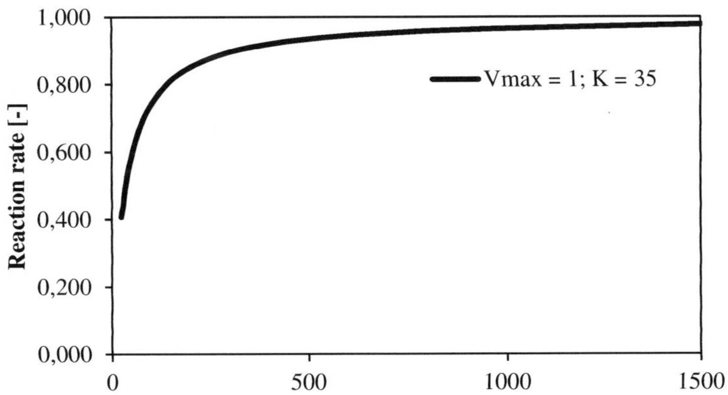

Figure 2. Illustration of Michaelis-Menten function.

Obtained values of $Ca(OH)_2$ consumption in time for single mixtures were processed by Matlab using Least Square Method (LSM). Present nonlinear solution allows acquisition of optimal fit and precision of the estimation. Besides acquisition of ultimate reaction capacity, it is also possible to monitor reaction kinetic for a single mixture. Amorphous content of ceramic powder is about 0.45, what is supposed to be the theoretically limit value.

3. Results and discussion

Performed experimental program was focused on the estimation and monitoring of pozzolanic reaction. The monitoring was conducted by using thermogravimetry and determination of actual content of $Ca(OH)_2$. Obtained results were processed by LSM to obtain estimation in terms of Michaelis-Menten equation.

Obtained values of Michaelis-Menten constant and ultimate reaction capacity are shown in table 3. There is obvious, that finesses of ceramic powder has significant role in the pozzolanic reaction of studied material and entire kinetic. The highest value of V_{max} was obtained, as expected, by the mixture CP-5, which contained the ceramic powder with the highest finesses. Mixtures CP-1 and CP-2 exhibited nearly similar results; following mixtures reached proved gradually increased reactivity.

Table 3. Obtained values from estimation process.

	V_{max}	K
	[%]	[hour]
CP-1	45.5	2768.7
CP-2	43.7	1763.7
CP-3	44.8	1672.1
CP-4	46.6	1748.5
CP-5	48.4	1409.9

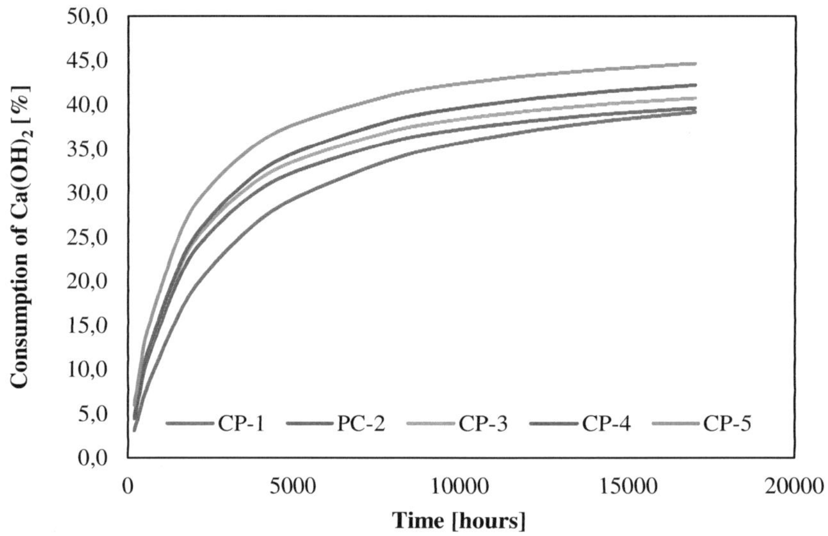

Figure 3. Estimation using Michealis-Menten equation.

However, the experiment was conducted in specific conditions excluding the process of carbonation, which plays an important role in case of lime mortars. The kinetics of particular mixtures are well visible in Figure 3. It is obvious, that mixture CP-1 (with grains up to 0.125mm) exhibited the slowest progress of pozzolanic reaction, on the other hand CP-5 exhibited rapid progress of pozzolanic reaction during initial phases of hydration. CP-5 during initial 14 days of hydration obtained more than two time higher consumption of $Ca(OH)_2$ in comparison with CP-1. Thorough experimental study was carried out by Cáchova et al. [13], who oriented the research on the monitoring of pozzolanic reaction kinetic of ceramic powder, which corresponds with CP-1. In this study was pozzolanic reaction compared with kinetics of carbonation, which are competitive process during the lime based mortars hardening, what is described by equation 2 and equation 3.

$$Ca(OH)_2 + Ceramic\ Powder \rightarrow hydration\ products \quad (2)$$

$$Ca(OH)_2 + CO_2 \rightarrow CaCO_3 + H_2O \quad (3)$$

Kinetic of $Ca(OH)_2$ consumption for carbonation was obtained from compressive strength evolution of lime based plasters published in [12-13], who studied similar systems with addition of pozzolans. These data shows, that carbonation plays a dominant role in the process of hardening. Obtained results were integrated to previous results, Figure 4. For better illustration of initial kinetic of all studied processes, it was transformed to logarithmic scale, Figure 5.

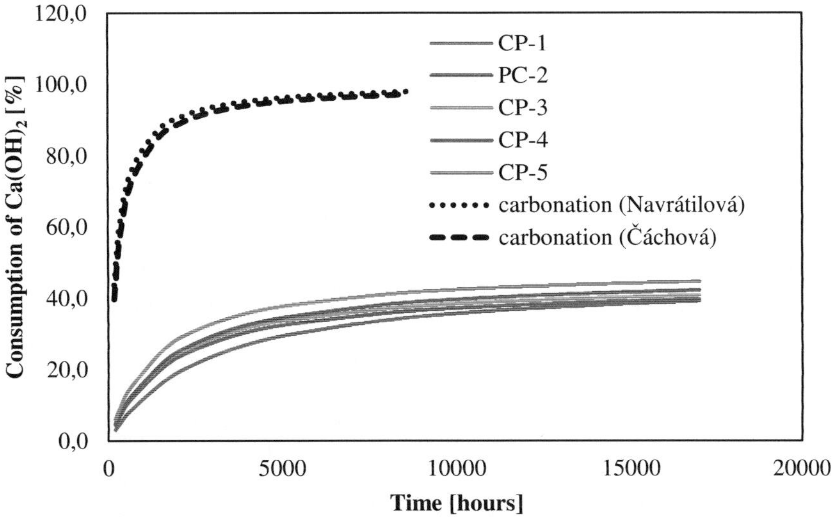

Figure 4. Comparison of $Ca(OH)_2$ consumption due to carbonation and pozzolanic reaction using Michaelis-Menten estimation.

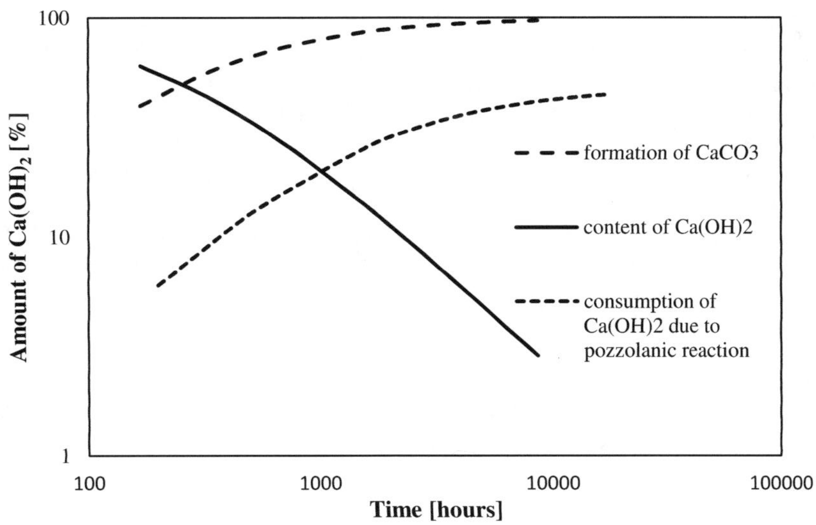

Figure 5. Comparison of $Ca(OH)_2$ consumption due to carbonation and pozzolanic reaction using Michaelis-Menten estimation.

It is obvious, that carbonation is the dominant process during hardening, so the reaction kinetic is controlled by diffusion. That means, that conversion of lime to calcite is much faster, than pozzolanic reaction. These conclusions confirms research published in [13-14], where thorough research of cement-ceramic powder system was realized. Cement based system is a specific, due to continuous formation of new Portlandite, that is why was applied model by Kondo [15].

4. Conclusion

Experimental program with aim at the reaction kinetic of various ceramic powder was realized in this work. Applied ceramic powders had different finesses, what caused various kinetic of pozzolanic reaction, which was estimated using Michaelis-Menten equation. Organizing of conducted experimental work eliminated the influence of carbonation. Performed work confirm original prerequisite, that finesses of the pozzolan has significant impact on the reaction progress. However, obtained results confirmed very slow progress of pozzolanic reaction in studied mixtures. Carbonation was considered as faster mechanism in comparison with obtained results on lime-ceramic powder pastes. These results also well correspond with current construction praxis, which applies the duration of hardening to approximately 7days per 10 mm of the plaster. It is necessary to note, that in this is not consumed entire lime for the carbonation, however this process considerably retards consequent pozzolanic reaction. That means that part of the used pozzolan is fixed in the system as active filler. This fact could be successively exploited during renovation and restauration work by lime water curing, which is often used as in conservation practice, despite that present treatment is based in the carbonation.

Acknowledgement

This work has been supported by Czech Science Foundation under the project No. 18-13525S.

References

[1] Hughes D C, Jaglin D, Kozlowski R and Mucha D 2009 *Cem. Concr. Res.* **39** 77-89
[2] Pintér F, Vidovszky I, Weber J and Bayer K 2014 *J Cult. Herit.* **15** 219-226
[3] Pavía S and Caro S 2008 *Constr. Build. Mater.* **22** 1807-1811
[4] Baronio G and Binda L 1997 *Constr. Build. Mater.* **11** 41-46
[5] Drdácký M, Fratini F, Frankeová D and Slížková Z 2013 *Constr. Build. Mater.* **38** 1117-1128
[6] Fridrichová M, Gazdič D, Dvořák K and Magrla R 2017 *Mater. Tehnol.* **51** 219-223
[7] Reiterman P. and Keppert M 2017 *Mater. Sci. Forum* **908** 40-44
[8] Monteagudo S M, Moregues A, Gálvez J C, Casati M J and Reyes E 2014 *Thermochim. Acta* **592** 37-51
[9] Grilo J, Santos-Silva A, Faria P, Gameiro, Veiga R, Velosa A 2014 *Constr. Build. Mater.* **51** 287-294
[10] Deboucha W, Leklou N, Khelidj A and Oudjit M N 2017 *Constr. Build. Mater.* **146** 687-701
[11] Keppert M, Urbanová M, Brus J, Čáchová M, Fořt J, Trník A, Scheinherrová L, Záleská M and Černý R 2017 *Constr. Build. Mater.* **148** 411-418
[12] Navrátilová E and Rovnaníková P 2016 *Constr. Build. Mater.* **120** 530-539
[13] Čáchová M, Scheinherrová L, Kobera L, Urbanová M, Brus J and Keppert M 2017 *Key Eng. Mater.* **722** 126-131
[14] Doleželová M., Čáchová M, Scheinherrová L and Keppert 2017 *AIP Conf. Proc.* **1906** 140008
[15] Kondo R, Lee K and Diamon M 1976 *J. Ceram. Soc. Japan* **84** 573-578

IOP Conf. Series: Materials Science and Engineering **385** (2018) 012047 doi:10.1088/1757-899X/385/1/012047

The bond strength by pull-off and direct tensile strength of concrete damaged by elevated temperatures

I Rozsypalova, P Danek and O Karel

Department of Building Testing, Faculty of Civil Engineering, Brno University of Technology, Veveri 95, Brno, 602 00, Czech Republic

Abstract. This paper presents the results of an experimental study of selected characteristics of concrete following its exposure to high temperatures. The monitored parameter was the influence of high temperatures on the bond strength and tensile strength of concrete. The bond strength was tested by a pull-off test, the tensile strength was determined as the direct tensile strength. Experimental concrete panels of 2300 × 1300 × 150 mm were exposed to high temperatures (corresponding to the temperatures in a normal fire) using a fire test furnace at the AdMaS Science Centre of the Faculty of Civil Engineering, Brno University of Technology. The direct tensile strength test was performed on cylinders (diameter of 100 mm and length of 100 mm) made from core samples taken from the panel. The pull-off test was performed on the surface exposed to high temperatures. The results of the experiment show a significant influence of high temperatures on the examined characteristics, and both characteristics were found to significantly decrease with a growing thermal load.

1. Introduction

At present, concrete ranks among the most commonly used building materials. The most important mechanical property of concrete, according to the current practice, is the compressive strength, according to which the concrete is classified into individual strength classes. Another important and no less significant role is played by the tensile characteristics of concrete, because they have a major influence on the occurrence and development of cracks in concrete structures [1,2]. When evaluating building structures made from this composite material, according to the recommendations given in ČSN ISO 13822 [3], the actual characteristics of the material are considered. It is also necessary to take into account material degradation caused by extraordinary influences which also include a load in the form of fire. Elevated temperatures cause damage of the concrete internal structure [1].

The direct tensile strength test of concrete is used very little in common practice due to its complicated execution. In order to achieve the required type of failure caused by direct tension, it is necessary to achieve a strict parallelism of the contact areas of the specimen (by grinding), and the specimens in the press must be perfectly centered [4,5].

For the testing, series of large-format experimental concrete panels were made as part of the experimental research, which were exposed with one side to the action of high temperatures in the furnace for fire tests of building materials and small-scale elements in the AdMaS Centre (Advanced Materials, Structures and Technologies), Faculty of Civil Engineering in the Brno University of Technology. Apart from the characteristics tested in this paper, also a number of other important properties of the thermally damaged concrete were assessed. The published partial outcomes include the results of both non-destructive [6,7] and destructive tests [8–12].

2. Methods

To determine the influence of high temperatures on the bond strength and tensile strength of concrete, a series of large-format experimental panels were made, which were subsequently thermally loaded.

2.1. Fire experiment

A total of 5 experimental panels (with dimensions of 2300 × 1300 × 150 mm) were made from standard concrete with a strength class of C30/37, with standard characteristics in terms of fire resistance. The mixture of fresh concrete consisted of Portland cement CEM I 42.5 R (335 kg/m^3), water (170 kg/m^3), silica sand 0–4 (840 kg/m^3), silica aggregate 8–16 (880 kg/m^3), fly ash (50 kg/m^3), and superplasticizer (2.67 kg/m^3). The concrete panels were reinforced with a mesh reinforcement near the bottom side.

In order to measure the temperatures achieved during the fire test, type K thermocouples were embedded into the panels. In each panel, a total of 10 thermocouples were placed along the longitudinal axis, three near the upper (unheated) side, three near the bottom (heated) side, three in the middle of the panel width, and one at a quarter of the panel width from the bottom side.

The test panels were placed on the upper side of the fire furnace, which guaranteed a one-sided heating of the panel from its bottom side. The upper side and lateral sides of the panel were left to have a spontaneous heat exchange with the surrounding environment. Loading with high temperatures was performed in accordance with EN 1992-1-2 [13] according to the standardized temperature curve specified in ISO 834 [14], because this chosen type of a sharp increase in the furnace temperature corresponds to the action of a real fire [13]. To achieve the prescribed maximum nominal temperatures (550, 600, 800 and 1000 °C), the level of thermal load specified for individual panels was always kept for the following 60 minutes. Then the panel was left on the furnace until cooling down to the environment temperature. One of the panels was left as a reference panel without thermal loading. The age of panels at the time of starting the fire tests was approximately 160 days, and thus the experiments were in accordance with the recommendations of RILEM TC 129-MHT [15] requiring the age of concrete of at least 90 days. Details of the fire test are given in the earlier publication [8].

2.2. Pull-off bond strength test

The strength test was performed in accordance with standard EN 1542 [16]. The procedure is based on a direct pull-off of a disc, which is glued onto the surface of the tested material. The pull-off area was defined in advance by making a core drill through the surface layer of the specimen. On each panel, 8 testing points were tested, and therefore, a total of 40 tests were performed.

At first, the tested surface was levelled by means of a grinding machine. A metal disc with a diameter of 50 mm was glued to the tested surface using a two-component adhesive. The metal disc provided with a thin continuous layer of an adhesive was attached to the surface of the drilled area concentrically with the core axis and pressed (air removal must be ensured). The potential adhesive running outside the disc area was immediately removed. The adhesive was left to set in accordance with the manufacturer's guidelines. Into the glued disc, a special screw with a small ball on one end was screwed, thanks to which the strength from the device is transferred to the pull-off area. The used Proceq DY-216 pull-off tester (Figure 1) is able to exert a maximum tensile force of up to 16 kN. The result of the test is then directly the value of bond strength in MPa, on the basis of the known area of the disc used.

Figure 1. Proceq DY-216: pull-off bond strength automated tester.

Figure 2. Pull-off bond strength test: a) selected panel (panel P7 loaded with 1000°C) after performing the pull-off test, b) detail of testing point P4-8 on the thermally unloaded panel P4 after performing the pull-off test, c) detail of testing point P7-3 on the thermally most loaded panel P7 after performing the pull-off test.

2.3. Direct tensile strength test

Specimens for the direct tensile strength test were obtained by core drilling from the experimental panels. The nominal diameter of the drilled core was 100 mm. Bases of the extracted cylinders were levelled from the side of the temperature action (therefore, we removed material with a width of several millimeters, so that the damaged part of the specimen was preserved as much as possible), and subsequently, the specimens were shortened also on the upper base, to 100 mm. The final test specimen was thus a cylinder with a diameter of 100 mm and a length of 100 mm. From each panel, 3 specimens were taken for this test. In total, we tested five series of three specimens, i.e. 15 specimens. Bases of the test specimens were glued with a fast-setting two-component Sikadur adhesive to steel plates, which have a thread for screwing a joint with a holding rod serving for transmitting load from the tensile machine through the steel plate to the test specimen.

The direct tensile strength is determined on specimens, which are loaded by an axial tension of the tensile machine in the longitudinal direction. The tensile test was performed using a Heckert FPZ 100/1 mechanical press (see Figure 3).

Figure 3. Heckert FPZ 100/1 mechanical press for the direct tensile strength test (specimen 8C_P3).

During the test, a maximum tensile force is recorded. The direct tensile strength is then determined using the equation (1):

$$f_t = F_{max}/A \quad (1)$$

where f_t represents the direct tensile strength of concrete [MPa], F_{max} represents the size of load at failure of the test specimen [N], and A represents the cross-sectional area of the test specimen [mm^2].

3. Results

The tensile characteristics were used to evaluate the conditions of the concrete after high temperature exposition simulating fire load of structures. Direct tensile strength and pull-off bond strength were determined.

The results of experimental work are summarized in the form of average values and standard deviations in Figure 4 for bond strength determined by a pull-off and in Figure 5 for direct tensile strength of concrete.

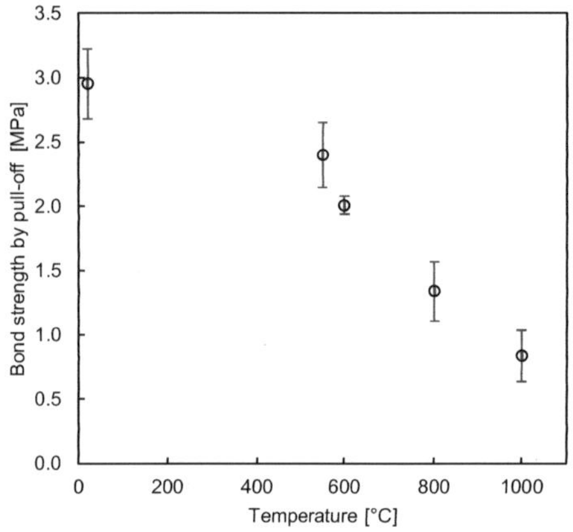

Figure 4. Dependence of the concrete bond strength determined by a pull-off test on temperature.

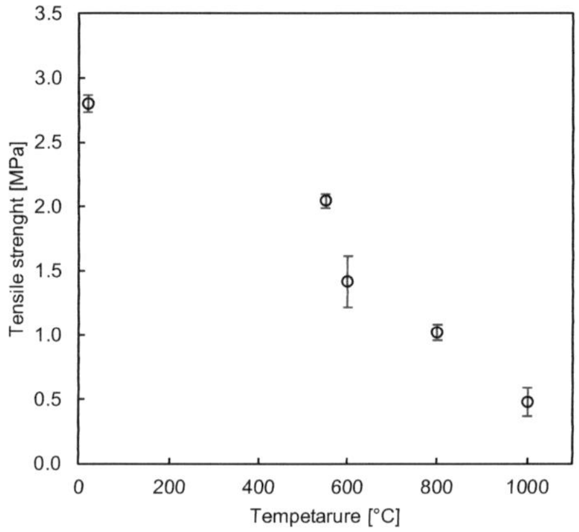

Figure 5. Dependence of the concrete direct tensile strength on temperature.

Table 1 and Figure 6 compares percentage changes of the tested characteristics, where the value determined on concrete without thermal loading (panel P4_T20) is considered a reference value (100 %). The rate of decrease of the direct tensile strength with increasing temperature is higher than for the pull-off bond strength. The characteristics declined the most significantly for temperature 1000 °C, the bond strength drops to 28.2 % and the direct tensile strength decreased to 17.1 % compared to the unloaded concrete.

Table 1. Percentage change in the characteristics of concrete depending on temperature – comparison of the bond strength by pull-off and the tensile strength.

Temperature (max.) [°C]	Bond strength by pull-off [%]	Tensile strength [%]
20	100.0	100.0
550	81.1	73.0
600	67.9	50.5
800	45.3	36.3
1000	28.2	17.1

In Figure 7, mutual comparison of the bond strength and the tensile strength values is shown. The strengths significantly decrease with increasing temperature. The direct tensile strength values are slightly higher than the bond strength by pull-off over the entire applied thermal load range.

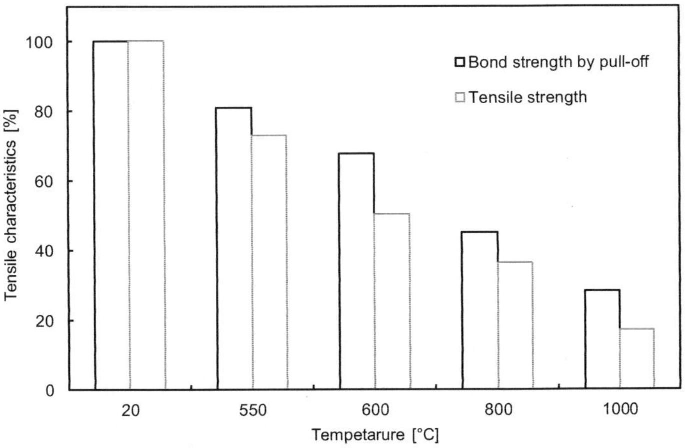

Figure 6. Percentage change in the tensile characteristics of concrete depending on temperature – comparison of the bond strength by pull-off and the direct tensile strength.

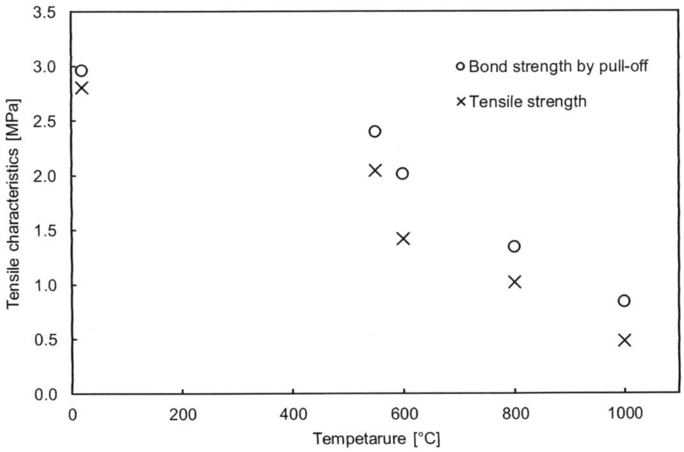

Figure 7. Mutual comparison of the bond strength and the tensile strength values.

4. Conclusion

The experiment examined the state of concrete after its thermal load with high temperatures. The values determined were the bond strength by a pull-off test and the direct tensile strength. The used methods proved a significant change in these characteristics due to exposure to high temperatures.

The bond strength of concrete determined by a pull-off test gradually decreases with a growing thermal load. In comparison with the reference specimens, there was almost a 72 % decrease in the most heavily loaded panel. No significant trend in the variability of individual measurements was observed, and its values were not particularly significant.

Similar behaviour was observed also for the direct tensile strength of concrete. The decrease in strength was even more significant, the strength for the highest temperature decreased by almost 83 %. The measurement variability also does not display significant changes caused by growing temperatures.

A mutual visual comparison of the determined characteristics suggests the possibility of a certain relation between the tensile strength of concrete and its bond strength. In order to arrive at valid conclusions, it would be appropriate to focus on examining also this option in future.

Acknowledgements
This outcome has been achieved with the financial support of specific research project No. FAST-S-18-4824 at the Brno University of Technology, Faculty of Civil Engineering.

References

[1] Guo Z and Shi X 2011 *Experiment and Calculation of Reinforced Concrete at Elevated Temperatures* (Oxford: Butterworth-Heinemann)
[2] Hudoba I 2008 *Vysokohodnotný betón: Materiály, vlastnosti, výroba, využitie* (Bratislava: STU)
[3] ISO 13822: 2010 *Bases for design of structures -- Assessment of existing structures*
[4] Kucharczykova B, Danek P and Zitt P 2008 Sborník příspěvků konference Zkoušení a jakost ve stavebnictví vol 1 (Prague: Kloknerův Ústav CTU in Prague) pp 283-290
[5] Rozsypalova I 2013 *Srovnání vybraných pevnostních tahových parametrů betonu* Bachelor thesis (Brno: BUT)
[6] Rozsypalova I, Danek P, Sachr J and Karel O 2017 *Proc. Int. Conf. on Rehabilitation and Reconstruction of Building* (Praha: ČVUT) p 117
[7] Rozsypalova I, Kumpova I, Danek P, Simonova H and Kersner Z 2018 *Proc. Int. Conf. Juniorstav* (Brno: BUT) pp 696–701
[8] Rozsypalova I, Danek P, Simonová H and Kersner Z 2017 *Proc. Int. Conf. on Dynamics of Civil Engineering and Transport Structures and Wind Engineering (MATEC Web of Conferences)* vol 107 ed J Melcer and K Kotrasova (Les Ulis: EDP Science)
[9] Simonová H, Halfar P, Rozsypalova I, Danek P and Kersner Z 2017 *Proc. Int. Conf. on Dynamics of Civil Engineering and Transport Structures and Wind Engineering (MATEC Web of Conferences)* vol 107 ed J Melcer and K Kotrasova (Les Ulis: EDP Science)
[10] Simonová H, Trcka T, Bejcek M, Rozsypalova I, Danek P and Kersner Z 2018 *Sol. St. Phen.* **272** 220–225
[11] Rozsypalova I, Simonova H, Danek P, Kersner Z and Vyhlidal M 2017 *Proc. Int. Workshop on Concrete Spalling* (RILEM: Paris) pp 105–111
[12] Simonova H, Rozsypalova I, Rovnanikova P, Danek P and Kersner Z 2018 *Sol. St. Phen.* **272** 47–52
[13] EN 1992-1-2: 2004 *Eurocode 2: Design of concrete structures - Part 1-2: General rules - Structural fire design*
[14] ISO 834: 2014 *Fire-resistance tests - Elements of building construction - Part 1: General requirements*
[15] RILEM TC 129-MHT 1997 Test methods for mechanical properties of concrete at high temperatures *Mater. Struct.* **33** 224–228
[16] EN 1542: 1999 *Products and systems for the protection and repair of concrete structures – Test methods – Measurement of bond strength by pull-off*

IOP Conf. Series: Materials Science and Engineering **385** (2018) 012048 doi:10.1088/1757-899X/385/1/012048

Technological characterisation of selected mineral additives

A Sicakova and M Kovac

Faculty of Civil Engineering, Technical University of Kosice, Vysokoskolska 4, 042 00, Kosice, Slovakia

Abstract. The paper was aimed on testing the technological properties of product "Zeoslag" which consists of 10% of zeolite and 90% of ground granulated blast furnace slag. Both the Perlite and Zeolite were tested for comparison, and also cement as reference material. Mineral additives were tested according to standard for testing the slag - the binder here consists of 50% of additive to be tested and 50% of cement. Normalized sand and water according to standard recipe were used to make a testing mixtures. Following technological parameters of additives were tested: water demand for standard consistency, flexural and compressive strength development (2, 7 and 28 days) and activity index. Parameters are discussed in the terms of chemical character of additives. At the given range of modulus of hydraulicity (0.02 – 2.44), the experiment showed tight nonlinear relationship between it and strength characteristics of additive-based mortars.

1. Introduction

Concrete additions are defined as finely divided materials used in cement-based mixtures in order to improve or to obtain desired fresh and hardened concrete properties. The EN 206 [1] lists two types of inorganic additions:

- Nearly inert additions (type I): Virtually inactive materials such as rock powders (lime fillers or quartz dust) and color pigments. Pigmented metal oxides (mainly iron oxides) are used to color concrete. They must remain color-fast and stable in the alkaline cement environment. With some types of pigment, the water requirement of the mix can increase.
- Pozzolanic or latent hydraulic additions (type II): [1] considers only fly ash and silica fume as well as ground granulated blast furnace slag as suitable additives for concrete production with the recommendations how to involve them to the water/binder ratio (k-value).

In the research and development phase of cement composites, a lot of other additives are considered for application in the mixture, while their influences on a wide range of parameters are evaluated. First of all, their pozzolanic or latent hydraulic ability is valuable. Solid industrial by-products (siliceous and aluminous), as well as some natural pozzolanic materials are being increasingly used in the cement and concrete industry, mainly due to environmental and economic effect. Moreover, the incorporation of these materials in concrete gives encouraging results with regards to the mechanical and durability properties of concrete [2].

Construction and demolition waste like glass, brick and concrete dust, seems to be valuable source of mineral fines [3-7]. Lot of studies present generally positive influence of powder additives on the properties of cement-based mixtures [8-10].

The stability or resistance to segregation of the plastic mixture is attained by increasing the total quantity of fines in the mixture and/or by using admixtures that modify the viscosity. Increased fines content can be achieved by increasing the content of cementitious materials or by incorporating mineral fines; they are used to improve the grading curve. Special requirements for aggregate grading are to be taken into account and especially fine particles (so called "micro filler") content should be controlled [11]. Micro fillers improve mix workability and provide particle dense packing in hydrated cement paste. High fluidity with relatively low water content is achieved by the use of a superplasticizing and air entraining admixtures.

Chemically active mineral additives (highly reactive pozzolan) increase the cohesiveness of concrete and require more water to maintain workability; however, the requirement of water may be offset by adding plasticizer. The workability of concrete with micro-filler mineral additives greatly depends on the particle size, specific surface area, particle shape, and replacement level. Higher fines content is recommended for easy workable concretes; however, there is also a risk of negative influence here, mainly due to shrinkage [12].

This paper presents following technological parameters of selected additives (Zeolite, Perlite and Zeoslag) – water demand for achievement of standard consistency, flexural and compressive strength development (2, 7 and 28 days) and activity index. Parameters are discussed in the terms of chemical character of additives. The paper is focused on the "Zeoslag" product, which consists of 10% of zeolite and 90% of ground granulated blast furnace slag, and its comparison with other fine-grain materials.

2. Experiment

The paper is aimed on the testing of technological properties of product "Zeoslag" and its comparison with other selected fine-grain materials. The Zeoslag is product being supposed to be a supplementary cementitious material (additive), while it consists of 10% of zeolite and 90% of ground granulated blast furnace slag (GGBFS).

For comparison, the cement, zeolite and perlite were involved to the experiment, too. Following testing was performed:

- chemical composition: using XRF analysis. The main oxides (MgO, Al_2O_3, SiO_2, CaO a Fe_2O_3) were determined by this testing, following by calculation of basic modulus to compare hydraulic activity of materials. The materials have been verified if they meet the typical character of the hydraulicity (Table 3),using the Rankin classification, given in Table 1,

Table 1. Rankin classification of hydraulicity.

Kind of material	CaO/SiO_2 ratio
Hydraulic	> 2
Latent hydraulic	1 – 1.5
Pozzolanic	< 0.5

Modulus of basicity (Mz) and modulus of hydraulicity (HM) were calculated using following formulas:

$$Mz = \frac{CaO + MgO}{SiO2 + Al2O3}$$

$$HM = \frac{CaO}{SiO_2 + Al_2O_3 + Fe_2O_3}$$

- determining the amount of water for standard consistency of binder paste (STN EN 196-3: Methods of testing cement. Part 3: Determination of setting times and soundness),
- initial setting time (STN EN 196-3: Methods of testing cement. Part 3: Determination of setting times and soundness)
- flexural strength after 2, 7 and 28 days of setting and hardening (STN EN 196-1: Methods of testing cement. Part 1: Determination of strength)

- compressive strength 2, 7 and 28 days of setting and hardening (STN EN 196-1: Methods of testing cement. Part 1: Determination of strength)
- activity index after 2, 7 and 28 days of setting and hardening (STN EN 15167-1 Ground granulated blast furnace slag for use in concrete, mortar and grout. Part 1: Definitions, specifications and conformity criteria)

As given above, samples of mixtures for testing the strength characteristics and calculating the activity index were prepared according to standard for testing the GGBFS. The binder here consists of 50 % of additive to be tested and 50 % of cement. Whole mixture is as follows: 225 g of cement, 225 g of additive, 225 g of water and 1350 g +/- 5 g of normalized sand. Next, the activity index is calculated as a ratio of compressive strength of this mixture and compressive strength of sample consisting of 100 % of cement.

Following criteria for the activity index are given in the standard: it should be higher than 45 % after 7 days and higher than 70 % after 28 days of setting and hardening.

3. Results

3.1. Chemical analysis

Chemical composition of tested materials is given in Table 2. Parameters defining the character of materials (Modulus of basicity, Modulus of hydraulicity and CaO/SiO_2 ratio) are evaluated in Table 3.

Table 2. Chemical composition of materials used in the experiment.

Material	MgO (%)	Al_2O_3 (%)	SiO_2 (%)	CaO (%)	Fe_2O_3 (%)
CEM I 42.5 N	9.03	2.97	20.34	64.1	2.99
Zeolite	1.16	11.61	78.75	3.50	1.73
Zeoslag	8.12	6.74	43.29	26.8	0.37
Perlite	0.93	12.06	73.11	1.59	2.39

Table 3. Modulus of basicity and modulus of hydraulicity of materials used in the experiment.

Material	Modulus of basicity	Modulus of hydraulicity	CaO/SiO_2	Character of materials
CEM I 42.5 N	3.14	2.44	3.15	hydraulic
Zeolite	0.05	0.04	0.04	pozzolanic
Zeoslag	0.69	0.53	0.62	pozzolanic/latent hydraulic
Perlite	0.03	0.02	0.02	pozzolanic

The results confirm the chemical character of individual materials; the Zeolite and Perlite are of pozzolanic one, while Zeoslag gains latent hydraulic character thanks to GGBFS.

3.2. Amount of water for standard consistency of binder paste and initial setting time

Results are given in Table 4.

Table 4. Water demand for standard consistency of binder paste and initial setting time.

Mixture	Amount of water (g)	Amount of water (%)	Initial setting time (min)
CEM I 42.5 N	140.1	100.0	190
Zeolite	177.4	126.6	320
Zeoslag	145.1	103.6	270
Perlite	171.9	122.7	250

Comparing the cement-based binder, the highest water demand was found for the Zeolite, following by Perlite. The Zeoslag product needs only a bit more water than the reference sample (by 3.6%). However, its initial setting time does not correspond exactly with this, being longer than that of the Perlite. Although the Zeolite and Perlite are of similar chemical composition, as well as the water demand, the initial setting time of Zeolite is much more longer than that of Perlite (by 70 minutes).

3.3. Flexural strength

Results of flexural strength after 2, 7 and 28 days of setting and hardening are given in Table 5.

Table 5. Results of testing the flexural strength.

Mixture	f_{f-2} [MPa]	f_{f-7} [MPa]	f_{f-28} [MPa]
CEM I 42.5 N	3.5	7.0	8.0
Zeolite	1.1	3.0	5.5
Zeoslag	2.7	5.5	7.0
Perlite	1.5	3.5	4.5

The flexural strength of samples corresponds with the assumption based on chemical character of tested additives. The highest values were achieved by mixture containing cement only, followed by Zeoslag product.

The lowest values were measured for Zeolite and Perlite, while they are close to each other (1.1 and 1.5 MPa; 3.0 and 3.5 MPa; 5.5 and 4.5 MPa - values after 2, 7 and 28 days respectively). The strength development is quite interesting; while Perlite gives little bit better results than zeolite in early ages (2 and 7 days), after 28 days the zeolite is better (by 1 MPa). The strength development is given in figure 1.

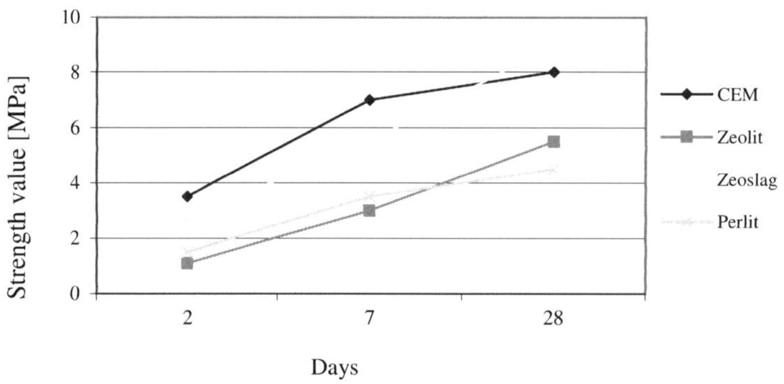

Figure 1. Flexural strength of standard mortars with additives.

3.4. Compressive strength and activity index

Results of strength and activity index after 2, 7 and 28 days of hardening are given in Table 6.

Table 6. Results of testing the compressive strength.

Mixture	f_{c-2} [MPa]	f_{c-7} [MPa]	f_{c-28} [MPa]	AI_2 (%)	AI_7 (%)	AI_{28} (%)
CEM I 42.5 N	14.0	46.5	51.5	100	100	100
Zeolite	4.0	9.5	25.0	28.6	20.4	48.6
Zeoslag	6.8	19.7	45.3	48.3	**42.4**	**88.0**
Perlite	4.5	10.5	19.0	32.1	22.6	36.9

Similarly, the same order of values was detected for compressive strength. The highest values were achieved by mixture containing cement only, followed by Zeoslag product. While values of zeolite

IOP Conf. Series: Materials Science and Engineering **385** (2018) 012048 doi:10.1088/1757-899X/385/1/012048

and perlite are close to each other again (4.0 and 4.5 MPa; 9.5 and 10.5 MPa; 25.0 and 19.0 MPa - values after 2, 7 and 28 days respectively), the difference between values of cement-based sample and Zeoslag are more significant than in the case of flexural strength. However, taking into account the rate of cement replacement (50%), the 28-day compressive strength of sample "Zeoslag" (45.3 MPa) is very promising.

The strength development of both the Perlite and Zeolite samples are the same as in the case of flexural strength: Perlite gives little bit better results than Zeolite in early ages (2 and 7 days), but after 28 days the Zeolite is better by 6 MPa).

The limit for activity index is reached only by Zeoslag sample after 28 days, however the value after 7 days lacks a little to reach the limit.

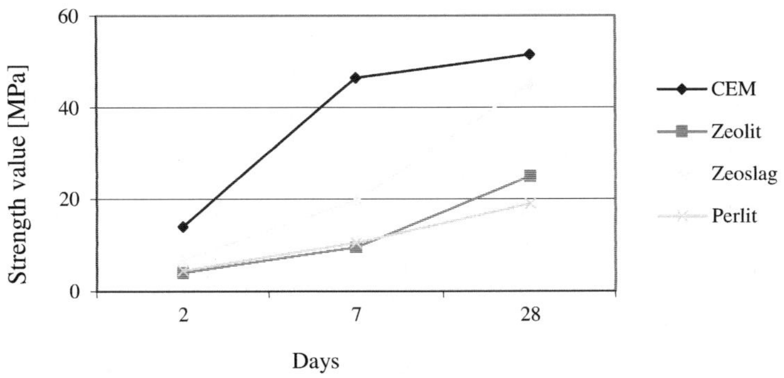

Figure 2. Compressive strength of standard mortars with additives.

The dependence between modulus of hydraulicity and strength characteristics is given in figure 3.

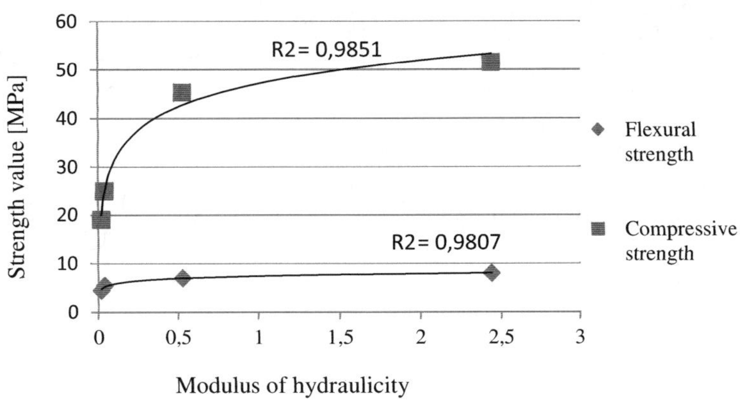

Figure 3. Dependence between modulus of hydraulicity and strength characteristics of standard mortars with additives.

Tight nonlinear dependence was found between modulus of hydraulicity and strength characteristics. R^2 value of 0.985 was found in case of compressive strength and R^2 value of 0.981 was found in case of flexural strength.

4. Conclusion

The paper was aimed on testing the technological properties of product "Zeoslag" which consists of 10% of zeolite and 90% of ground granulated blast furnace slag. Both the Perlite and Zeolite were tested for comparison, and also cement as reference material. Mineral additives were tested according to standard for testing the slag - the binder here consists of 50 % of additive to be tested and 50 % of cement. Normalized sand and water according to standard recipe were used to make a testing mixtures. The chemical analysis confirmed the chemical character of individual additives; the zeolite and perlite are of pozzolanic one, while zeoslag gains latent hydraulic character thanks to GGBFS. To have a standardized consistency, the Zeoslag needs only a bit more water than the reference sample (by 3.6%). Both the Zeolite and Perlite need for more water. Zeolite has the longest initial setting time (320 min). Perlite has higher strength characteristics than zeolite in early ages (2 and 7 days of setting and hardening), but lower after 28 days. After 28 days, the strength characteristics of Zeoslag are close to the reference sample. This is positive finding, taking into account the rate of cement replacement (50%). Regarding dependences between modulus of hydraulicity and strength characteristics, experiment showed tight nonlinear dependence at the given range of modulus of hydraulicity (0.02 – 2.44).

Acknowledgements
The results presented in this article have been carried out within the project of Slovak Scientific Grant Agency VEGA (grant number 1/0524/18) and were supported also by Slovak Cultural and Education Grant Agency (contract No. 073TUKE-4/2018).

References
[1] STN EN 206: 2015 *Concrete. Specification, performance, production and conformity*
[2] Roziere E, Samara M, Loukili A and Damidot D 2015 *Constr. Build. Mater.* **81** 1-10
[3] Papadakis V G and Tsimas S 2002 *Cement Concrete Res.* **32** 1525-1532
[4] Korjakins A, Shakhmenko G and Bumanis G 2009 *Academic Journal Constructioin Science* **10** 131
[5] Singh R, Kaushik R and Singh G 2013 *IJERA* **3**1283-1286
[6] Ma X, Wang Z and Wang Z 2013 *Adv. Mater. Sci. Eng.* **2013** 5
[7] Draganovska M and Sicakova A 2015 *Pollack Periodica* **10** 153-163
[8] Kovac M, Sicakova A and Draganovska M 2016 *Proc. Int. Conf. SGEM 2016 (Albena)* vol 2 (Sofia: STEF92 Technology Ltd.) pp 151-158
[9] Kala T F 2013 *International Journal of Engineering and Science* **2** 36-50
[10] Uncik S and Kmecova V 2013 *Concrete and Concrete Structures* **65** 51-56
[11] Beeralingegowda B and Gundakalle V D 2013 *International Journal of Innovative Research in Science* **2** 4996
[12] Khan S U 2014 *Sci. World J.* **2014** 11

IOP Conf. Series: Materials Science and Engineering **385** (2018) 012049 doi:10.1088/1757-899X/385/1/012049

The effect of mixing method on capillary moisture content of concretes with recycled concrete aggregates

A Sicakova[1] **and K Urban**[2]

[1] Institute of Environmental Engineering, Faculty of Civil Engineering, Technical University of Kosice, Vysokoskolska 4, 042 00, Kosice, Slovakia
[2] Institute of Construction Technology and Management, Faculty of Civil Engineering Technical University of Kosice, Vysokoskolska 4, 042 00, Kosice, Slovakia

Abstract. Recycled concrete aggregate represents a major environmental burden in the affected area and its use as an aggregate in concrete production is widespread in many countries. However, such aggregates have worse characteristics and strength than those of natural aggregate. In particular they have a higher absorption rate than natural aggregates, need much more water for mixing, and cause a high slump loss rate depending on the elapsed time. In this context, it is possible to monitor globally extensive research aimed on the elimination of improper properties of alternative aggregates, which are an obstacle to their application in concrete. The sorptivity expressed by capillary moisture content is a characteristic of moisture transport into the material, and recently it has been recognized as an important performance characteristic of durability. The paper is focused on the specific approach to concrete mixing as a way of the improvement of recycled aggregate´s surface using 3 variations of powders as coating materials and 2 kinds of aggregate. The changes in the capillary moisture content are compared and discussed. It is clear that the kind of coating material can influence this performance parameter significantly, so the careful design of mixing process together with the selection of powdery materials is essential. The triple mixing method was found to be promising for recycled concrete aggregate concrete, when fly ash or recycled concrete powder are used in the first stage of mixing. It is a favourable result, taking into account the perspective of the application of secondary raw materials for concrete production.

1. Introduction

Many variables need to be considered for producing the high-quality concrete; they can be considered in the terms of constituent material variables and in the terms of production parameters (mixing course, transportation and handling, as well as temperature).

As for concrete components, the alternative materials play an important role in current concrete technology. Almost three quarters of the volume of concrete is composed of aggregates. To meet the global demand of concrete in the future, it is becoming a more challenging task to find suitable alternatives to natural aggregates for preparing concrete. Therefore, the use of alternative sources for natural aggregates is becoming increasingly important [1]. However, such aggregates have worse characteristics and strength than those of natural aggregates. In particular they have a higher absorption rate than normal aggregate, need much more water for mixing, and cause a high slump loss rate depending on the elapsed time [2–4].

Recycled aggregates are composed of original aggregates and adhered mortar [5]. The major problems with the use of recycled aggregates in structural concrete are their high water absorption capacity, porosity and lower strength. The presence of RCA and the porous nature of the old cement mortar affect the bond between the RCA and cement paste when used in new concrete. That is why the poorer quality of RCA often limits its utilization [6].

The mixing procedure includes the type of mixer, the order of introduction of the materials into the mixer (loading method), the energy of mixing (duration and power) and the discharge method. Those parameters of mixing procedures are generally under wide investigation, evaluating their impact on the concrete properties [7–9].

Various loading methods are investigated worldwide, for the improvement of concrete quality. One of them is triple mixing method (TM); as given in [10], its principle lies in dividing the mixing operation into three steps to further enhance of ITZ and thus the properties of the RCA by surface-coating of pozzolanic materials, such as fly ash, slag and silica fume. It was designed as follows: coarse and fine aggregates are mixed with the addition of a certain amount of water to obtain wet aggregates. Then the additive is added to get the surface-coated aggregates, following by cement. Finally, the rest water together with superplasticizer is added to obtain fresh concrete. Also other authors involve this mixing technique into the investigation process and present the positive effect on the properties of hardened concrete [11,12]. The method primarily works with recycled aggregates and cementitious materials, but possibilities are wider, as presented in [13], when geopolymer slurry was used to coat recycled concrete aggregate, or in [14] when recycled bricks powder was used to coat air cooled blast furnace slag aggregate.

The sorptivity is an index of moisture transport into unsaturated specimens, and recently it has also been recognized as an important index of concrete durability [15]. During the sorptivity process, the driving force for water ingress into concrete is capillary suction within the pore spaces of concrete [16]. Martys and Ferraris have shown that the sorptivity coefficient is essential to predict the service life of concrete as a structural material [17]. Water absorption through capillarity is a phenomenon that occurs due to the difference between the capillary motion force (depending on the surface tension of liquid and wettability of capillary pores) and the gravity force acting on the column of elevated water (the weight of water column), which induces water movement until the balance is established. The height of capillary suction (action) increases with decreasing the capillary diameter of a capillary pore. The process is particularly visible in dry–wet conditions and has the most relevance near the element's surface [18].

In this study, the own course of TM method is presented, as well as its influence on the capillary water content of concrete. For finding the advantages of triple mixing technology, the adequate samples and tests were performed using the normal mixing approach. Two kinds of aggregates were tested (recycled concrete aggregate and natural aggregate), as well as two kinds of additives (fly ash and recycled concrete powder) were used to coat the coarse fraction of RCA in the first step of mixing. While fly ash is considered worldwide to be a standard additive, recycled concrete powder is not widely used; even it represents a problematical part of recycled concrete. It was involved into the experiment from environmental reasons to check whether it would be suitable to play the role of the coating material within the triple mixing technology. For comparison, cement as a coating material was used, too. The paper is focused on the evaluation of changes in capillary water content of concrete from three points of view: mixing technique, kind of aggregate and kind of coating material.

2. Materials and methods

Six kinds of concrete mixtures were tested, depending on the kind of aggregate and the kind of coating powder, while samples were prepared in two series: using triple mixing (TM) and normal mixing (NM). While the standard mixing involves the pre-mixing of dry components first, following by the addition of water with plasticizer, the principle of triple mixing lies in dividing the mixing process into three steps, differing in the order and timing of concrete´s components addition. It results in coating the aggregate in the first stage of mixing by specific coating material, thus improving the surface character of aggregate. In this experiment, the own course was applied (see figure 1). The materials and their parameters were as follows:

- Aggregates:

- Natural aggregate (NA): fraction 0/4 was used in all mixtures (water absorption 1.2% apparent density ρ_0 = 2650 kg/m^3).
- Recycled concrete aggregate (RCA): fractions 4/8 (water absorption 6.8% and apparent density ρ_0 = 2200 kg/m^3) and 8/16 (water absorption 5.3% and apparent density ρ_0 = 2300 kg/m^3).

- Coating materials:
  - Cement CEM I 42.5 R, ρ_0 = 3100 kg/m^3.
  - Fly ash (FA): the energy segment of the steel-making factory. The original granulometry of fly ash was $d_{(0.9)}$ = 95 μm; ρ_0 = 2100 kg/m^3.
  - Recycled concrete powder (RCP): Particles under 125μm were separated from unsorted C&DW by sieving.
- Binder: Cement CEM I 42.5 R, ρ_0 = 3100 kg/m^3.
- Admixture: polycarboxylate type of superplasticizer.

The triple mixing method given by Kong [10] was slightly modified. The actual mixing method (see Figure 1) mainly differs in the following aspects: the sequence of adding aggregate (coarse aggregate in the first stage of mixing, fine aggregate in the third one), the method of calculation of the coating layer volume, and the method of calculation the water amounts (W_1 and W_2). Two kinds of powdery materials were tested for coating of RCA in the first stage of mixing: fly ash and recycled concrete powder. As a control mixture, concrete of standard composition was tested too, having natural aggregate (NA) and cement (CEM); in the TM cement meets two functions: coating the coarse aggregate and filling the voids between grains. To see the effect of triple mixing, samples of the same compositions were prepared by normal mixing (Figure 2).

The compositions of concrete were designed keeping the limiting amounts of cement and water for specific class of exposure, in accordance to standard for concrete production (min 300 kg of cement and max w/c ratio = 0.5). The amounts of real mixing water (W_1 and W_2) were adjusted taking into account the actual absorption capacity of aggregates, i.e. the effective amount of water W_{ef} was increased by water absorption value during mixing. For calculation of the additive amount, the thickness of coating layer was considered as δ = 0.150 mm. The compositions were calculated taking into account densities of individual components to keep the constant volume 1m^3. The compositions of concrete mixtures are given in Table 1.

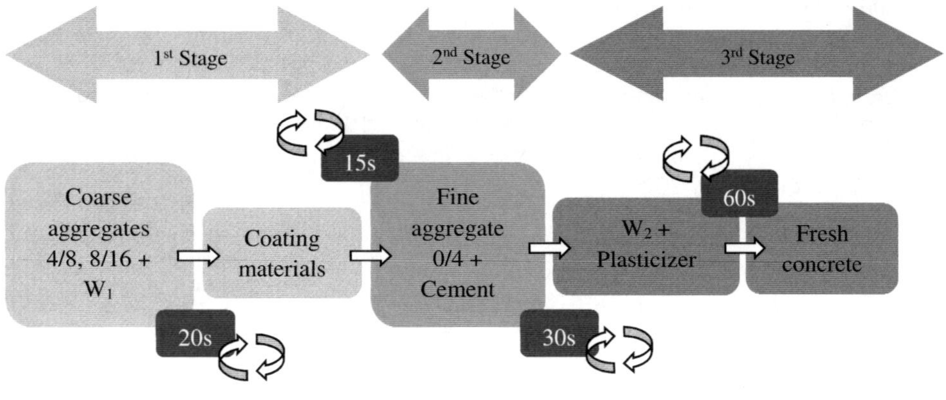

Figure 1. Experimental triple mixing procedure.

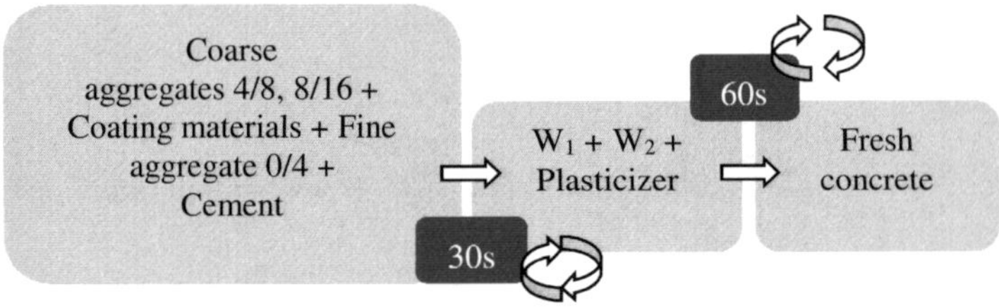

Figure 2. Experimental normal mixing procedure.

Table 1. Mix proportions of concrete samples for TM approach.

Component [kg.m^{-3}]		RCA - coarse aggregate			NA - coarse aggregate		
		Coating material					
		CEM	FA	RCP	CEM	FA	RCP
		RCA_{CEM}	RCA_{FA}	RCA_{RCP}	NA_{CEM}	NA_{FA}	NA_{RCP}
CEM I 42.5 R		310	310	310	336	336	336
NA	0/4	898	898	898	896	896	896
	4/8	-	-	-	269	269	269
	8/16	-	-	-	627	627	627
RCA	4/8	224	224	224	-	-	-
	8/16	545	545	545	-	-	-
Material for coating		80	68	68	55	47	47
Plasticizer		2.5	2.5	2.5	2.7	2.7	2.7
Water	W_{ef-1}	39.8	33.8	33.8	27.4	23.2	23.2
	W_{ef-2}	155	155	155	168	168	168
	Total	232	226	226	203	198	202

Cube samples of 100x100x100 mm were prepared and cured under standard conditions. After 28 days, the capillary moisture (water) content was tested. The capillary absorption test allows the characterization of the porous structure of concrete and it is an indicator of the concrete durability. Samples were dried to constant mass, and then one face of the specimens is immersed in water at a depth of 5–10 mm for a specific period of time (normally 10 and 90 min). For this evaluation, the values in 90 min are discussed. Capillary moisture content m_c is characterized by water absorption coefficient A_w according to [19,20].

$$m_c = A_w \cdot \sqrt{t} \tag{1}$$

m_c	capillary moisture content (kg/m^2),
A_w	water absorption coefficient (kg/m^2 . s$^{1/2}$),
t	time (s).

Water absorption coefficient (A_w):

$$A_w = \left(\frac{m_t - m_i}{A\sqrt{t}}\right) \tag{2}$$

A_w	water absorption coefficient (kg/(m^2 . s$^{1/2}$)),
m_t	mass of the specimen after time 't' (kg),
m_i	initial mass of the specimen (kg),
A	liquid contact area of the specimen (m^2),
t	time (s).

3. Results and discussion

The comparison of capillary moisture content of the tested concretes prepared by various mixing methods after 28 days of curing is shown in Figure 3. The figure is organized in terms of mixing method, kind of aggregate and kind of coating material.

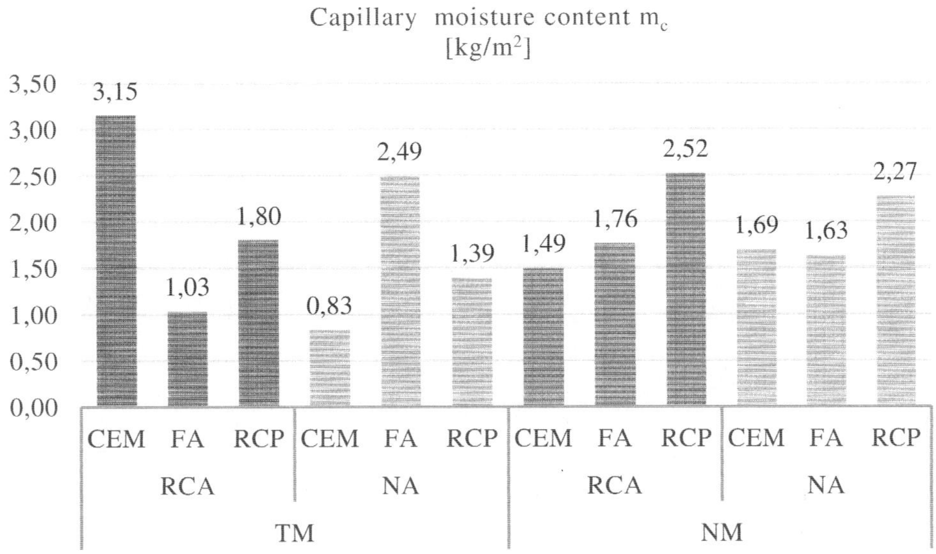

Figure 3. Capillary moisture content of concrete samples.

Evaluation in terms of mixing method:

In most cases, the triple mixing method gives better results than that of normal mixing. Looking at samples with RCA, the triple mixing process gives better results when aggregate is coated by FA and RCP. For coating with cement, the normal mixing seems to be more favorable.

For samples with NA, the triple mixing process gives better results when aggregate is coated by CEM and RCP while the normal mixing seems to be more favorable for coating with FA.

Evaluation in terms of kind of aggregate:

The samples with RCA have usually higher capillary water content than that of the samples with NA. The only exceptions are samples having RCA when coated with FA and prepared by TM, and the samples having RCA when coated with CEM and prepared by NM.

Evaluation in terms of kind of coating material:

There is no clear dependence in capillary moisture content in terms of coating material. Those additives affect differently in individual cases of the mixing method and the kind of aggregate. In 2 of 4 cases (when normal mixing was applied), the RCP gives the highest values; this corresponds with the character of the material, as it has no premise to have some pozzolanic/hydraulic character. However, the values themselves are so far from the ones of other coating materials.

4. Conclusion

The concretes prepared by two mixing methods (triple and normal) with 2 kinds of aggregate and 3 kinds of coating materials were tested. The capillary water content was measured as a durability parameter. The triple mixing method is promising for recycled concrete aggregate concrete, when fly ash or recycled concrete powder are used in the first stage of mixing to coat the coarse grains. It is a favourable result, taking into account the perspective of the application of secondary raw materials for concrete production.

Acknowledgements
The results presented in this article have been carried out within the project of Slovak Scientific Grant Agency VEGA (grant number 1/0524/18) and by the project NFP 26220220051 Development of progressive technologies for utilization of selected waste materials in road construction engineering (European Union Structural Funds).

Reference

[1] Hiraskar K G and Patil C 2013 *IJSER* **4** pp 95–98
[2] Ryou J S and Lee Y S 2014 *Int. J. Concr. Struct. Mater.* **8** 165–172
[3] Tabsh S W and Abdelfatah A S 2009 *Constr. Build. Mater.* **23** 1163–1167
[4] Eguchi K, Teranishi K, Nakagome A, Kishimoto H, Shinozaki K and Narikawa M 2007 *Constr. Build. Mater.* **21** 1542–1551
[5] Etxeberria M, Vázquez E, Marí A and Barra M 2007 *Cement Concrete Res.* **37** 735–742
[6] Spaeth V and Djerbi Tegguer A 2013 *Int. J. Sustain. Built Environ.* **2** 143–152
[7] Ferraris C F 2001 *J. Res. Natl. Inst. Stand. Technol.* **106** 391–399
[8] Dehn F 2006 *Proc. Int. RILEM Workshop HPFRCC Struct. Appl. (Honolulu)* (Paris: RILEM Publication) pp 1–8
[9] Prasittisopin L and Trejo D 2015 *ACI Mater. J.* 399-107
[10] Kong D, Lei T, Zheng J, Ma C, Jiang J and Jiang J 2010 *Constr. Build. Mater.* **24** 701–708
[11] Lee S-H, Hong K-N, Park J-K and Ko J 2014 *Materials* **7** 4739–4754
[12] Surya M, Vvl K R and Lakshmy P 2013 *Procedia - Soc. Behav. Sci.* **104** 1158–67
[13] Junak J and Sicakova A 2017 *Build. - Open Access J.* **8** 1–11
[14] Urban K and Sicakova A 2017 *IOP Conf. Ser. Earth Environ. Sci.* **92** 012069
[15] Dias W P 2000 *Cement Concrete Res.* **30** 1255–1261
[16] Hall C 1989 *Mag. Concr. Res.* **41** 51–61
[17] Martys N S and Ferraris C F 1997 *Cement Concrete Res.* **27** 747–60
[18] Evangelista L and de Brito J 2010 *Cem. Concr. Compos.* **32** 9–14
[19] Kumaran M K 1999 *J. Therm. Envel. Build. Sci.* **22** 349–355
[20] Mukhopadhyaya P, Kumaran K, Normandin N and Goudreau P 2002 *J. Therm. Envel. Build. Sci.* **26** 179–195

IOP Conf. Series: Materials Science and Engineering **385** (2018) 012050 doi:10.1088/1757-899X/385/1/012050

Examination of potential radiation hazard of concrete composites with different slags as additives

E Singovszka and A Estokova

Department of Material Engineering, Technical University of Kosice, Faculty of Civil Engineering, Institute of Environmental Engineering, Kosice, Slovakia

Abstract. Present day steelmaking slags are being successfully used as a high quality mineral aggregate for the building materials. With this, it is of vital importance to be familiar with the technical significance of the secondary application of slag, because some slag might contain increased concentration of substances harmful to human health. In terms of slag impact on the environment, radionuclides are the least researched of all pollutants emitted from the metallurgical processes. The paper presents the results of the measurement the natural radioactivity levels in concrete composites with incorporated slags in various proportions. The radionuclides´ activity index, dose rate and annual effective dose were evaluated to assess the potential radiological hazard associated with composites with slag and ferromanganese slag.

1. Introduction

Most building materials of natural origin contain small amounts of naturally occurring radioactive material (NORM), mainly radionuclides from the ^{238}U and ^{232}Th decay chains and the radioactive isotope of potassium, ^{40}K. The activity concentration of the radionuclides in building materials varies considerably, depending on both the nature and the origin of the raw material compounds. Generally, natural building materials reflect the geology of their site of origin. The average activity concentrations of ^{226}Ra, ^{232}Th and ^{40}K in the Earth's crust are about 40, 40 and 400 $Bq.kg^{-1}$ respectively [1-2]. Available literature shows typical and maximum activity concentrations in common building materials and industrial by-products used for building materials in the EU (max. values of activity for slag of ^{226}Ra is 2100 $Bq.kg^{-1}$, for ^{232}Th $Bq.kg^{-1}$ is 340 and for ^{40}K is 1000 $Bq.kg^{-1}$). The radioactivity concentrations found in certain kinds of industrial by-products (fly ash, phosphogypsum, etc.) can often be significantly higher in comparison with most common building materials [1-3]. Metallurgical industry influences the environment directly with its by-products, i.e. various hazardous and non-hazardous technological wastes (slag, refractories, sludge, dust, mill scale, etc.) which are most commonly disposed of at their inadequate landfills. The most common technological waste inadequately disposed of in the said manner is unprocessed steel slag. Through awareness of environmental considerations and more recently, the concept of sustainable development, extensive research and development has removed slag from industrial waste into modern industrial product which is effectively and profitably used for many industrial purposes, especially as raw material in the numerous building applications [4-6]. Slags from different metallurgical processes contain many useful components (metals and oxides) used for various industrial and construction purposes. The properties of slag including mineralogical composition play important roles in determining specific applications. Steel-making processes produce significant volumes of waste, which is a problem both from the economical and environmental point of view. In order to find solutions for exploitation of steel slag, as well as due to increasingly demanding legal regulations in environment protection, physical, chemical and radiochemical properties of this material are more and more frequently subjected to systematic research. The amounts of radionuclides involved are noteworthy. US, Australian, Indian and UK coals contain up to about 4 ppm uranium, those in Germany up to 13 ppm, and those from Brazil and China range up to 20 ppm uranium. Thorium concentrations are often about three times those of uranium [7-8].

2. Experimental

2.1. Materials

Research on radioactivity of building materials was performed using 12 mixtures of cement composites with different portions of cupola furnace slag (TH) and ferromanganese slag (TO). Recipes of cement composites are given in table 1.

Table 1. Recipes of cement mortars per $1m^3$.

The components of mortar mix [kg]	TH 0	TH 10	TH 20	TH 30	TO 0	TO 10	TO 20	TO 30
Cement CEM I 42.5 R [kg]	450	405	360	315	450	405	360	315
Cement replacement [wt.%]	0%	10%	20%	30%	0%	10%	20%	30%
Cupola furnace slag [kg]	0	45	90	135	-	-	-	-
Ferromanganese slag [kg]	-	-	-	-	0	45	90	135
Fine aggregate [kg]	1350							
Water [L]	225							

2.2. Radiological measurements

The mass activities of radionuclides (^{226}Ra, ^{232}Th and ^{40}K) in cement composites were measured using gamma ray spectrometry. This technique is an important tool in field of environmental radioactivity measurements due to its high resolution, large photo peak efficiency and it can measure different radionuclides in a single spectrum. Measurements were carried out using an EMS-1A SH (Empos, Prague, Czech Republic) detection system equipped with a NaI/Tl scintillation detection probe and a MC4K multichannel analyzer with optimized resolution of 818 V, 4.096 channel and with 9 cm of lead shielding and internal lining of 2 mm tinned copper.

The specific activity concentrations of ^{226}Ra, ^{232}Th and ^{40}K were determined in Bq.kg−1 using the count spectra. The ^{40}K radionuclide was measured directly through its gamma ray energy peak at 1461 keV, while activities of ^{226}Ra and ^{232}Th were calculated based on the mean value of their respective decay products. Activity of ^{226}Ra was measured using the 351.9 keV gamma rays from ^{214}Pb and the activity of ^{232}Th was measured using the 238.6 keV gamma rays of ^{212}Pb. The same counting time of 86,400 s (24 h) was used for all measured samples.

2.3.Methods for radiological analysis

Gamma index Iγ

Within the European Union effective doses exceeding 1 mSv.y^{-1} should be taken into account from the radiation protection point of view. Since several radionuclides contribute to the overall dose, in order to assess whether the dose criterion is met, an activity concentration index Iγ have been established. Radionuclides´ activity index, or shortly gamma index, Iγ was defined according to the following equation (1):

$$I_\gamma = \frac{A_{Ra}}{300} + \frac{A_{Th}}{200} + \frac{A_K}{3000} \quad (1)$$

where A_{Ra}, A_{Th}, A_K are the radium, thorium and potassium, activity concentration (Bq.kg^{-1}) in the building materials. The gamma index is derived to identify whether a dose criterion is met. The gamma index of materials used in building should not exceed limit values depending on the dose criterion ($I_\gamma \leq 1$) [9,10]. The gamma index should be used only as a screening tool for identifying materials which might be of concern.

Dose rate and annual effective dose
The absorbed dose rates in air in a room can be calculated by using the specific dose rates given in [9-10]. The specific dose rates for radionuclides are given for different screening tool of identifying materials which might be of concern. Indoor dose rates for a model room (dimension of 4m x 5m x 2.8 m, thickness of 20 cm, density of 2,35 g/cm^3, and the background of 50 nGy.h^{-1}) are calculated with different structures in a building causing the irradiation as follows:
All structures:

$$D = 0.92.A_{Ra} + 1.1.A_{Th} + 0.08A_K \quad (2)$$

Floor and walls:

$$D = 0.67.A_{Ra} + 0.78.A_{Th} + 0.057A_K \quad (3)$$

Floor only:

$$D = 0.24.A_{Ra} + 0.28.A_{Th} + 0.02A_K \quad (4)$$

The annual effective dose D_E (mSv), due to gamma radiation from building materials with the annual exposure time of 7000 h [9,10] was calculated by next equation:

$$D_E = 0.7\ (Sv.Gy^{-1})x7000\ (h)x10^{-6}\ xD(nGy.h^{-1}) \quad (5)$$

3. Results and discussion

Table 2 presents the measured radionuclides´ mass concentrations in input materials and studied cement composites with slag additives as well as the calculated gamma indexes.

Table 2. The mass activities of radionuclides and gamma indexes in input materials and composites samples.

Sample	Mass activity [Bq.kg^{-1}] ^{226}Ra	^{232}Th	^{40}K	Gamma index Iγ [-]
TH 0	5.72	20.74	80.45	0.149
TH10	3.17	8.54	5.78	0.055
TH20	3.54	7.53	3.74	0.049
TH30	4.06	8.013	22.31	0.053
TO 0	5.72	20.74	80.45	0.149
TO10	3.36	6.66	14.37	0.044
TO20	2.96	6.08	4.987	0.040
TO30	4.07	8.26	4.794	0.055
Cupola furnace slag	16.95	48.95	97.04	0.345
Ferromanganese slag	14.82	48.93	155.28	0.333
Cement	3.03	11.63	57.27	0,09
Fine aggregates	4.11	4.86	33.23	0.06

The highest values of individual ^{226}Ra, ^{40}K and ^{332}Th mass activities were found for cupola furnace slag followed by ferromanganese slag. Measured activities of radionuclides in cement were lower than measured previously [11]. In [11], the radionuclide activities in CEM I cements ranged from 3.69 – 36.8 Bq·kg^{-1}, 11.8 – 24.9 Bq·kg^{-1} and 36.98 – 331.4 Bq·kg^{-1} for ^{226}Ra, ^{232}Th and ^{40}K, respectively. The analyzed mass activities in aggregates were also low. According to Terpakova [12], the aggregates radioactivity strongly depends on the geological sources and ranges from 21.5 to 29.9 Bq·kg^{-1} for ^{226}Ra, 5.0 to 11.4 Bq·kg^{-1} for ^{232}Th and 106.5 to 270.6 Bq·kg^{-1} for ^{40}K.

Surprisingly, the highest gamma radiation was observed in cement composites without any slag additions. Cabanekova [13] presents a very wide range of radionuclides activities regarding concretes. In spite of the highest radiation in slags themselves, no increasing trend was observed with the increasing portion of slag in composites. The calculated gamma indexes for input materials have been also extremely low when compared to Vietnamese and Indian study [10,14] where for example the Iγ of cement was four times higher. On the other hand, the results are similar to the Egyptian ones, as reported in [15].

The gamma index ranged from 0.04 (TO 20) – 0.35 (slag). The limit value of gamma index ($I_\gamma \leq 1$) was not exceed in any sample. Surprisingly, the highest values of gamma index have been calculated for the composites without any additives (TH0, TO0). In descending, the gamma index of 12 samples are of the sequence: Cupola furnace slag >ferromanganese slag> TH0=TO0 > cement >fine aggregates>TH10=TO30>TH30 >TH20>TO0>TO20. In particular, the I_γ values of cupola furnace slag and ferromanganese slag are significant to be considered in the assessment of radiological risk linking to standards and regulators on natural radioactivity in composites.

The values of dose rates depended on the structures in building causing the irradiation in samples are compared in figure 1.

The annual effective dose, calculated in accordance with equation (5), is given in table 3.

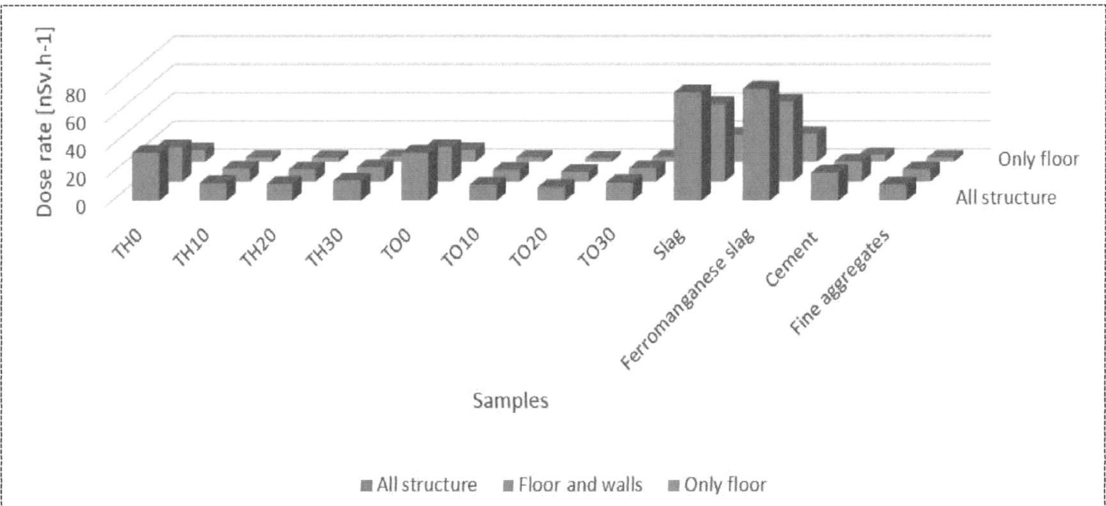

Figure 1. The dose rates of analyzed samples in $nSv.h^{-1}$.

Table 3. The annual effective dose for composite samples.

Sample	Annual effective dose DE [mSv]
TH 0	0.17
TH10	0.06
TH20	0.06
TH30	0.07
TO 0	0.17
TO10	0.06
TO20	0.05
TO30	0.06

The obtained data show that there are many samples with annual effective doses not exceeding the criterion of 1mSv. The highest values of rate dose calculated for slag-free composites were significantly lower than recommended dose criterion of 84 nSv.h^{-1}. The calculated values for dose rate for input materials have been lower when compared to Turkish study [16] where for example the D of cement was twice higher. This was the case for all three applications of composites: in floors, in floors and walls, and even in all structures.

4. Conclusion

The natural radioactivity of composites with slag additives was investigated to estimate the potential radiological hazard of the waste-based materials. The results showed that the mass activities of radionuclides or gamma indexes of the composites did not exceed the limit values recommended by UNSECAR. In addition, the values of the annual effective dose have been found within the safe limit. Addition of slag materials has not proved any significant positive or negative effect.

Another concern was to study the correlation between the amount of the slags and radioactivity´s increasing. However, no correlation was observed in this study. The present study will be helpful to understand radiation level in the research of cement composites with additives for further studies.

Acknowledgments
This work has been supported by the Slovak Grant Agency for Science (Grant No. 1/0648/17).

References

[1] Radiological Protection Principles concerning the natural radioactivity of building mateials, Radiation Protection 112, European Commision, Directorate Geeneral, Environment, Nuclear Safety and Civil Protection, 1999 11 p 16
[2] Mustonen R Pennanen M Annanmaki M Oksanen E 1999 *Enhanced radioactivity of building materials. Final Report of the Contract No 96-ET-003 for the European Commission*, Radiation and Nuclear Safety Authority, STUK: Finland p 16
[3] Soflic T Barisic D Sofilic U 2011 *Arch. Metall. Mater* **56** 627-634
[4] Huang Y Bird RN Heidrich O 2007 *Resour. Conserv. Recycl.* **52** 58-79
[5] Ducman V and Mirti B 2009 *Waste Manage.* **29** 2361-2368
[6] Svyazhin AG Shakhpazov E KH Romanovich DA 1998 *Metallurg* **42** 25-27
[7] Naturally-occuring radioactive materilas (NORM), Avaliable on internet: http://www.world-nuclear.org/information-library/safety-and-security/radiation-and-health/naturally-occurring-radioactive-materials-norm.aspx (Access: 18.12.2017)
[8] Liu H et al. 2013 Characteristics and disposal categorization of solid radioactive waste from the front end of the uranium fuel cycle. *In Naturally occuring radioactive material (NORM VII) Proceedings of an International Symposium Beijing China*
[9] EC (European Commission) 2000 Radiological Protection Principles Concerning the Natural Radioactivity of Building Materials. Directorate-General Environment, Nuclear Safety and Civil Protection; Luxembourg, Belgium: (Radiation Protection 112)
[10] Sieu LN et al. 2011 Natural radioactivity in commonly building materials used in Vietnam In *37th Annual Radioactive Waste Management Symposium* pp 1–9
[11] Estokova A and Palascakova L 2013 *Int J Env Res Pub He* **10** 7165–7179
[12] Terpakova E 2000 *Zeszyty naukowe Politechniki Rzeszowskiej* **32** 555-562
[13] Cabanekova H 1996 *J Radioanal Nucl Ch* **209** 301–306
[14] Raghu Y 2017 *J. Taibah Uni. Sci* **11** pp 223-233
[15] El-Taher A 2010 *Radiat Prot Dosim* **138** 158–165
[16] Mehmet A et al 2017 *J Radioanal Nucl Ch* vol **314** 941-948

IOP Conf. Series: Materials Science and Engineering **385** (2018) 012051 doi:10.1088/1757-899X/385/1/012051

Usability of alternative admixtures for the potential production of blended cements

D Sklenarova, K Dvorak, D Gazdic and V Hladik

Faculty of Civil Engineering, Brno University of Technology, Veveri 331/95, Brno 602 00, Czech Republic

Abstract. This paper studies how pozzolan materials can be mechanically activated. For this, two major grinding technologies and different types of artificial pozzolans, such as red brick dust, granulated blastfurnace slag and sheet glass, were used. Observed properties were chemical and mineral composition, amorphousness, particle size and specific surface area. Pozzolan activity of selected materials was compared by evaluation of the reaction with lime using X-Ray diffraction (XRD) analysis and differential thermal analysis (DTA).

1. Introduction

Pozzolans are materials, which do not harden themselves, nor by the effect of the exciter, but because of their high amorphous siliceous or siliceous and aluminous content, it is capable to react with calcium ions to form similar hydration products (calcium silicate hydrates and calcium silicate aluminate hydrates) such as alite or belite.

The use of natural pozzolans, such as trass, tuffs, volcanic ash or pumices, is known from the Neolithic period [1]. Contrary to this, artificial pozzolans as an admixture with exceptional properties was used in the construction since Classical Antiquity [2]. It is known that the Roman Empire used both types of pozzolans. Natural ones such as those from city Pozzuoli located nearby volcano Mount Vesuvius, but they they purposefully created a binder with pozzolan properties by mixing crushed bricks with lime [3].

Recently, the frequently discussed topic is about the tendency to reduce CO_2 emissions from cement production. There are many options that directly reduce emissions. One of the options, how to reduce the emissions of CO_2, NO_x, SO_x and other substances is replacing a certain amount of cement with secondary raw materials with pozzolan properties such as microsilica or metakaolin, however, these materials are expensive today, so materials of similar properties with a lower purchase price are sought. Therefore, we cannot directly reduce emissions in cement production, but we can indirectly reduce them in relation to the unit quantity of resulting blended cement of the same or better properties than pure Portland cement. This has a significant impact on the ecology of cement production [4].

In this experiment, we had monitored whether the mechanical activation had occurred and the conditions during this process. One way to prove the effect of mechanical activation is to compare the grinding methods. The effort was to achieve the same specific surface area for both grinding methods and to monitor outputs of XRD and DTA analysis.

2. Materials and methods

Three materials were selected for testing - Red Brick Dust (RBD) made from grinded brick (Heluz 2012), Sheet Glass (SG) from the old window glass and Granulated Blastfurnace Slag (GBS) from power station Dětmarovice (2017). On each of them a chemical analysis and X-Ray diffraction (XRD)

analysis was performed. For research, two grinding methods were used – grinding in a vibratory disc mill and in a laboratory planetary mill. The specific surface area was measured by using a ZEB PC-Blaine Star. The milled samples were mixed with lime immediately after milling, and pozzolan activity was then monitored by using X-Ray diffraction (XRD) analysis and differential thermal analysis (DTA) after 1, 7 and 28 days.

To ensure grinding efficiency, all materials were first crushed to a maximum grain size 2 mm. Granulated Blastfurnace Slag did not have to be prepared in any way, because it was delivered already milled (specific surface area was 280 m^2/kg). For measurement of specific surface area of samples, it is necessary to specify the density of the tested materials. Thus, the density of all materials was measured on the Micromeritics AccuPyc II 1340 helium pycnometer.

2.1. *Method of grinding in a planetary mill*

At first, the grinding was done in one-position planetary mill PULVERISETTE 6 (FRITSCH). For grinding was used 500 ml grinding bowl with 25 grinding balls. Rotational speed of main disk was 500 rpm. The actual grinding process was dry, so no liquid was added. Total batch of 200 g was ground to a specific surface area 500 m^2/kg. From the beginning, 1–2 minutes grinding intervals were selected and then the grinding balls and bowl were cleaned. Later, the final specific surface area correction was done by using shorter, 30 seconds grinding intervals. Immediately after grinding, all samples were mixed with water and reactive CaO.

For this experiment, lime was specially prepared by burning $CaCO_3$ in a grade of purity analytical reagent. The temperature increase during burning process was 8 °C/min up to 1050 °C, where the temperature stayed for 4 hours. To determine the correct batch of CaO, it was assumed that the material was only reactive SiO_2, and so a gel of $C_3S_2H_3$ could form from a 100% of material. The ratio of CaO and materials was then determined from the molar masses.

M(Ca) = 40.08 g/mol	M(CaO) = 40.08 + 16.00 = 56.08 g/mol
M(O) = 16.00 g/mol	M(SiO2) = 28.09 + 2(16.00) = 60.09 g/mol
M(Si) = 28.09 g/mol	$3 \cdot C \rightarrow 3 \cdot M(CaO) = 3 \cdot 56.08 = 168.24$ g/mol
	$2 \cdot S \rightarrow 2 \cdot M(SiO2) = 2 \cdot 60.09 = 120.18$ g/mol

While maintaining a ratio of 168.24 g/mol of CaO to 120.18 g/mol of pozzolan (test materials), we provide sufficient amount of CaO to produce the ideal conditions for the $C_3S_2H_3$ gel formation. When we adjust this ratio, we come to a ratio of 1.4 g of lime per gram of test material.

2.2. *Method of grinding in a vibratory disc mill*

The second grinding method was done in vibratory disc mill RETSCH RS 200 with grinding bowl (size 100 ml). Each batch weighed 180 g and it was ground to a specific surface area 500 m^2/kg. As with the planetary mill, it was necessary to do short grinding cycles (up to 1 minute) followed by cleaning the grinding bowl and tools, otherwise the grinding process became very inefficient. Rotational speed was 900 rpm. Next steps were the same as for grinding in the planetary mill and during the addition of CaO, the same calculated ratio was maintained (1.4 g : 1 g of test material).

2.3. *Sample preparation for X-Ray diffraction (XRD) analysis and differential thermal analysis (DTA)*

A batch for XRD and DTA analysis was 5 g (grinded material and CaO). The samples were ground for 3 minutes in a McCrone Micronizing Mill and subsequently dried at 55 °C. Samples of each material were taken after 1 day, 7 days and 28 days from mixing.

3. Results

The material characteristics of the samples were determined for the specification of input raw materials. In the first part, before grinding, chemical analysis and XRD analysis were done.

Table 1. Chemical composition of the pozzolans.

	Granulated Blastfurnace Slag (GBS)	Sheet Glass (SG)	Red Brick Dust (RBD)
SiO_2 [%]	40.37	71.74	70.38
MgO [%]	2.01	0.85	0.25
MnO [%]	1.21	0.00	0.05
CaO [%]	47.09	10.80	4.88
Al_2O_3 [%]	7.85	1.06	14.54
TiO_2 [%]	0.41	0.07	0.81
Fe_2O_3 [%]	0.19	0.03	5.17
Na_2O [%]	0.57	15.02	0.90
K_2O [%]	0.61	0.16	2.12
SO_3 [%]	0.27	0.02	0.17

From the chemical analysis, it is evident that the brick dust contains the highest amount of SiO_2. From the diffractogram was identified β-quartz, which coincided with the chemical composition of the material. The sheet glass had a high content of SiO_2 and Na_2O, which is typical for glass. Other oxides are present in very small amounts. XRD analysis of the glass was not performed because it is particularly amorphous material. The granulated slag compared with other observed materials contained the highest amount of CaO and the lowest amount of SiO_2.

3.1. Results of differential thermal analysis (DTA)

On the grounds that the chemical formula of C-S-H gels cannot be accurately determined, it was necessary to observe the total content of CaO in the samples. In figure 1 shows an example of a DTA analysis output, where two important reactions are marked.

- Reaction 1 - Decomposition of portlandite $Ca(OH)_2 \rightarrow CaO + H_2O$
- Reaction 2 - Decomposition of carbonates $CaCO_3 \rightarrow CaO + CO_2$

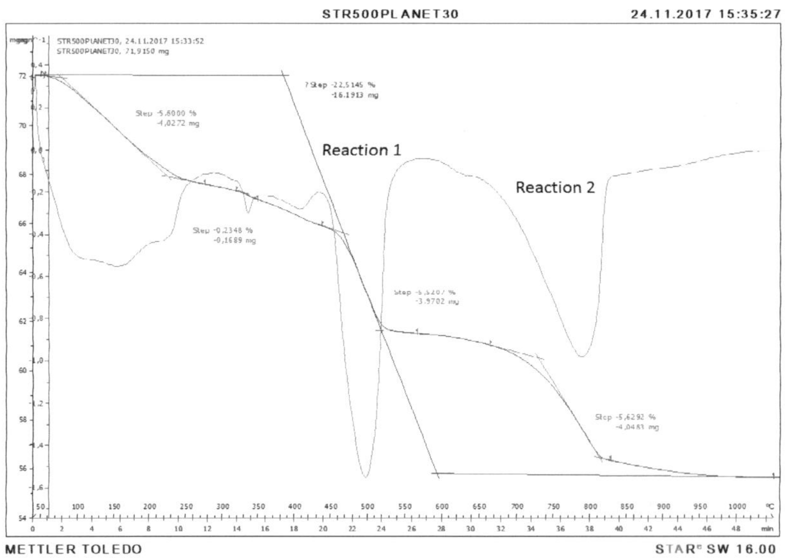

Figure 1. Example of output from DTA analysis.

The next part was to express input and output substances from the reaction in molar amounts.

- Reaction 1 - Decomposition of portlandite $M(Ca(OH)_2) \rightarrow M(CaO) + M(H_2O)$
- Reaction 2 - Decomposition of carbonates $M(CaCO_3) \rightarrow M(CaO) + M(CO_2)$

Calculation coefficients were also used to simplify the calculation. In Reaction 1, H_2O is exited, which is reflected by weight loss of the sample.

$$\text{Calculation coefficient } (Ca(OH)_2) = \frac{M_{(Ca(OH)_2)}}{M_{(H_2O)}} = \frac{74.0958}{18.0158} = 4.1128232$$

In Reaction 2, CO_2 is exited, which is reflected by weight loss of the sample.

$$\text{Calculation coefficient } (CaCO_3) = \frac{M_{(CaCO_3)}}{M_{(CO_2)}} = \frac{100.09}{44.01} = 2.2742558$$

For further evaluation, it was also necessary to determine the actual CaO content in the samples. Calculated content of portlandite and calcium carbonates was further recalculated to their net CaO content.

$$\text{CaO content in } (Ca(OH)_2) = \frac{M_{(CaO)}}{M_{(Ca(OH)_2)}} = \frac{56.08}{74.0958} = 0.7568631$$

$$\text{CaO content in } (CaCO_3) = \frac{M_{(CaO)}}{M_{(CaCO_3)}} = \frac{56.08}{100.09} = 0.5602957$$

Thus, if the sample weight loss was 10% in the temperature range 450 °C - 550 °C in reaction 1, it means that the sample contained 10% · 4.1128232 = 41.1% $Ca(OH)_2$ and that contained 41.1% · 0.7568631 = 31.12844109% CaO. It was similar for reaction 2. If the sample weight was 10% in the temperature range 650 °C - 900 °C, it means that the sample contained 10% · 2.2742558 = 22.7% $CaCO_3$ and that contained 22.7% · 0.5602957 = 12.74255745% CaO. The CaO content, which was calculated from both reactions together, indicates the total CaO content in the samples.

For the sake of clarity, the total CaO content calculated from both reactions obtained from DTA analysis of all samples (Granulated Blastfurnace Slag (GBS), Red Brick Dust (RBD) and Sheet Glass (SG)) is shown in figure 2 (grinded in the vibratory disc mill) and in figure 3 (grinded in the planetary mill).

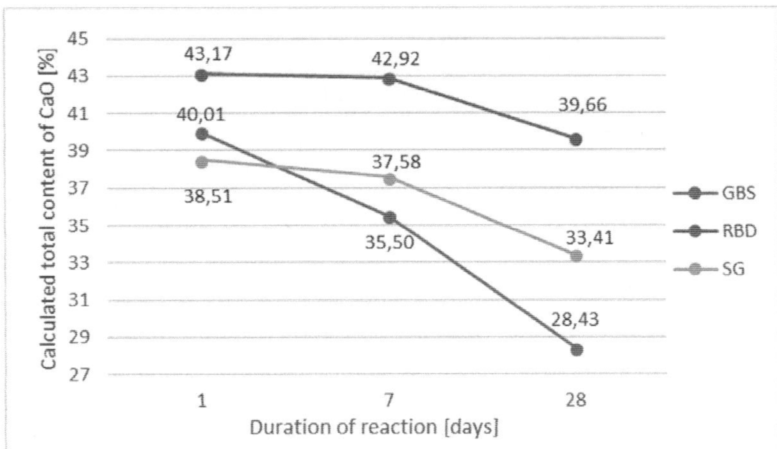

Figure 2. The calculated total content of CaO in the samples grinded in the vibratory disc mill after 1, 7 and 28 days of reaction.

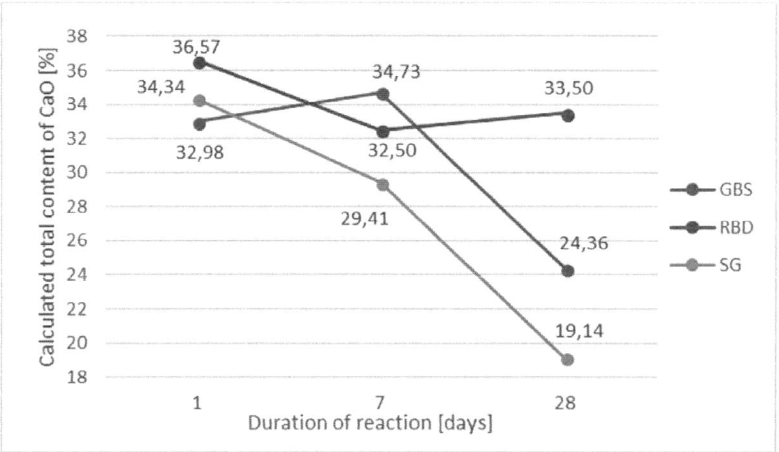

Figure 3. The calculated total content of CaO in the samples grinded in the planetary mill after 1, 7 and 28 days of reaction.

3.2. Results of X-Ray diffraction (XRD) analysis

In this part, the size and shapes of the $Ca(OH)_2$ lines were mainly examined in X-Ray diffractograms.

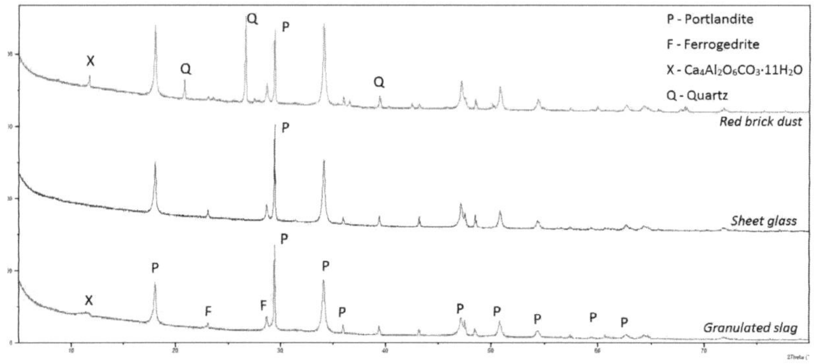

Figure 4. Diffractogram of the samples grinded in the vibratory disc mill after 28 days (pink line - Red Brick Dust (RBD), blue line - Sheet Glass (SG) and red line - Granulated Blastfurnace Slag (GBS)).

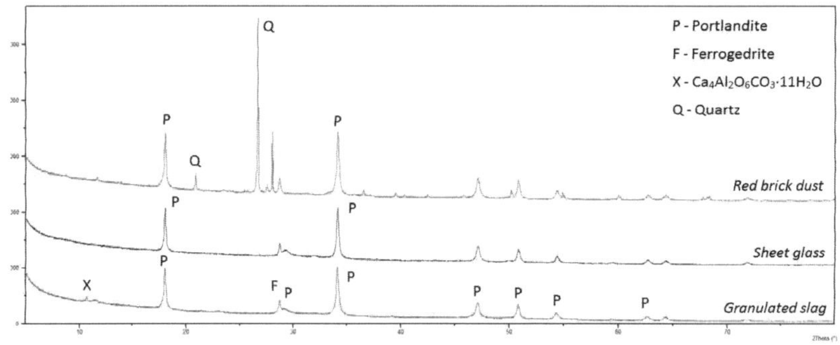

Figure 5. Diffractogram of the samples grinded in the planetary mill after 28 days (pink line - Red Brick Dust (RBD), blue line - Sheet Glass (SG) and red line - Granulated Blast furnace Slag (GBS)).

As can be seen in the figure 4 and figure 5, the lowest intensity was reached by Granulated Blast furnace Slag grinded in the vibratory disc mill. Similarly, low intensity can be observed for Sheet Glass grinded in the planetary mill. On the contrary, Red Brick Dust grinded in the planetary mill reached very high intensity.

4. Discussion of results and conclusion

This paper deals with mechanical activation of pozzolans. The main result is the evaluation of the pozzolan activity of Sheet Glass, Granulated Slag and Brick Dust by DTA analysis. The amount of CaO in the samples mixed with lime in a given ratio was subsequently determined for different time intervals. A chemical analysis was performed and actual CaO content in the samples was found.

The highest actual CaO content was measured in samples of Granulated Slag (47.09%.), which also had the largest loss of CaO during the measurement, which meant it had high pozzolanic activity. Mechanical activation due to grinding methods could also occur in the slag because a greater decrease in CaO on the samples grinded in the planetary mill was observed on the same specific surface (500 m^2/kg).

The Sheet Glass showed very interesting results. The loss of CaO was very variable depending on used grinding technology. The glass grinded in the vibratory disc mill had a significantly higher CaO content and unchanged consistency after 28 days while the glass grinded in the planetary mill had after 28 lower CaO content and hard cracked structure. This pozzolan activity was caused by the high amount of SiO_2 and method of grinding in planetary mill, which apparently allowed to reach the limit of the plastic fracture, which means that we can consider this as a mechanical activation [5]. Recycled glass as an admixture for the production of blended cements can be a suitable alternative. Pozzolanic activity increases for high specific surface area and the high fineness of grinding ensures that the concrete does not degrade due to the alkali–silica reaction [6].

From all of the examined materials the losses of CaO were lowest in a Brick Dust sample with approximately 3% loss after 28 days of reaction. However, the total CaO content was very low and it is not possible to assume a pozzolanic reaction for this sample. By the experiment we have confirmed that the sample did not harden after 28 days of reaction.

In conclusion, it can be said that Granulated Slag and Sheet Glass can be mechanically activated, and they indicate pozzolan properties. So, it is possible to use such materials in the production of blended cements as an alternative admixture. For Brick Dust no such properties were observed.

5. Acknowledgements

This work was financially supported by project number: GA17-24954S “The Conditions of Thermodynamic Stability and Transformation of AFt Phases” and project No. LO1408 "AdMaS UP - Advanced Materials, Structures and Technologies", supported by Ministry of Education, Youth and Sports under the „National Sustainability Programme I".

References

[1] Malinowski R and Garfinkel Y 1990 *Concrete International* **13** 62-68
[2] Moropoulou A, Carmak A, Labropoulos K C, Van Grieken R and Torfs K 2004 *Cement Concrete Res.* **34** 1–6
[3] Rovnaníková P 2000 *Keim Info* **5**
[4] Sobolev K 2016 *Cement Concrete Comp* **71** 53-62
[5] Boldyrev V V, Pavlov S V and Goldberg E L 1996 *Int. J. Miner. Process.* **44-45** 181-185
[6] Dvorak K, Dolak D and Dobrovolny P 2017 *Procedia Engineer.* **180** 1229-1236

Determination of durability of mortar with slag exposed to bacterial environment

M Smolakova[1], A Estokova[1], V Vaclavik[2], K Soucek[3] and V Zajicova[3]

[1] Institute of Environmental Engineering, Faculty of Civil Engineering, Technical University of Kosice, Vysokoskolska 4, 042 00 Kosice, Slovakia
[2] Institute of Environmental Engineering, Faculty of Mining and Geology, VSB-Technical University of Ostrava, 17. Listopadu 2172/15, 708 33 Ostrava-Poruba, Czech Republic
[3] Institute of Geonics of the CAS, Studentska 1768, 708 00 Ostrava-Poruba, Czech Republic

Abstract. Leaching of cement based materials and dissolution of cement hydrates under biogenic acid attack result in an increase of the porosity and changes in the pore structure of the building materials. The impact of certain physicochemical concrete/mortar parameters such as cement chemistry, water/binder (w/b) ratio, porosity and fluid permeability of cementitious materials has been recognized from a material sciences perspective. In this work the durability of cement composite made with blast furnace slag exposed to sulphur oxidizing bacteria A. thiooxidans during the period of 12 months was investigated. Leaching of calcium and silicon ions, water absorbability and porosity of the slag based composite using X-ray computed tomography was evaluated.

1. Introduction

For many years, cement-based materials have been used as structural or filling materials in a great variety of applications, such as building materials. The durability of these materials is affected as they are altered in contact with water or exposed to various environmental conditions [1]. Durability of cement based materials has become a main concern mainly for safety, economic and environmental reasons. It is well known that the durability issues, such as chloride or sulphate-induced corrosion, carbonation, freeze-thaw action are closely linked with fluid and ion transport in cementitious materials [2].

Pore network in cement pastes plays a fundamental role in the transport processes of elements (major and trace elements) within the matrix and hence their potential release in the environment [1]. In the hardened cement paste, there are complex pore structures, including gel pores, capillary pores, and air voids [3]. Pore size distribution in composite matrix is ranging from few nanometers (nm) up to tens of micrometers (μm) [1,4]. Such a diverse pore system is closely associated with the mechanical performance and transport properties of the matrix. To a large extent the pore structure determines the ingress of surrounding water molecules into the cement material. Entrained aggressive substances, such as sulphate, carbonate and chloride ions, penetrate the cement matrix through pore transportation, resulting in neutralization of cover of cement-based material and possible generation of chemical stress. In addition to the external load, the porosity of cement composites downgrades their mechanical properties [3].

Concrete corrosion is a complex process which involves the physical adsorption of hydrogen sulphide, chemical and biological oxidation of sulphide to various sulphur species, including sulphuric acid, and reaction between the sulphuric acid and the cement matrix. During active sewer corrosion sulphur oxidizing bacteria (SOB) oxidize the sulphur compounds resulting in sulphuric acid production which then diffuses into the inner corrosion layer and reacts with alkaline compounds released from the material. Ultimately, the loss of cement mass occurs and the microbial conversions of sulphur compounds are critical for the process [5].

Recently, the utilization of waste materials and supplementary cementitious materials (SCMs) with fine size has been found to compensate both physically and chemically the deficiency in manufacturing denser and more compact cement-based structures [3,6,7,8]. Granulated blast furnace slag (GBS) and silica fume (SF) are two cement replacements which have grown in popularity in recent years. Slag is a by-product from blast furnaces used to make iron. There are many advantages of GBS as: greater compressive and flexible strengths, greater workability and the ability to produce a denser matrix [9,10,11]. GBS provides greater mechanical and durability properties as it significantly decreases both the content and the size of $Ca(OH)_2$ crystals in the aggregate-paste interface, which makes the microstructure of the transition zone dense and strong [9,12].

This paper presents an attempt to determine the durability of mortar sample made with blast furnace slag affected by the action of *Acidithiobacillus thiooxidans* after long-term, 12-month exposition. Evaluated characteristics were leaching of calcium and silicon ions, water absorbability and porosity of composite sample which was characterised by X-ray computed tomography.

2. Material and Methods

2.1Mortar sample

Two mixtures where designed for the experiment, whereas S0 sample represents the reference sample made with ordinary CEM I Portland cement (OPC) and sample S1 was made with 35 wt.% of ordinary Portland cement (OPC) and with 65 wt.% of blast furnace slag, respectively. The mortar specimens with dimensions approximately 160 mm × 40 mm × 40 mm were designed in cooperation with cement factory plant which we cooperate with. Chemical composition of OPC and blast furnace slag is shown in table 1.

Table 1. Chemical composition of ordinary Portland cement and blast furnace slag.

Content of main components	OPC (wt.%)	Slag (wt.%)
CaO	57.15	39.55
SiO_2	18.11	38.95
Al_2O_3	4.02	8.33
Fe_2O_3	2.69	0.54
SO_3	1.49	0.57
MgO	1.37	10.11
K_2O	1.12	0.48
TiO_2	0.18	0.37
P_2O_5	0.33	0.04
Cl	0.06	0.02

2.2 Microbiological exposure

In the experiment, sulphur-oxidising bacteria *Acidithiobacillus thiooxidans* was used to create the aggressive acidic environment, because of their ability to produce sulphuric acid. *A. thiooxidans* were isolated from an acid mine drainage from Pech shaft (locality of Smolnik, Eastern Slovakia).The nutrient medium by Waksman and Joffe [13] was used for the preparation of the active bacterial culture of *A. thiooxidans* in the experiment. The composition of the nutrient medium was as follows: 0.25 g/L $CaCl_2.6H_2O$, 0.2 g/L $(NH_4)_2SO_4$, 3.0 g/L K_2HPO_4, 0.5 g/L $MgSO_4.7H_2O$, traces of $FeSO_4.7H_2O$, 10.0 g/L elementary S and up to 1000 mL supplemented by distilled water. Elemental sulphur was added to the nutrient medium.

The exposure of mortar samples to acidic bacterial environment has taken place in glass containers under aerobic laboratory conditions over a period of 12 months. Optimal growth temperature of bacterial culture (28 - 30°C) and pH interval 2.0-3.5 were controlled during the experiment [14,15].

2.3 X-ray fluorescence analysis
X-ray fluorescence analysis (XRF) was used to analyze the chemical composition of leachates. After placing samples into the bacterial medium, a chemical composition of mortar sample leachates was investigated every 2 months. SPECTRO iQ II (Ametek, Germany) with SDD silicon drift detector with a resolution of 145 eV was used for the analysis. The samples were analysed for 300 s (solid specimens) and 180 s (liquids) using X-ray tube (25 kV and 50 kV, 0.5 mA and 1.0 mA) combined with a Highly Ordered Pyrolytic Graphite (HOPG) crystal for polarization of the primary beam. The measurement proceeded under helium atmosphere. The concentrations of the elements were calculated by using the calibration method for cements and liquids, respectively [14].

2.4 Water absorbability test
The water absorbability test was performed on mortar specimens in accordance with a corresponding national standard [16]. All samples were dried at 105 ± 5°C to the constant mass and consequently weighted (m_D). After that, the reference samples were immersed in water for 72 hours and weighted again (m_W). Based on these two mass values, the original absorbability of samples, before the experiment, was determined. The final absorbability of the exposed samples was calculated using wet mass of the samples (m_W) measured after 12 months when the samples were removed from the liquid medium. Water absorbability for both samples was determined using equation (1) [16]:

$$v = \frac{m_W - m_D}{m_D} \,.\, 100\% \qquad (1)$$

where m_W refers to the weight of wet sample and m_D is weight of dried sample.

2.5 Micro X-ray computed tomography
X-ray computed tomography (CT) is a non-destructive method for obtaining a large number of consecutive sectional images of the internal microstructure of specimens. It has been used in several studies to characterize the engineering properties of cement-based materials in terms of such parameters as pore network, air-void space, or spatial distribution of air content under axial loading [17,18,19].

To study the inner structure of mortar sample, the Nikon Metrology XT H 225 industrial micro X-ray CT system with reflex X-ray source and flat detector with resolution of 4000x4000 pixels was used. It is a fully automated apparatus with a rotating scanning system equipped with a micro focal X-ray source which generates cone-shaped beams. Studied volumes were reconstructed using CT Pro 3D and CT Pro 2D software (by Nikon Metrology NV). Visualization was provided by a VGStudio Max software (by Volume Graphics), version 2.2, with additional modules for proper image analysis [20].

3. Results
The changes in Ca and Si ion concentrations in leachates of studied mortar sample are given in figure 1. The highest concentration of Ca ions was observed for sample S1 at the end of the experiment with the value of 841.6 mg/L. Highest amount of leached Ca ions (578.1 mg/L) for reference sample was also observed at the end of the experiment. The leaching of Si ions for specimen S1 was most intensive with maximum reached after 6th month of bio-corrosion experiment (603.4 mg/L). For the reference sample S0 the highest concentration of leached out Si ions was measured after the 12 month period with the value of 663.6 mg/L. The increase in the values of leached out calcium and silicon ions are linked with gradual deterioration of sample caused by dissolving the composite's hydrates.

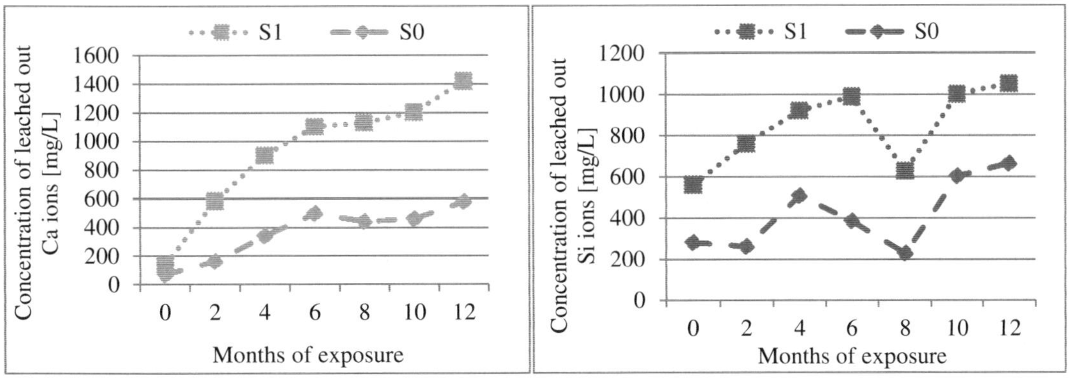

Figure 1. Concentration of leached out Ca and Si ions from mortar sample during bacterial exposition.

Water absorbability of mortar composites before and after the bio-corrosion experiment is reported in table 2.

Table 2. Water absorbability of mortar composite.

Mortar sample	Water absorption capacity [%]		Change [%]
	Before the experiment (72 hours)	After 12 months	
S0	7.98	10.99	3.01
S1	8.55	8.76	0.21

It can be seen that water absorbability of both samples increased due to the sulphuric acid exposure. Considerably higher change in water absorption capacity was observed for reference sample S0, which increased by 3.01%, while the change in water absorption capacity for the sample S1 made with blast furnace slag was only 0.21%. The lower water absorbability of sample S1 is likely linked to the presence of blast furnace slag in the mortar, which resulted in denser matrix of the sample with reduced porosity.

Nevertheless the sample S1 made with blast furnace slag had higher leached out amount of Ca and Si ions, what is in contrary to the water absorbability results. This contradiction can be linked to the different chemical compositions of the cement mortars.

The pore size and porosity of the sample S1 with slag, measured using X-ray computed tomography, is illustrated in figure 2. Different colours on the images indicate the different size of the pores in the sample.

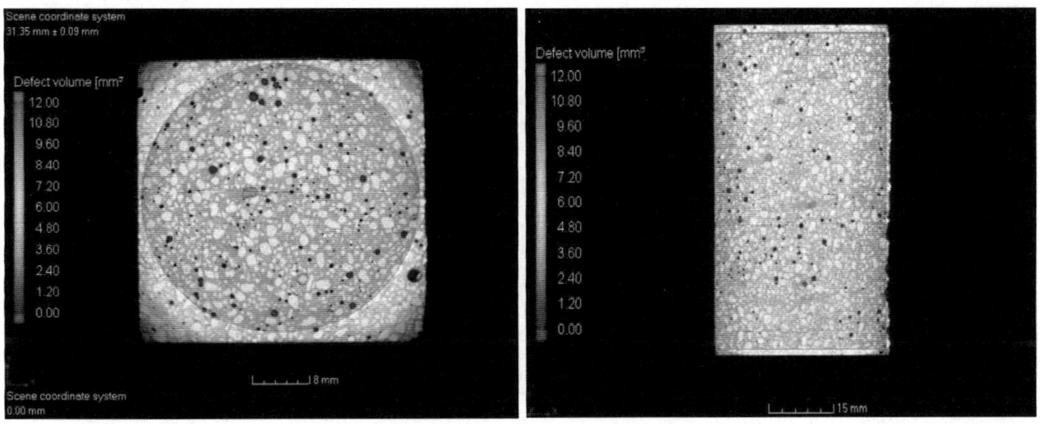

IOP Conf. Series: Materials Science and Engineering **385** (2018) 012052 doi:10.1088/1757-899X/385/1/012052

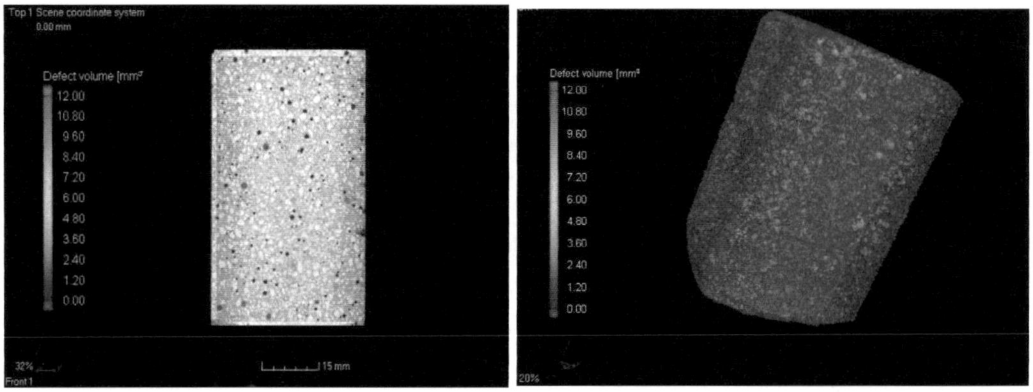

Figure 2. Micro X-ray CT images of mortar sample made with blast furnace slag.

The size of pores in sample varied from 0.01 mm to 11.54 mm, the smallest pores such as gel or capillary pores were not measured by micro computed tomography which is accordance with the study by Bossa N. et al [1]. The dimensions of tested cylinder were, the diameter 38.7 mm with height 63.6 mm. The total number of pores in the tested area of specimen were 16 690. Micro computed tomography measurements lead to a calculated porosity of 0.8% in the composite sample, in agreement with studies by Chengcheng D. et al. [19], where the cement composite samples had porosity ranging from 0.044% to 1.486%. Some authors state, that porosity of cement composites is ranging from 1.8% [1]. This result confirms that micro X-ray CT is not sensitive enough to detect all the smallest pores that represent up to 70–90% of the whole pore volume of the cement matrix [1,4].

4. Conclusion

In order to predict the biodeterioration of the material it is necessary to evaluate various physico-chemical parameters. The objective of this paper was to determine the durability of cement composite made with the addition of blast furnace slag under the 12 months long exposition of aggressive sulphate environment represented by bacteria *A. thiooxidans*. Evaluated characteristics were leaching of basic inorganic elements (calcium and silicon ions), water absorbability, porosity and pore size of the sample, which was measured by micro computed tomography. Obtained results of leaching of basic inorganic elements and increase in water absorbability of sample confirmed the degradation effect of sulphur-oxidizing bacteria, however the pore network and the overall porosity could not be properly measured due to low sensitivity of the micro X-ray computed tomography.

Acknowledgements

This paper has been prepared within the Grant No. 2/0145/15 of the Slovak Grant Agency for Science.

References

[1] Bossa N, Chaurand P, Vicente J, Borschneck D, Levard C, Aguerre-Chariol O and Rose J 2015 *Cem. Concr. Res.* **67** 138-147
[2] Zhang M 2017 *Cement Concrete Res.* **95** 18-29
[3] Wang Y S and Dai J G 2017 *NDT&E Int.* **86** 28-35
[4] Cook R A, Hover K C 1999 *Cement Concrete Res.* **29** 933–943
[5] Sun X, Jiang G, Bond P L, Keller J and Yuan Z 2015 *Water Res.* **70** 279-287
[6] Luquot L, Rodriguez O, Gouze P 2014 *Transp. Porous. Med.* **101** 507–32
[7] Dvorsky T, Vaclavik V, Simicek V and Brenek A 2015 *Inz. Mineral.* **16** 51-56
[8] Gola L, Vaclavik V, Valicek J, Harnicarová M, Kusnerova M and Dvorský T 2015 *Adv. Struct. Mat.* **70** 155-165
[9] Limbachiya V, Ganjian E and Claisse P 2016 *Constr. Build. Mater.* **113** 273-279
[10] Atici U, Ersoy A 2008 *Constr. Build. Mater.* **22** 1507–1514
[11] Stevulova N, Vaclavik V, Junak J, Grul R and Bacikova M 2008 *8th Int. Mul. Sci. Geo. SGEM 2008 (Albena)* vol. 2 (Sofia: STEF92 Technology Ltd.) pp 193-200

[12] Pavía S and Condren E 2008 *J. Mater. Civ. Eng.* **20** 313–320
[13] Waksman S A and Joffe J S 1922 *J. bacterial.* **7** 239-256
[14] Kovalcikova M, Estokova A and A. Luptakova 2015 *IOP Conf. Ser.: Mater. Sci. Eng.* **96** 1-7
[15] Estokova A, Smolakova M and Luptakova A 2016 *41st IAHS World congress Sustainability and Innovation for the Future (Algarve)* 1-9
[16] STN 73 1316: 1989 Determination of moisture content, absorptivity and capillarity of concrete
[17] Kim K Y, Yun T S, Choo J, Kang D H and Shin H S 2012 *Constr. Build. Mater.* **37** 93-101
[18] Promentilla M A B, Sugiyama T and Shimura K 2008 *3rd Asian concrete federation (ACF) international conference (Ho Chi Minh)* 940–7
[19] Wong R C K and Chau K T 2005 *Cem. Concr. Res.* **35** 1566–76
[20] Zajicova V, Stas L, Soucek K and Georgiovska L 2016 *16th Int. Mul. Sci. Geo. SGEM 2016 (Albena)* vol 3 (Sofia: STEF92 Technology Ltd.) pp 355-362
[21] Du C, Zou D, Liu T and Lv H 2017 *Measurement* **122** 611-619

IOP Conf. Series: Materials Science and Engineering **385** (2018) 012053 doi:10.1088/1757-899X/385/1/012053

The fly ash of class C for ceramic technology

R Sokolar and M Nguyen

Institute of Technology of Building Materials and Components, Brno University of Technology, Faculty of Civil Engineering, Veveri 331/95, 602 00 Brno, Czech Republic

Abstract. The properties of different Czech fluidized fly ashes FFA (class C fly ash according to ASTM C618 12a with high content of CaO in the form of anhydrite primarily) and emisions of SO_2 in flue gas during the firing of FFA are discussed. The decomposition of anhydrite CaSO4 was observed during the soaking time at 1200 °C during the firing of dry pressed bodies based on pure fluidized fly ash. The effect od fluidized fly ashes admixtures on the properties of ceramic bodies is as fluxing agent and the source of CaO for the creation of anorthite in the fired ceramic body. The properties of domestic fluidized fly ashes in connection with their utilization in ceramic technology are described.

1. Introduction

Fluidized fly ash FFA is a secondary raw material - energy by-product from the process of combustion of coal in fluidized bed boilers. Fluidized ash is produced by fluid combustion of finely ground coal together with limestone or dolomite at lower temperatures (about 850 °C) compared with tradition combustion of pure solid fuel (coal) at about 1400 °C – classical high-temperature fluid fly ash (CFA) is produced. The term "fluidized fly ash" is unknown in foreign scientific literature. Fluidized fly ash is generaly called as "class C fly ash" according to the American Standard ASTM C618-12a - Standard Specification of Coal Fly Ash and Raw Calcined Natural Pozzolan for Use in Conrete, which classifies fly ash based on the content of selected oxides and divides them into two basic groups. "Class F" – classical high temperature fly ash and "class C" - fluidized fly ash, see table 1.

Table 1. The classification of fly ashes according to ASTM C618.

Class	$\Sigma(SiO_2+Al_2O_3+Fe_2O_3)$	SO_3	LOI
C	min 50 %	max 5 %	max 6 %
F	min 70 %	max 3 %	max 6 %

Total production of by-product energy products in the Czech Republic in 2014 was approximately 13.7 million tonnes. The total production of all types of fly ash was 9.3 million tonnes - high-temperature ash production about7.9 million tonnes (85 %) and fluidized fly ash 1.4 million tonnes (15 %). The fluidized technology is one of the most popular methods for burning of coal and other sorts of fuel in thermal power plants. In conjunction with desulphurization, this is the most efficient method for the limitation of harmful emissions (especially sulphur dioxide) in the air. Fluidized fly ashes are generated during burning a fine grain mixture of coal powder, fly ash and limestone or dolomite in fluidized-bed boilers, which are burning the air-borne coal dust at lower temperatures (usually up to 900 °C) in comparison with the classic burning on fire grates where the burning temperature is up to 1450 °C.

The fly ash-clay mixtures for the single-firing process technology for the dry pressed ceramic tiles were developed experimentally, using kaolinic stoneware clay as the basic raw material and classical high temperature fly ash [1-2]. The bodies prepared by this method show a high shrinkage after firing, often even in the reduction cores, in comparison with standard bodies based on natural resources. The limited shrinkage after firing the fly ash-clay mixture may be achieved by the addition of limestone [3-4], which however decreases the bending strength of the body [3]. Blast furnace slag [5] with a high content of CaO in the mixture and with fly ash may also decrease the firing shrinkage of body at firing temperatures up to 1150 °C, much like the low alkali pyrophyllite [6]. Tincal ore waste [8], talc [9] and metal finishing wastes [10] in the fly ash mixture behave as a flux, and, on the contrary, the firing shrinkage of the body increases with increased proportions of those materials in raw materials mixture. Fluidized fly ash was used successfully during the preparation of glass-ceramic materials [11-12]. Fluidized fly ash shows a worse sinterability (higher water absorption, porosity) than classical high temperature fly ash [13]. It is possible to improve the sintering activity of fly ash mixtures by milling - their water absorption after firing decreases [1, 14].

The use of fluidized fly ash in concrete technology, geopolymers, solidification of dangerous substances, or stabilization of unstressed layers for roads is studied abroad especially. The studies about the application of FFA in ceramic technology are rare. The recycling of three different fly ashes (FFA) obtained from the coal fired thermal power plants has been studied for the glass materials [15]. There is evident that all used fly ashes are typ FFA (class C) according to their chemical composition. The significant difference in CaO content (6,76 – 37.80 %) is interesting (table 2).

Table 2. Chemical composition of used fly ashes [15].

Fly ash	SiO_2	Al_2O_3	CaO	MgO	Fe_2O_3	Na_2O	K_2O	SO_3	LOI
Cayirhan	41.53	17.77	12.52	4.46	9.93	2.57	2.43	7.25	0.68
Seyitomer	44.58	22.54	6.76	8.98	9.85	0.22	0.60	2.52	3.83
Afsin-Elbistan	18.11	7.63	37.80	3.50	5.23	0.22	0.60	18.22	8.40

All presented results of scientific research of FFA utilization in ceramic (fired) materials did not solve the problem of sulfur dioxide emissions – this is the aim of Czech Science Foundation project GACR 18-02815S Elimination of Sulphur oxide emission during the firing of ceramic bodies based on fly ashes of class C (2018 – 2020). The aim of the article is an evaluation of properties of Czech fluidized fly ashes (mineralogical and chemical composition, granulometry, morphology of grains) compared with traditional classic high-tempertature fly ash. The problem of sulphur oxide SO_2 emissions in flue gas during the firing of pure FFA and pure stoneware clay dry pressed bodies is discussed. Main goal of the research is to define conditions when the FFA can be used in ceramic technology as a unexpensive source of CaO – necessary in the modern production of porous wall tiles (group BIII according to EN 14411) for example.

2. Properties of fluid fly ash (FFA)

Three Czech fluidized fly ashes FFA (from Hodonin, Tisova and Ledvice Thermal Power Plant) compared with classical high-temperature fly ash CFA (from Melnik Thermal Power Plant) as non-plastic raw materials for the preparation of laboratory samples were used.

2.1. Chemical and mineralogical composition

The basic difference between CFA and FFA consists in the mineralogical composition (figure 1). CFA contain up to 80 % glassy phase as the main component, and up to 20 % mullite. SO_3 content is usually below 1 %, because calcium sulphate decomposes at temperatures above 1150 °C. FFA are characterized by their higher SO_3 content (in the form of anhydrite $CaSO_4$), free calcium oxide CaO (up to 15 %) and calcite $CaCO_3$. Fluidized fly ashes do not contain either glass phase or mullite. The chemical composition of FFA therefore typically has a high content of CaO (table 3). FFA shows usually hydraulic properties (they solidify and harden after the mixing with water, without addition of any other components). Quartz is presented in both types of fly ashes (figure 1). A comparison of the mineralogical composition of high temperature and fluid fly ash is shown in figure 1 (powder X-Ray diffraction). The creation of anhydrite in FFA during the combustion of the milled coal and limestone mixture inside the fluidized boiler we can described according to chemical equations:

1. Decomposition of limestone: $CaCO_3 \rightarrow CaO + CO_2$ (1)
2. Reaction of SO_2 released from coal with CaO: $2CaO + 2SO_2 + O_2 \rightarrow 2CaSO_4$ (2)

The following table (table 3) shows the chemical composition of the typical Czech class C fla ashes from three thermal power plants in Hodonin, Ledvice and Tisova, as well as the classical high temperature fly ash from the Mělník power plant. The chemical composition of all fluidized fly ashes is very similar with slightly higher amount of calcium oxide (over 20 %) and a lower amount of alumina and SiO_2. On the other hand, classical high-temperature fly ash from the Mělník plant contains only 2.2 % CaO and higher content of Al_2O_3 and SiO_2.

Table 3. Chemical composition of different Czech fluid class C fly ashes (Hodonin, Tisova, Ledvice) compared with traditional high temperature fly ash (Melnik).

	SiO_2	Al_2O_3	Fe_2O_3	TiO_2	CaO	MgO	MnO	K_2O	Na_2O	SO_3	LOI
Hodonin	32.5	16.0	6.7	0.6	24.5	3.4	0.1	0.8	0.1	4.0	4.5
Tisova	35.0	23.3	5.5	5.4	21.5	1.6	0.1	0.5	0.1	2.5	4.5
Ledvice	35.7	25.4	6.5	1.8	20.5	1.1	0.1	1.5	0.3	3.3	3.8
Melnik	57.3	29.3	5.1	1.7	2.2	1.4	0.0	1.6	0.1	0.1	1.2

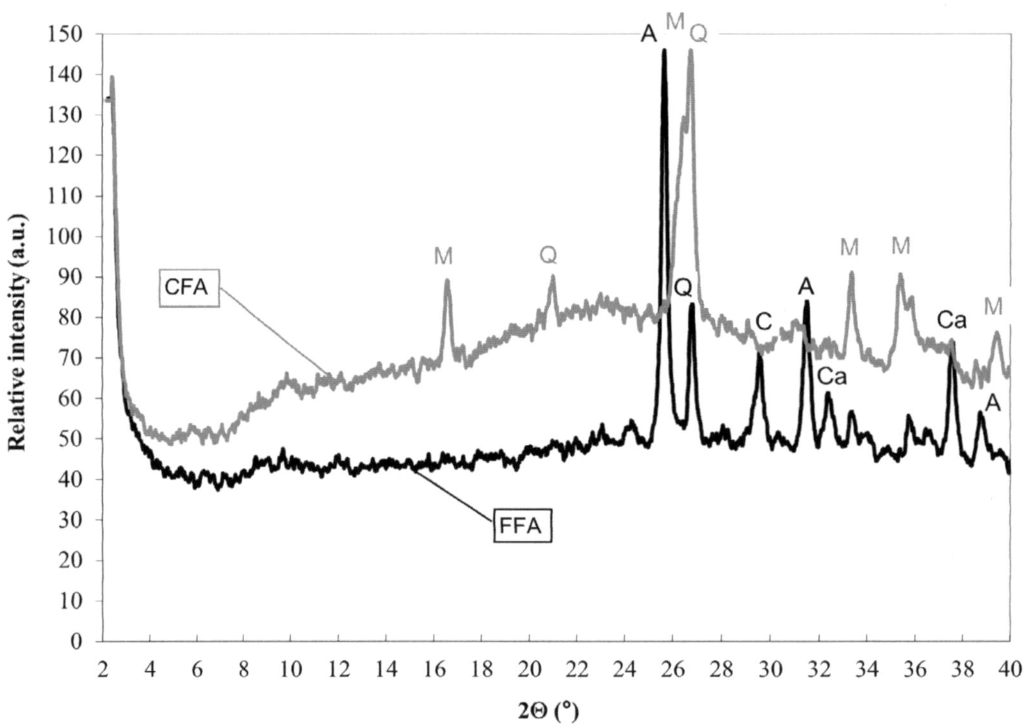

Figure 1. X-Ray diffraction of different types of fly ash – CFA and FFA: M-mullite, Q-quartz, A-anorthite, C-calcited, Ca-CaO.

2.2. Granulomnetry

The difference in morphology between classical high temperature fly ash CFA (left) and fluidized fly ash (right) in the images from the scanning electron microscope SEM is evident (figure 4). High-temperature of combustion creates spherical shape of CFA grains with different diameters. Grains of fluidized fly ash are irregular shaped with sharp edges due to lower temperature of combustion.

Granulometry of the fly ashes used was determined according to the residue on a screen with a size of 63 μm (R63). From the granulometric point of view, the fluidized fly ash from the Ledvice power plant differs. It contains the highest proportion of grains above 63 μm which is approximately twice number compared to the other analysed FFA. The grain size comparison is shown in table 4.

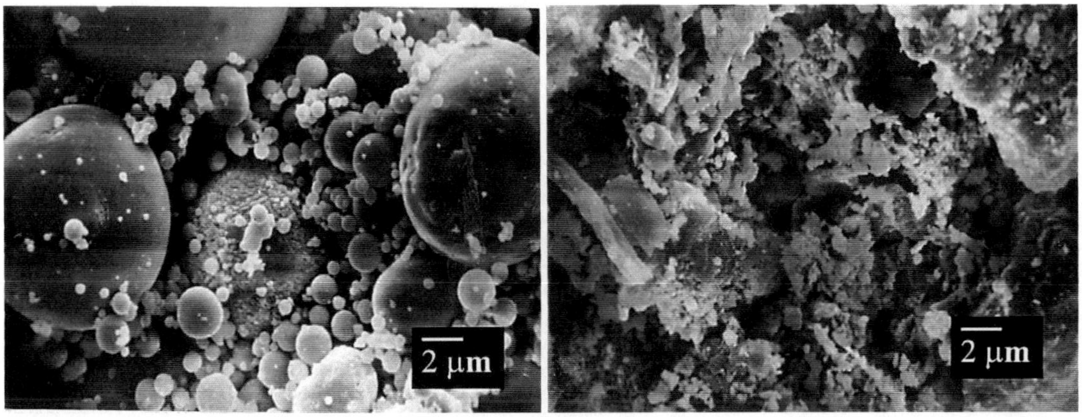

Figure 2. Morphology of classical high-temperatuire fly ash CFA (left) and fluidized fly ash – class C FFA (right), SEM 1500×.

Table 4. Granulometry of the fly ashes according to the residue on a screen with a size of 63 μm - R_{63}.

FFA	Hodonin	Ledvice	Tisova
R_{63} [%]	32.2	54.2	25.1

2.3. SO_2 emissoins during the firing

Dry pressed body based on pure FFA (Hodonin) has been fired with continual analysing of flue gas (content of SO_2) by use of the TESTO M-I 300 flue gas analyzer, with a maximum measurable limit of 4000 ppm. Decomposition of anhydrite during the soaking time at 1200 °C (figure 3) produces calcium oxide, sulfur dioxide gas and oxygen very intensively according to the equation (reverse reaction (2):

$$2CaSO_4 \rightarrow 2CaO + 2SO_2 + O_2 \tag{3}$$

The same decomposition we can observe during the firing of pure ceramic clays (for example stoneware clay MM – figure 4.), but from the lower temperatures (about 700 °C). The source of SO_2 in fuel gas during the firing of ceramic bodies is content of pyrite (marcasite) in clays – the decomposition of pyrite (marcasite) is evident about the temperature of 450 °C (figure 4) via equation (4):

$$4FeS_2 + 11O_2 \xrightarrow{asi\ 450\,^{\circ}C} 2Fe_2O_3 + 8SO_2 \tag{4}$$

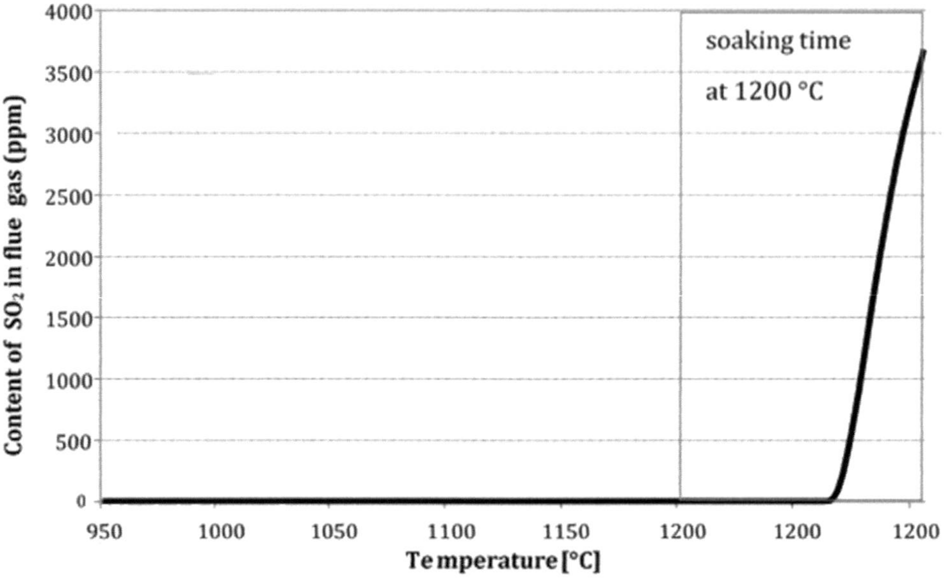

Figure 3. The content of sulfur dioxide in flue gas during the firing of class C fly ash (FFA – Hodonin).

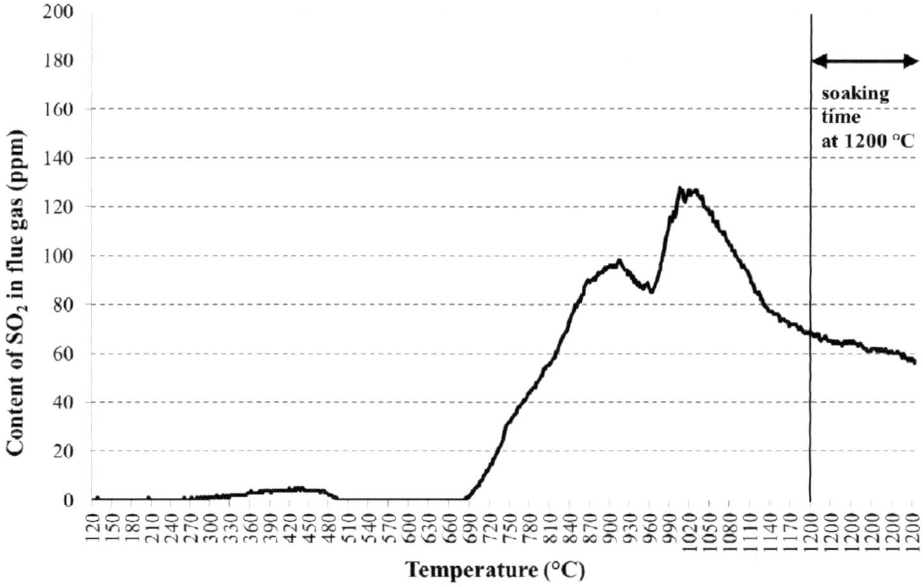

Figure 4. The content of sulphur oxide in flue gas during the firing of stoneware clay (MM).

3. Conclusion

The aim of the paper was to have a think about the possibility of application of fluidized fly ash (class C fly ash according to ASTM C618 12a) in the production of fired (ceramic) materials as a source of CaO in raw materials mixture. Fluidized and classical high-temperature fly ash were compared in terms of mineralogical and chemical composition, grain morphology and granulometry and their different firing behavior.

Every year about 1.5 million tonnes of fluidized fly ash are produced in Czech Republic but there has not been found a wider application ofr FFA for example in the production of fired building materials because the decomposition of anhydrite and thus releasing the emissions of sulfur dioxide have not been solved.

The aim of research within the Czech Science Foundation (via acknowledgement) solution is to define the temperature interval in which the anhydrite decomposes, the determination of the SO_2 content in the flue gas and the design of the treatment of the raw material mixture for the binding of SO_2 into the ceramic body without emissions.

Acknowledgement
This article was supported under the Czech Science Foundation project GACR 18-02815S Elimination of Sulphur oxide emission during the firing of ceramic bodies based on fly ashes of class C.

References
[1] Sokolar R and Smetanova L 2010 *Ceram. Int.* **36** 215
[2] Queralt I, Querol X, López-Soler A and Plana F 1997 *Fuel* **76** 787
[3] Zimmer A and Bergmann C P 2007 *Waste Management* **27** 59
[4] Haiying Z, Youcai Z and Jingyu Q 2007 *Journal of Harazdous Materials* **141**106
[5] Dana K, Dey J and Das S K. 2005 *Ceram. Int.* **31** 147
[6] Chandra N, Sharma P, Pashkov G L, Voskresenskaya E N, Amritphale S S and Baghel N S 2008 *Waste management* **28** 1993
[7] Furlani E., Brückner S, Minichelli D and Maschio S 2008 *Ceram. Int.* **34** 2137
[8] Olgun A, Erdogan Y, Ayhan Y and Zeybek B 2005 *Ceram. Int.* **31** 153
[9] Chandra N, Agnihotri N, Bhasin S and Khan A F 2008 *J. of the Eur. Cer. Soc.* **25** 81
[10] Little M R, Adell V, Boccaccini A R. and Cheeseman C R 2008 *Resources, Conservation and Recycling* **52** 1329
[11] Erol M, Genç A, Öveçolu M L, Yücelen E, Küçükbayrak S and Taptik Y 2000 *J. of the Eur. Cer. Soc.* **28** 1245
[12] Barbieri L, Lancellotti I, Manfredini T, Queralt I, Rincon J Ma and Romero M 1999 *Fuel* **78** 271
[13] Erol M, Küçükbayrak S and Ersoy-Meriçboyu A 2008 *Fuel* **87** 1334
[14] Ilic M, Cheeseman Ch, Sollars Ch and Knight J 2003 *Fuel* **82** 331
[15] Erol M, Kucubayrak S and Ersdoy-Mericboyu A 2006 *Fuel* **86** 706

Degradation of composites containing alkali-resistant glass fibres by ammonium nitrate solutions

Z Stefunkova, M Ledererova and V Gregorova

Department of Material Engineering, Faculty of Civil Engineering, Slovak University of Technology, Radlinskeho 11, 810 05 Bratislava, Slovakia

Abstract. The aim of the research program was a degradation of composites containing alkali-resistant glass fibres (ARGF) by ammonium nitrate solutions. The effect of 3 and 5% ammonium nitrate solutions and storage long time on physical properties and structure of cement composites is presented. Application of glass fibre had a positive effect on the properties of cement pastes, most notably, this was reflected on the compressive strength and transverse tensile strength. Despite the positive effects but also the application of fiberglass there was a significant loss of strength cement mortars exposed to prolonged exposure to ammonium nitrate solutions.

1. Introductions

History of fiberglass applications is relatively young. In the past were not available suitable sources of alkali-resistant glass fibers appropriate to the concrete. Glass fibers that were used initially in cement composites were disturbed with alkalis in pore solution of cement stone. Durability of cement stone and composites decreased as a result of degradation of fibers.

Electrical grade glass (E-Glass) was originally developed for standoff insulators for electrical wiring. It was later found to have excellent fiber forming capabilities and was used as the reinforcing phase in the material commonly known as fiberglass. With further research proved to be unsuitable for use in fiber reinforced concrete. Fiber E-glass has a low resistance to acids and alkalis contained in the cement stone. It is characterized by high electrical resistance and good deformation characteristics.

Glass fibers currently used to produce reinforcing bar are made of glass resistant to the effects of alkali. They were developed from the 1960th in England and Spain. It was found that the content of zirconia in the glass fiber has a positive impact on the resistance against alkali. Fibers initially contained 16% ZrO_2; gradually increasing doses up to 20%. The alkali-resistant fibers are suitable to produce diverse types of cement composites with wide application in the construction industry [1,2].

Glass fibers Cem-FIL are produced from molten glass with temperature up to 1 560°C flowing through perforated plate. Diameter of glass fibers that are produced by this method is 14μm. The resulting monofilament cool quickly, and the surface lubrication is directly sprayed onto and connects them to the springs. Lubrication must ensure resistance to high fiber co-alkaline environment of cement paste (pH 12 to 14). In addition, it provides a straight-line basis stirring fiber composite in the shape, which is most appropriate for the reinforcement. In the production of continuous glass fibers, it is used a variety of sizing according to technological requirements specified for use springs. When applied to concrete, strands of fiberglass must withstand the friction when mixed with aggregate and friction in the production of dry mixtures. If there was a breakage of spring during mixing, may produce clusters of broken fibers, which reinforcement ability is virtually nil. Lubrication shall allow the flexibility of springs, to avoid breaking of springs during mixing [3].

The object of the research was to verify the impact of alkali-resistant glass fibers Cem-FIL AntiCrak and the properties of cement pastes exposed to aggressive action of ammonium nitrate solutions. This medium can cause intense leaching of cement composites, under certain circumstances and expansion processes of concrete corrosion (rust Moskvin by type III [4]).

2. Materials and methods

Samples were made of cement CEM I 42.5 R and potable mains water. Samples were blended with alkali-resistant glass fibres Cem-FIL AntiCrak (ARGF) up to 6mm at 2.5wt.% and 5wt.%.

As aggressive media were used ammonium nitrate solutions with concentration at 3% and 5%. In the production of samples, it was important to determine the optimal water cement ratio. Based on tests of consistency of cement slurries was elected a single factor for all water produced composites 0.33. With this factor it reached a generally good workability and mixing water was not separating.

Cylindrical samples with a diameter and height of 30mm were made of cement paste blended with fibers and as a comparative sample without fibers additives. Samples were prepared by a standard manner in a standard mixer used to make a standard cement mortar according to EN 196-1 [5]. Samples were treated 24 hours in a humid environment, and for another 27 days were stored in water. Then they were treated with water or were placed in a solution of ammonium nitrate.

Aggressive media volume was ten times the volume of samples. NH_4NO_3 solution was exchanged with decreasing concentration of the aggressive component of 10%. Water was exchanged at regular 28-day intervals.

The samples were determined by the following properties: appearance and weight change, porosity, water absorption, density, compressive strength, splitting tensile strength. The microstructure and the interaction between the fibers and the cementitious stone were examined on selected samples using a scanning electron microscope (SEM), specifically TESCAN VEGA3 (SEM HV: 30.0kV, SEM MAG: 2.40kx, View field: 100μm).

Properties of the samples treated in water were tested at 28, 56, 308 and 1488 days. Properties of the samples stored in the aggressive solutions were tested at 1, 28, 280 and 1460 days of storage in aggressive solutions (the prior 28-day storage in water).

3. Results and discussion

3.1. Appearance and weight change subsection

Samples of hardened cement composites stored in water showed almost no change in appearance during the reporting periods.

The samples appearance in 3% and 5% solutions NH_4NO_3 has changed over time. The first 28 days samples without fibers and samples with 2.5wt.% glass fibers were no changes. Continuing storage samples in solutions started to emerge cracks on the surface of the samples, which incurred due to the ongoing corrosion processes. After 280-days storage in solution, samples have already been cracked. After 1460 days of storage in solution, it led to destruction of the surface structure of samples (figure 1).

Samples with addition of 5wt.% fiberglass at all concentrations throughout the period showed no change in appearance.

Weight of test samples hardening cement composites stored in water during treatment increased. The increase has caused ongoing cement hydration and absorbed water pore structure of the samples. After the 308-day and 1488-day storage in water, there were no major difference in weight between the samples without additives and samples blended with glass fibers.

Weight of the samples stored in cement composites aggressive NH_4NO_3 solutions (due to the ongoing corrosive processes in the cement stone) decreased with storage time of the samples and with increasing concentration of the solution. This is consistent with literature data [4]. For samples without additives fiberglass was the decline after 1460 days of storage at 3% solution of NH_4NO_3 20%, in 5% to 27% nominal. Samples blended with 5% glass fibres had a lower decline in the 3% - a solution of

9.5% and 5% - 19% of nominal, which can be attributed to the positive impact of the presence of glass fibres.

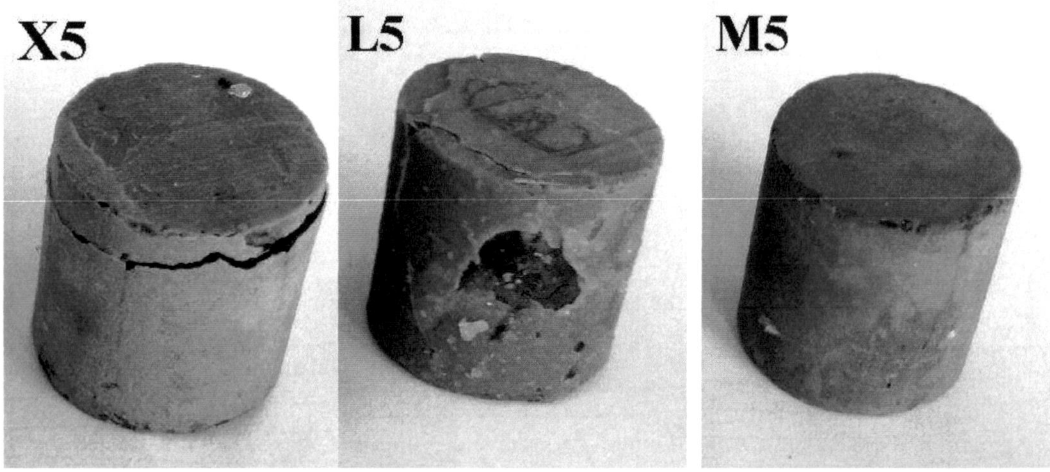

Figure 1. Appearance of samples of cement composites after 1460 days of storage at 5% solution ammonium nitrate without additives (X5) with admixture of 2.5wt.% (L5), and 5wt.% (M5) glass fibers.

3.2. Porosity
Porosity of the samples of cement composites stored in water treatment decreased with time because of progressive hydration structure and compacting samples. In the samples, stored in 3% and 5% NH_4NO_3 solutions, had gone in due course of corrosion II. type with storage time to porosity increased. A higher concentration of the solution generally means higher porosity of cement stone. The porosity was reflected positively by the impact of fiberglass. With increasing dose glass fibers are generally decreasing porosity (table 1).

3.3. Bulk density, compressive strength and splitting tensile strength
Bulk density, compressive strength and splitting tensile strength of samples of cement composites stored in water and aggressive solutions NH_4NO_3 in terms of the test set out in table. 1

Bulk density samples stored in water without additives and with addition of glass fibres with time of treatment grew larger. This corresponds with an increase in weight and a decrease in porosity. For samples stored at 3% and 5% NH_4NO_3 solutions out there was a reduction of their density with time saving, while for samples stored in aggressive environment was more pronounced decline.

The compressive strength of samples without additives blended with glass fibers deposited in water treatment has risen over time (figure 2) [6,7]. Compressive strength of the samples stored in the out of 3% and 5% solutions of NH_4NO_3-out with time and storage generally decreasing concentration of the solution.

Application of glass fibers has generally positive impact on reducing the strength of cement slurry stored in aggressive solutions. Strength loss was mitigated mainly by prolonged storage in aggressive solutions. For example, in the case of cement paste without fibers decreased compressive strength after 1460 days of storage at 3% solution of NH_4NO_3 from about 79 MPa to 13.5 MPa, while in the case of slurries containing 2.5% glass fibers was about the decline of 80 MPa to 34.6 MPa. Increasing doses of fiberglass from 2.5% to 5%-lo contributed to an increase in strength during prolonged storage in solution.

The same downward trend strength is confirmed also in the case of testing transverse tensile strength (table 1). For samples stored in aggressive solutions NH_4NO_3 there was a reduction in transverse tensile strength with time in storage solutions, and this decrease is highlighted with increasing concentration of aggressive media.

Table 1. Bulk density, porosity, compressive strength and splitting tensile strength.

Dose glass fiber (wt.%)	Medium	Age of samples /time saving in solution (days)	Bulk density (kg/m³)	Porosity (%)	Splitting tensile strength f_{csp} (MPa)	Compressive strength f_c (MPa)	$\frac{f_{ci}}{f_{c28(29)i}} x100$ (%)
0	water	28/0	2120	31.3	7.1	77.7	100.0
		56/0	2130	31.0	7.9	85.0	109.4
		308/0	2140	28.6	8.4	93.8	120.7
		1488/0	2180	28.4	9.1	94.9	122.1
	3% NH_4NO_3	29/1	2100	32.1	6.6	78.7	100.0
		56/28	2020	37.1	5.6	68.3	86.8
		308/280	1950	40.8	3.8	48.7	61.9
		1488/1460	1550	57.0	0.5	13.5	17.2
	5% NH_4NO_3	29/1	2150	34.5	7.0	78.1	100.0
		56/28	1930	41.6	5.3	50.6	64.8
		308/280	1850	49.2	2.9	38.0	48.7
		1488/1460	1460	65.4	0.04	5.0	6.4
2.5	water	28/0	2150	31.7	12.4	80.9	100.0
		56/0	2160	29.1	13.3	88.8	109.5
		308/0	2160	26.8	10.6	96.2	118.9
		1488/0	2170	26.5	7.4	97.8	120.9
	3% NH_4NO_3	29/1	2150	31.1	11.9	80.8	100.0
		56/28	2060	35.3	10.6	74.8	92.6
		308/280	2000	37.6	8.4	67.1	83.1
		1488/1460	1730	47.7	4.6	34.6	42.8
	5% NH_4NO_3	29/1	2150	29.6	12.5	77.0	100.0
		56/28	2030	37.1	10.4	57.1	74.2
		308/280	1960	41.1	7.5	40.0	51.9
		1488/1460	1560	62.0	3.1	15.6	20.3
5	water	28/0	2120	29.1	9.6	67.7	100.0
		56/0	2130	28.6	10.5	78.4	115.8
		308/0	2150	27.5	8.9	95.8	141.5
		1488/0	2170	26.1	6.4	97.4	143.9
	3% NH_4NO_3	29/1	2130	29.3	9.6	71.4	100.0
		56/28	2080	33.2	8.6	68.7	96.2
		308/280	2010	34.0	7.3	64.2	89.9
		1488/1460	1920	42.8	5.5	37.4	52.4
	5% NH_4NO_3	29/1	2120	31.2	10.5	75.2	100.0
		56/28	2020	39.0	9.3	59.9	79.7
		308/280	1920	43.8	7.0	44.6	59.3
		1488/1460	1680	58.8	4.6	29.0	38.6

Loss transverse tensile strength with time saving samples in aggressive solutions was also quite significant, but more modest than the compressive strength. For example, long-term storage for the sample in a 5% solution of NH_4NO_3 to achieve a decrease of 75.2 MPa to 29.0 MPa, a decrease to 38.6% of the original strength, the transverse tensile strength, the decrease from 10.5 MPa to 4.6 MPa, which represents 43.8% of the original strength.

In the case of transverse tensile strength more effectively showed a positive impact on reducing fiberglass tensile specimens stored in aggressive media. The achieved results are consistent with literature data [3,4,8,9].

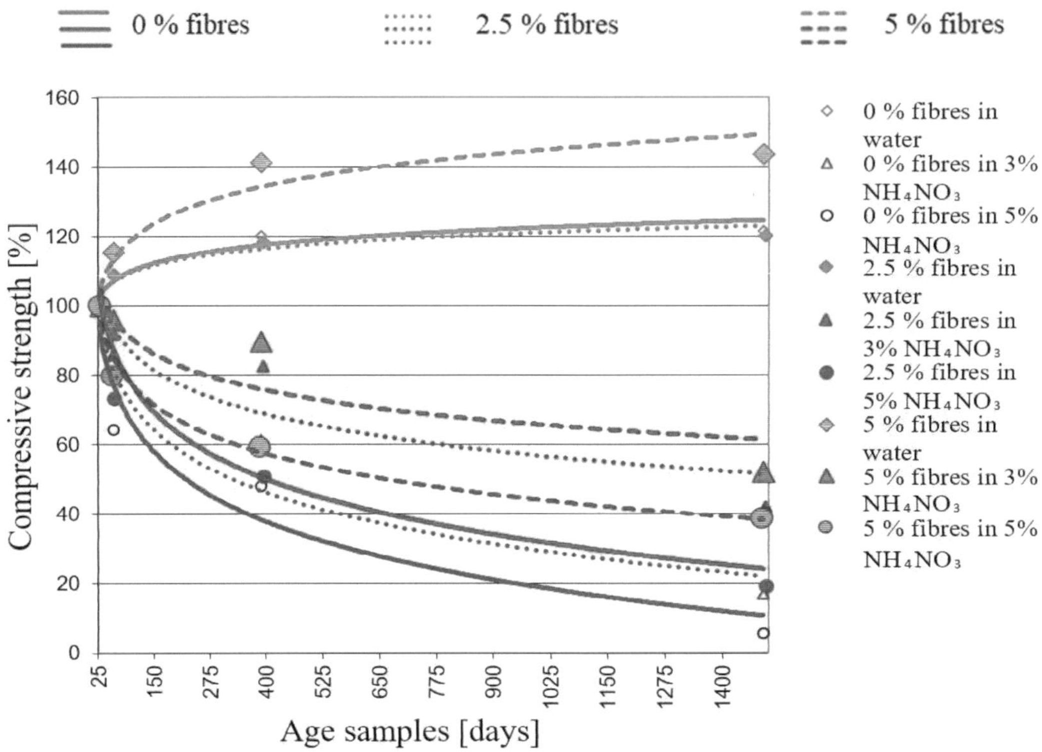

Figure 2. The compressive strength of the samples with the addition of 0, 2.5 and 5 wt.% of the glass fibres as a function of time, represented by a logarithmic trend line.

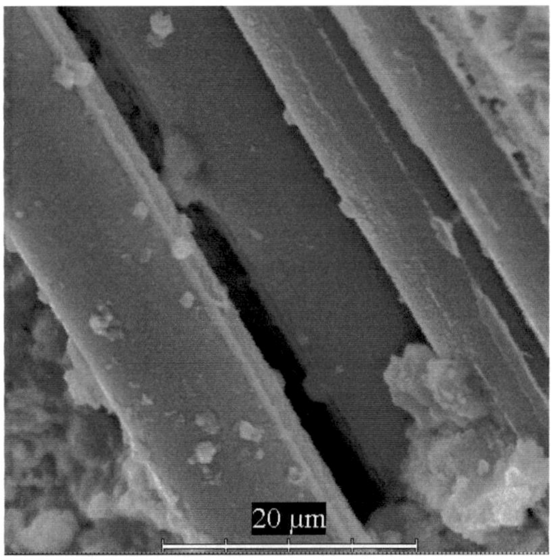

Figure 3. SEM micrograph of the sample with cement composites admixture 2.5wt.% fiberglass stored 1460 days in 3% NH_4NO_3 solution.

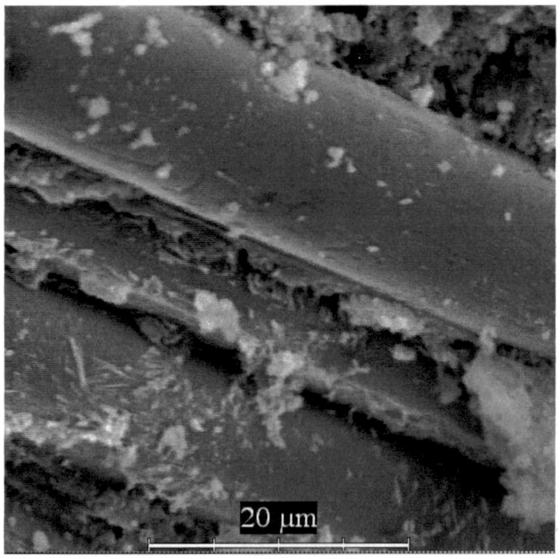

Figure 4. SEM micrograph of the sample with cement composites admixture 2.5wt.% fiberglass stored 1460 days in 5% NH_4NO_3 solution.

Figure 3 and 4 are SEM micrographs of cement composites samples with 2.5wt.% alkali-resistant glass fibers stored 1460 days in 3% and 5% NH_4NO_3 solutions. The microstructure is significantly disturbed by the effect of aggressive influence of ammonium nitrate solution, which are corresponded with decreasing of the compressive strength and increasing of the porosity (table 1). Surface of the glass fibers is smooth and without damage.

4. Conclusion

The achieved results confirm the significant effect of aggressive NH_4NO_3 solutions, which appeared to pre-weight loss, increase porosity, decrease in density, compressive strength and transverse tensile strength. These results point to a leaching test samples and their corrosion II type. Application of glass fibre had a positive effect on the properties of cement pastes, most notably, this was reflected on the compressive strength and transverse tensile strength. Despite the positive effects but also the application of fiberglass there was a significant loss of strength cement mortars exposed to prolonged exposure to ammonium nitrate solutions. The obtained results are showed that the application of glass fiber can mitigate or slow down the process of corrosion of concrete leaching, respectively corrosion II type. Fiber application can extend the durability of structures exposed to such action but cannot prevent the corrosion process.

Acknowledgements
This article was created with the support of the Ministry of Education, Science, Research and Sport of the Slovak Republic within the Research and Development Operational Program for the project "University Science Park of STU Bratislava", ITMS 26240220084, co-funded by the European Regional Development Fund and by the Slovak Research and Development Agency under the contract No. APVV-15-0681.

References

[1] Bartos P J M 2017 *IOP Conf. Ser. Mater. Sci. Eng.* **246** 012002
[2] Brandt A M 1995 *Cement-Based Composites. Materials, Mechanical Properties and Performance* (London: E and FN SPON) p 470
[3] Benes T 2002 *Glass Fibre Chopped Strand Reinforcement*, Concrete TSR, vol 4 http://www.betontks.cz/sites/default/files/BETON_TKS_2002-04.pdf, pp 31 -33
[4] Bajza A and Rousekova I 1990 *Il Cemento* **87** 29-46
[5] STN EN 196-1: 2016 *Methods of testing cement Part 1: Determination of strength*
[6] Bharathi Murugan R et al 2017 *IOP Conf. Ser. Earth Environ. Sci.* **80** 012004
[7] Ďubek M et al 2017 *Detecting fibers in the cross sections of steel fiber-reinforced concrete* (London: Taylor&Francis Group) pp 383-88
[8] Bentur A and Mindess S 2006 *Fibre Reinforced Cementitious Composites* (London: CRC Press) p 624
[9] Horakova A et al 2017 *IOP Conf. Ser. Mater. Sci. Eng.* **246** 012038

Permeation and diffusion tests for estimation of durability of fibreconcretes with dense aggregate and recycled concrete

M Stehlik, J Knapova and V Hermankova

Department of Building Testing, Faculty of Civil Engineering, Brno University of Technology, Veveri 95, Brno, Czech Republic

Abstract. Mechanical and deformational properties of concrete are markedly improved by the addition of long structural fibres. This addition, however, not always improves the durability of concrete. One important factor for an estimation of concrete durability is to assess the permeation and diffusion characteristics of the surface layer. The research was performed by means of the TORRENT permeability method and the method for determination of the non-steady-state migration coefficient of chlorides in concrete – NORDTEST. The addition of long polypropylene fibres contributes to increasing the permeability of the surface layer for air and chloride solution, but no too significantly. The higher permeability was determined in concrete with concrete recyclate and polypropylene fibres. The measured results by both used methods are partially comparable and can be used for raw estimation of durability of concretes. The higher values of permeability indicate the expected lower durability of tested concretes.

1. Introduction

According to the Guidance Paper F to the European Construction Products Directive of December 2004, the durability of concrete can be defined as the ability of a product to maintain its required performance over a given or long time, under the influence of foreseeable actions. An important indicator in this respect is the surface layer of concrete constituting a "gateway" for various harmful substances [1,2]. The research of durability is mainly concerned with the study of the physico-mechanical and chemical damage to concrete.

Fibre concretes are modern building materials with the sufficiently proven mechanical and deformation properties. Long polymer fibres purposefully improve the mechanical properties of the concrete composite. However, the presence of fibres in the concrete composite can also have a negative impact [3]. The diffused fibres are able to improve the low tensile strength of concrete, however, at the cost of exceeding the critical shear stress of the fibre-cement contact. Poor adhesion of fibres to the cement matrix, the exceeding of the shear stress, or corrosion can lead, according to some authors, to an increase in permeability and subsequently to a decrease in the durability of the cement composite. On the other hand, however, some experts believe that the addition of long and short fibres to concrete increases its durability on the grounds of preventing the occurrence of microcracks at all stages of concrete curing [4]. If we accept the fact that the durability of concrete is considerably determined by the permeability of its surface layers [5], it seems to be reasonable to determine the permeability of the surface of the tested concretes by one of the modern non-destructive methods, e.g. TORRENT for air and NORDTEST for a solution of chlorides in water.

2. Experimental part

Focus is given to the determination of the durability properties of fibre concretes made from natural dense aggregate and polymer fibres, and alternatively, in one formula, from the concrete recyclate with a fraction of 0-16 mm by Dufonev, substituting the coarse fraction of the dense aggregate.

From among the surface permeability tests [6,7], the TORRENT test [8] was chosen (permeability of the surface layer for air). This test was carried out on the concrete tiles of 300x300x80 mm. For each formula, 6 measurements were taken. Another method used in the tests was the NORDTEST method for a determination of the coefficient of the non-steady-state migration of chlorides in concrete. This test was carried out on the segments separated from the core samples. Each segment had a diameter of 100 mm and a height of 50 mm. For each formula, the measurements were taken on two separate segments.

2.1 Formulas of concrete

We made a total of 12 tiles (300x80x300 mm) according to 4 concrete formulas (the 1st formula – reference, with dense aggregate, „O“, the 2nd formula - dense aggregate with 1% of PP fibres, „B“, the 3rd formula – dense aggregate with 0.15% of PP fibres, „HV“, and the 4th formula – concrete recyclate of 0-16 mm with 1% of PP fibres, „C“), see table 1. The type of cement used was CEM II/B-S 32.5R. The concretes were prepared from the natural extracted dense aggregate with a fraction of 0 – 4 mm from the Bratčice sand pit, the natural extracted dense aggregate with a fraction of 4 – 8 mm from Tovačov, and the natural crushed dense aggregate with a fraction of 8 – 16 mm from Olbramovice. The concrete made from concrete recyclate was prepared by substituting the coarse fraction of the natural aggregate of 8-16 mm with the raw concrete recyclate with a fraction of 0-16 mm by Dufonev s.r.o. All the four formulas were plastified by means of the CHRYSOPLAST 760 plasticizer, and the type of long construction polymer fibres used was FORTA FERRO with a length of l = 50 mm and aspect ratio of l/d = 110.

Table 1. Formulas of concrete.

Labelling of concrete	Amount of cement	Aggregate 0-4 mm	4-8 mm	8-16 mm	Water-cement ratio	Plastisizer	PP fibres
	[kg/m^3]	[kg/m^3]	[kg/m^3]	[kg/m^3]	-	[%] of m_c	[kg/m^3]
O	490	890	100	745	0,34	1	0
B	490	890	100	745	0,36	1	9,1 (1%)
HV	490	890	100	745	0,35	1	1,37 (0,15%)
C (recyclate)	490	890	100	633 (recyclate)	0,43	1	9,1 (1%)

2.2 Principles of TORRENT and NORDTEST testing methods

The TORRENT testing method is suitable for determining the air permeability of the surface layer of concrete and it can be used both on the test specimens, and on the construction site. The principle of determining the air permeability consists in creating vacuum on the surface layer of concrete and in measuring the penetration of air through the concrete to the measuring device during the specified period of time. A cross-section of the TORRENT apparatus can be seen in figure 1.

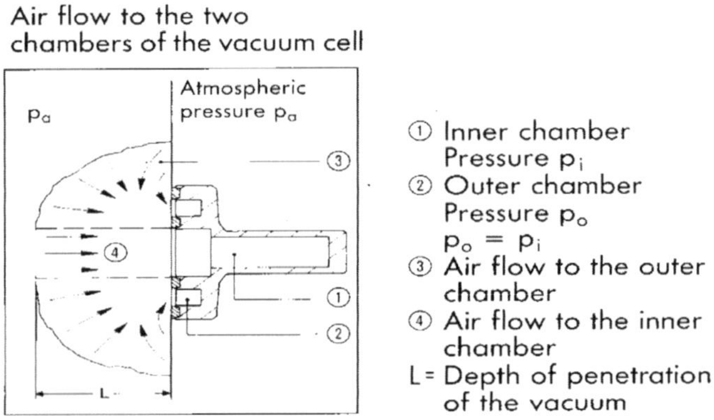

Figure 1. Cross-section scheme of the TORRENT device.

In the direction of the specimen axis by the testing method NORDTEST, an external electrical potential is applied which forces the outer chloride ions to pass inside the specimen. When the specified test time interval expires, the specimen is split axially, and the freshly split surfaces are covered with silver nitrate $AgNO_3$. In the areas of penetration, it reacts into the white silver chloride AgCl, according to which it is easy to recognize and measure the depth of penetration. The migration coefficient is subsequently determined from this measured depth. The position of the specimen in the apparatus is shown in figure 2.

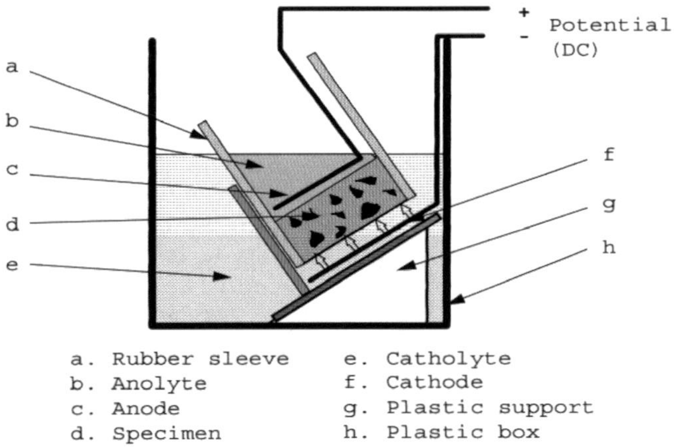

Figure 2. Cross-section scheme of the NORDTEST device.

3. Results and discussion

Both permeability tests were carried out after 1 month of the concrete specimens' curing in a humid environment and the subsequent drying to the 3% moisture content. Table 2 and table 3 give the values of the determined coefficients of permeability and of the migration of chlorides using both methods.

Table 2. Coefficients of air permeability of concretes k_T from the TORRENT method.

Labelling of concrete	Current moisture content w average [%]	Correction of value k_T for w=3% average [*10^{-16} m^2]	[%]	Depth of vacuum intrusion L average [mm]	Quality of concrete layer for w=3% average
O (concrete with dense aggregate without fibres)	2,92	0,003	100	4,1	Very good
B (concrete with dense aggregate + 1 % PP fibres)	3,25	0,007	233,3	4,2	Very good
HV (concrete with dense aggregate + 0,15 % PP fibres)	3,42	0,011	366,6	4,5	Good
C (concrete with concrete recyclate + 1% PP fibres)	3,68	0,035	1166,6	6,3	Good

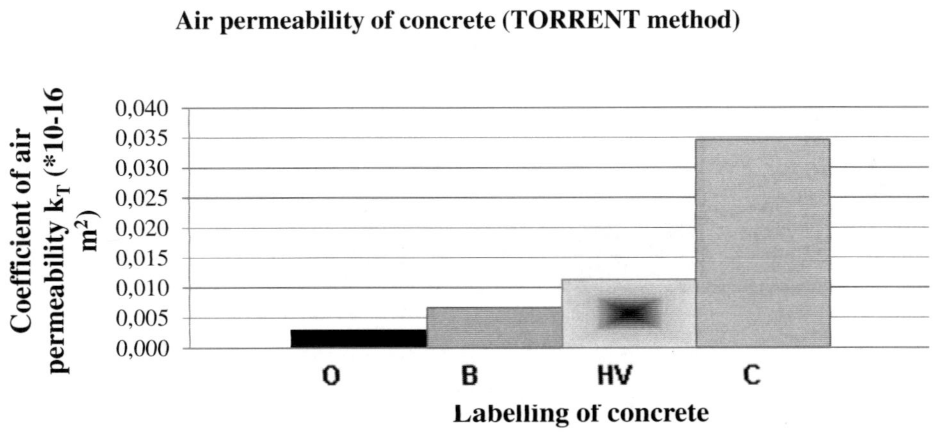

Figure 3. Dependence of coefficients of air permeability k_T on types of concrete.

Table 3. Measured non-steady state chloride migration coefficients by the NORDTEST method.

Labelling of concrete	Initial current	Applied voltage	Chloride penetration height	Non-steady state chloride migration coefficient	
	I30v	U	X_d (individually)	D_{nssm} (individually)	D_{nssm} (average)
	[mA]	[V]	[mm]	[m²/s]	[m²/s]
O1	45	30	10,929	4,815	4,625
O2	45	30	10,141	4,434	
B1	50	30	17,714	8,137	8,067
B2	50	30	17,429	7,997	
HV1	40	35	22,107	8,843	9,064
HV2	60	25	16,931	9,286	
C1	65	25	23,285	13,095	12,838
C2	60	25	22,431	12,581	

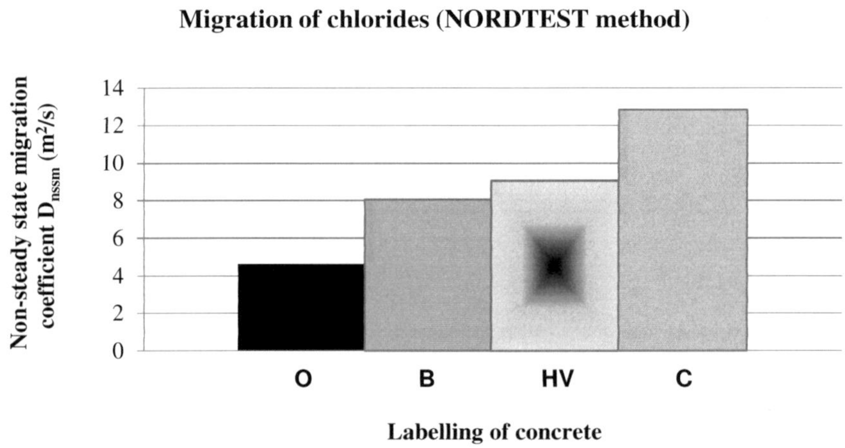

Figure 4. Dependence of non-steady state migration coefficient of chlorides D_{nssm} on types of concrete.

It is apparent from both charts that polymer fibres slightly increase the permeability of concrete for air and for chlorides, namely, inversely to their amount. It is possible to observe a close correlation dependence between the results of both the TORRENT and NORDTEST methods. The permeability coefficients of the concretes made from concrete recyclate determined by both the methods are the highest, which is caused by the higher porosity of the concrete recyclate compared to the dense aggregate. The overall permeability is probably also worsened by a combination of the concrete recyclate and PP fibres. The higher permeability is characteristic of the concretes with 0.15% of fibres in comparison to the concretes with 1% of fibres. This effect is caused by an insufficient amount of fibres in the concrete volume which do not effectively suppress the occurrence of microcracks during the stages of hydration and subsequently shrinkage of concrete.

4. Conclusion

Durability of concrete is from great amount a function of physical-mechanical properties of concrete layer. In this article the results of two basic methods for concrete layer testing were described. The coefficients of permeability and chloride migration determined by means of the TORRENT and NORDTEST methods are relatively closely correlationally dependent and can be used separately or in combination for a quick informative estimate of the durability properties of fibre concretes made from dense aggregate as well as from concrete recyclate. Generally, the higher value of permeability of concrete decreases its durability, or in other words, its expected life time.

If we compare the permeability and diffusion properties of the surface layer of concrete with fibres with that without fibres, it is possible to say that fibres contribute to increasing the permeability of the surface layer for air and chloride solutions, but the increase is not too marked, and according to the tables, the concretes fall under the same class. The higher permeability, especially for the gaseous medium, was determined in the concrete made from the concrete recyclate with fibres, probably due to the higher porosity of the concrete recyclate compared to the dense aggregate. In the case of the migration of chlorides, this difference is not very important.

In conclusion it is possible to say that the concrete made from concrete recyclate shows slightly worse permeability results in terms of the formulas tested, and we can expect its lower durability, but this concrete can definitely be used even in demanding structures with a low aggressiveness of the environment of class X and XC.

Acknowledgement
This paper was created under the support of project No. LO1408 AdMaS UP – Advanced Materials, Structures and Technologies, supported by the Ministry of Education, Youth and Sports of the Czech Republic within the scope of the “National Sustainability Programme I”.

References

[1] Adamek J, Jurankova V 2010 *Transactions on Transport Sciences* **2** 188-195
[2] CEP-FIP Model Code Final Draft Section 5.1.13. 2010 *Properties related to durability* (Germany: Ernst und Sohn) p 106-110
[3] Kim B, Boyd A J and Lee J Y 2011 *J. Compos. Mater.* **45** 2379-2389
[4] Cangiano S, Plizzari G A, Gadoni E, Frigeri G and Teruzzi T 2005 *Proc. Int. Conf. on Cement Combinations for Durable Concrete (Dundee)* (London: Thomas Telford Ltd) p 477-486
[5] Adamek J, Jurankova V and Kucharczykova B 2009 *Proc. Int. Conf. on Fibre Concrete (Prague)* (Prague: CTU Praque) p 9-14
[6] Adamek J, Juranková V, Kadlecova Z and Stehlík M 2012 *Three NDT methods for the assesment of concrete permeability as a measure of durability* (Paris: Springer in RILEM Bookseries) p 32-38
[7] Collepardi M, Marcialis A and Turriziani R 1972 *Journal of American Ceramic Society* **55** 534-535
[8] Romer M 2005 *Mater. Struct.* **38** 541-547

Study of a different carbonation intensity on the internal surface of cracks and inside concretes with dense aggregate and concrete recyclate

M Stehlik[1], **J Knapova**[2], **P Macová**[3]

[1,2]Department of Building Testing, FAST BUT in Brno, Veveri 95, 602 00 Brno, CR
[3]UTAM AV CR v.v.i., Centre of Excellence Telc, Batelovska 485, 588 56 Telc, CR

Abstract. The paper focuses on the experimental physical - chemical analysis of different carbonation intensity by TG/DTG and FTIR method on the selected surfaces of concretes with aggregate and with concrete recyclate, with and without polymer and steel fibres. By means of thermal analysis and infrared spectrometry, carbonation intensity from the specific amount of calcium carbonate and portlandite was tested on specimens scratched from the internal surface of cracks and on specimens from the inside of concrete. It seems that bending macrocraks do not significantly accelerate the process of carbonation in the depth of concrete. But they bring various corrosion agents to concrete. As regards the evaluation of carbonation intensity of different concrete samples, fibre concretes with concrete recyclate are more suitable for structural purposes than fibre concretes with dense aggregate.

1. Introduction

The best way of preventing occurrence of cracks in a system that is bound by hardened cement, is to use fibres. Cracks in concrete can occur due to tensile stress caused by plastic shrinkage, drying shrinkage or autogenous shrinkage. At the same time, fibres can also improve other properties, such as bending or impact strength, or abrasion resistance. Fibre concretes are modern building materials with sufficiently proven mechanical and deformation properties. Long polymer fibres purposefully improve the mechanical properties of the concrete composite. However, the presence of fibres in the concrete composite can also have a negative impact [1]. The dispersed fibres are able to improve the low tensile strength of concrete, however, at the cost of exceeding the critical shear stress of the fibre - cement contact. Poor adhesion of fibres to the cement matrix, the exceeding of the shear stress or corrosion can lead, according to some authors, to an increase in permeability and subsequently to a decrease in the durability of the cement composite. On the other hand, however, some experts believe, that the addition of both long and short fibres to concrete increases its durability on the grounds of preventing the occurrence of microcracks at all stages of concrete curing.

The aim of the paper is to assess the influence of macrocracks on the intensity of carbonation in the loaded structures of fibre concretes. Under macrocracks we understand cracks with a width of 0.1 mm and more, which are visible to the naked eye. The paper seeks to answer the question whether macrocracks created when loading fibre concrete with steel or polymer fibres, accelerate the process of carbonation in the depth of concrete or not. The paper compares fibre concrete with dense natural aggregate with fibre concrete with its coarse aggregate fraction substituted with concrete recyclate.

2. Experimental setup

The paper focuses on examining the carbonation of fibre concretes with dense natural aggregate and fibre concretes with the coarse aggregate fraction substituted with concrete recyclate. It also studies combinations of steel and polymer fibres added to concretes.

Beams containing flexural cracks from the previous flexural strength testing were kept in a carbonation chamber with a CO_2 concentration of 98% for a period of 9 months.

For the purposes of the experiment, three specimens were chosen and labelled B2/2 (1% of polymer fibres), C2/1 (1% of polymer fibres + concrete recyclate), and DA3/2 (0.15% of steel fibers). The three specimens were tested for carbonation intensity by means of thermal analysis and infrared spectrometry with Fourier transformation. Material for the analysis of the studied specimens was taken by scraping the surface layer A) from the surfaces of cracks in the middle depth of sectional area of beams dimensions 100x50x400 mm , B) from the surfaces split off from the inside of concrete (distance from the beam surface is min. 25 mm).

2.1. Process of carbonation

Carbonation of concrete is a chemical process caused by a reaction of carbon dioxide with the components of hardened cement paste in concrete. Portlandite present in the hardened cement paste reacts with carbon dioxide to form calcium carbonate (note: in the form of calcite or aragonite).

The equation (1) of the process of concrete carbonation:

$$CO_2 + Ca(OH)_2 \rightarrow CaCO_3 + H_2O \quad (1)$$

With a decreasing amount of portlandite, also the pH of the covering layer of concrete is decreased. In case the pH decreases below 9, the so-called depassivation of the reinforcement occurs, and its corrosion begins. Carbonation damage to concrete can be detected by means of a 1% solution of phenolphthalein. In case the pH decreases below 9.4, the typical violet colouration does not occur. That part of concrete which does not change colour after the solution application, shows signs of carbonation [2].

2.2. Concrete formulas

The tested specimens were made according to the following three formulas: 1st formula marked "B" – dense aggregate with 1% of PP fibres, 2nd formula marked "C" – concrete recyclate of 0-16 mm with 1% of PP fibres, 3rd formula marked "DA" – dense aggregate with 0.15% of steel fibres. See table 1.

Table 1. Concrete formulas.

Labelling of concrete	Amount of cement	Aggregate 0-4 mm	4-8 mm	8-16 mm	Water-cement ratio	Plasticizer	PP + steel fibers
	[kg/m^3]	[kg/m^3]	[kg/m^3]	[kg/m^3]	-	[%] of m_c	[kg/m^3]
B	490	890	100	745	0.36	1	9.1 (1%) PP
C (recyclate)	490	890	100	633 (recyclate)	0.43	1	9.1 (1%) PP
DA	490	890	100	745	0.36	1	11.7 (0.15%) steel

The type of cement used was CEM II/B-S 32.5R. The concretes were prepared from the natural extracted dense aggregate with a fraction of 0-4 mm from the Bratčice sand pit, from the natural extracted dense aggregate with a fraction of 4-8 mm from Tovačov, and from the natural crushed dense aggregate with a fraction of 8-16 mm from Olbramovice. The concrete made from concrete recyclate was prepared by substituting the coarse fraction of the natural aggregate of 8-16 mm from Olbramovice with the raw concrete recyclate with a fraction of 0-16 mm by Dufonev s.r.o.

All the three formulas were plastified by means of the CHRYSOPLAST 760 plasticizer, the type of polymer fibres used was FORTA FERRO with a length of 50 mm, the type of steel fibres used was Dramix with a length of 60 mm.

2.3. Thermal analysis (TG/DTG)

Thermal analysis involves experimental methods by means of which we analyze changes in the composition and properties of the analyzed substances or mixtures under thermal load.

The basic method of thermal analysis is thermogravimetry (TG), which studies changes in mass over time depending on the temperature. From the resulting thermogravimetric curve, it is possible to read temperatures of mass changes.

A method derived from this is the derivative thermogravimetry (DTG), in which the resulting thermogravimetric curve shows the dependence of the mass change rate on temperature. This method is suitable for a differentiation of consecutive effects [3,4].

2.4. Infrared spectrometry with Fourier transformation (FTIR)

Infrared spectrometry is an analytical technique used for identification and structural characterization of organic compounds and determination of inorganic matters. The principle of infrared spectrometry is the absorption of infrared radiation in passing through the specimen. The infrared radiation includes the area of wavenumbers of 12,500 – 10 cm^{-1}.

Passage of radiation through a substance causes rotational-vibrational transitions of molecules. In case a change in the given rotational or vibrational states is connected with a change in the dipole moment, the radiation characteristic of a specific bond in the molecule is absorbed. Each organic compound has its own region, i.e., no two different organic compounds have the same spectral expression.

The most frequently used band of the infrared spectrometry is the so-called central region given by the wavelength range of 4000 – 400 cm^{-1}. In the range of 4000 – 2500 cm^{-1}, there are valence vibrations of hydrogen bonds. The loose O-H bond is absorbed at wavelengths of about 3600 cm^{-1} [5,6,7].

3. Results and discussion

3.1. Results of TG/DTG

Table 2. Amount of calcium carbonate and portlandite measured by thermal analysis of specimens (A – internal surface of the crack, B – the inside of concrete).

Sample	Mass % c / hm. % $Ca(OH)_2$ (350-550°C)	$Ca(CO)_3$ (550-850°C)
B2/2 A	6.3	19.4
B2/2 B	5.3	15.8
C2/1 A	-	21.5
C2/1 B	-	18.5
DA3/2 A	-	22.4
DA3/2 B	-	18.7

Specimens were measured by a STA504 instrument (TA Instruments) in the temperature range of 50 – 1050°C. The analysis was performed with 15 – 20 mg of a specimen spread into ceramic crucibles, the heating was performed in the N_2 atmosphere with a heating rate of 20°C per 1 minute. Figure 1 shows the sample plot of the thermal analysis with graphs of TG, DTG and DSC of the internal surface of the crack of specimen B2/2 (A). Table 2 shows the representation of portlandite and calcite in the individual specimens.

The measured values of all specimens show a higher amount of calcium carbonate on the surface of the crack (A) than inside concrete (B). The higher amount of calcium carbonate indicates a higher extent of concrete carbonation in the area of the crack.

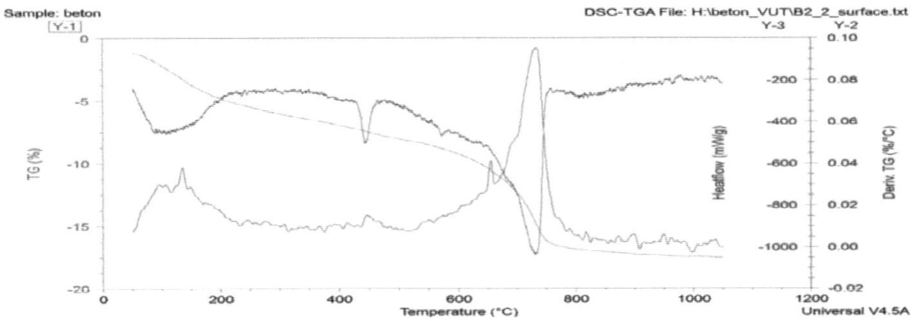

Figure 1. Specimen B2/2 – Plot of the thermal analysis of the internal surface of the crack (A).

3.2. Results of FTIR

Infrared spectrometry allows for measuring specimens of all physical states. One of the basic solid state measuring techniques uses tablets of potassium bromide, which transmits a wide region of the infrared spectrum. A small amount (1 – 10 mg) of the crushed specimen is homogenized with 300 – 400 mg of KBr. Subsequently, it is pressed into a tablet, giving rise to the so-called glassy modification of KBr, in which the loss of the intensity of the passing infrared radiation caused by reflection is small [9].

Using infrared spectrometry with Fourier transformation, we determined presence of portlandite in the specimens. Always two specimens from each formula were compared: specimen (A) from the crack surface, specimen (B) from the inside. 1 mg of the spread specimen was homogenized with 150 mg of KBr, and the tablet prepared from it was measured on the neighbouring iZ10 FTIR module of the Nicolet iN10 microscope (Thermo Scientific) by means of a transmission technique in the spectrum range of 4000 to 400 cm-1, using resolution of 4 cm-1. Attention was given especially to the band of 3643 cm-1, which corresponds to the vibration of OH bond in portlandite. A comparison of spectra of the measured specimens is shown in the following charts in figure 2-4.

In specimen B2/2, a higher intensity of band 3643 cm-1 was measured in the part of specimen taken from the inside (B) than in the area of the crack surface (A). In specimen C2/1, portlandite occurs only inside (B) and in a very low concentration. Specimen DA3/2 was fully carbonated in both the monitored part (A and B), and calcium carbonate does not occur only in the form of calcite, but also in the form of aragonite, especially on the surface of the crack.

Values obtained from the thermal analysis show a higher amount of calcium carbonate present in the fibre concrete in the area of the surface of cracks than in the inside of concrete, in all the three specimens tested.

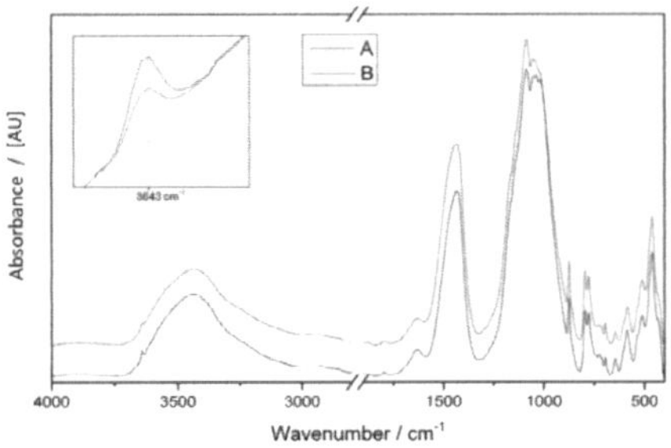

Figure 2a. Specimen B2/2: Infrared spectrum (A – internal surface of a crack, B – the inside of concrete).

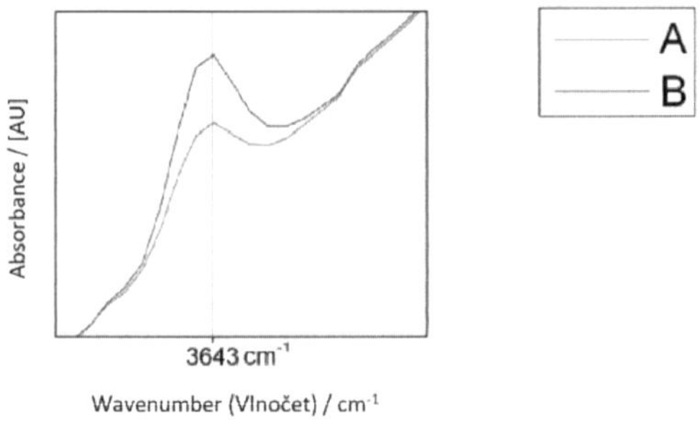

Figure 2b. Detail of a chart of a wavenumber of the vibration of the OH bond in portlandite. Specimen B2/2: Infrared spectrum (A – internal surface of a crack, B – the inside of concrete).

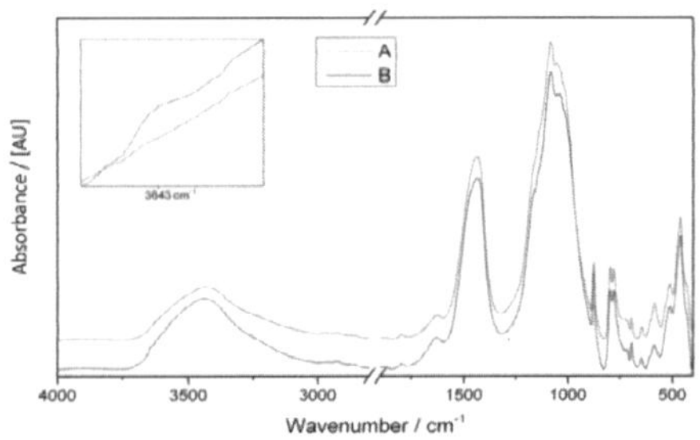

Figure 3. Specimen C2/1: Infrared spectrum (A – internal surface of a crack, B – the inside of concrete).

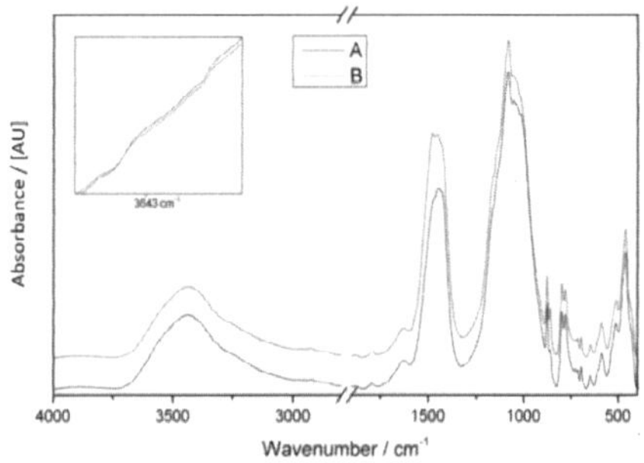

Figure 4. Specimen DA3/2: Infrared spectrum (A – internal surface of a crack, B – the inside of concrete).

The measurement with Fourier transformation infrared spectrometry (FTIR) showed that specimen B2/2 with an addition of 1 % of PP fibres had the best result in terms of the lowest extent of carbonation. Specimen C2/1 with concrete recyclate and 1% addition of PP fibres showed a better result than specimen DA 3/2 with coarse aggregate and an addition of 0.15% of steel fibers. The specimen of fibre concrete with steel fibres (DA 3/2) was fully carbonated on the crack surface as well as in the inside of concrete. This higher susceptibility to carbonation is probably caused by an occurrence of a high number of microcracks during the hydration shrinkage of concrete. Considering the different elasticity moduli of polymer and steel fibers, tensile microcracks can occur in the solid steel-fiber skeleton during shrinkage of concrete at the initial stage of setting [8]. In this case, steel fibres seem to be less advantageous for their low plasticity than the polymer ones. Fibre concrete made from concrete recyclate with an addition of polymer fibres resisted depth carbonation in the experiment better than steel fibre reinforced concrete with dense aggregate, in which carbonation occurred with a greater extent and intensity.

4. Conclusions

On the basis of the TG/DTG and FTIR tests performed on fibre concrete specimens of three formulas, it is possible to make the following conclusions:

a) Bending macrocracks (more than 0.1 mm wide) do not significantly accelerate the process of carbonation in the depth of concrete due to low ventilation of air with a natural content of CO_2. On the internal surface of the crack, carbonation occurs with approximately the same intensity as on the free surface of the concrete of an open structure.

b) If durability of fibre concretes is evaluated narrowly, only from their susceptibility to carbonation, then fibre concretes with concrete recyclate are more suitable for structural purposes than fibre concretes with dense aggregate. The bearer of a higher durability of fibre concretes with concrete recyclate, however, is not the concrete recyclate itself, but the higher plasticity of polymer fibres in comparison with the steel ones.

c) Although macrocracks themselves do not significantly accelerate the process of carbonation inside concrete, they bring CO_2, water and other substances to concrete and in the case of crossing steel reinforcement and a crack, accelerated corrosion occurs.

Acknowledgement

This paper was created under the support of project No. LO1408 AdMaS UP – Advanced Materials, Structures and Technologies, supported by the Ministry of Education, Youth and Sports of the Czech Republic within the scope of the "National Sustainability Programme I".

References

[1] Adamek J, Jurankova V and Kucharczykova B 2009 *Proc. Int. Conf. on Fibre Concrete* (Prague) (CTU Praque: Czech Republic) pp 9-14

[2] CEP-FIP Model Code, Final Draft 2010 *Properties related to durability* (Ernst und Sohn: Germany) p 106-110

[3] CSN EN 14630: 2008 *Products and systems for the protection and repair of concrete structures – Test methods – Determination of carbonation depth in hardened concrete by the phenolphthalein method*

[4] Brown M E 1998 *Handbook of Thermal Analysis and Calorimetry* [online] (Elsevier B V)

[5] Blazek A 1974 *Thermal analysis* (Prague: SNTL) (in Czech)

[6] Bohm A and Smrckova-Voltrova S 1995 *Structural analysis of organic compounds* (Prague: VŠCHT) (in Czech)

[7] Vaculikova L, Plevova E 2005 *Acta Geodyn. Geomater.* **2** 165-175

[8] Kania P 2007 *Infrared spectroscopy* (Prague: VŠCHT) (in Czech)

[9] Stehlik M, Hermanková V, Vitek L 2015 *J.Civ. Eng. Manag.* **21** 177-184

[10] Otyepkova E, Prucek R, Kvitek L, Panacek A and Otyepka M 2013 *Basics of selected experimental chemical methods* (Olomouc: UPOL) (in Czech)

IOP Conf. Series: Materials Science and Engineering **385** (2018) 012057 doi:10.1088/1757-899X/385/1/012057

The potential utilization of the rubber material after waste tire recycling

J Svoboda[1,2], V Vaclavik[1,2], T Dvorsky[1], L Klus[1,2] and R Zajac[1]

[1] Department of Environmental Engineering, Faculty of Mining and Geology, VSB-Technical University of Ostrava, 17. listopadu 2172/15, 708 33 Ostrava-Poruba, Czech Republic
[2] Institute of Clean Technologies for Mining and Utilization of Raw Materials for Energy Use, Faculty of Mining and Geology, VSB-Technical University of Ostrava, 17. listopadu 2172/15, 708 33 Ostrava-Poruba, Czech Republic

Abstract. This article describes the method of production of rubber granulate after the recycling of waste tires and the potential utilization of this material. The aim of this study was the utilization of rubber in the building industry, especially the use of granulate as rubber filler in concrete. It has been found that the use of rubber granulate as a substitute for natural aggregates in a concrete mix has a significant impact on the properties of concrete. Although studies on the utilization of rubber granulate in a concrete mix have been conducted in different sectors and in different research areas, it is clear that we do not know much about it yet. That is why it is necessary to continue and to obtain more information about the properties of concrete based on recycled rubber.

1. Introduction

With the development of the automotive industry, tires have become an integral part of everyday life of people. We use them almost daily, whether it is when driving a passenger car or on a public transport bus. They are also widely used in air transport, but they represent an important part of goods traffic as well.

As a result of the development of the society in the world, the demand for tires has been going up, which is associated with higher production of waste tires [1]. Waste tires are a global problem and an increasing risk to the environment, because they are not biodegradable, and they are often incorrectly stored and disposed of. These stocks represent a threat of uncontrolled fires and other environmental risks. It is estimated that almost 1 000 million tires reach the end of their lifetime each year and more than 50% of them are discarded without any further use [2–4].

The storage of tires at landfills after the end of the lifetime has been banned by the European Commission in 1999. The recycling and utilization of waste tires is therefore studied in many countries [5]. This step has forced us to rethink the view and attitude towards waste materials in order to increase their utilization and to dramatically reduce their amount required by their disposal or storage [6].

The simplest and the cheapest way of disposing of waste tires is their combustion. When burning tires, i.e. during the oxidation reaction, the largest part is converted into carbon oxides and soot. The other released substances include butadiene and styrene, aliphatic and aromatic hydrocarbons, benzene, toluene, phenylacetylene etc. The sulphur contained in tires reacts to sulphur dioxide and sulphur derivatives. We can find heavy metals, such as lead, in the combustion product as well [7]. The use of tires as a fuel is not economically attractive and, with regard to the environment, their use as a secondary raw material is more profitable. Waste tire as a raw material has a high potential in terms of its further processing and utilization. Products made from waste tires can be used in various civil and industrial sectors, such as highway construction, agriculture, seashore civil engineering, breast walls, etc. [8].

Civil engineering has been trying to face the challenge of improving the environment by looking for more environmentally friendly raw materials or by using waste materials, such as filler in concrete for several years. Metallurgical industry, which has been trying to process the produced waste in other metallurgical aggregates, as shown by research [9, 10], can serve as a role model in the utilization of waste. A possible solution for the use of natural rubber from waste tires is its use in concrete as a substitute for some natural aggregates. Concrete composites based on recycled tires have the potential to reduce the total greenhouse gas emissions, provide economic benefits, reduce the environmental problems associated with the end-of-life of tires in transport, and they also bring an eco-friendly way of recycling natural rubber from waste tires [3,8,11,12].

Studies, in which scientists examined the properties of concrete containing crushed natural rubber, have been conducted. Scientists have suggested that a concrete mix containing natural rubber granulate can increase the toughness and ductility, and it improves the acoustic absorption and resistance to heat changes. A reduction of the unit weight and better durability, when compared to conventional concrete, have been found as well. The studies have also revealed a decrease in the tensile strength due to low natural rubber stiffness and poor binding power between aggregates of natural rubber and cement articles. Pre-treatment of the surface of natural rubber granulate using NaOH solution could improve the adhesion of natural rubber and cement articles in concrete mix. This modification of the natural rubber surface has been presented by the research of "Development of waste tire modified concrete" [12,13].

2. Methods and potential of the utilization of waste tires

Tire recycling makes it possible to obtain rare materials, to save energy in relation to primary production, and to reduce the amount of waste. Used tires represent the largest share of the total volume of recycled natural rubber. A worn tire contains a number of chemicals with an energy potential, but it is also a source of secondary raw material. Tires are made of a flexible rubber material whose construction is reinforced with textile and metal materials. The composition of tire includes materials containing natural rubber, soot, steel reinforcements, oils and vulcanizing agents, synthetic rubber and textiles. Figure 1 shows the percentage composition of tires for passenger and freight transport [6,14].

When recycling tires, it is necessary to take the following measures [6,15]:

- reduce the risk of contamination during temporary storage of used tires,
- increase the usability of used tires,
- reduce the consumption of raw materials, especially those originating from non-renewable resources, and utilize used tires as secondary raw materials.

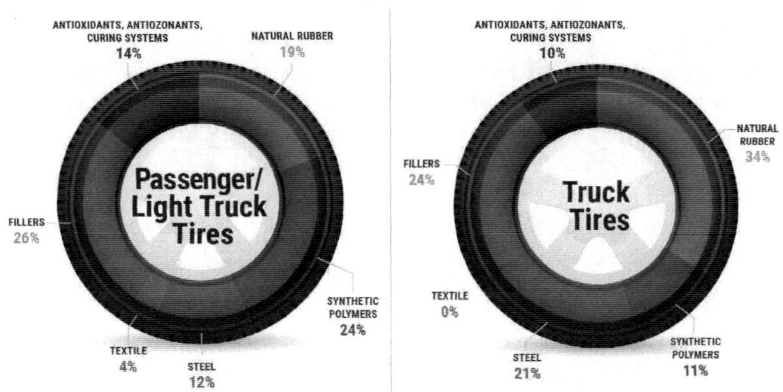

Figure 1. Composition of tires for passenger and freight transport [13].

The use of tires as a source of raw materials leads to three product categories. They are rubber granules, steel wires, and nylon and textile fibres. These materials can be retroactively separated from crushed rubber using modern technologies, so the materials can be recycled or further utilized [6,14].

Waste tires are fed to a primary crusher with fixed blades that cuts them to a size ranging from 50 to 300 mm. At this stage, the tire cuttings can be used as filler, for example in a drainage system, a biofilm reactor, etc., or they are collected for secondary crushing. After the primary phase of crushing, the rubber cuttings proceed into a granulator that cuts and shears the rubber to the dimensions of 10-50 mm. Subsequently, the material passes through a sieve with a fixed hole size. In the next step, the rubber material passes through is a magnetic separator, where steel wire particles are separated. An air-gravity separator is used to remove the textile and nylon fibres. Air is blown into the rubber material and gravity and the weights of the individual particles lead to their separation. To obtain materials with a finer fraction <10 mm, it is necessary for the material to pass through a progressive series of granulating mills (see figure 2). The resulting material is distributed under the name of rubber dust. There may be other process phases, such as a secondary collection of metallic elements, or wet grinding process resulting in a production of particles with the size of up to ≤100μm [16,17].

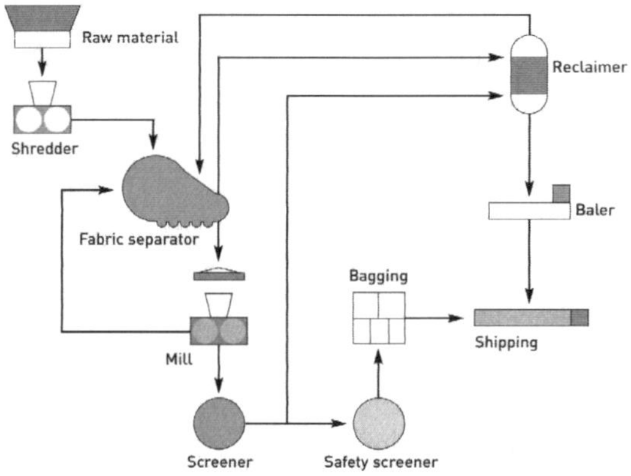

Figure 2. Tire crushing and material separation process [16].

Natural rubber can be manufactured in various sizes, ranging from the smallest powder to granules that are used for various purposes. It is very important to make sure that the quality of recycled rubber is high so that the recycled rubber can be used as a substitute for new natural rubber. It must be completely clean and the particle size must be uniform. If these quality requirements are met, the industry will guarantee that recycled rubber is fully capable of replacing the original raw material. The outcome of the follow-up processing of rubber crumb is the material cut in different sizes and types depending on its intended use [14,18]:

- rubber chips (size of 20-50 mm)
- rubber granulate (0.8-20 mm),
- rubber dust (<0.8 mm).

The utilization of the materials from tires that are used repeatedly has been developed for many years, including the applications on road surfaces, the foundations of children playgrounds, sports playgrounds and noise barriers. There has been an increase in the number of companies that use material from tires to produce various rubber products. The utilization of crushed rubber in a mix of asphalt paving has proved to be successful due to the good compatibility and interaction between the rubber particles and the asphalt binder. A research on the use of rubber particles in concrete composites has already been documented and tested, however the information about the size of the rubber particles or their distribution in concrete, which may also influence the specific properties, is limited [14,19].

3. Use of rubber granulate in concrete

One of the possible solutions for the use of natural rubber from waste tires is the incorporation of natural rubber into concrete as a substitute for natural aggregates [8].

Eldin and Senouci (1993) have carried out primary studies on the possibility of using rubber particles from waste tires in concrete. In their study, the part of the aggregate was replaced with rubber aggregate. They have found that the concrete has lower workability, compressive strength and tensile strength. Worsening the mechanical properties was caused by the low adhesion between the surface of the rubber particles and the cement articles. The loss of compressive strength has corresponded with the size of the rubber aggregate. Concrete, however, has demonstrated the ability to absorb large amount of plastic energy when exposed to pressure and tensile load [20,21].

A research on the strength and penetration of chlorides into concrete with rubber granulate has been carried out by Gesoğlu and Güneyisi [22]. They have pointed out to the fact that the unit weight of rubberised concrete decreases with the increasing share of added rubber, which reduced the unit weight by up to 18%. The development of rubberised concrete strength between 3 and 7 days was relatively high, with a slower hardening rate between days 7 and 28. The compressive strength was reduced depending on the percentage share of the used natural rubber. When rubber was used in concrete production, chloride ion penetration in concrete was considerably worse, but the degree of chloride penetration rate depended mainly on the amount of used natural rubber [20,22].

Ganjian et al. (2009) have examined the use of tires as a substitute for filler in concrete. They used rubber granulate as a substitute for coarse aggregate of the concrete components and rubber dust in 5%, 7.2% and 10% weight substitute was used to replace cement. Replacement of rubber increases water permeability depth in the concrete mixtures.The substitution of 5% and 7.5% tyre rubber showed mechanical properties similar to the control sample and it's classified as low permeability according to DIN standard but the mixture with higher percentage substitution of 10% tyre rubber replacement is classified as medium [20,23]. The compressive strength of rubberised concrete can be increased through the effect of carbon tetrachloride and sodium hydroxide on the rubber aggregates in concrete [20,24]. Yilmaz and Degirmenci (2008) have found a decrease in water absorption when the size of the rubber particles in concrete was increased [25].

In paper [26], the acoustic properties of materials made using recycled rubber were tested. It has been found that these materials can effectively absorb sound when the size of the rubber aggregate and the binder content are carefully selected and its thickness is set to the desired frequency range.

4. Summary and conclusions

The objective of this article is to present the potential of the utilization of rubber granulate from waste tires as an introduction to our research. The utilization of crushed rubber in the building industry is becoming more and more frequent, and it is necessary to find ways to effectively apply this filler. The processing of crushed rubber in the building industry can be beneficial to this industry, but the most important benefit is a relief to the environment. Increasing living standards also mean more waste, which is why it is important to think about how to further process these waste materials.

The examination of the studies clearly shows that the mechanical properties of concrete with the addition of crushed rubber are rather deteriorating. The reason for decreasing compressive strength is the percentage share of crushed rubber in the concrete mix. The modulus of elasticity of rubber is many times lower than the modulus of elasticity of concrete. Rubber granulate in concrete acts as a pore, and higher content of rubber particles increases the amount of pores. This is the reason why the mechanical properties of concrete mix deteriorate. However, some physical properties of concrete are also improved for example as acoustic properties of concrete, or the weight of concrete decreases with the amount of added tire recycle.

Although concrete containing crushed rubber is not a completely new feature in the field of construction, we do not know much about it yet. That is why more research is needed in this field in order to study the properties of rubberised concrete. Further research will be focused on the design of concrete formulations where natural aggregate is replaced with rubber granulate with the fractions of 1-3 mm and 0-1 mm, which was obtained from tire recycling from Bonus-CB spol. s.r.o. company. The physical-mechanical properties of the concrete based on recycled rubber will be tested as well.

Acknowledgement

- This article was written in connection with project Institute of clean technologies for mining and utilization of raw materials for energy use - Sustainability program. Identification code: LO1406. Project is supported by the National Programme for Sustainability I (2013-2020) financed by the state budget of the Czech Republic.
- This research work has been supported by the student grant competition project SGS SP2018/12.

References

[1] Adamcova D 2011 *Research of selected types of municipal waste with focus on waste tyres* (Brno: Mendel University in Brno, Faculty of AgriSciences,) Dissertation Thesis
[2] Rawahi Z A and Waris M B 2017 *International Journal of Structural and Civil Engineering* **5** 273-276
[3] Thomas B S, Gupta R CH and Panicker V J 2016 *J. Clean. Prod.* **112** 504–513
[4] Xue J and Shinozuka M 2013 *Cons. Build. Mat.* **42** 196–204
[5] Asaro L, Gratton M, Seghar S and Hocine A N 2018 *Res. Cons. Rec.* **133** 250–262
[6] Bulei C, Todor M P, Heput T and Kiss I 2018 *IOP Conference Series: Mat. Sci. Eng.* 294
[7] Toutanji H A 1996 *Cem. Con. Comp.* **18** 135–139
[8] Thomas B S and Gupta R Ch 2016 *Ren. Sust. En. Rev.* **54** 1323–1333
[9] Jursova S, Honus S and Pustejovska P 2016 *Inter. Con. Met. Mat.* 1838–1843
[10] Pustejovska P, Jursova S, Brozova S and Sousek J 2013 *Metallurgist.* **56** 908–911
[11] Kashani A, Ngo D T, Mendis P, Black R J and Hajimohammadi A 2017 *J. Clean. Prod.* **149** 925–935
[12] Si R, Wang J, Guo S, Dai Q and Han S 2018 *J. Clean. Prod.* **180** 823–831
[13] Li G, Stubblefield A M, Garrick G, Eggers J, Abadie Ch and Huang B 2004 *Cement Concrete Res.* **34** 2283–2289
[14] Bulei C 2017 *Opportunities for low investment recycling waste tire derived materials into rubber pavers.* 5
[15] Raffoul S, Garcia R, Pilakoutas K, Guadagnini M and Medina F N 2016 *Const. Build. Mat.* **124** 391–404
[16] Waste & resources action programme 2007 Tyre-derived rubber materials, A technical report on the manufacture of tyre-derived rubber materials B.m.: Environment Agency.. https://www.pharosproject.net/uploads/files/sources/1828/1330040519.pdf
[17] Sienkiewicz M, Borzedowska-Labuda K, Zalewski S and Janik H 2017 *Const. Build. Mat.* **154** 144–154
[18] Godlewska J 2017 *Proc. Eng.* **182** 229–234
[19] Medina F N, Medina F D, Hernandez-Olivares F a Navacerrada M A 2017 *Const. Build. Mat.* **144** 563–573
[20] Thomas B S and Gupta R Ch 2015 *J. Clean. Prod.* **102** 78–87
[21] Eldin N and Senousi A 1993 *Cem. Con. Aggr.* **215** 74–84
[22] Gesoglu M and Guneyisi E 2007 *Mat. Struct.* **40** 953–964
[23] Ganjian E, Khorami M and Maghsoudi A A 2009 *Con. Build. Mat.* **23** 1828–1836
[24] Guleria S P and Dutta R K 2016 *Proc. Envi. Sci.* **35** 629–638
[25] Yilmaz A and Degirmenci N 2009 *Was. Man.* **29** 1541–1546
[26] Swift M J, Bris P and Horoshenkov K V 1999 *App. Acous.* **57** 203–212

Assessment of damp and salinity of historical building in Markušovce

E Terpakova and A Estokova

Institute of Environmental Engineering, Faculty of Civil Engineering, Technical University of Kosice, Vysokoskolska 4, 042 00 Kosice, Slovakia

Abstract. This article deals with diagnostics and current state of historical building in the grounds of the manor house Markušovce, eastern Slovakia. In the introduction, the problems of historical building, materials and structures are discussed. The next part is devoted to assessment of the selected structures, moisture survey and salinity above all. Determination of salinity shows high concentrations of chlorides, in part due to maintenance of local roads, which is in the immediate vicinity of the building.. Also, the high damp content of the masonry was confirmed due to the lack of waterproofing as well as high levels of groundwater. Near the historical building is a river Hornád that this fact suggests. The proximity of the park with trees and grassing turn causes shading of western and northern facade parts. The damp and shading of façade parts contribute to bio-corrosive manifestations, increased growth of moss and algae on the façade, and also confirm corrosion of masonry evidenced by higher sulphate content due to biogeneous sulfuric acid. The results of diagnostics tests show, that the samples from damaged parts of façade have a higher degree of salinity, this fact confirms the need for a reconstruction.

1 Introduction

In today´s era, historical buildings are frequently having problems with damp. There are four main sources of dampness in building, rising damp, penetrating damp, water from leaks (pipes bedded) new buildings an additional problem is trapped construction water. Structural movement and consequential cracking can form obvious paths for water to penetrate a wall. Similarly, frost erosion caused by atmospheric impurities can reduce a wall's resistance to damp penetration. These phenomena are influenced by chemical attack from environment.

In the times of their construction there were no moisture insulation used to insulate the ground from the construction systems. Moisture is a major source of deterioration of technical state and corrosion processes in historic masonry constructions. Rising damp is a natural phenomenon which occurs when groundwater flows into the base parts or pore structure of a masonry construction. From practical experience it is known that many other factors may play a role regarding permeability problems in masonry. The amount of possible causes of moisture problems in historic masonry underlines the complexity of this phenomenon. Evaporation is an important factor in rising damp. The surface of an affected wall contains moisture that has risen from the ground and this moisture is then subject to evaporation. The factors controlling evaporation include: temperature, humidity, air movement and surface [1].

Also, damp in constructions has significant effect on indoor climate when the relative humidity of indoor air rises. High relative humidity creates molds at the wall surfaces, most of them at the lower places, where are the lowest surface temperatures.

Historical buildings not commonly used, such as manor house, have serious problems with molds because of poor ventilation. The subject of the article is the assessment of the current state of the former technical building in the grounds of the manor house Markušovce. Methodical procedures of diagnostics this historical building are discussed in contribution.

IOP Conf. Series: Materials Science and Engineering **385** (2018) 012058 doi:10.1088/1757-899X/385/1/012058

2. Material and Methods

2.1 Historical building in Markušovce- insights from the input inspection

Markušovce is a village and municipality in the Spišská Nová Ves District in the Košice Region of central-eastern Slovakia. Markušovce was founded in the 12th century before the Tatar invasions of the region. The village was owned by the Mariassy family from the 13th century, and many members of the family are buried in the village's Church of St. Michael. The manor house of Markušovce was built in 1643 and is now a museum, together with the Rococo belvedere (or garden house) „Dardanely", dating from 1778, which stands in its grounds; this contains a collection of musical instruments and is frequently used for concerts [1].

There is also a historical building in the grounds of the manor house, but is without making public visits available. This damaged historical building (see figure 1 - in midle part of photo), is at present time registered in the list of Monuments Board as Technical Building. The assessed building dates from the 18th century and was built in the Baroque style [1].

During the visual survey of object it was found that damp content of the internal masonry pillars reaches up to the height at around 1700 – 1900 mm. The situation was the worst at the north and west side of the technical building. On the inner surface of the plasters there were visible efflorescence and mildew occurrences especially in the higher parts of the plinth. On the outside of the walls there are still visible lichens, mosses and algae.

The biggest problem was the partially destruction of roof of building, which is source of negative influence of the water towards the walls, mortars, plasters and reason of foundations disrepair. The neighboring building was renovated aproximately 5 yers ago (see figure 1 - right side in the photo).

Figure 1. Manor house Markušovce in region Spiš (Slovakia) and assessed building *(owner photo)*.

Usually natural damp of building construction is not a technical problem. Every material in natural environment even after application contains certain amount of water on the surface or inside structure. Total content of water is closely connected with character of structure of materials, but also with the way and location of their build in. When we take this into consideration, water chemical composition due to real circulation in environment (the presence of water pollutants etc.), in the structure of building materials appears chemical and chemical - physical processes. These processes are initiated by pollutants by specific thermodynamic conditions and by time links [2]. In building constructions the most important are physical processes as soaking, damping and drying. The main types of such processes in building are soaking, damping and drying, he related degradation processes as well as effect lies in the formation of mold, growth of algae, etc. [3]. Dumped part of construction is very favorable for mushrooms.

When chemical pollutants are present in water in soluble form, after evaporation of liquid phase, on the surface is likely to build up their concentration or inside the construction in the form of crystals or new products - efflorescences. The pollutants may also participate in chemical reactions in structure of materials. Production of new products can be connected with volume changes, which are the most clear at crystallization processes [3]. All volume changes are linked with mechanical consequences which are visible in increased tension, decreasing of strength parameters or even destruction.

Figure 2. Damaged part of east side of façade, damp, salinity and biocorrosion *(owner photo)*.

2.2 Methods of diagnostic of historical building

Various diagnostics methods can be used to study parameters such as damp, mechanical properties of construction, salinity (presence of salts and efflorescences), thermal parameters and other. This way, the methods are classified according to their functioning principle, allowing choosing each technique independently of the failure diagnostics.

An investigation of the effects of humidity and salt crystallisation on masonry has been undertaken at the historical building in Markušovce. The chemical - analytical methods for the assessment of mortars and plasters of facades due to biocorrosion will be under consideration. Assessment of degradation masonry consisted of determination of moisture and salinity as combination of gravimetric and chemical analysis of obtained samples from most damaged part. It's important to understand that one method can be used for multiple problems that are present on the building. The article discusses in particular the results of the material test in terms of degradation. However, a static survey was also carried out showing that part of the building was in an emergency.

2.2.1 Determination of damp. Regarding the results of the diagnostic survey the mass of humidity (w_{mh}) samples were determined by gravimetric method, samples were weighed and placed into a laboratory drier ELOP 1200, at 105±5°C. The drying process was carried out in 3 stages. In the first step, samples are dried for 2 hours, in the second stage 6 hours, in the third stage individually up to the constant weight of the samples. After each drying, the samples were weighed. The mass humidity (w_{mh}) calculation was according to the equation (1):

$$w_{mh}=\frac{(m_w-m_d)}{m_d}\cdot 100\% \quad (1)$$

where: w_{mh} - mass humidity, m_w - mass of wet sample, m_d - mass of dray sample.

To determine the water content in structure of building material we can used different portable devices as well, but gravimetric determination is more accurate [3].

2.2.2 Determination of salinity. Masonry within the building has been contaminated with chloride salts above all, which are undergoing cycles of crystallisation - dissolution in response to fluctuations in relative humidity.

This problem which affects the durrability is related to the salt migration processes in the building construction. The presence of aggressive saltes is limiting criterion for production novadays silicates too [3]. Presence of salts in mortars is also observe and within evaluation according to WTA rules for reconstruction purposes [7] concentration of salts is expressed as salinity.

To gain quick results In situ, portable kits are used to semiquantitative determination of chlorides, nitrates and sulphates. Determination of chlorides in laboratory conditions is based on either on partial decomposition of sample by mixture of acids and following determination, or by determination of chlorides from liquid leaching of the silicate sample. Range and accuracy of these determinations is influenced by several factors. For accurate analytical determination is best to used potentiometric determination with use of specific ion selective electrode [2]. Potentiometric determination of chlorides is more useful from practical point of view in compare with clasical determination such as gravimetric method or argentometric titration.

To the determination of sulfates, the gravimetric method (precipitation of sulphates from the dissolved form of silicate samples into the insoluble form of $BaSO_4$ according to the chemical equation (2) was used most frequently. Sulphate content was calculated after ignition of $BaSO_4$.

$$BaCl_2 + SO_4^{2-} \rightarrow BaSO_4 + 2\ Cl^- \qquad (2)$$

However, the determination is time-consuming and requires, in particular, lower analytical capacity and precise monitoring of the method, especially in the case of lower sulphate concentrations, therefore, in particular, direct powder determination using the XRF method is currently used. To determine precise structural changes RTG analysis and SEM or EDX microscopy are used.

Circulation of nitrogen in environment is greatly affected by anthropogenic activities, by which is its concentration increased even in waters. Influence of nitrogen salts is observed in connection with silicates for the reasons of corrosion by leaching. According to WTA rules [7] the presence of nitrates increases salinity of mortars. At periodic changes of humidity inside construction, concentration of nitrate salts, formation new corrosive product and crystals in finale phase even destruction signs will happen. Increased concentration is probably likely for oxidation- reduction reaction of certain forms of nitrogen in natural environment reasons [2].

Nitrates can be quite fast determine by potenciometric method, with the help of ion selective electrode but disadvantage is likelihood of interference of some ions. Another analytical method for nitrates determination is spectrophotometry.

3. Results and discusion

Several physical factors affect the humidity in the masonry. The presence of salts in massonry increases the w_{mh}, because salts have hygroscopic properties [4] and these consequently also enhances the chemical degradation of the masonry. The obtained values w_{mh} (see table 1) were different for the individual materials and also according to the orientation on the world sides of the façade, essentially reflecting the long-term negative effects on the masonry construction.

Based on the experimental determination of humidity expresed by results w_{mh} and according to the criteria given in [6], was confirmed that the content of the water phase in the structure of the masonry is individually distributed. The obtained w_{mh} values in the range from 7.5% to 10% means high humidity, values above 10% very high humidity respectively, which are created in any case suitable conditions for masonry biodegradation. Biocorosion of facades effect the quality of building envelope causes the changes with integrity, chemical composition, mechanical properties, thermal characteristics, fire resistance and changing aesthetic properties and appearance. These serious issues need to be monitored.It is possible to conclude, that the long-term negative effects of the environment as well as the way of maintenance appear on the present technical state of masonry structure.

Determination of chlorides c(Cl) concentration, (as selected representative salts) was provided by potentiometric method using selective chloride electrode. Alkality changes of samples were expressed as pH values, the potentiometric method was chosen for their estimation. Table 1 shows the comparison of the average results of stone samples and bricks taken from facade at the same height from the ground about 300 mm.

The results of diagnostics tests show, that the samples from damaged parts of façade have a higher degree of salinity, this fact confirms the need for a reconstruction. The samples taken from the stone-redominantly porous sandstone are able to absorb humidity and consequently salts are deposited into wall of pores. The presence of salts has also been observed at the surface of the sampling points as white efflorescences. Increased salt concentrations are related to the location of historical building, as we can see from figure 1 and figure 2, the building is located in the immediate vicinity of local communication where salt is used for winter maintenance.

Except to chlorides, higher sulphate concentrations were also detected in the samples using semiquantitative analysis by $BaCl_2$ precipitation of sulfate anions. In addition to above mentioned method was made FTIR spectral analysis of sample by ATR FTIR spectrometer Alpha Brucker, in range 4000-360 cm^{-1}, with 24 scans. The results of white salts sample taken from the east part of the facade is shown in figure 3.

Table 1. Results of analytical determination of mass humidity and salinity.

Orientation of facade	Type of sample	w_{mh} ±0.01%	pH ± 0.01 [-]	c(Cl)± 0.01%	Degree of salinity [7]
east	stone	5.64	9.11	0.285	high
	brick	10.64	8.95	0.334	high
south	stone	3.64	10.42	0.276	high
	brick	5.58	9.14	0.380	high
west	stone	8.16	7.83	1.74	very high
	brick	11.71	8.25	1.15	very high
north	stone	7.42	6.95	0.85	very high
	brick	16.45	8.02	2.47	very high

To simplify the analysis of the results of the vibration spectrum is presented in three main regions relating to: vibrations of the sulfate ions 500-1300 cm^{-1}, bending vibrations of water molecules 1610–1660 cm^{-1} and the water stretch region 3000-3800 cm^{-1}[2]. IR spectrum shows the intensive peak of a wave number of 1086 cm^{-1} corresponds to $CaSO_4$ due to sulphate degradation [4].

In figure 3 three characteristic FTIR transmitance spectra samples obtained from east facade are presented. The spectra from sample brick (2) and sample mortar (3) have dominated the characteristic peaks of calcite (1400 cm^{-1} and 880 cm^{-1}) while a weak broad band of approximately 1100 cm^{-1} is an indication of silicate (-Si-O) vibrations, which are overlapped the peak of the anion $-SO_4^{2-}$ in both cases.

Finally the spectrum obtained from sample salt (1) has dominated the characteristic peaks about 1650cm^{-1}, 1064cm^{-1} respectively at 620cm^{-1} which are an indication of sulhpates [5], guiding us to conclude that the sample contains products of bio-corrosion processes, because on the historical masonry cement was not used as a binder.

Although IR spectroscopy is one of the most widely used physical methods of investigation of molecular structure, information about the crystal's structure can be useful to interprete the IR spectra established by X-ray diffraction methods.

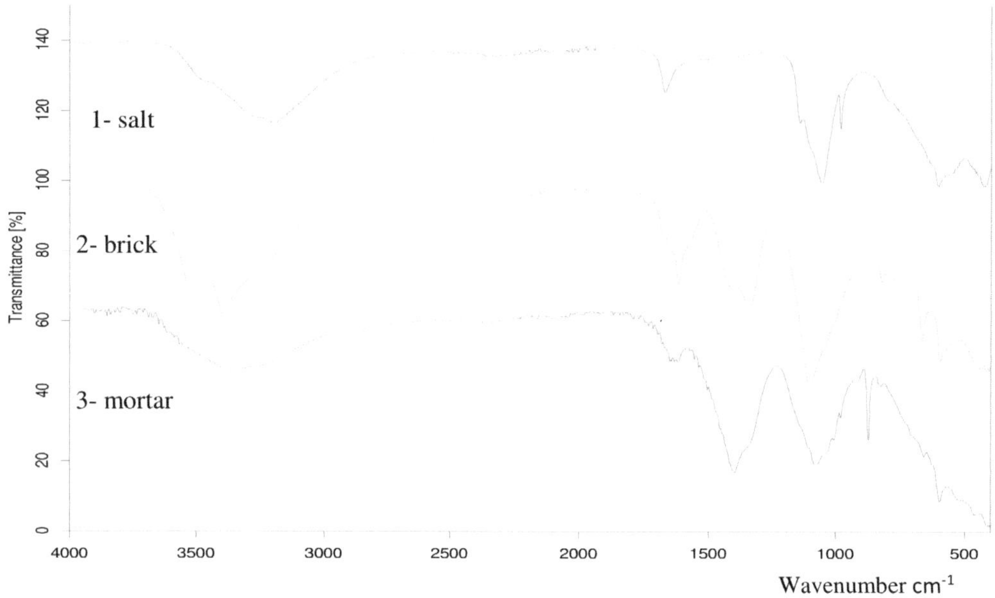

Figure 3. FTIR spectra of samples.

4. Conclusion

The results of diagnostic assays show, that the samples from damaged parts of façade have a higher degree of salinity, this fact confirms the need for a reconstruction. Determinations of humidity and salinity of construction parts are very important aspects from durability and reliability building. Damaged parts necessarily require urgent reconstruction because the object is in several parts of statically disturbed. When choosing technologies and materials, it must also be taken into account that it is a historic building, each other step should be in accordance with the requirements of the monumental authority under which the object belongs.

Refurbishment of historical building also must be taking into account:

- the general requirements for the protection of other historical monuments and environment in areal of castle Markušovce,
- selection and application of suitable remedial materials compatible with historical masonry.

At the same time, account must be taken on the technical needs and social requirements for the use of building during the reconstruction, cultural and tourist activities and events in areal of manor house of Markušovce.

Acknowledgements

The work was supported by Project VEGA 2/0145/15.

References

[1] online: http://www.pamiatky.net/kastiele-a-kurie/markusovce-kastiel
[2] Balik M 2008 *Dehumidification of buildings* (Praha: Grada)
[3] Terpakova E 2016 *Diagnostics in civil construction* (Košice: TUKE)
[4] Matousek M and Drochytka R 1998 *Atmospheric corrosion of concretes* (Praha: Ikas)
[5] Wasserbauer R 2000 *Biological damaged of buildings* (Prague: ABF a s)
[6] ČSN P 730610:2000 *Waterproofing of buildings The rehabilitation of damp masonry and additional protection of buildings against ground moisture atmospheric water- Basic provision* (In Czech)
[7] WTA 4-6-98:2004 *Sealing of structural elements on contact with soil at a later state* (in Czech)

The study of capillarity properties of bio-composites

E Terpakova and N Stevulova

Institute of Environmental Engineering, Faculty of Civil Engineering, Technical University of Kosice, Vysokoskolska 4, 042 00 Kosice, Slovakia

Abstract. Composite materials demanded in various technical areas as well as civil engineering, have important influence on the progress in research and development of materials. Among other things, composites materials based on organic fibers, obtained from the processing of natural raw or as a product of recycling, have come to the fore in the last years. The results of the comparative study of the physical parameters (density, thermal conductivity coefficient and water absorption behaviour) of the bio-based composite are presented after 28 days of setting and hardening. The ability of the material to receive and retain water in the pore system is also very important, so parameters such as absorption and capillary absorption are also studied. Results are discussed in terms of impact of individual content of cellulosic fibers in each other properties of composite materials. Density results showed an improvement in the values with the increasing weight % fiber in composites for both types of fiber. Also, an improvement in thermal conductivity has been demonstrated, confirming the possibility of applying composites to the insulating elements. However, the results of absorption and capillary absorption imply the need to optimize the technology of preparation and surface treatment of composites.

1. Introduction

The increased quality requirements in the construction industry are also associated with the requirements for new materials. Every year, thus it is expanding the range of products that are designed to meet the requirements of consumers. In the current civil engineering, efforts are also being made to use non-traditional materials, or attention is drawn to renewable sources as hemp, cotton, line, coconut fibers etc.. Renewal of scientific as well as industrial interests in the use of cellulosic fibers as reinforcing in lightweight composite materials relates to a need of progress of environmental friendly products with high use value in term of sustainable development [1]. Due to low density, bio-degrability, interesting thermal, mechanical, acoustic and other properties of natural fibers, low cost and eco-friendly raw material, this natural fibrous material is used as a (partial) replacement of synthetic fibers, such as glass, carbon or metallic fibers [2]. The applying of natural fibers in cement based composites is narrowly linked to the ecological building sector, where a choice of materials is based on components including recyclable, renewable raw materials and low-resource manufacture [3]. Application of cellulosic fibers into cement composites must be thoroughly studied, because as it is generally known, the surface of the cellulosic materials has a hydrophilic character [4]. The hygroscopic properties expressed by absorbability and capillary water content are as basic characteristics of water transport processes into material, and recently they has been recognized as an important performance characteristic of durability materials as well [5]. This fact can be reflected both in the preparation and mixing of fresh composites, also after hardening of mixtures to their final shape. This paper is focused on specific approach to composite preparing as a way of improvement of organic filler as particulary substitution of natural sand. The aim of present study is to investigate the utilisation of cellulosic fibers from wood pulp and waste paper as reinforcing element with insulating properties into cement composites f. e. as mortars for plastering. The major problems with the use of cellulosic fibers (both natural and recycled types) in cement composites are their high water absorption capacity, porosity and lower strength. Several research reports can be recently found worldwide, dealing with various surface treatments of fibers [2] in order to improve their surface properties. The results of the comparative study of the physical parameters (density, thermal conductivity coefficient and water absorption behaviour) of the bio-based composite are presented. Changes in capillary moisture content of composites are compared and discussed. The ability of the material to receive and retain water in the pore system is also very important, so parameters such as absorption and capillary absorption are also studied. In article these parameters were investigated after 28 days of setting and hardening. Results are discussed in terms of impact of individual content of cellulosic fibers in each other composite material. The composite durability will also be derived from the values obtained.

2. Materials and methods

2.1 Materials

The cement used in this study was Portland cement CEM I 42.5 N (Cement factory Ladce, Slovakia). Silica sand (as filler), was obtained from Šaštín - Stráže, Slovakia, (fraction size: 0/1mm). Its properties were in accordance with European standard STN EN 196-1 [7]. The two types of cellulosic fibers used in this study were - bleached wood pulp (typ GW 500), indicated in the text as sample "B" and unbleached - recycled fibers (typ G-500T), indicated in the text as sample "W" obtained from different waste papers. Both cellulosic fibers were provided by company Greencel Ltd. (Hencovce, Slovakia). Table 1 shows the basic characteristics of used cellulosic fibers. Deionised water was used in all experiments as mixing water as well as curing water. Parameters of water were in accordance with STN EN 1008 [8].

Table 1. Physical and chemical characteristics of cellulosic fibers (Greencel).

	Type of cellulosic fibers	
Characteristics	GW-500 (bleached wood pulp)	G-500 T (unbleached recycled fibers)
Cellulose content [%]	99.5	80
Bulk density (kg/m^3)	60-80	50-100
Max. length (μm)	500	400
Dry matter (%)	93	93
Ash (%)	0.5	0.5
pH value (-)	6 ± 1	7 ± 1

2.2. The course of experiments

The experimental cellulosic fibers reinforced composites were prepared using normal type Portland cement CEM I 42.5 N, silica sand and water. Cellulosic fibers from wood pulp and waste paper were used as particularly substitution of inorganic filler (5, 10, 15, and 20 wt. % from weight of filler). The mix proportions of specimens were designed with Cement/Sand (C/S) weight ratio of 1:3 and Water/Cement (W/C) ratio of 0.70. Reference cement composite was manufactured without using of cellulosic fibers. Cube samples of 100mm x100mm x100 mm were prepared and cured under standard conditions. All of the experimental mixtures were obtained in a laboratory by using a laboratory mixer.

The density, coefficient of thermal conductivity of composites with various content of cellulosic fibers (5, 10, 15 and 20 wt. %) were investigated after 28 days of hardening. Density was calculated from the weight and the determined dimensions of the individual samples.

The thermal conductivity coefficient of samples as the main parameter of heat transport λ ($W.m^{-1}.K^{-1}$) was measured by the commercial device ISOMET 104 (Applied Precision Ltd., Germany). The determination is based on the analysis of the temperature response of the analyzed material to heat flow impulses. The heat flow is induced by electrical heating using a resistor heater having direct thermal contact with the surface of the sample.

Determination of absorbability was performed by immersing the dried samples into a container with deionised water at 20 ±2°C. The evaluation of absorbability was carried out gradually for 1, 6, 12 and 24 hours, when the pores of the composites were filled completely with water. Total water absorbability of composites is represented in table 3 (evaluation after 24 hours testing). This way is in contrast to the determination of total water absorbability (when samples are immersed completely into water bath).

In addition, the capillary water content was tested after 28 days of hardening. Samples (cubic type 100mm x 100mm x 100mm) were dried up to constant mass. In case of capillarity determination, samples are immersed in water at a depth of 5–10 mm for a specific period of time only by defined area of sample (exactly specified in mm^2). The test procedure consists of recording, at defined time intervals, the mass increment by capillary suction of a specimen with contact surface of samples with water, up to constant weight difference between two consecutive weightings less than 0.1%.
For this experiment, measurements were done in one step in the following intervals: 300s, 600s, 900s, up to 3600s, consequently at next measurements to show the development of this parameter in time up to 7 days. All procedures were performed in accordance with standards rules [9], [10]. The values of the coefficient of the capillary water absorption were calculated by the equation (1):

$$c = \frac{(m_{ti} - m_{t0})}{A_s} \quad (1)$$

where:

c - coefficient of the capillary water absorption
m_{ti} - specimen mass after the process of soaking (g) in time t = i (s)
m_{t0} - specimen mass after the process of drying (g) in time t = 0 (s)
A_s - cross area of the specimen surface immersed in water (m^2)

3. Discussion of results

3.1. *Density and thermal conductivity of fibers cement composites*

Table 2 reports the values of important parameters such as density and thermal conductivity of fiber cement composites as well as reference composite. While density of reference sample after 28 days of hardening reached 2110 kg/m^3, specimens with cellulosic fibers GW-500 (sample B) in an amount of 0.5 wt. % to 20 wt. % as a filler substitute, showed the decrease in density 6.16% up to 8.53% compared to the reference sample. The lowest values of density composites were determined in case of composites on the base G500-T (unbleached recycled fibers) (sample W). There was observed reduction in density of composites from 3.32% up to 4.98% compared to the reference sample (see table 2). Reduction in density values is a positive fact, it confirms the possibility of composites using as insulating element of constructions. To reduce the density of composite has also technological importance, because it can improve the workability of fresh mixture, when using composite as mortar for construction.

Regarding the evaluation of thermal conductivity values it can be stated that there is in all the samples a decrease in average values compared to the reference sample. The lowest value of thermal conductivity (0.291W/m.K) and its maximal reduction (30.38 %) of 28 days hardened fiber cement composites was found in case of the sample 20% W in comparison with reference sample (0.418 W/m.K). The remaining detail results of samples "W" and their comparison with reference sample are in table 2. Reducing the coefficient of thermal conductivity with the increasing content of cellulose fibers was also observed in samples based on bleached cellulose samples (B), however this reduction, compared with reference sample was 12.44 % (sample 5%B) in case sample 20B% it was 21.77% . Generally, the fiber additions into matrix have positive impact on density and thermal conductivity of fibers cement composites, as well. Incorporation of cellulosic fibers into structure of composites creates the voids and the heat-insulating properties are enhanced when density decreases.

Table 2. Density and thermal conductivity values of cement composites.

Sample	Density		Coefficient of thermal conductivity	
	Average values ±5 ($kg.m^{-3}$)	Change (%) / decrease of values	Average values ± 0.001(W/m.K)	Change (%)/ decrease of value λ
Reference	2110	100/0	0.418	100/0
5% B	1980	93.84/ -6.16	0.366	87.56/ - 12.44
10%B	1945	92.18/ -7.82	0.361	86.36/ -13.64
15%B	1935	91.71/ -8.29	0.358	85.64 / -14.36
20%B	1930	91.47/ -8.53	0.327	78.23 / -21.77
5% W	2040	96.68/-3.32	0.350	83.73/ -16.27
10%W	2030	96.20/-3.80	0.332	79.43 / -20.57
15%W	2015	95.49/-4.51	0.304	72.73 /-27.27
20%W	2005	95.02/ -4.98	0.291	69.62 / -30.38

3.2 Assessments of hygroscopic properties of composites

Generally, the assessments of the hygroscopic properties of building elements - composites are of considerable importance also from the service life point of view [6]. They also relate to parameters such as frost resistance or durability, and hence the location and method of composites application in building constructions. From hygroscopic parameters has focused attention on assessment of water absorbability of composites, being for technologically application of composites is important, whether it is the total water absorption or capillarity.

3.2.1 Determination of absorbability. The results of total absorbability show the expected increasing of amount of water in the composites, both over time and with the increasing content of cellulose fibers as well (table 3). Increasing the water absorbability also may be caused by rising of fiber amount in cement matrix because cellulosic fibers have hydrophilic character [1].

3.2.2 Capillary absorption testing of composites. The data on the capillarity properties obtained from all of the produced composites (with 5%, 10%, 15% and 20% GW-500 fibers mixes (B), and 5%, 10%, 15% and 20% of G-500T fibers mixes (W) are shown graphically. The results of the determination of capillary absorption at the time up to 1 hour for individual mixes composites (W) are shown in figure1 and composites (B) in figure 2. A comparison of the development capillary absoption results shows, that composites (B) with a lower % fiber replacement of filler exhibit lower values over the reference sample except sample 20% B (with density 2005 kg/m^3).

Regarding the results of the capillary absorption for the composites (W), see figure 1, that a growing share of fibers as a substitute filler occurred seal the pores, thereby affecting the ability of soaking. This knowledge is very important in terms of the intended use of composites as a plaster. In further experiments, long-term capillary absorption also will be tested from the point of view of the durability of the composites.

Table 3. Water absorbability of composites after 24hours testing.

Sample	Absorbability ±0.01 %	% changes
Reference	8.85	0
5% B	9.85	+11.29
10% B	11.21	+26.66
15% B	11.24	+27.05
20% B	11.93	+34.80
5% W	10.01	+13.10
10% W	10.36	+17.06
15% W	10.42	+17.18
20% W	10.83	+22.37

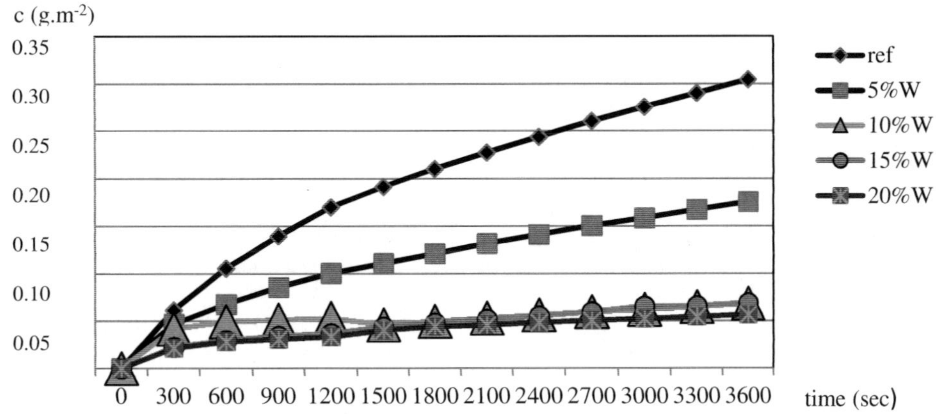

Figure 1. Results of the capillary absorption test (1 hour) mixes of composites (W).

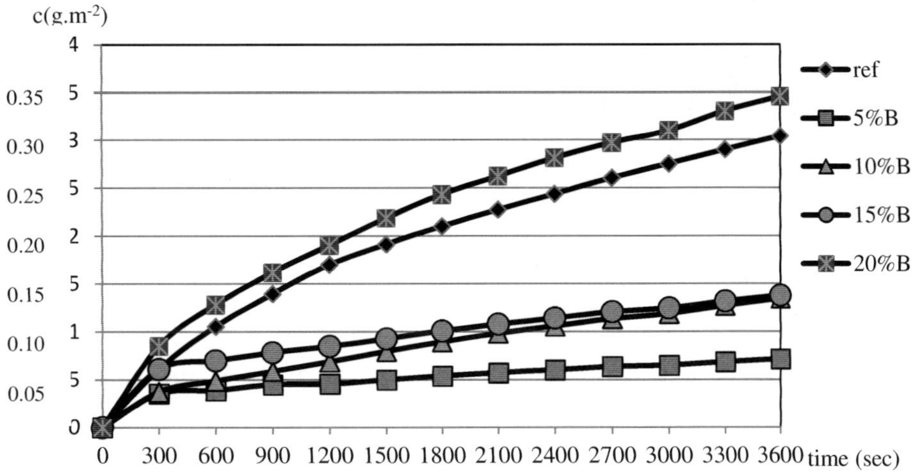

Figure 2. Results of capillary absorption test (1 hour), mixes of composites (B).

4. Conclusion

The quality and durability of composites is strongly dependent on the kind, properties and dosage of the basic components. The cellulosic fibers due to their low density, thermal, mechanical and other technical properties are used as a partial replacement of inorganic filler or another synthetic fiber into cement composites. It is clear, that the technological procedures of composites preparing or the way of coating material can influenced the terminal properties of products from durability point of view. The capillary water content as well as absorbability are important parameters which must by take in count, because of increasing water content in construction threatens to create disorder, failures, up to an emergency.

Acknowledgment

The results presented in this article have been carried out within the project of Slovak Scientific GrantAgency VEGA grant number 1/0277/15.

References

[1] Stevulova N, Cigasova J, Estokova A, Terpakova E, Geffert A, Kacik F, Singovszka E and Holub M 2014 *Materials* **7** 8131-8150
[2] Terpakova E, Stevulova N, Kidalova L and Estokova A 2012 *Proc. Eng.* **42** 931 - 941
[3] Hospodarova V, Stevulova N, Vaclavik V, Dvorsky T and Briancin J 2017 *Proc. Int. Conf. on Environmental Engineering (Vilnius)* (Vilnius: Gediminas Technical University Lithuania) pp 1-8
[4] Terpakova E, Stevulova N, Sicakova A and Schwarzova I 2014 14th *Proc. Int. Conf. Multidisciplinary Scientific GeoConference SGEM (Albena)* vol 2 (Sofia: STEF92 Technology Ltd) pp 193-198
[5] Kidalova L, Stevulova N and Terpakova E 2015 *Pollack Periodica* **10** 123-132
[6] Urban K, Sicaková A 2018 *International Journal of Research in Engineering and Science* **6** 35-39
[7] STN EN 196-1:2016 *Methods of testing cement Part 1: Determination of strength* (in Slovak)
[8] STN EN 1008:2003 *Mixing water concrete. Specification for sampling, testing and assessing the suitability of water, including water recovered from processes in the concrete industry, as mixing water for concrete* (in Slovak)
[9] STN EN 480-5: 2006 *Admixtures for concrete, mortar and grout Test methods Part 5: Determination of capillary absorption* (in Slovak)
[10] STN EN 1015-18: 2003 *Methods of test for mortar for masonry Part 18: Determination of water absorption coefficient due to capillary action of hardened mortar* (in Slovak)

IOP Conf. Series: Materials Science and Engineering **385** (2018) 012060 doi:10.1088/1757-899X/385/1/012060

Differences in the accuracy of test methods in the field of testing aggregate and determination of their causes

R Uxa and T Vymazal

Department of Building Testing, Faculty of Civil Engineering, VUT in Brno, Veveri 95, Brno, Czech Republic

Abstract. A The article is based on the authors' many years of experience in providing laboratory proficiency testing including organization of proficiency testing by interlaboratory comparison. In the introduction are compared the basic tests in the field of aggregates, based on their coefficient of variation. Another part details the influence of each testing stage (sampling, sample preparation and the test itself) on the accuracy of the test results and proposes possible measures to reduce their variance.

1. Introduction

The main goal of laboratory testing is to determine with adequate accuracy the desired properties of materials and structures. This accuracy then decides subsequent designs, calculations, evaluations, or result interpretations, which affect the general outcome of a construction or its part. This is why it is necessary to limit or eliminate all parts of a testing procedure that could have a negative impact on the accuracy of the results.

The proportioning of the stages of a particular testing procedure can be ascertained from data obtained in proficiency testing programmes aimed at testing aggregates and organised by ASPK, s.r.o. Brno and by the Proficiency Testing Provider at SZK FAST, Brno University of Technology between 1995 and 2017. Their statistical evaluation is performed in accordance with CSN ISO 5725 "Accuracy (trueness and precision) of measurement methods and results"[11].

The paper uses the following definitions:

- Accuracy: the closeness of agreement between a test result and an accepted reference value.
- Trueness: the closeness of agreement between the m value of a great number of test results and an accepted reference value μ.
- Bias: the difference between the mean value of test results m and an accepted reference value μ,
- Precision: the closeness of agreement between independent test results under pre-defined conditions (reproducibility).
- Reproducibility: precision under reproducibility conditions.
- Reproducibility conditions: conditions under which test results are arrived at by the same method, using the same test specimens, in different laboratories, and by different testers using different equipment.

These definitions allow the acceptably simplified statement that the quantity which determines the accuracy of a test method is its reproducibility. It is, however, not possible to compare the values of reproducibility without taking the reference value into account.

A comprehensive evaluation was thus performed using the reproducibility coefficient of variation defined by CSN ISO 3534-1 "Statistics - Vocabulary and symbols - Part 1: General statistical terms and terms used in probability" [13] as a ratio of reproducibility to the mean value of the quantity being measured.

Figure 1 below shows the values of variation coefficient for selected aggregate tests measured as part of proficiency testing programmes from 1996 to 2017.

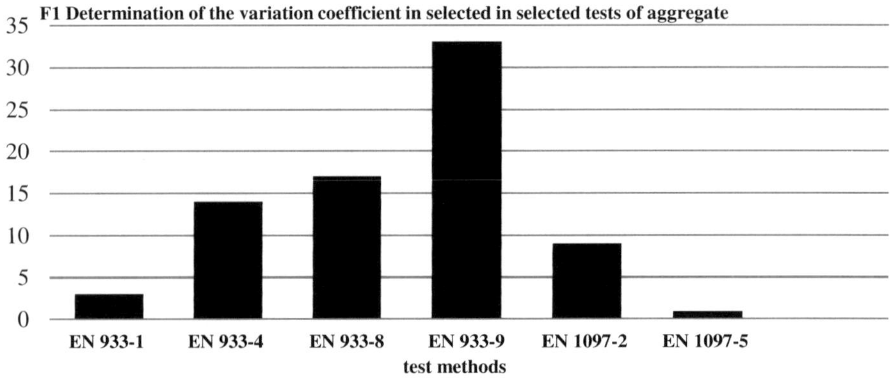

Figure 1. Determination of the variation coefficient in selected tests of aggregate.

The following section describes the influence that each stage of testing (sampling, specimen preparation, and the testing itself) has on the accuracy of test results.

2. Division of testing activities

Laboratory testing encompasses all activities connected with carrying out tests including sampling and specimen preparation.

2.1. Sampling

Sampling is a standard procedure, during which a part of the material or product is taken so that it provides a representative sample for the purposes of testing CSN EN 932-1 [4].

2.2. Sample preparation and adjustment

Sample preparation is a procedure during which the representative sample is adjusted (homogenized and reduced) so as to achieve a test (laboratory) sample, see CSN EN 932-2 [5].

Table 1. Percentage of accredited sampling and sample preparation in testing laboratories.

Activity	Standard	Percentage of testing laboratories to perform said activity
Sampling	Tests for general properties of aggregates Part 1: Methods for sampling CSN EN 932-1	75 %
Sample preparation	Tests for general properties of aggregates Part 2: Methods for reducing laboratory samples CSN EN 932-2	10 %

2.3. Testing

Testing is a procedure during which the test sample is tested for the required or observed properties. This activity has been extensively explored and documented, which is why this paper does not address it in detail.

3. Determination of the influence of individual operations on the overall test result

According to CSN ISO 11648-1 [12] each process has its own variance component:

- the variance component of sampling caused by the taking of partial samples,
- the variance component of sample preparation produced during adjusting and reducing the sample,

- the variance component of measurement characterising the precision of the method used.

The percentage of each component for the overall result of a test method is quantitatively determined using differences between the proficiency testing programmes from each of the above-listed providers. The differences lie in the way of sample preparation, making it possible to estimate the percentage of each component in the overall result. The specification of the influence of sample preparation by homogenization and reduction on the determination of particle size distribution was supplemented by the authors' own measurement in cooperation with laboratories at COLAS CZ, a.s. and M.I.S. a.s. The results are discussed in section 4.

3.1. Determination of particle size distribution - sieving method

The testing of particle size distribution according to CSN EN 933-1 [6] underwent a change in 2005 in the way samples are prepared within proficiency testing programmes, where the test sample is centrally prepared from narrow fractions of aggregate and supplied to the laboratories at an amount equal to a standard testing batch (2.63 kg for fraction 0/16). Thanks to this, the variance components of a) sampling and b) sample preparation are eliminated. Figure 2 shows the difference.

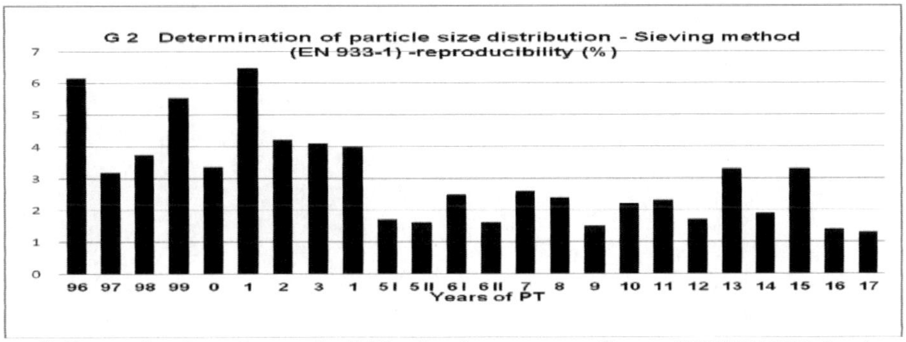

Figure 2. Determination of reproducibility for the test according to CSN EN 933-1 using a 4 mm sieve.

3.2. Determination of particle shape - shape index

The determination of particle shape - shape index according to CSN EN 933-4[7] entailed comparing the results of standard IC with IC from 2008 and 2017, which used identical samples - the variance components a) and b) were thus eliminated. Figure 3 shows the difference.

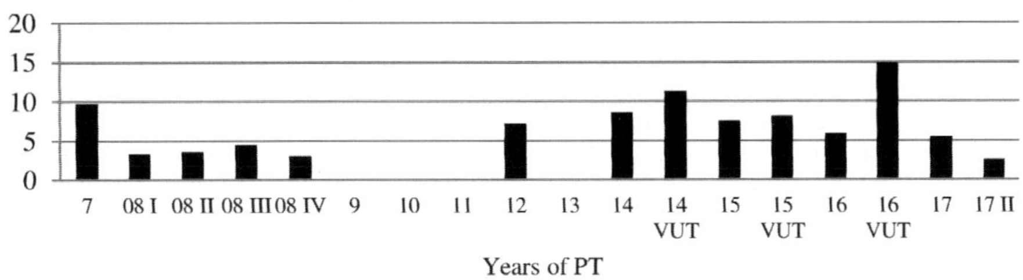

Figure 3. Determination of reproducibility for the test according to CSN EN 933-4 using fraction 8/16

Note: Proficiency testing programmes implemented thus far still lack sufficient knowledge and documentation for a possible determining of the influence of the components of the testing process for tests according to CSN EN 933-8 [8], CSN EN 933-9 [9], CSN EN 1097-2 [10] and CSN EN 1097-6 [11].

IOP Conf. Series: Materials Science and Engineering **385** (2018) 012060 doi:10.1088/1757-899X/385/1/012060

4. Determination of the influence of sample preparation by homogenization and reduction for determining particle size distribution

The next step for eliminating the influence of variance components is ascertaining the percentage of influence that the variance component of sampling and variance component of sample preparation have. This step was carried out only for the determination of particle size distribution CSN EN 933-1[6]. Two laboratory samples (fraction 0/16, mass 2.640 kg) were mixed to form a single sample (fraction 0/16, mass 5.280 kg). A laboratory sample was then taken from it by two ways:

- reduction and homogenization according to CSN EN 932-2 [5],
- by simply taking a sample.

4.1. Homogenized samples

Table 2. Sieve analysis of the homogenized laboratory sample, fraction 0/16.

Lab. no./sieve	16	8	4	2	1	0.50	0.25	0.125	0.063
A/1	100	59	30	20	12	8	4	2	1.4
A/2	100	59	31	19	12	8	5	3	1.9
A/3	100	58	31	20	12	9	5	3	1.9
A/4	99	59	30	19	11	8	4	3	1.6
m	99.8	58.8	30.5	19.5	11.8	8.3	4.5	2.8	1.7
s. d.	0.50	0.50	0.58	0.58	0.50	0.50	0.58	0.50	0.24
B/5	99	59	30	19	13	8	4	3	*1.5*
B/6	99	59	30	20	12	8	4	3	*1.4*
B/7	99	59	31	20	12	8	4	3	1.7
B/8	99	58	30	20	12	8	4	3	1.5
m	99.0	58.8	30.3	19.8	12.3	8.0	4.0	3.0	1.5
s. d.	0.00	0.50	0.50	0.50	0.50	0.00	0.00	0.00	0.13
C/9	100	59	30	20	12	9	5	4	*2.1*
C/10	100	58	31	19	12	8	5	3	*1.9*
C/11	100	59	31	19	12	8	5	3	1.9
C/12	100	58	31	20	12	8	5	3	1.9
m	100.0	58.5	30.8	19.5	12.0	8.3	5.0	3.3	2.0
s. d.	0.00	0.58	0.50	0.58	0.00	0.50	0.00	0.50	0.10

Legend m general mean value s.d. standard deviation

4.2. Non-homogenized samples

Table 3 Sieve analysis of a non-homogenized sample, fraction 0/16.

Lab. no./sieve	16	8	4	2	1	0.50	0.25	0.125	0.063
A/1	97	57	31	19	12	8	5	4	*1.3*
A/2	98	58	28	20	13	8	5	3	1.5
A/3	100	57	30	18	12	7	3	2	0.6
A/4	97	56	29	19	12	8	5	3	1.5
m	98.0	57.0	29.5	19.0	12.3	7.8	4.5	3.0	1.2
s. d.	1.41	0.82	1.29	0.82	0.50	0.50	1.00	0.82	0.43
B/5	100	57	30	19	12	8	5	3	*1.6*
B/6	97	57	29	19	11	8	5	3	*1.8*
B/7	99	58	31	20	13	7	5	2	1.4
B/8	98	58	30	19	12	8	5	3	0.5
m	98.5	57.5	30.0	19.3	12.0	7.8	5.0	2.8	1.3
s. d.	1.29	0.58	0.82	0.50	0.82	0.50	0.00	0.50	0.57
Lab. no./sieve	16	8	4	2	1	0.50	0.25	0.125	0.063

C/9	99	54	26	18	11	7	4	3	2.0
C/10	99	58	30	20	13	8	5	4	1.0
C/11	98	54	26	18	11	7	4	3	1.4
C/12	99	56	28	20	13	8	5	4	1.6
m	98.8	55.5	27.5	19.0	12.0	7.5	4.5	3.5	1.5
s. d.	0.50	1.91	1.91	1.15	1.15	0.58	0.58	0.58	0.42

Legend m general mean value s.d. standard deviation

5. Result discussion

Available literature [16-18] does not address this question. We believe that this is largely due to the fact that the test sample batch was approximately twice the size; in the case of fraction 0/16, it was at least 5 kg, see CSN 72 1172 [15] compared to 2.6 kg according to CSN EN 933-1 [6]. The number of measurements is also different: CSN 72 1172 [15] requires two, CSN EN 933-1[6] considers one measurement to be enough.

Until quite recently, one test of aggregate particle size distribution (fraction 0/16) thus required a minimum of 10 kg, today it is only 2.6 kg; i.e. about one quarter. Such a significant reduction in testing sample size causes both preparatory stages (sampling and later homogenisation) to have a much more dramatic effect on the test result. Data shown in graphs G2, G3, and G4 can be used to partially determine the percentage of each variance component in the overall result; at the moment, however, only the influence of the variance component of the test itself and the compound effect of the variance component of sampling and sample preparation. The results are shown in table 4.

Table 4. Determination of the influence of each variance component on the overall result.

Test method	R_{normal}	$R_{reduced}$	SR_{a+b+c} %	SR_{c} %	SR_{a+b} %
Particle size distribution ČSN EN 933-1	4.5	2.1	100	**46**	**54**
Determination of the shape index SI ČSN EN 933-4	9.1	3.6	100	**39**	**61**

Legend
R normal reproducibility during normal testing %
R reduced reproducibility with elimination of sampling and laboratory sample preparation %
SR a+b+c influence of variance components a+b+c (total) = 100%
SR c influence of variance components c (test itself)
SR a+b influence of variance components a+b (sampling and sample preparation)

Data shown in tables 2 and 3 can be used to determine the percentage of the influence of sample homogenization on the overall result. The result is in table 5.

Table 5. Determination of the influence of sample homogenization on the overall result.

Sieve (mm)	8	4	2	1	0.5	0.25	0.125	0.063	average
R_H	1.59	1.48	1.65	1.20	1.17	1.60	1.39	0.71	1.35
R_N	4.15	4.98	2.64	2.65	1.56	1.90	1.85	1.43	2.65
R_H/R_N %	38	30	63	45	75	84	75	50	51

Legend
R_H reproducibility of the homogenized sample %
R_N reproducibility of the non-homogenized sample %

The average reproducibility of tests performed on the non-homogenized sample is 2.65 %, while tests with the homogenized/reduced sample returned a value of 1.35 %.

Using this data helps to clarify the influence (variance components) of each operation on the overall test result. The results are listed in table 6.

Table 6. Determination of the influence of each operation on the test result according to EN 933-1[6].

Process	methodology	influence %
Sampling	CSN EN 932-1	27
Laboratory sample preparation	CSN EN 932-2	27
Testing	CSN EN 933-1	46

6. Conclusion

The above-presented data shows that the influence of sampling and subsequent preparation (homogenisation) of a test sample on the test result is considerable and is often neglected in contemporary laboratory testing. The actual testing receives far more attention, which, however, takes only a 50 % part in the accuracy of the result. It is probable that in other tests, or areas of testing, this ratio could be different, but in no case is it strongly significant (100 %). It must be noted that making an effort to improve this (i.e. reducing the variance components of reproducibility for the whole process) carries no extra material costs and requires no time-consuming and complicated testing procedures. All that is necessary is to thoroughly follow existing standards. As concerns meeting the requirements of [18] and [19]; i.e. standards on Factory Production Control, the scope of further required tests may be limited. Consistent application and compliance with the proposed model will improve the accuracy of test results. This will allow for more precise design of components (such as road layers) or entire structures. With the annual cost of construction in the order of CZK 100 billion, improving the precision of a testing/monitoring activites, resulting savings of 1% can be considered very significant.

Acknowledgements
This paper was written as part of the project FAST-S-17-4736.

References

[1] Uxa R and Kralikova M 2015 *Proc. Conf on Asphalt Road* (Ceske Budejovice) (Prague: Sdruzeni pro vystavbu silnic) (in Czech)
[2] Uxa R 2016 Proc. *Conf. on Testing and Quality in Construction* (Brno) (Brno: VUT Brno) (in Czech)
[3] ASPK 1996 - 2017 *Results of proficiency testing programs* Report
[4] CSN EN 932-1:1997 *Tests for general properties of aggregates - Part 1: Methods for sampling*
[5] CSN EN 932-2:2000 *Tests for general properties of aggregates - Part 2: Methods for reducing laboratory samples*
[6] CSN EN 933-1:2012 *Tests for geometrical properties of aggregates - Part 1: Determination of particle size distribution - Sieving method*
[7] CSN EN 933-4:2008 *Tests for geometrical properties of aggregates - Part 4: Determination of particle shape - Shape index*
[8] CSN EN 933-8+A1:2016 *Tests for geometrical properties of aggregates - Part 8: Assessment of fines - Sand equivalent test*
[9] CSN EN 933-9+A1:2013 *Tests for geometrical properties of aggregates - Part 9: Assessment of fines - Methylene blue test*
[10] CSN EN 1097-2:2010 *Tests for mechanical and physical properties of aggregates - Part 2: Methods for the determination of resistance to fragmentation*
[11] CSN ISO 5725-2:1996 *Accuracy (trueness and precision) of measurement methods and results - Part 2: Basic method for the determination of repeatability and reproducibility of a standard measurement method*
[12] CSN ISO 11648-2:2003 *Statistical aspects of sampling from bulk materials - Part 2: Sampling of particulate materials*
[13] CSN ISO 3534-1:2010 *Statistics - Vocabulary and symbols - Part 1: General statistical terms*

and terms used in probability
[14] CSN 72 1170:1967 *Testing of aggregates for building purposes. Basic regulations*
[15] CSN 72 1172:1967 *Determination of grain size distribution of aggregates*
[16] Kloboucek B1979 *Road Laboratory* (Prague: SNTL)
[17] Spurek J 1979 *Road Construction* (Prague: SNTL)
[18] Hanzik V 2015 *Laying Of Compacted Asphalt Mixtures* (Prague: Sdruzeni pro vystavbu silnic)
[19] CSN EN 12620+A1:2008 *Aggregates for concrete*
[20] CSN EN 13043: 2004 *Aggregates for bituminous mixtures and surface treatments for roads, airfields and other trafficked areas*

Mechanical and thermal properties of concrete suitable for radioactive waste disposal sites

E Vejmelkova[1], M Cachova[1], L Scheinherrova[1], P Konvalinka[2], M Keppert[1], P Bezdicka[3] and R Cerny[1]

[1] Czech Technical University in Prague, Faculty of Civil Engineering, Department of Materials Engineering and Chemistry, Thakurova 7, 166 29 Prague 6, Czech Republic
[2] Czech Technical University in Prague, Faculty of Civil Engineering, Experimental Centre, Thakurova 7, 166 29 Prague 6, Czech Republic
[3] Institute of Inorganic Chemistry, Academy of Sciences of the Czech Republic, Husinec-Rez 1001, 250 68 Rez, Czech Republic

Abstract. Secondary casing as a part of a multi-barrier system serves to meet the safety requirements on the radioactive waste repositories. Such casing needs to be durable, radiation- and thermal resistant, with good mechanical properties; these requirements are fulfilled by concrete. However, the casing should also serve as a barrier limiting the leaching of radionuclides, in case of failure of primary package. This problem is addressed in the article by the addition of a sorption material (natural zeolite rock) in the concrete mixture. Mechanical and thermal properties of concrete containing different amount of natural zeolite rock are studied with the link to concrete porosity. The addition of natural zeolite is found to cause a decrease in thermal conductivity and specific heat capacity. Both compressive and flexural strength decrease as well when natural zeolite is added in the mixture, which correlates with the increase of open porosity of studied concrete mixtures.

1. Introduction

Radioactive waste repositories require a multi-barrier system of protection to ensure safety of any living species within the repository neighborhood and to preserve the surrounding environment. Such a hazardous waste has to be encapsulated in a solid and durable matrix (cementitious, glass, ceramics etc.) placed in a durable container (usually metal) – this serves as the first protection and it is called a primary package. Primary packages are further protected by secondary engineering barriers, which should further limit the leaching of radionuclides into the final protection barrier, which is comprised by the surrounding rock [1]. However, the last barrier is in direct contact with groundwater, which cannot be contaminated; in that case, the radionuclides would spread in a large environmental area.

For low and intermediate level waste, the secondary barriers are made of concrete for its mechanical properties, durability, low corrosion and low cost. The secondary casing is usually reinforced to have sufficient mechanical parameters in order to withstand the transport, manipulation on site and ground pressure after the final emplacement in the repository [2,3]. Besides the mechanical properties, there are requirements on the radiation stability of concrete, its thermal stability and resistance to aggressive water and eventual organisms. Moreover, the concrete barrier should limit the leaching of radionuclides into the environment in case there is an accidental leach from the primary package. Well known sorption materials for radionuclides such as ^{137}Cs or ^{90}Sr etc. are zeolites, both natural or synthetic [4,5]. As zeolites present pozzolanic activity, these can be used as additive in concrete taking part in the cement hydration and/or it can act as a filler [6,7]. The article aims at studying the properties of concrete incorporating natural zeolite as an additive to enhance the sorption properties of secondary casing in rad-waste repositories. The paper is focused on the change in mechanical and thermal properties of concrete when different amount of zeolite is added in the concrete mixture.

2. Materials and methods

2.1. Materials

Table 1 presents the composition of studied materials. The concrete mixtures were designed to have high strength with continuous granulometry curve made of three fractions of aggregates Dobříň (Quartz, Spar) and Zbraslav (Andezite) complemented with fine filler – silicate flour ST6 (specific surface – 0,4 m^2g^{-1}, d50 = 16 μm). The principal binder component was Portland cement CEM I 42,5 R, which was the only binder component for the reference mixture RZR. In the next four mixtures, natural zeolite rock was added in different amounts. 10% of cement weight (40 kg in 1 m^3) was substituted by natural zeolite (ZeoBau 200, specific surface – 24,5 m^2g^{-1}, d50 = 34 μm), which takes part in the hydration of the binder system, then the addition of zeolite was dosed increasingly by 40 kgm^{-3}, while the cement dosage was kept constant and zeolite rock served as an inert mineral additive. Superplasticizer Stachement 2000 (based on polycarboxylate) was added to achieve good rheology of the fresh mixture and the amount of water was dosed in order to meet the prescribed slump of the fresh mixture. Table 2 gives the chemical composition of the input materials.

Table 1. Composition of studied materials.

	RZR	RZ1	RZ2	RZ3	RZ4
Portland cement CEM 42,5 R [kg/m^3]	400	320	320	320	320.00
Natural zeolite ZeoBau 200 [kg/m^3]		80	120	160	200
Silicate flour ST6 [kg/m^3]	120	120	80	40	-
Aggregates 0-4 mm Dobříň [kg/m^3]	760	760	760	760	760
Aggregates 4-8 mm Zbraslav [kg/m^3]	455	455	455	455	455
Aggregates 8-16 mm Zbraslav [kg/m^3]	483	483	483	483	483
Superplasticizer - Stachement 2000 [kg/m^3]	6.9	6.9	6.9	6.9	6.9
Water [kg/m^3]	195	160	160	160	160

Table 2. Chemical composition of the input materials (wt. %).

	Portland cement 42.5 R	Natural zeolite	Silicate flour
SiO_2	18.7	74.5	99.7
Al_2O_3	4.5	15.4	0.2
Fe_2O_3	3.4	1.6	-
CaO	65.9	3.3	-
MgO	1.3	0.7	-
K_2O	0.8	3.5	-
Na_2O	0.2	0.6	-
TiO_2	0.3	0.2	-
SO_3	4.3	4.9	-
P_2O_5	0.1	0.2	-

2.2. *Experimental methods*

2.2.1. Basic physical properties. Bulk density, ρ [kg·m^{-3}], matrix density, ρ_{mat} [kg·m^{-3}], and open porosity ψ [%] were measured using the helium pycnometry (Pycnomatic ATC) and water vacuum saturation method [8].

For the determination of the pore distribution, mercury intrusion porosimetry was used. The experiment was carried out by Pascal 140 and Pascal 440 devices. Results were plotted as the cumulative curves of pore specific volume in dependence on the pore diameter and distribution curves of incremental pore volumes in dependence on the pore diameter.

2.2.2. Mechanical properties. Compressive strength [MPa] and flexural strength [MPa] as the main mechanical parameters were determined according to CSN EN 12390-3 [9] and CSN EN 12390-5 [10] respectively; device EU 40 was used. The strength was determined for 28-days cured samples.

2.2.3. Thermal properties. The thermal conductivity, λ [W·m^{-1}·K^{-1}], and the specific heat capacity, c [J·kg^{-1}·K^{-1}], in dependence on the weight fraction of water in the material, were determined to assess the thermal properties. The pulse method was used with the use of portable device ISOMET 2114.

3. Results and discussion

3.1. *Basic physical properties*

Measured parameters of basic physical properties are given in table 3. Results obtained by pycnometry and gravimetric method correlated well with the results gained from vacuum saturation, only in a little bit higher absolute values. Bulk density as well as matrix density decreased with the increasing amount of zeolite in the mixture, which is the result of increasing open porosity. This is caused by the very porous nature of the input zeolite with channels of pores. The decrease of bulk and matrix density is within 15%, which is insignificant, but the increase in open porosity is by 45% when comparing RZ4 with the reference mixture. However, the increasing trend is not proportional as there is high rise in values for RZ1 compared to RZR (by 37%), but the further increase is lower. This means that even the initial 10% of natural zeolite which should have taken part in the pozzolanic reaction, resulted in increase of porosity of the hardened structure of concrete.

Table 3. Basic physical properties of studied materials.

	Water vacuum saturation method			Pycnometry and gravimetric method		
	Bulk density [kgm^{-3}]	Matrix density [kgm^{-3}]	Open porosity [%]	Bulk density [kgm^{-3}]	Matrix density [kgm^{-3}]	Open porosity [%]
RZR	2 319	2 579	10.1	2 361	2 638	10.5
RZ1	2 111	2 514	16.0	2 166	2 595	16.5
RZ2	2 097	2 505	16.3	2 157	2 594	16.8
RZ3	2 046	2 456	16.7	2 152	2 595	17.1
RZ4	2 002	2 448	18.2	2 082	2 606	20.1

Figures 1 and 2 present the pore size distribution. It can be seen that there is higher amount of larger pores (sizes around 10 - 100 μm) for the reference mixture in comparison to RZ1 and RZ2. However, the total pore volume is lower in case of RZR. For RZ3 and RZ4 there are higher amounts of all sizes of pores. It can be concluded that the presence of natural zeolite increases the total porosity, but there is a shift in the average size of pores from macro-pores (10 – 100 μm) to micro-pores (around 0.1 μm).

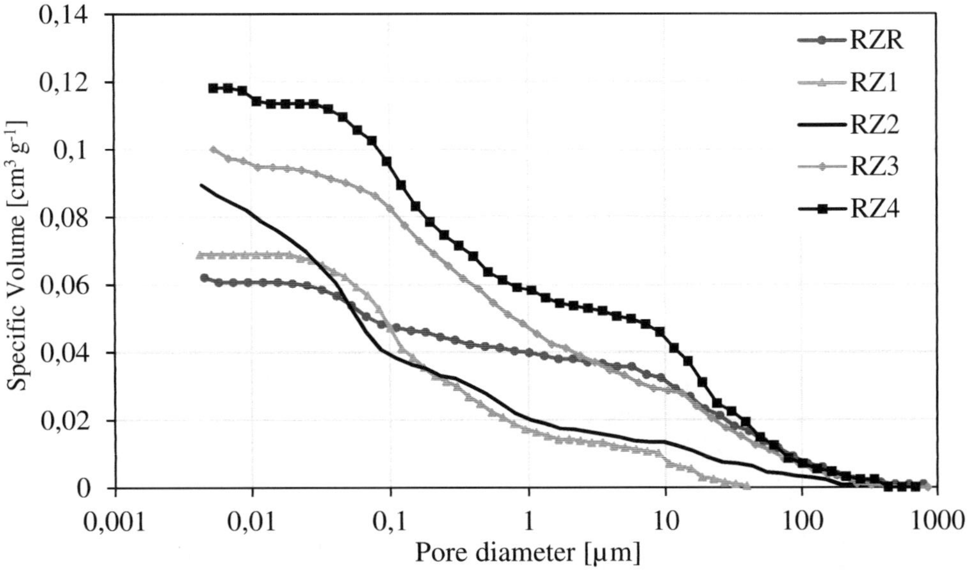

Figure 1. Cumulative curve of pore size distribution of studied materials.

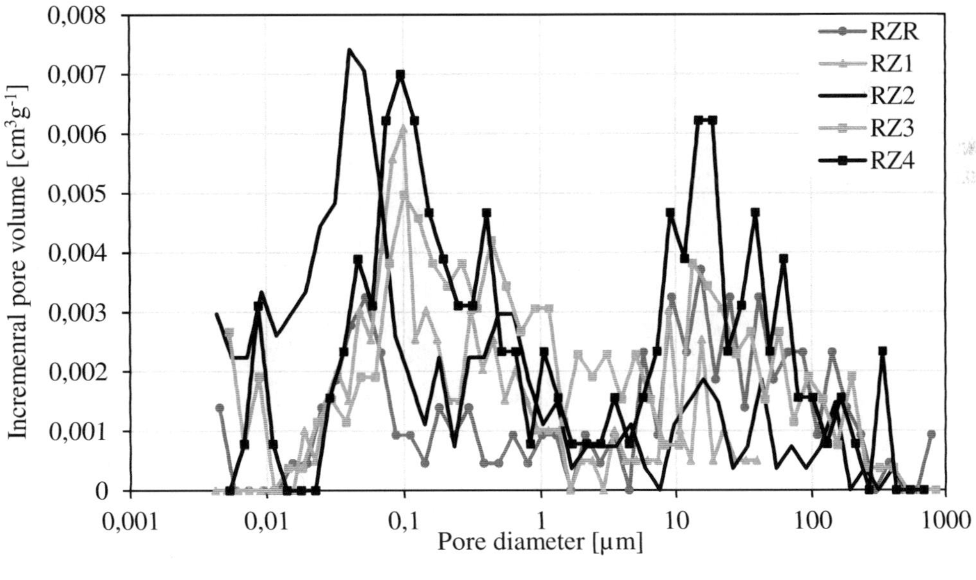

Figure 2. Pore size distribution curves of studied materials.

3.2. Mechanical properties

Results of mechanical properties are given in table 4. There is a clear decreasing trend for both compressive and flexural strength. The drop of compressive strength for RZ1 compared to reference concrete is by 35% and in case of RZ4 the residual strength is at about half the reference value. In case of flexural strength, the decrease for RZ1 is lower than in case of compressive strength – “only” by

12%; but for RZ4 again the flexural strength showed only half the reference value. The results correlate well with the values of measured porosities, which indicates that the worsened mechanical properties were caused by the increase in open porosity.

Table 4. Mechanical properties of studied materials.

Material	Compressive strength [MPa]	Bending strength [MPa]
RZR	165.73	8.04
RZ1	107.67	7.06
RZ2	101.73	6.02
RZ3	87.50	4.31
RZ4	85.75	4.17

3.3. Thermal properties

Figures 3 and 4 present the values of measured thermal properties in dry and saturated state. There is a decrease of thermal conductivity of concrete when natural zeolite is present. The drop is for RZ1 in comparison to reference mixture is by 30% and 23% for dry and saturated state respectively, however further decrease with increasing amount of zeolite is negligible in both dry and saturated state. The only exception is RZ4, where a further drop in the dry state by 15% was recorded.

Values of specific heat capacity in the dry state of RZ4 concrete were 40% below the reference. A decreasing trend was recorded also for the saturated state, however the rate of the decline was lower, as the value of RZ4 decreased "only" by 20% in comparison to the reference concrete.

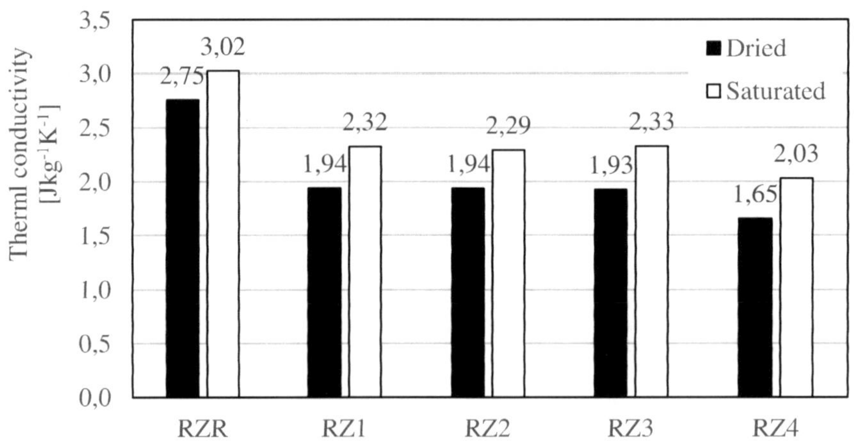

Figure 3. Thermal conductivity of studied materials.

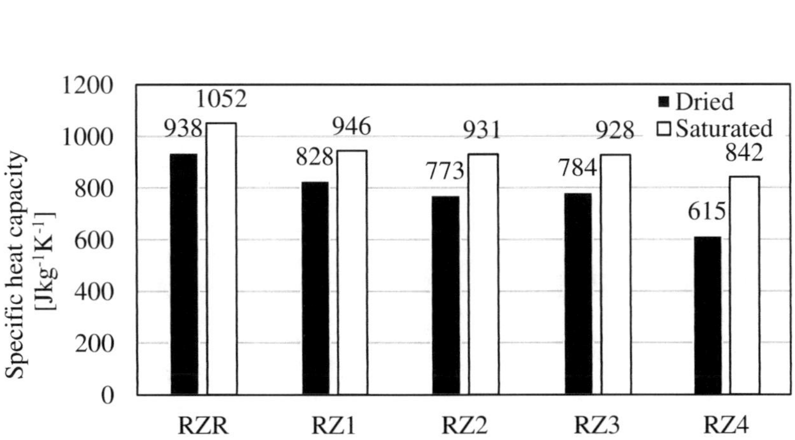

Figure 4. Specific heat capacity of studied materials.

4. Conclusions

The need to ensure safety criteria within the radioactive waste repositories calls for finding the best way how to limit any leaching of radionuclides into the environment. Although this is the main task of primary packages in the multi-barrier system of the rad-waste repository, it is always better to have double protection. Therefore, there is ongoing research on studying concrete for secondary engineering barrier incorporating natural zeolite as a good sorbent of radionuclides. The paper studied the mechanical and thermal properties in connection to the porous system. However, despite the good sorption properties of natural zeolite, its presence in concrete for secondary barriers result in significant worsening of mechanical properties due to increased porosity. And also in the view of thermal properties, the thermal conductivity and specific heat capacity decreased with the incorporation of natural zeolite. Further experiments and numerical simulations need to be made to consider the advantages and disadvantages of the zeolite presence in secondary barrier concrete.

Acknowledgment

This work was supported by the Czech Science Foundation, under project No 17-11635S and by the project no. SGS16/199/OHK1/3T/11.

References

[1] Kotatkova J, Zatloukal J, Reiterman P, and Kolar K 2017 J. Environ. Radioactiv. 178–179 147–155
[2] CEA 2009 Nuclear waste conditioning. A Nuclear Energy Division Monograph (Paris) pp 78-82
[3] Ojovan M I 2011 *Handbook of advanced radioactive waste conditioning technologies* (Cambridge: Woodhead publishing)
[4] Elkamash Q, Elnaggar M and Eldessouky M 2006 *J. Hazard. Mater.* **136** 310-316
[5] Plecas I, Dimovic S and Smiciklas I 2006 *Prog. Nucl. Energ.* **48** 495-503
[6] Karakurt C, Kurama, H and Topçu I B 2010 *Cement Concrete Comp.* **32** 1-8
[7] Najimi M, Sobhani J, Ahmadi B and Shekarchi M 2012 *Constr. Build. Mater.* **35** 1023-1033
[8] Roels S, Cerny R, Pavlik Z et al 2004 *Journal of Thermal Envelope and Building Science* **27** 307–325
[9] CSN EN 12390-3: 2002 *Testing of hardened concrete – Part 3: Compressive strength*
[10] CSN EN 12390-5: 2009 *Testing of hardened concrete - Part5: Flexural strength*

IOP Conf. Series: Materials Science and Engineering **385** (2018) 012062 doi:10.1088/1757-899X/385/1/012062

Influence of the fracture toughness on durability of the silicate structures

J Vyborny, M Petriková, L Fiala and J Koci

Department of Materials Engineering and Chemistry, Faculty of Civil Engineering, Czech Technical University in Prague, Thakurova 7, 166 29 Prague 6, Czech Republic

Abstract. Generally, the crack propagation is a problem of equilibrium and stability of silicate structures. The critical crack length depends on fracture toughness. The relationships between fracture toughness, bulk density, modulus of elasticity, and compressive strength are depicted in silicate material diagrams. Those diagrams have numerous application in engineering structural design.

1. Introduction

The basic imperfection of classical theory of calculation of material strength is a fact, that those calculations neglect material defects and discontinuities. Usually, the materials are comprehended as a homogenous continuum, which is getting damaged suddenly after reaching limit conditions in the whole exposed body. In other words, the failure is detected according to highest average stress (sheer, compressive, tensile, etc.) or highest strain energy in the vicinity of the critical cross-section [1]. However, such an interpretation is insufficient for a description of complex processes of cracks formation and propagation. The propagated crack will reach its critical length eventually implying fracture of the whole body. The illustration of the dependence between crack length and durability is given in figure 1.

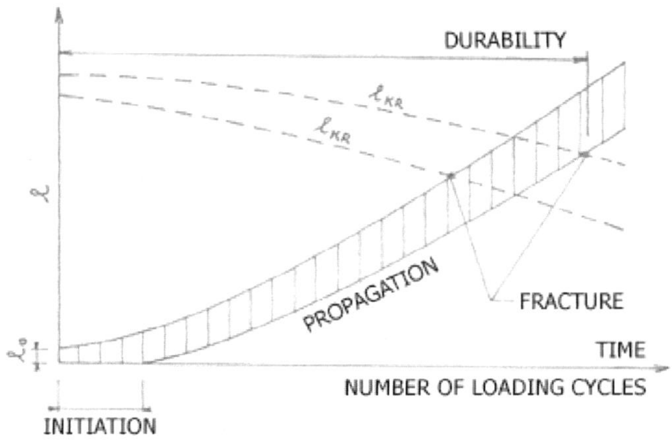

Figure 1. Relationship between durability of silicate materials and crack development.

In figure 1, the critical crack length l_{CR} varies in time. The length is shortening during material's service life period due to decreasing of fracture toughness as the material ages. The aging is usually due to various operational loadings and their intensities given by the environmental conditions. Typically, the material properties are changing in complex processes taking place across the whole body such as strain or precipitation aging, degradation due to solar radiation, moisture effects, intercrystalline corrosion etc. For that reason, a significant spread is apparent in the critical crack length l_{CR}. The static material fatigue is accompanied by the loss of the ability of plastic strains, which is apparent after long-lasting loading periods especially when the material is exposed to the effect of high temperatures. The final fracture comes then from small macrocrack or system of cracks in the form of brittle fracture as shown in figure 2. The formation of major macrocrack can be observed also in brick columns or pillar loaded by concentric pressure (see figure 3).

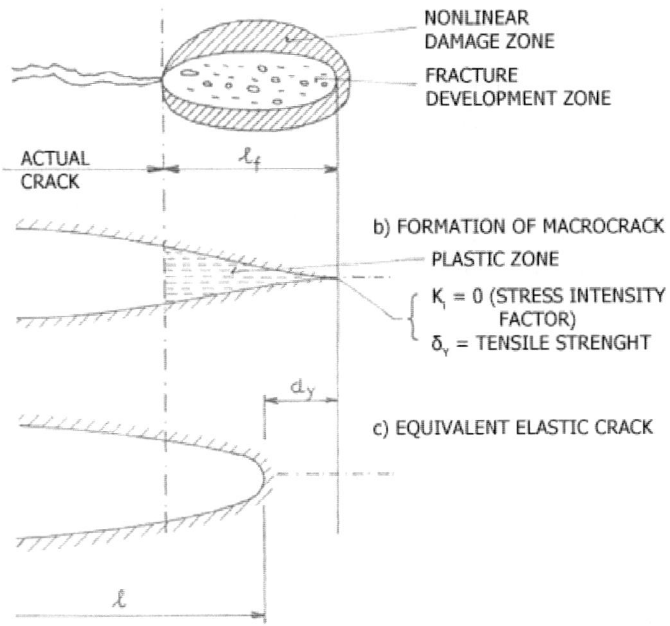

Figure 2. Connecting of small cracks into a macrocrack in concrete, stone, and bricks.

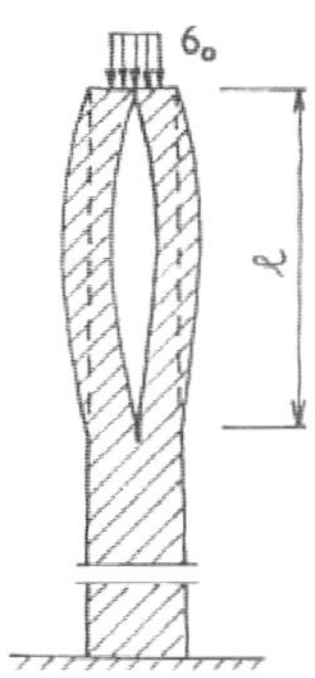

Figure 3. Formation of major macrocrack in columns loaded by concentric pressure.

Therefore, the description should not be limited to the description of stress fields from point to point only, but it should be also aimed, regarding the limit state, at the respective quantities in the vicinity of the initial inhomogeneity, macroscopic defect or developed cracks after loading (figure 3). Thus, practically new type of material is defined in continuum mechanics. Its deformation is given not only by elastic strain but also the local strain of ideal zone having size d_y (figure 2).

The stress intensity factor K_I is an important variable in continuum mechanics of deformable bodies containing macrocracks and narrowly bounded fracture development zones.

The critical value of stress intensity factor K_I with respect to the dynamic character of fracture process is referred to as fracture toughness K_{IC}.

The size of fracture development zone d_y, defined as a distance between the end of opening microcrack (where δ_y is tensile strength and $K_I = 0$) and to the root of equivalent elastic crack for the type of loading *I* (crack opening), is defined as:

$$d_y = \frac{1}{\pi} \cdot \frac{K_{IC}^2}{\delta_y^2}. \tag{1}$$

According to the theory of nonlinear fracture mechanics, the stress concentrated in the root of actual crack creates a development zone where the material is fatigued and fractured. The crack that is to be developed (if tensile stress equals the tensile strength of the material with a fragile matrix, i.e. silicates or thermosetting plastics) has the length according to (1), which means that the critical crack length l_{CR} is equal to the size of fracture development zone d_y.

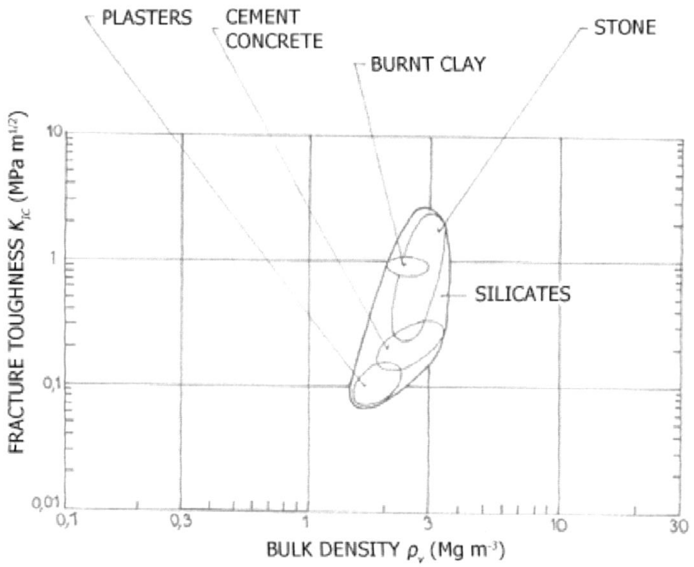

Figure 4. The dependence of fracture toughness on bulk density.

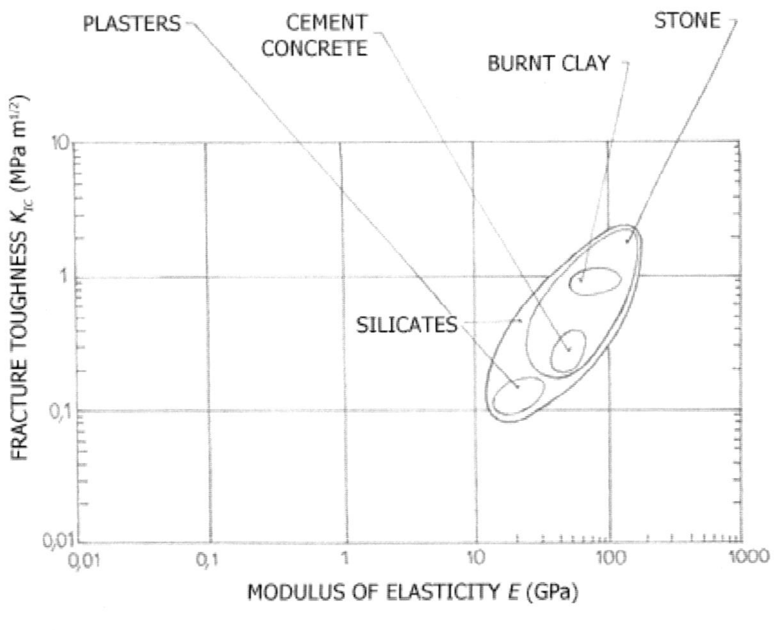

Figure 5. The dependence of fracture toughness on modulus of elasticity.

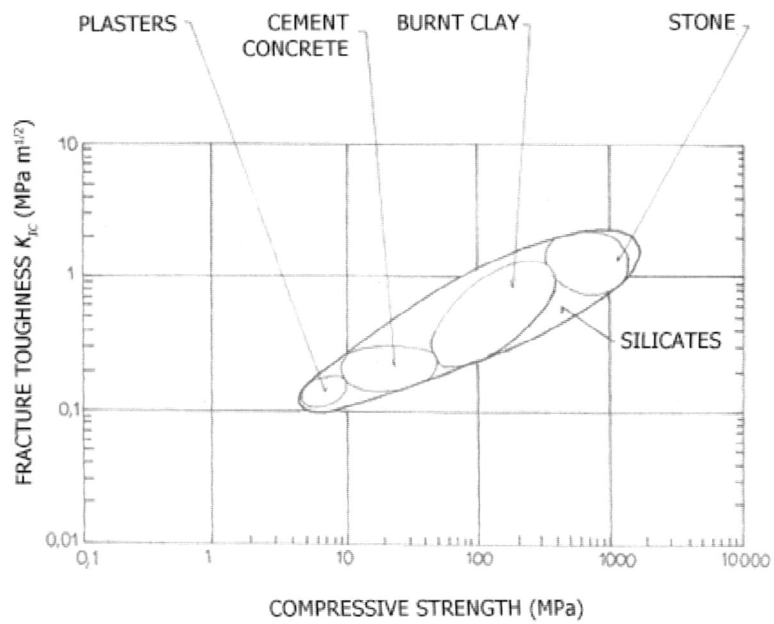

Figure 6. The dependence of fracture toughness on compressive strength.

2. Discussion

It generally applies for silicates that plastic strain in the top of the crack reduces the stress peak by rounding the sharp edge of the crack (figure 2) as the energy absorbed in the plastic zone makes the crack development more difficult. The plastic strain in the root of crack in burnt clay-ceramics is very limited because the stress peak is reduced by a network of cracks. Thus, both the absorbed energy and fracture toughness are low. This results in the fact that fracture toughness K_{IC} for ceramic materials is approximately 15 times lower than fracture toughness of metals.

Due to low fracture toughness and presence of cracks and voids in the material structure the strength of ceramic is quite low. The voids weaken the material, however, if they are spherical in the ideal case, the stress concentrated in the vicinity of such voids is quite low. More dangerous for the material integrity is the presence of small cracks, which can result from thermal expansion (during wet clay burning, wetting of unprotected brick masonry by driving rain etc.) or isotropy in modulus among grains. Nowadays, there exist two ways of producing ceramics with increased strength:

- reducing the length of crack by technological solutions (for example by adding fine brick grinding powder), precision in manufacturing, curing etc.,
- increasing K_{IC} by using fine aggregates (from 1.0 to 0.4 μm), utilization of phase transformation, designing of composite structures.

Note that for a reasonable value of fracture toughness K_{IC} = 2.0 MPa is the highest length of crack approximately 60 μm, which corresponds to the size of aggregate. Therefore the use of ceramic grinding from brick block production seems to be an ideal solution [6].

The values of fracture toughness as a function of bulk density (figure 4), modulus of elasticity (figure 5) and compressive strength (figure 6) are shown in diagrams of silicate materials [4]. The diagrams show the limiting values of particular quantities. If the combination of values lies inside the zone in the diagram, it can be considered as safe. Position outside of that zone is questionable.

3. Conclusions

The damage of structural masonry walls is one of the most widespread deterioration and reason of service life shortage and reduction of building seismic capacity. Synergic effects are considered as a key element in the assessment of the durability and structural performance of brickworks. New observation highlight the importance of rain penetration effect on the durability and structural integrity of brick masonry [5].

Acknowledgement

This research has been supported by the Ministry of Industry and Trade, Czech Republic under Project No. FV 10036 and by Czech Ministry of Education, Youth and Sports, under project No SGS16/199/OHK1/3T11.

References

[1] Flinn A R and Trojan P K 1990 *Engineering Materials and Their Applications* (Boston: Houghton Mifflin)

[2] Bazant Z P and Cedolin A 1991 *Stability of Structures: Elastic, Inelastic, Fracture and Damage Theories* (New York, Oxford: Oxford Univesity Press)

[3] Emmons H P, Drochytka R and Jerabek Z 1999 *Saving and Maintenance of Concrete in Illustrations* (Brno: CERM s.r.o.)

[4] Ashby F M 1999 *Acta Metall. Mater.* **37** 1273-1293

[5] Cacciotti R 2018 *Brick Masonry Response to Wind-driven Rain* (Prague: Czech Technical University)

[6] Heluz s.r.o., available at http:www.heluz.cz

IOP Conf. Series: Materials Science and Engineering **385** (2018) 012063 doi:10.1088/1757-899X/385/1/012063

Rheological and mechanical properties of lime-brick powder mortars with expanded clay aggregate

M Vysvaril, T Zizlavsky and P Rovnanikova

Faculty of Civil Engineering, Brno University of Technology, Veveri 331/95, Brno 602 00, Czech Republic

Abstract. The rheological behaviour and mechanical properties of lightweight lime-brick powder mortars with combined expanded clay-sand aggregates are presented in this paper. The rheological investigation included the characterization of flow properties and viscoelastic properties of the fresh mortars. Flexural and compressive strengths were determined at 7 and 28 days. These mortars are Non-Newtonian thixotropic liquids with conversion from shear-thickening to shear-thinning behaviour depending on expanded clay aggregate (ECA) content. Yield stress and consistency coefficient of mortars significantly increase with decreasing amount of ECA; the mortars become more rigid and more viscous. Some of the mortars reached higher flexural strengths than similar lime mortars with metakaolin and higher compressive strengths than pure lime mortars. The lightweight lime-brick powder mortar with the mass ratio of 1:1 between gravelous sand and ECA was identified as the best among the analyzed mortars.

1. Introduction

Recently, the use of lightweight aggregates (LWA) in mortar composition to improve acoustic and thermal insulation, or fire resistance has been developed [1–5]. However, lowering the density produces a strength reduction, which can limit the direct exposition of the mortar to the environmental agents, requiring an external protection [6]. Expanded perlite, expanded glass, expanded clay and hollow micro-spheres are some lightweight materials that, because of their composition based on aluminosilicates, are potentially able to give pozzolanic reactions. They can therefore provide additional advantages in mortars by improving their mechanical strength. Expanded clay aggregate (ECA) has been successfully used as a component of lightweight concrete mixtures in the past [7–10]. The application of ECA as the LWA in lime mortars was newly reported [11].

Brick powder as a waste material is available from the development of building elements for the construction of low-energy houses, or from a demolition waste. Bricks waste is usually rich in glass and burned clays, consisting in dehydrated aluminosilicates not only in crystalline phase but also in amorphous state which is important for its pozzolanic reaction.

In this study, the influence of quantity of ECA on the rheological and mechanical properties of blended lime-brick powder mortars is monitored. As their supposed applications include the renovation of surface layers of historical buildings, brick powder is utilized as the pozzolanic admixture replacing cement. The rheological properties of lime or cement-lime mortars containing LWA was reported in common literature sources just in the very few cases [12]. The main ambition of this study is to partially contribute to filling the gap in the scientific knowledge in that respect. The knowledge of mortars rheology may contribute to understand the behaviour of fresh materials and can allow predicting their flow properties that are important in a rendering.

2. Materials

Hydrated lime CL 90-S (Carmeuse Czech Republic, s.r.o.) was used as the main binder in all studied mortars. The brick powder from skiving of heat-insulating bricks from the HELUZ Brickworks factory, v.o.s. was added as a material with good pozzolanic properties [13] which was supposed to improve the mechanical characteristics. Natural gravelous sand (the fine fraction 0/2 mm from Českomoravský štěrk, a.s., Hulín) and expanded clay (Lias Vintířov, LSM k.s.) were used as aggregates. The ECA was produced using clay expanded in a rotary kiln at the temperature of 1150 °C. According to the producer information, the final product has spherical grains and its internal porosity system is uniform, while the surface of spheres is sintered. Basic properties of the applied ECA, as given by the producer are presented in figure 1. The chemical composition of all raw materials is given in table 1. The phase compositions obtained by the X-ray diffraction analysis are presented in table 2. The granulometry of aggregates and binders is shown in figure 1.

Mortar mixtures were prepared using the correct amount of water required to obtain normal consistency and a good workability of the mortars (160 ± 5 mm; measured by the flow table test). The proportioning of the mortar mixtures is given in table 3. Fresh mixtures were cast into prismatic moulds of size 40 × 40 × 160 mm. Standard conditions of sample storage were 22 ± 2 °C and relative humidity of 50 ± 5%. After 7 and 28 days the bulk density, the flexural and compressive strengths were determined.

Table 1. Chemical composition of initial materials examined by XRF spectroscopy (mass %).

	SiO_2	Al_2O_3	Fe_2O_3	CaO	MgO	K_2O	Na_2O	P_2O_5	TiO_2	SO_3	MnO
Lime	–	–	–	99.2	0.6	–	–	–	–	0.1	–
Brick powder	57.4	14.6	5.5	9.8	4.0	2.8	1.4	0.1	0.3	1.2	–
Sand	86.8	6.5	1.2	0.5	0.3	2.3	0.9	–	0.2	0.1	–
ECA	42.9	30.6	11.3	4.4	2.4	2.5	0.8	0.3	3.4	0.8	0.2

Table 2. Phase composition of initial materials (mass %).

Mineral	Lime	Brick powder	Gravelous sand	ECA
Albite	–	13.5	–	–
Anatase	–	2.8	–	–
Anorthite	–	–	–	7.8
Biotite	–	–	3.0	–
Calcite	1.8	–	–	–
Diopside	–	–	–	3.0
Epidote	–	–	3.6	–
Gypsum	–	–	–	0.8
Hematite	–	3.1	0.7	–
Hercynite	–	–	–	12.2
Microcline	–	12.4	3.9	–
Mullite	–	–	–	7.9
Muscovite	–	8.9	–	–
Orthoclase	–	2.9	3.0	–
Portlandite	98.1	–	–	–
Quartz	–	31.2	75.6	2.9
Rutile	–	3.1	0.1	–
Sanidine	–	2.4	–	–
Staurolite	–	–	6.1	–
Amorphous phases	–	19.6	–	64.7

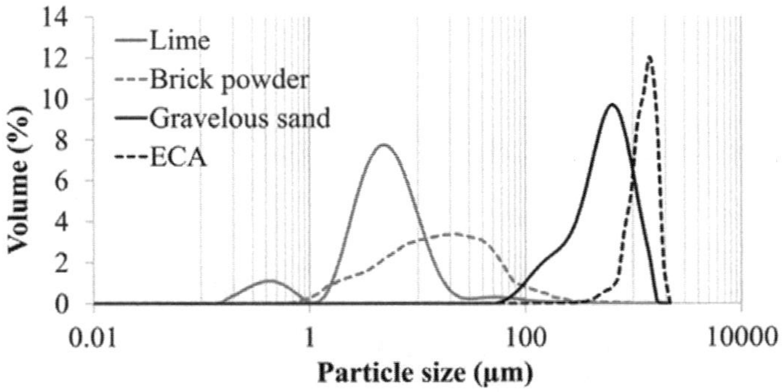

Property of ECA	Value
Fraction (mm)	0/2
Loose bulk density (kg m^{-3})	575
Particle bulk density (kg m^{-3})	1050
Thermal conductivity (W m^{-1}K^{-1})	0.12
Water absorption (% by mass)	5
Flammability (DIN 4102)	A1
Crush resistance (MPa)	4

Figure 1. Particle size distribution of initial materials and properties of the expanded clay aggregate.

3. Experimental procedures

Rheological tests were carried out on the hybrid rheometer Discovery HR-1 (TA Instruments). The experimental conditions, the technique of flow and oscillation measurements and the data evaluation were identical as in the previous studies [14,15]. The Herschel-Bulkley model was applied to descending branches of flow curves to fit the experimental data and used to describe the mortars rheological behaviour. Thixotropy of the mortars was determined by the TRIOS software as an area between flow curves [16]. The strain of 0.003% used in the frequency sweep oscillation tests was lower than the critical strain for all studied mortars (table 4; γ_c).

Flexural strengths of the samples were determined using a standard three-point-bending test and compressive strengths were measured on the far edge of both residual pieces obtained from the flexural test according to EN 1015-11. The flexural and compressive strengths were obtained after curing times of 7 and 28 d. The coefficients of variation were below 15%. The water absorption of mortars was measured according to EN 13755 after curing times of 28 d.

Table 3. Composition of mixtures.

Mixture	Lime (g)	Brick powder (g)	ECA (g)	Gravelous sand (g)	Water (ml)
LV-I	100	100	100	0	240
LV-II	100	100	75	25	210
LV-III	100	100	50	50	180
LV-IV	100	100	25	75	150

4. Results and discussion

4.1. Rheological properties

4.1.1. Flow properties.

Figure 2 illustrates the flow curves of fresh mortars. The shape of curves indicates the lime-brick mortars as Non-Newtonian liquids with gradual conversion from shear-thickening to shear-thinning suspension for decreasing ECA content. It is also confirmed by diminishing flow index values, n, (table 4). The shear-thickening behaviour of lime mortars is very unusual especially with such high doses of mixing water and spherical grains of ECA, conversely to typical shear-thickening behaviour of lime-metakaolin mortars where the dilatant properties can be explained by considering the plate like and angular shape of metakaolin particles. All of the mortars exhibited a thixotropic (time-dependent shear thinning) character with an increasing tendency for decreasing ECA content in the mortar (table 4). It is apparent that the yield stress and the consistency coefficient of mortars significantly increased with decreasing amount of ECA (table 4). This was expected because the lower mixing water amount in mortar drops off the solubility of the lime and brick particles, the structure of the mortar is denser, and the mortar is less fluid with higher plasticity.

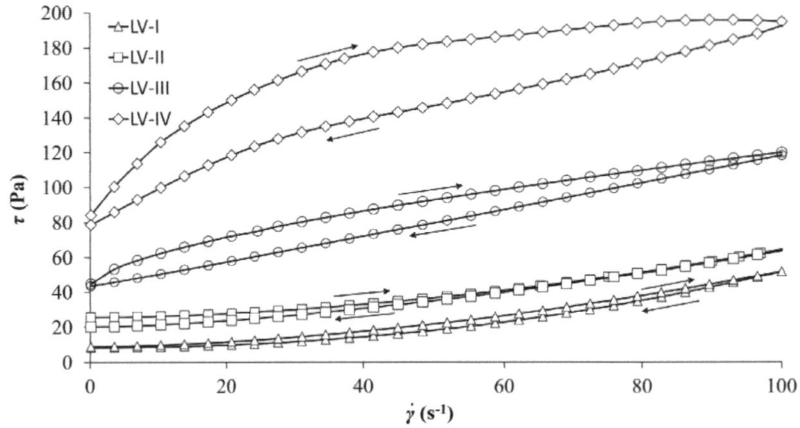

Figure 2. Flow curves of lime-brick powder mortars.

Table 4. Comparison of bulk density (28 d) and rheological parameters of lime-brick powder mortars.

Mixture	ρ_b (kg m^{-3})	τ_0 (Pa)	k (Pa s)	n (-)	Thixotropy (Pa s)	γ_c (%)	γ_f (%)	G^* (Pa)	η^* (Pa s)	tan δ (-)
LV-I	790	8.35	0.002	2.14	242.5	0.005	0.359	14501	2308	0.296
LV-II	930	20.00	0.036	1.54	226.5	0.005	0.841	13872	2208	0.280
LV-III	1120	43.45	0.602	1.05	1043.0	0.006	5.778	39135	6229	0.212
LV-IV	1250	75.44	6.152	0.63	2670.2	0.008	15.08	57750	9191	0.138

4.1.2. Viscoelastic properties

A linear viscoelastic (LVE) range was determined for mortars by applying varying strain amplitude oscillatory test and measuring the dynamic moduli, G' and G'', (figure 3). The LV-IV mortar exhibited the largest LVE range ending at a critical strain, γ_c, at which the decrease of G' is observed, but the variations were quite small (table 4). All mortars have G' greater than G'' in the LVE range, where the structure of the system is not disturbed; this indicates that the mortars are composed of a strongly multi-connected network of particles. Decreasing amount of ECA in the mortars leads to an increase of the storage modulus (G'), indicating an increase of the material cohesion.

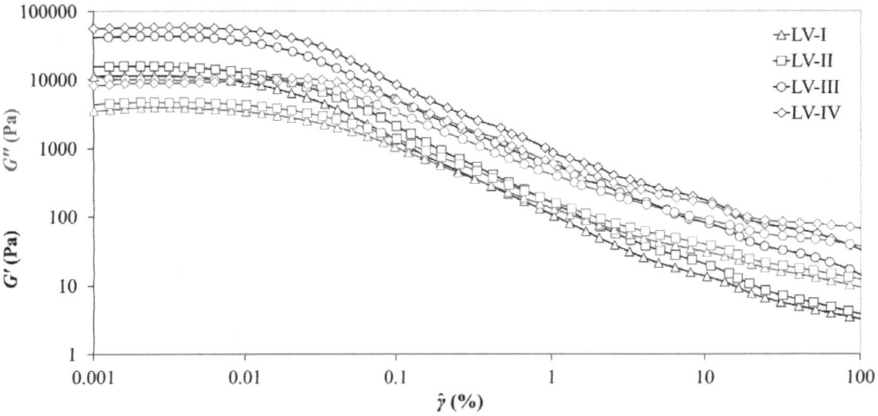

Figure 3. The evolution of G' and G'' as a function of strain amplitude.

A flow strain γ_f at a flow point (when G′ = G″, the gel-like character of the sample is changed to the liquid state) increases with decreasing ECA content; this confirms the more fluid character of LV-I and LV-II mortars.

The results of frequency sweep tests proved the relationship between the amount of ECA and the stiffness of lime mortars; the ratio between dissipation and elasticity increased (growing G^*) with decreasing amount of ECA and the mortars became more rigid and more viscous (figure 4). Since the ratio between viscosity and elasticity (loss factor, tan δ) varies below 1 (table 4), the mortars show a higher proportion of the elastic component (G') than the viscous one (G''), suggesting that the mortars have a strong structure resistant to external interference.

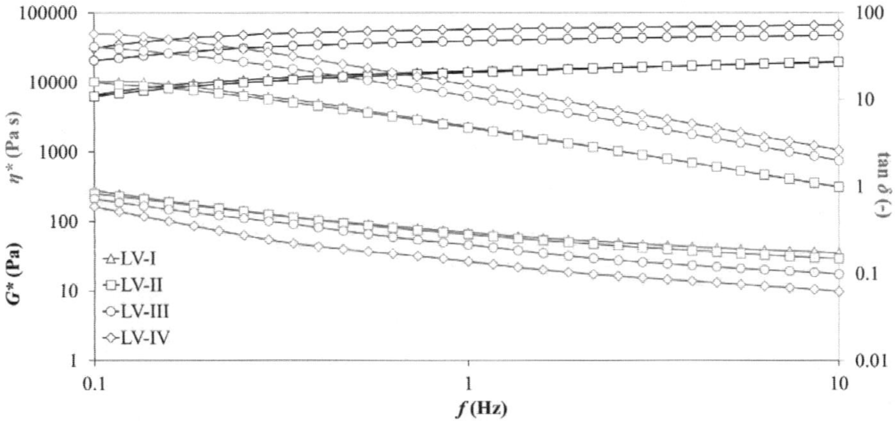

Figure 4. Evolution of G^*, η^* and tan δ as a function of frequency.

4.2. Mechanical properties

Obviously, because of mortar hardening, the mechanical strengths of mortars increased with curing time (figure 5). It can be stated that the mortars with higher content of ECA (LV-I and LV-II) are much weaker than LV-III and LV-IV. The latter mortars reached higher flexural strengths than similar lime mortars with metakaolin [11]. The effect of ECA amount on the values of compressive strength was contradictory to the previous findings [11]. The compressive strength of LV-III and LV-IV was much higher than for pure lime mortars and if the reduced bulk density (table 4) is taken into account, the LV-III mortar represents a suitable variation of lightweight lime-brick powder mortars for practical utilization. The differences in strengths properties and water absorption between LV-III and LV-IV are insignificant but the difference in bulk density is fundamental.

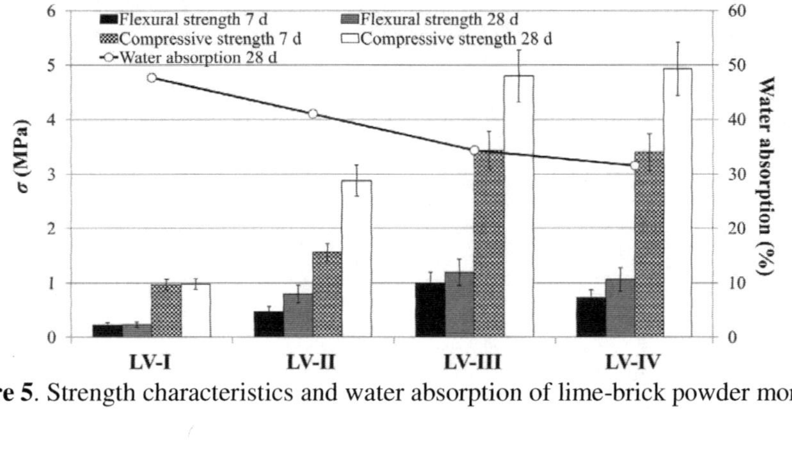

Figure 5. Strength characteristics and water absorption of lime-brick powder mortars.

5. Conclusions

Lightweight lime-brick powder mortars with combined expanded clay-sand aggregates were investigated in this paper. The effect of quantity of expanded clay aggregate (ECA) on the rheological and mechanical properties of mortars was determined. The main conclusions are as follows:

- The lime-brick powder mortars were Non-Newtonian liquids with gradual conversion from shear-thickening to shear-thinning behaviour for decreasing ECA content. All of the mortars exhibited a thixotropic character with an increasing tendency for decreasing ECA content in the mortar.
- The yield stress and the consistency coefficient of mortars significantly increased with decreasing amount of ECA because of the diminishing amount of mixing water required for the same consistency.
- Decreasing amount of ECA in the mortars led to the growth of the storage modulus (G') and complex modulus (G^*) indicating the material cohesion increase; the mortars became more rigid and more viscous.
- The mortars with high content of ECA were weak in strengths. Some of the mortars reached higher flexural strengths than similar lime mortars with metakaolin and higher compressive strengths than pure lime mortars.
- The lightweight lime-brick powder mortar with the mass ratio of 1:1 between gravelous sand and ECA, which was identified as the best among the analyzed mortars, has a very good potential for practical applications because its properties are significantly enhanced, as compared with the common lime plaster with silica aggregates.

Acknowledgments

This work has been financially supported by the GA ČR research project 18-07332S and by Ministry of Education, Youth and Sports under the „National Sustainability Programme I" – the project No. LO1408 „AdMaS UP – Advanced Materials, Structures and Technologies".

References

[1] Silva L M, Ribeiro R A, Labrincha J A and Ferreira V M 2010 *Cem. Concr. Compos.* **32** 19
[2] Torres M L and García-Ruiz P A 2009 *Cem. Concr. Compos.* **31** 114
[3] Zach J, Hela R, Hroudová J and Sedlmajer M 2011 *Adv. Mater. Res.* **335–336** 1199
[4] Brás A, Gonçalves F and Faustino P 2014 *Energy Build.* **72** 296
[5] Mazhoud B, Collet F, Pretot S and Chamoin J 2016 *Build. Environ.* **96** 206
[6] Palomar I, Barluenga G and Puentes J 2015 *Constr. Build. Mater.* **75** 306
[7] Lo T Y and Cui H Z 2004 *Mater. Lett.* **58/6** 916
[8] Kong L, Zhang B, Ge Y and Yuan J 2009 *J. Wuhan Univ. Technol. Mater. Sci. Ed.* **24/1** 145
[9] Dilli M E, Atahan H N and Şengül C 2015 *Constr. Build. Mater.* **101** 260
[10] Ke Y, Beaucour A, Ortola S, Dumontet H and Cabrillac R 2009 *Constr. Build. Mater.* **23/8** 2821
[11] Konakova D, Cachova M, Vejmelkova E, Keppert M, Jerman M, Bayer P, Rovnanikova P and Cerny R 2017 *Cem. Concr. Compos.* **83** 374
[12] Pereira de Oliveira L A, Castro Gomes J P and Nepomuceno M C S 2013 *Appl. Rheol.* **23** 15505
[13] Navratilova E and Rovnanakova P 2016 *Constr. Build. Mater.* **120** 530
[14] Vysvaril M and Zizlavsky T 2017 *IOP Conf. Ser.: Mater. Sci. Eng.* **251** 1
[15] Vysvaril M, Hegrova M and Zizlavsky T 2018 *Key Eng. Mater.* **760** 257
[16] Bohac M and Necas R 2016 *Procedia Eng.* **151** 34

IOP Conf. Series: Materials Science and Engineering **385** (2018) 012064 doi:10.1088/1757-899X/385/1/012064

Comparing of the external bearing wall using three cultural perspectives in the life cycle impact assessment

M Wolfova, A Estokova, M Ondova and A Monokova

Department of Environmental Engineering, Faculty of Civil Engineering, Technical university of Kosice, Vysokoskolska 4, 042 00 Kosice, Slovakia

Abstract. The main aim of this research is to perform cultural perspectives analysis of a Life Cycle Assessment (LCA) case study to understand if the use of different Life Cycle Impact Assessment (LCIA) approaches may lead to a very different conclusions. A LCA was applied to analyse the environmental impacts of external bearing wall of two different compositions. In this study, the environmental comparison of the constructions was performed using the ReCiPe method in software SimaPro while considering three approaches of cultural theory: the hierarchist, the individualist, and the egalitarian. The results were compared using midpoint and endpoint indicators. The endpoint LCIA showed that each approach provided a different conclusion except for the hierarchist and egalitarian approach in the Resources indicator. The results for midpoint LCIA were in all impact categories the same with the exception of the terrestrial acidification and marine acidification.

1. Introduction

Life Cycle Assessment (LCA) is a methodology for evaluation the environmental impacts of service, product or a process all through its life cycle (ISO, 2006a) [1]. ISO characterizes between four phases in LCA framework: the goal and scope definition, the inventory analysis, the environmental impact assessment and the interpretation. Product life cycle includes all processes that may be related to the production, use and disposal of the product [2].

The building sector is considered to be the largest contributor to environmental pollution and resource depletion because it produces globally for 33% of greenhouse gases (GHG) emissions and consumes for 40% of the primary energy. Nevertheless, in line with the United Nations Environment Program, construction has the greatest potential to significantly reducing GHG emissions compared to other sectors that also emit emissions into the environment, in particular by reducing the operating energy of buildings, especially in cities [3]. The global urbanization trend, which is predicted to exceed 70% of all urban people by 2050, requires more comfort in the living space. Nowadays, cities represent approximately 67% of the global energy consumption, accounting for over 70% of all GHG emissions and consuming about 75% of all natural resources [4]. Another way to reduce the environmental loads of buildings is connected to the selective choice of the building materials which are environmentally suitable. LCA is the optimal method to evaluate the environmental impacts of the materials and constructions in order to provide their environmental performance.

This paper is focused on the LCA analysis of the environmental impacts of external bearing wall of two different compositions from three cultural perspectives. Both mid-point and end-point indicators have been taken into consideration.

2. Material and methods

2.1 Evaluated construction
The external bearing wall consisting of two different core materials, light clay bricks and aerated concrete block, was evaluated regarding its environmental impacts. The material compositions of the walls, expressed in function unit of 1 m^2, are given in tables 1 and 2.

Table 1. The materials used in the bearing wall consisting of light clay bricks.

Material	Density [kg/m^3]	Thickness [mm]	Mass per 1 m^2 [kg]
Base plaster	600	10	6.0
Light clay brick	680	440	299.2
Polystyrene foam	10	100	3.0
Lime mortar	400	30	12.0

Table 2. The materials used in the bearing wall consisting of aerated concrete blocks.

Material	Density [kg/m^3]	Thickness [mm]	Mass per 1 m^2 [kg]
Base plaster	600	10	6.0
Aerated concrete block	500	300	150
Polystyrene foam	160	160	25.6
Cellulose fibre	230	40	2.0
Acrylic binder	1850	3	5.6

Schemes of the analysed walls are presented in figure 1.

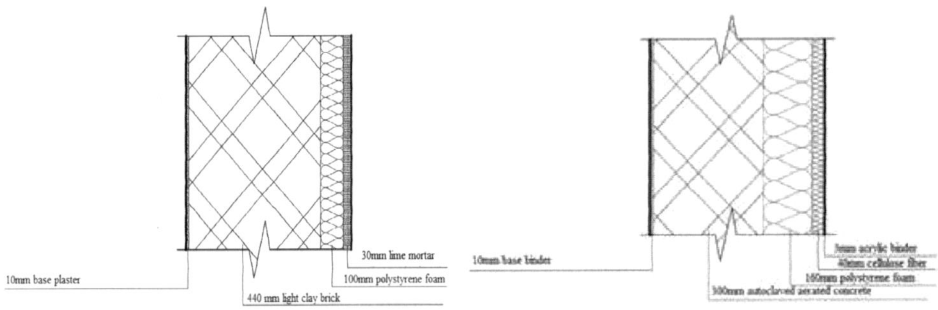

a) b)

Figure 1. a) Light clay brick wall composition, b) Aerated concrete block wall composition.

2.2 The ReCiPe method
The ReCiPe method (SimaPro software) was used to evaluate the environmental impacts of the constructions. Characterization models can use two types of impact indicators, midpoint indicators and endpoint or damage indicators. Figure 2 shows the characterization framework addressing both midpoint and endpoint and illustrates the impact categories for the ReCiPe methodology [5].

Endpoints (or damages) typically refer to three areas of protection: human health, ecosystem health and resources. Damage to human health caused by different types of environmental stressors are quantified by changes in mortality and morbidity. The indicator commonly used to quantify human health is disability-adjusted life years. The DALY concept in LCA was introduced by Hofstetter (1998) on the base of Murray and Lopez (1996) for the WHO (World Health Organization) [6]. The unit for ecosystem quality is local relative species loss in terrestrial, freshwater and marine ecosystems, respectively, integrated over space and time (potentially disappeared fraction of species×m^2×year or potentially disappeared fraction of species×m^3×year). To aggregate the impacts of terrestrial, freshwater and marine ecosystems into one single unit (species×year), we included species densities for these three types of ecosystems in the same way. The unit for resource scarcity is dollars ($), which represents the extra costs involved for future mineral and fossil resource extraction. Endpoint characterisation factors (CFe) are derived from midpoint characterisation factors (CFm) with a constant mid-to-endpoint factor per impact category [7].

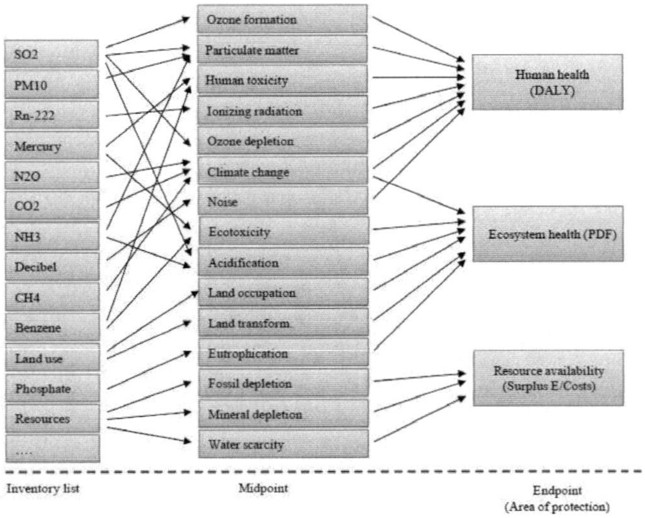

Figure 2. List of impact categories for characterization at midpoint and endpoint level [5]

2.3 Approaches of Cultural Theory

Cultural Theory has evolved over the last 20 years to become an important framework for understanding how groups in society interpret dangers and build confidence or mistrust in institutions that create and regulate risk. Cultural Theory has been used as a tool within framework LCA. This Cultural Theory reflects vision of society and view on nature. Cultural theory distinguishes five perspectives generaly three of these perspectives are used in environmental decision making. They are: individualist, hierarchist and egalitarian perspectives. Each archetype reflects the composition of ideologies, cultural prejudices, social relationships or moral convictions. The individualist is characterized by weak group coherence and prescriptions for social relations, and considers nature to be stable and able to recover from any intervention. In contrast with the individualist, the hierarchist is characterized by strong group coherence with obligatory prescriptions for social relations. In this perspective, nature is considered to be equilibrium. The egalitarian perspective has strong group coherence and considers nature to be breakable and unstable [2]. Overview of the different visions among the individualist, hierarchist and egalitarian perspectives as used in the evaluation are presented in table 3.

Table 3. Different visions in LCA.

	I (Individualist)	H (Hierarchist)	E (Egalitarian)
Time perspective	Balance between short and long term (20 yers)	Balanced time perspective (100 years)	Very long term (infinite)
Vision on society	Market - driven economic production	Development within natural limits	Equality and social governance
Manageability	Flexible management	Preventive and complex style of leadership	Control and limitation style of leadership
Vision on nature	Considers nature robust	Considers nature tolerant	Considers nature vulnerable
Level of proof	Ultimate reserve	Only proven effects	All possible effects

The individual agrees with the view that humanity has a high adaptive capacity through technological and economic development and that the short-term perspective is justified. The hierarchical view coincides with the view that the effect of regular management can be avoided and that the choice of what is to be included in the model is based on the (scientific) consensus level. The egalitarian is consistent with the view that nature is strictly responsible, that the long-term perspective is justified and that the worst scenario is needed [2].

3. Results and discussion

3.1 Midpoint indicators

Modeling choices were analyzed by defining three coherent scenarios, based on Cultural Theory perspectives. Figure 3 presents the results of the environmental impact of 1m^2 of the external bearing wall materials using light clay brick and aerated concrete block, calculated by ReCiPe method, in midpoint categories considering three cultural approaches I, H, E as mentioned above. Individual, hierarchist and egalitarian perspectives had their own contextual and preferential values.

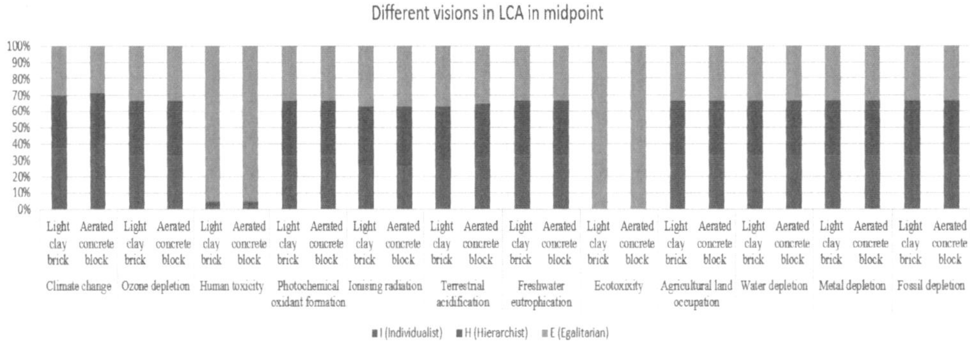

Figure 3. Comparison of the percentage share of the material on the midpoint category.

The significant differences in the environmental impacts, evaluating I, H, and E cultural approaches, have been found for two indicators: human toxicity and ecotoxicity (figure 3). In our study, the overall ecotoxicity indicator included terrestrial, freshwater and marine ecotoxicity. In individualist and hierarchist perspective, the values calculated were almost the same, however, in egalitarian perspective the results achieved were 200 times higher for light clay brick and even 600 times higher for aerated concrete block. In egalitarian approach, ecotoxicity refers to the potential for biological, chemical, or physical stressors to affect ecosystems in infinite time perspective.

Small differences in the results, considering the cultural approaches, have been also noticed for climate change and terrestrial acidification. In all others environmental indicators, the calculated values were comparable for all approaches and the choice of the time horizon was negligible.

3.2. Endpoint indicators (damage assessment)

In this section, a comparison between the three cultural approaches is made using the results from the calculations of the endpoint indicator in three damage categories (figure 4).

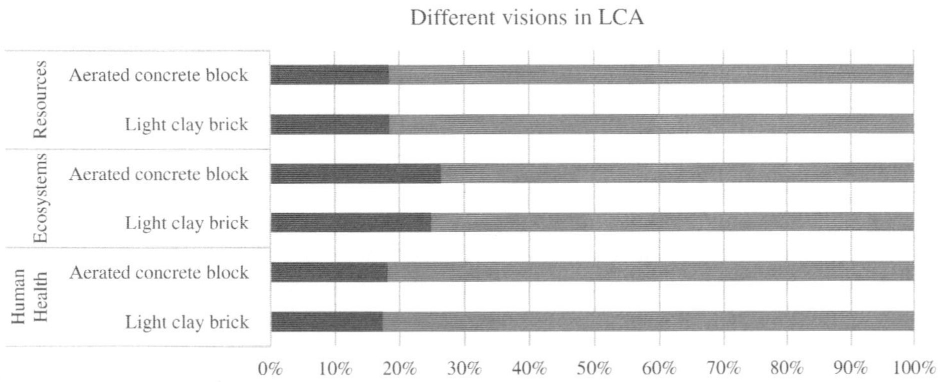

	Human Health		Ecosystems		Resources	
	Light clay brick	Aerated concrete block	Light clay brick	Aerated concrete block	Light clay brick	Aerated concrete block
■ Individualist	0,000111	0,000322	0,000000817	0,00000182	191	542
■ Hierarchist	0,000116	0,000307	0,000000785	0,00000148	422	1200
■ Egalitarian	0,000413	0,00115	0,00000168	0,00000359	422	1200

Figure 4. Comparison of the percentage share of the material on the total impacts.

The individualist and hierarchist approaches proved very similar and comparable results for both material compositions in two damage categories: human health and ecosystems (figure 4). Significant difference was observed when comparing these results to the results using egalitarian approach. Similarly as for the midpoints, the egalitarian-based calculation provided much higher score and thus much more negative environmental impacts.

Surprisingly, in the resources indicator, the hierarchist and egalitarian approaches resulted in the same environmental scores (422 for light clay brisk and 1200 for the aerated concrete) which were several times higher than those of calculated for the individualist approach.

3.3 Comparison of both material compositions of the bearing wall

The impact assessment results for the external bearing wall of both material compositions calculated by ReCiPe are presented in figure 5. Hierarchist method was applied to compare the environmental impacts of the constructions since this approach is most often used in the LCA analysis [8,9].

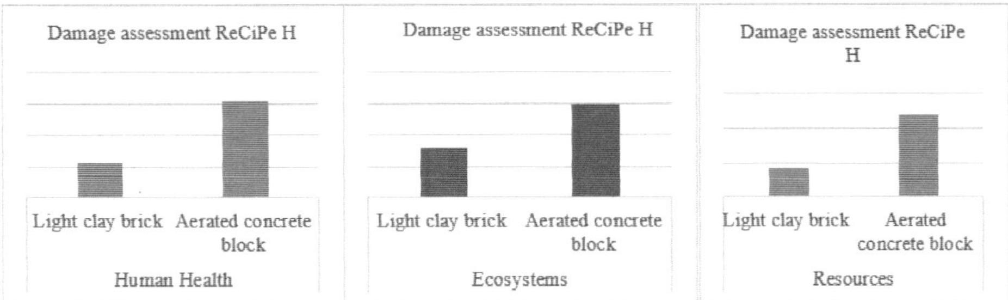

Figure 5. Comparison of the environmental impacts of two alternatives of the external bearing wall

As it can be seen in figure 5, the wall with the aerated concrete blocks reached worse environmental score than the wall with light clay brick in all damage categories. The calculated values were higher 2.6, 1.9, and 2.8 times for human health, ecosystems and resources, respectively. Considering only these results without any additional technical parameters, use of light clay brick is a better environmental alternative than use of aerated concrete blocks what is in accordance with [10]. However, to make a clear conclusion, the multi criteria analysis, including other parameters, is of importance.

4. Conclusion

The paper presented a comparative LCA study of the environmental impacts of the construction of external bearing wall with light clay brick and aerated concrete block. Special attention was paid to evaluation of the results when applied three different cultural approaches (individualist, hierarchist and egalitarian). In order for the different LCA cultural theory to be compared, the results of the midpoint and endpoint impact categories have been calculated by ReCiPe. The external wall with the light clay brick was evaluated to be the better alternative than the wall with the aerated concrete blocks. The results point to the importance of choosing the optimal approach for the environmental evaluation in building industry. The proper choice of method or perspective should be dependent on the purpose of the evaluation and way of utilization of the results. Another point is that not only the results calculated by the same method but by the same cultural perspective as well can be compared each other.

Acknowledgements
This paper has been prepared with a support of the Grants No. 1/0648/17 of the Slovak Grant Agency for Science.

References

[1] Loiseau E, Aissani L, Le Féon S, Laurent F, Cerceau J, Sala S and Roux P 2018 *J. Clean. Prod.* **176** 474-485
[2] De Schryver A N 2010 *Value choices in life cycle impact assessment* (Nijmegen: Ipskamp Drukkers BV) p 2-10
[3] Hoxha E, Habert G, Lascaux S, Chevalier J and Le Roy R 2017 *J. Clean. Prod.* **144** 33-47
[4] Weiler V, Harter H and Eicker U 2017 *Energy. Build.* **134** 319-328
[5] Hauschild M Z, Goedkoop M, Guinée J, Heijungs R, Huijbregts M, Joliet o and Sala S 2013 *Int. J. Life Cycle Assess.* **18** 683-697
[6] Arnesen T and Kapiriri L 2014 *Health policy* **70** 137-149
[7] Huijbregts M A, Steinmann Z J, Elshout P M, Stam G, Verones F, Viera M and van Zelm R 2017 *Int. J. Life Cycle Assess.* **22** 138-147
[8] Weidema B 2015 *J. Ind. Ecol.* **19** 20-26
[9] De Schryver A N, Geodkoop M J, Leuven R S and Huijbregts M A 2010 *Int. J. Life Cycle Assess.* **18** 682-691
[10] Estokova A and Porhincak M 2012 *Theor. Found Chem. Eng.* **46** 704-712

IOP Conf. Series: Materials Science and Engineering **385** (2018) 012065 doi:10.1088/1757-899X/385/1/012065

Study of the behaviour of thermal insulation materials made with recycled textile fibers

J Zach, V Novak and Jitka Peterkova

Institute of Technology of Building Materials and Components, Faculty of Civil Engineering, Brno University of Technology, Veveri 331/95, 602 00 Brno, Czech Republic

Abstract. The requirements placed on the thermal protection of buildings are becoming stricter worldwide, which causes an increased demand after thermal insulation materials. One of the ways of achieving sustainable development in this area is the use of secondary and renewable raw materials. This paper describes the results of research aimed at developing and studying the behaviour of thermal insulation materials made with recycled textile fibers.

1. Introduction

Thermal and acoustic insulation materials have become a standard part of basically all buildings. The worldwide usage of insulation materials has been rising every year (source EEB) [1]. For reasons of sustainable development and meeting contemporary requirements for the energy efficiency of buildings it is necessary to seek ways of further increase the production of thermal and acoustic insulations without putting excessive strain on material resources and increasing primary energy intensity (PEI) together with CO_2 emissions (GWP) as a consequence.

One of the possible ways of addressing this is the utilization of secondary raw materials, which can be used as a full substitute for primary materials [2, 3, 4, 5, 6, 7]. One potentially interesting option is the use of recycled textile fibers produced by various branches of industry. These fibers are currently available in good enough quality practically anywhere in the world, are affordable, and insulations made with them can have very good utility properties.

2. Recycled textile fibers

The research involved a study of various secondary fibers produced by different industrial branches. These were PES-based polymer fibers, cotton, and bast fibers. The fibers were microscopically examined and tested for thermal conductivity using specimens prepared from free layers of fibers with the bulk density of 35 kg/m^3. The measurements were made with a total 14 types of recycled fibers (including fiber mixtures). It was found that the thermal conductivity of the samples ranged within 0.038 – 0.049 W/(m.K). Thermal conductivity was proved to be markedly lower in samples that contained badly defibered clusters (e.g. recycled carpets, etc.). The thermal conductivity of samples made from pure fibers ranged within 0.038 – 0.042 W/(m.K). The subsequent research presented herein was performed with PES fibers with the thermal conductivity of 0.039 W/(m.K), average thickness of 20 μm and length of 61.2 mm, which appear to be useful thanks to their good properties and broad availability.

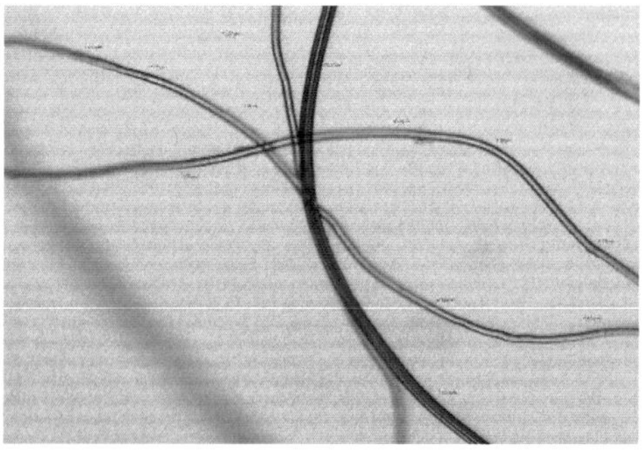

Figure 1. Microscopic image of the PES fibers.

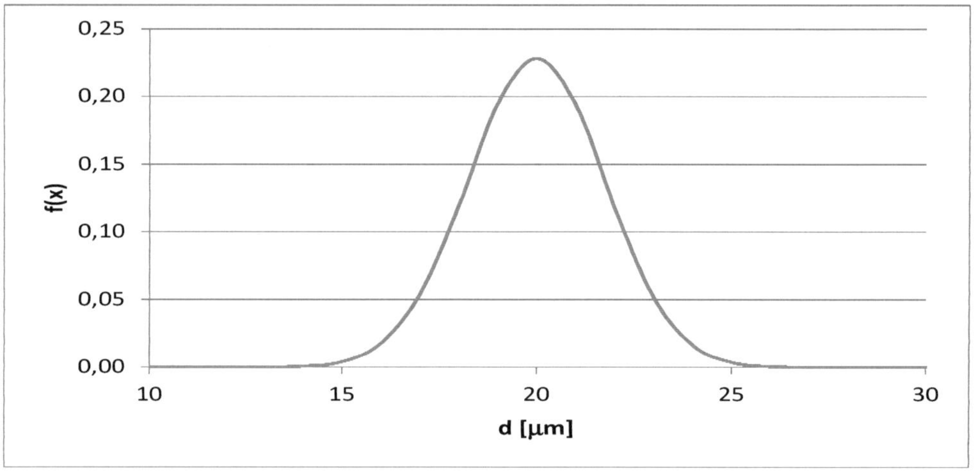

Figure 2. Diagram of the normal distribution of the fiber thickness in the PES sample (probability density).

3. Test samples

The test samples were made from a sample of recycled PES fibers (see above) and an addition of 15 % bicomponent PES fibers combined by thermal bonding. Two series of samples were made:

- Samples with the bulk density of 50 kg/m^3 with a variable thickness of 20, 40, 60, 80 and 100 mm,
- Samples with a thickness of 35 mm and varying bulk density 25, 50, 75, 100 and 125 kg/m^3.

It was not always possible to maintain the desired parameters during production. The actual thickness and bulk density values are listed below.

The samples of insulation were then made into mat specimens with the dimensions of 200 x 200 mm or 300 x 300 mm. Prior to further testing the specimens were stored at a temperature of +23°C and relative humidity 50%.

IOP Conf. Series: Materials Science and Engineering **385** (2018) 012065 doi:10.1088/1757-899X/385/1/012065

Figure 3. Photograph of a specimen with density of 100 kg/m^3.

4. Methodology
The fiber samples or mat specimens were measured for basic physical and thermal insulation properties. This involved the determination of:

- thickness according to EN 823 [8],
- length and width according to EN 822 [9],
- bulk density according to EN 1602 [10],
- thermal conductivity according to EN 12667 [11] and ISO 8301 [12].

The specimens had the shape of a mat, and, depending on thickness, had the dimensions of 200x200 mm or 300x300 mm. Thermal conductivity was measured on specimens with a stable laboratory moisture content ($u_{23,50}$), mean temperature + 10 °C, and temperature gradient 10 °C.

5. Results and discussion
The first step was the measurement of thickness, bulk density, and linear dimensions. It was found that the real parameters of the specimens diverted in some cases from the design, which is due to the system of prototype production and a relatively small volume of samples produced.

Table 1. Thickness and bulk density of the first series of samples with varying bulk density.

Sample No.	d [mm]	ρ_v [kg/m^3]
1 - a	21.043	48.58
1 – b	30.124	66.69
1 – c	55.438	56.53
1 – d	76.154	47.86
1 - e	90.161	52.84

Table 2. Thickness and bulk density of the second series of samples with varying thickness.

Sample No.	d [mm]	ρ_v [kg/m^3]
2 - a	35.928	26.33
2 – b	31.490	57.06
2 – c	33.247	77.22
2 – d	37.736	89.35
2 - e	38.049	104.89

Next, thermal conductivity was measured using the hot plate method. The tables below show the results.

Table 3. Thermal conductivity of the first series of samples with varying bulk density.

Sample No.	λ [W/(m.K)]
1 - a	48.58
1 – b	66.69
1 – c	56.53
1 – d	47.86
1 - e	52.84

Table 4. Thermal conductivity of the second series of samples with varying thickness.

Sample No.	λ [W/(m.K)]
2 - a	26.33
2 – b	57.06
2 – c	77.22
2 – d	89.35
2 - e	104.89

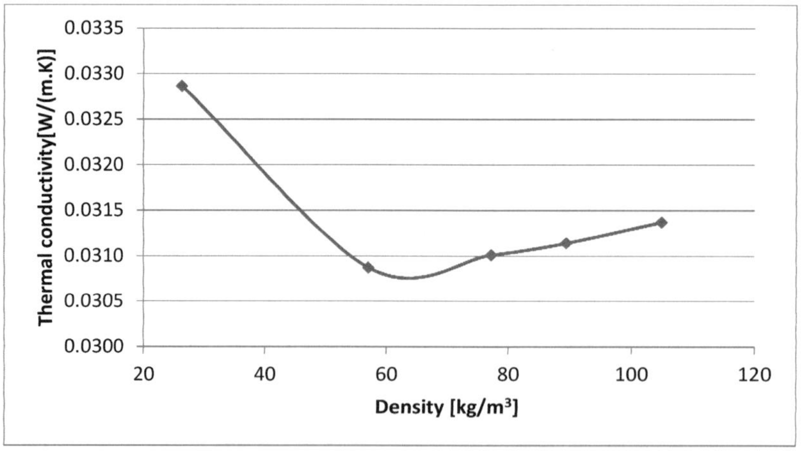

Figure 4. Diagram showing the dependence of thermal conductivity on bulk density.

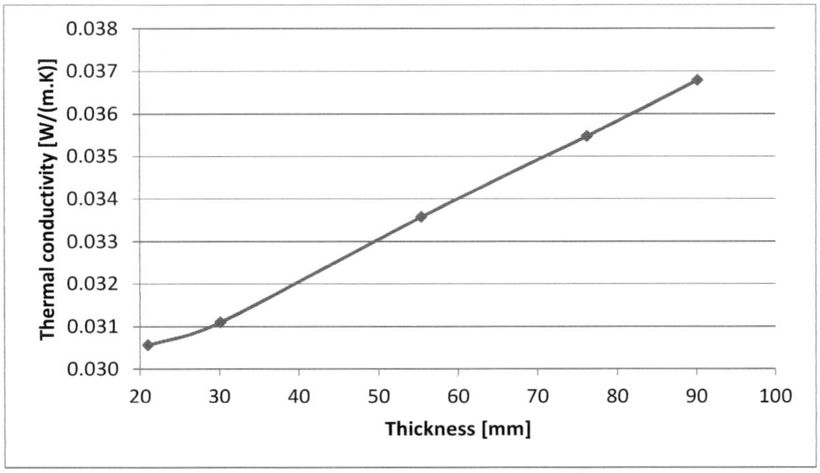

Figure 5. Diagram showing the dependence of thermal conductivity on thickness.

6. Conclusion

Thermal insulation materials made with recycled textile fibers show great promise for development and usage in civil engineering. The experiments show that the thermal insulations made with recycled PES fibers can compete with products based on mineral wool or expanded polystyrene, which currently dominate the market. When designing these materials, it is, however, important to correctly set the bulk density of the insulations and to know all the factors that influence the final properties.
The results also show the following.

- Insulations made with PES fibers have excellent thermal insulation properties; their thermal conductivity can reach values of less than 0.031 W/(m.K).
- The optimal bulk density of the samples is 50 – 80 kg/m^3.
- The key parameter of these insulations is their thickness and fiber orientation. This property is more important than bulk density (as demonstrated by diagrams in figures 3 and 4).

Acknowledgements
This paper has been worked out under the project No. LO1408 "AdMaS UP - Advanced Materials, Structures and Technologies", supported by Ministry of Education, Youth and Sports under the „National Sustainability Programme I" and FV20127 Development and research advanced thermal and acoustic insulation based on waste textile and natural fibers.

References

[1] Directive 2012/27/EU of the European Parliament and of the Council of 25 October 2012 on energy efficiency
[2] Latif E, Lawrence R M H, Shea A D and Walker P 2018 *Energ. Buildings* **165** 76-91
[3] Battegazzore D, Alongi J, Duraccio D and Frache A 2018 *J. Polym. Environ.* **26**1652-1660
[4] Lau KT, Hung PY, Zhu MH and Hui D 2018 *Composites Part B-Engineering* **136** 222-233
[5] Lacoste C, El Hage R, Bergeret A, Corn S and Lacroix P 2018 *Carbohyd. Polym.* **184** 1-8
[6] Zheng C, Li D F, Ottenhall A and Ek M 2017 *Holzforschung* **71** 633-639
[7] Zach J, Hroudova J and Korjenic A 2016 *J. Chem. Technol. Biot.* **91** 2156-2161
[8] EN 823: 2013 *Thermal insulating products for building applications. Determination of thickness*
[9] EN 822: 2013 *Thermal insulating products for building applications. Determination of length and width*
[10] EN 1602: 2013 *Thermal insulating products for building applications. Determination of the apparent density*

IOP Conf. Series: Materials Science and Engineering **385** (2018) 012065 doi:10.1088/1757-899X/385/1/012065

[11] EN 12667: 2001 *Thermal performance of building materials and products. Determination of thermal resistance by means of guarded hot plate and heat flow meter methods. Products of high and medium thermal resistance*

[12] ISO 8301: 1991 *Thermal insulation - Determination of steady-state thermal resistance and related properties - Heat flow meter apparatus*

IOP Conf. Series: Materials Science and Engineering **385** (2018) 012066 doi:10.1088/1757-899X/385/1/012066

Mortars with crystalline additive in aggressive environments

N Zizkova, L Nevrivova, M Ledl and S Keprdova

Department of Technology of Building Materials and Components, Faculty of Civil Engineering, Brno University of Technology, Veveri 95, 602 00 Brno, Czech Republic

Abstract. The paper describes how a crystalline additive influences the properties of cement-based composites. The investigation was focused on the degradation of mortars with crystalline additive exposed to selected aggressive environments. The specimens were tested for compressive and flexural strength. The influence on the microstructure is also discussed (high pressure mercury intrusion porosimetry, SEM, DTA). The results confirmed the positive influence of a crystalline additive on the values of compressive strength after exposure to all aggressive environments.

1. Introduction

Nowadays, concrete is the most widely used construction material, and therefore improving its properties remains an important issue [1,2,3].Cement composites usually consist of cement, water, aggregates, admixtures, additives and air. The binder is a continuous phase in the cement composite, thus the fact that it is porous is critical in terms of the transport of water and chemical substances in or out of the concrete [4]. One of the progressive technologies for cement based composites protection is the use of coatings, sprays or additives on so-called "secondary crystallization bases" also known as "crystalline technology" base [3]. The dry crystalline products are powder compounds that consist of Portland cement, quartz sand, and a compound of active chemicals [3,5].

In principle, crystalline materials work so that the chemical components react with the cementitious matrix during hydration, temporarily forming $Ca(OH)_2$ followed by the formation of disilicate and polysilicate anions. Available knowledge suggests that the cumulated process is accompanied by the formation of $3CaO \cdot 2SiO_2 \cdot 3H_2O$ together with the production of $4CaO \cdot Al_2O_3 \cdot 13H_2O$ [5].

Laboratory study [2] has shown, using X-ray fluorescence spectrometry (XRF) analysis, the resulting needle-like crystals likely comprise calcium or silicon [2]. The fine insoluble crystals seal the pore structure [6]. Authors [7] described a general process, where a general crystalline promoter M_xR_x reacts with tricalcium silicates and water to produce modified calcium silicate hydrates and a pore-blocking precipitate $M_xCaR_x - (H_2O)_x$ by the following equation (1):

$$3CaO - SiO_2 + M_xR_x + H_2O \rightarrow Ca_xSi_xO_xR - (H_2O)_x + M_xCaR_x - (H_2O)_x \qquad (1)$$

Similar process may exist involving the calcium aluminates. The resulting crystalline deposits become integrally bound with the hydrated cement paste and become a permanent part of the cement-based composites [7].

The objectives of the work was to study the degradation of mortars with crystalline additive exposed to selected aggressive environments.

2. Experiment

The experiment was focused on evaluating the influence of crystalline additive on physical-mechanical properties and microstructure of mortars containing a crystalline additive and polypropylene fibres. In order to observe the degradation of the cementitious composite, specimens were exposed to an aggressive environment.

2.1. Materials

This study used Portland cement CEM I 42.5 R as binder, crystalline additive, quartz sand and polypropylene fibres. The chemical composition of the crystalline additive (CA) is shown in table 1. CA was added in an amount of 0.5 and 1.5 % of cement weight, see table 2.

Table 1. Chemical composition of crystalline additive.

Chemical composition of CA [% wt.]									
SiO_2	Al_2O_3	Fe_2O_3	CaO	MgO	SO_3	K_2O	Na_2O	Cl^-	Organic Substances
11.5	2.34	1.74	45.7	0.734	2.050	0.387	6.61	0.026	7.03

Table 2. Composition of the mortars per m^3.

Mixture ID	Cement CEM I 42.5 R [kg]	Sand [kg]	CA [kg]	PP Fibres [kg]	Water [L]
CA0	512	1535	0	0	256
CA0.5	511	1533	3	0	256
CA1.5	510	1531	8	0	255
FCA0	511	1533	0	1	256
FCA0.5	511	1532	3	1	255
FCA1.5	510	1539	8	1	255

2.2. Methods

Specimens of mortars with the dimension of 40×40×160 mm were made. Porosity was determined by high pressure mercury intrusion porosimetry using a Thermo Finnigan POROTEC of the Pascal 140 – 240 type. DTA analysis was performed with a Mettler Toledo TGA/DSC 1 and SEM imaging with a TESCAN MIRA3 XMU. All specimens were stored 28 days at a temperature of 23 °C and relative humidity of 95 % and then under selected environments (A, B, C, D) which are shown in table 3.

Table 3. Curing environments.

Environment ID	Curing characteristic
A (23/95)	90 days stored at a temperature of 23 °C and relative humidity of 95 % (23/95)
B (Na_2SO_4)	90 days in sodium sulphate solution (concentration of 36 000 $mg \cdot L^{-1}$ of SO_4^{2-})
C (NH_4Cl)	90 days in ammonium chloride solution (concentration of 3 000 $mg \cdot L^{-1}$ of NH_4^+)
D (CO_2)	90 days in gaseous CO_2 at the concentration of φ 65 % and 75 % relative humidity

3. Results and Discussion

Figure 1 and 2 below show the results of compressive and flexural strength tests performed at an age of 118 days (23/95), 28 days (23/95) + 90 days in sodium sulphate environment, 28 days (23/95) + 90 days ammonium chloride environment and 28 days (23/95) + 90 days in gaseous CO_2.

IOP Conf. Series: Materials Science and Engineering **385** (2018) 012066 doi:10.1088/1757-899X/385/1/012066

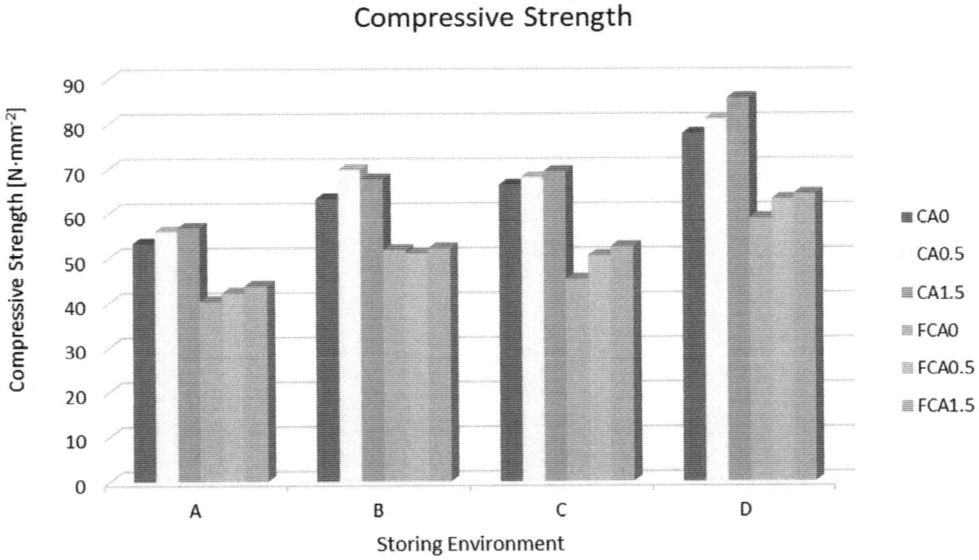

Figure 1. Determination of 118-day compressive strength of the mortars.

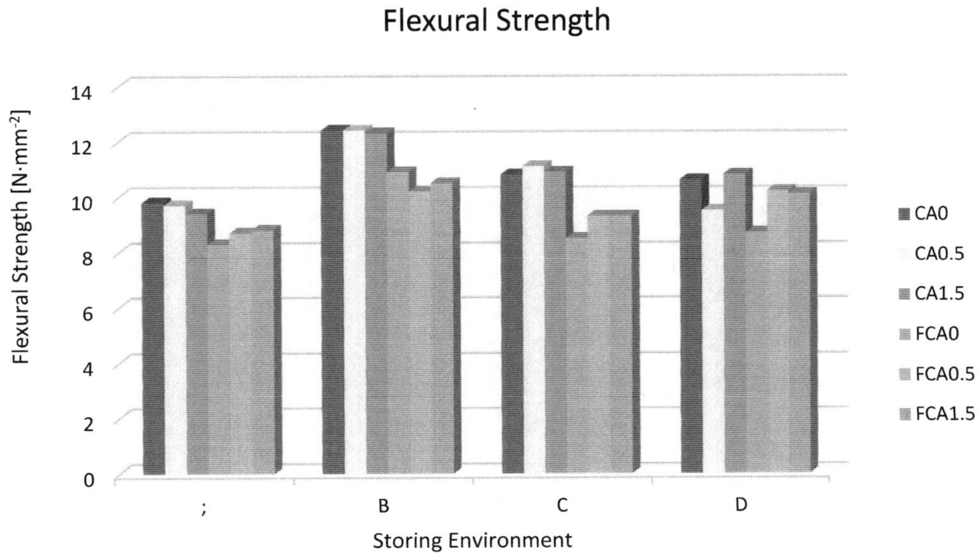

Figure 2. Determination of 118-day flexural strength of the mortars.

Figure 1 and 2 show that the addition of crystalline additive has a positive effect on the compressive strength values of mortars both with and without polypropylene fibres.

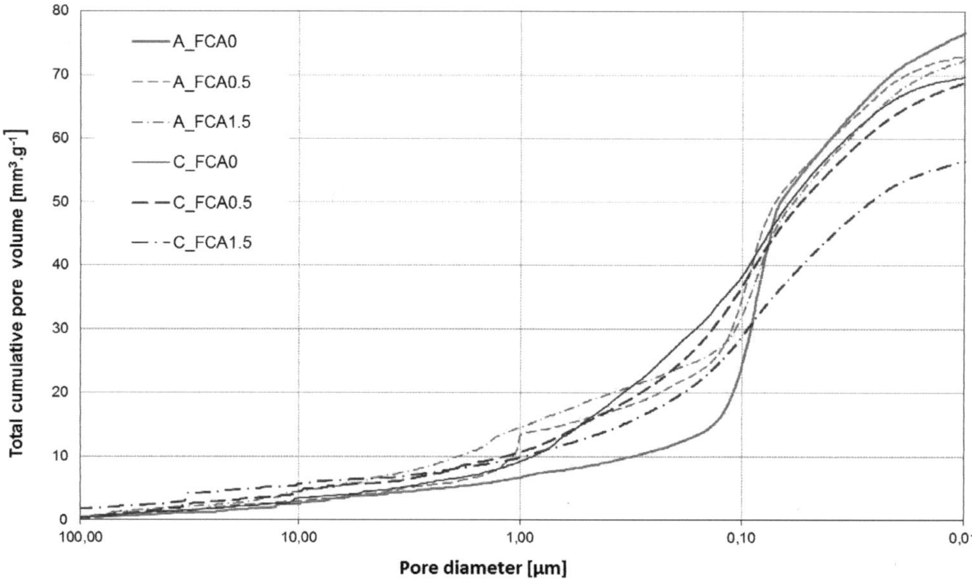

Figure 3. Pore size distribution of mortars stored at a temperature of 23 °C + relative humidity of 95 % (environment A) and in ammonium chloride solution (environment C).

Figure 3 shows the influence of the crystalline additive on the porosity of mortars containing polypropylene fibres stored in environment A and C. The data shows that the greatest reduction in porosity as a result of CA addition was in specimen C_FCA1.5, which contained 1.5 % of CA and was exposed to an ammonium chloride solution.

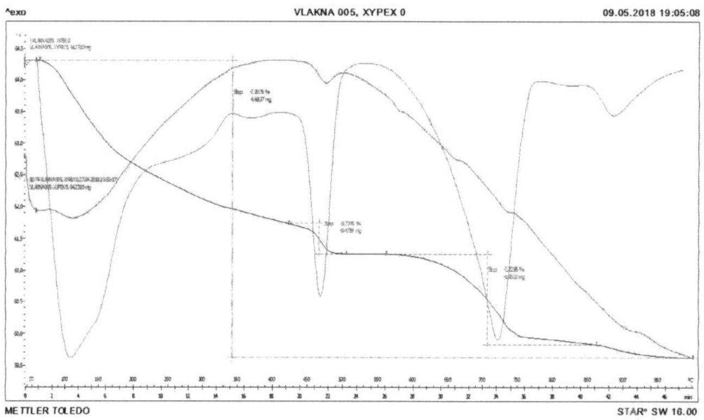

Figure 4. DTA record of the mixture FCA0 stored in ammonium chloride solution.

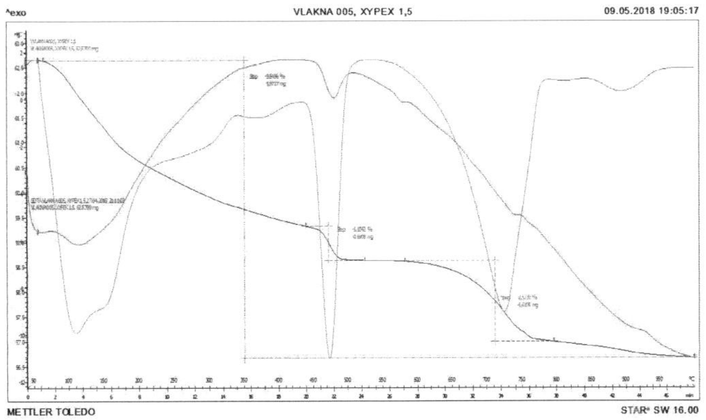

Figure 5. DTA record of the mixture FCA1.5 stored in ammonium chloride solution.

Figures 4–5 show DTA (heating rate 20 °C per minute; air atmosphere) records of mortars containing polypropylene fibres without CA (mixture FCA0) and with CA (mixture FCA1.5). Both mortars stored in ammonium chloride solution underwent three basic reactions: decomposition of C-S-H phase, $Ca(OH)_2$ and $CaCO_3$. The results indicate that the crystalline additive has an influence on the increase of the amount of portlandite. Mass content of $Ca(OH)_2$ in the mixture FCA0 was 3.03 % and in the mixture FCA1.5 was 4.54 %, see table 4.

Table 4. Results of DTA – mass loss and mass content.

Mixture ID	Mass loss corresponding to decomposition of $Ca(OH)_2$		Mass content of $Ca(OH)_2$	
	[%]	mg	[%]	mg
FCA0	0.74	0.47	3.03	1.95
FCA1.5	1.10	0.69	4.54	2.84

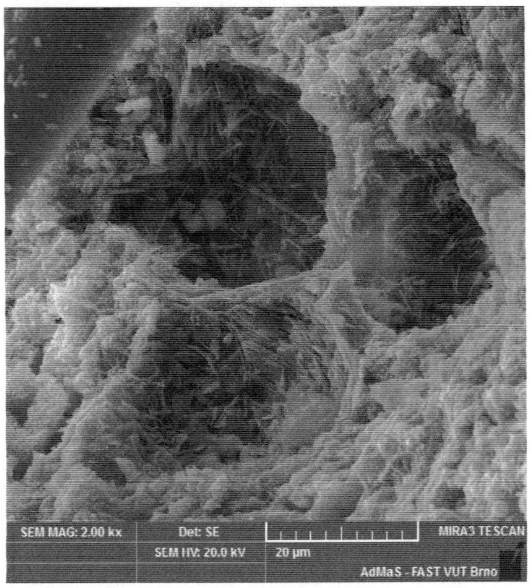

Figure 6. Picture of a specimen with CA (FCA1.5) stored in ammonium chloride solution.

Figure 7. Detail of newly formed CA (FCA1.5) in a specimen stored in ammonium chloride solution.

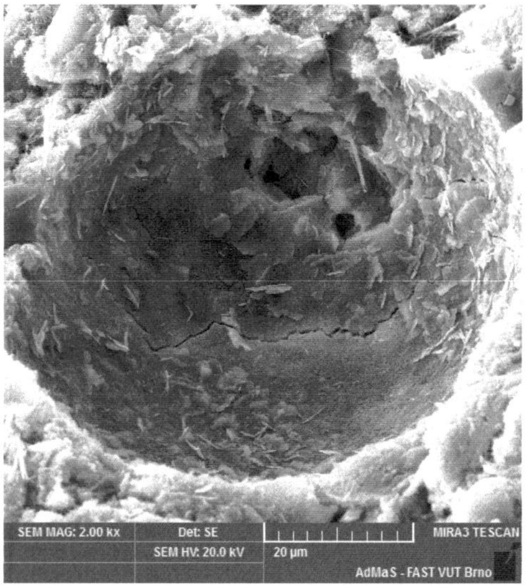

Figure 8. Picture of a specimen without CA (CA0) stored in ammonium chloride solution.

Figure 9. Detail of ettringite needles found in a specimen (CA0) stored in ammonium chloride solution.

Figure 6–9 show differences in the microstructure of specimens stored in an ammonium chloride solution. Mortars containing CA have clearly visible needle-like products, see figure 7.

4. Conclusion

The tests confirmed the positive influence of a crystalline additive on the values of compressive strength after exposure to all aggressive environments. Mortars which contained polypropylene fibres also saw an increase in flexural strength. The addition of polypropylene fibres can negatively influence the porosity of hardened mortars, while a higher CA content, i.e. 1.5 %, can bring about a reduction in porosity in mortars exposed to aggressive environments. The research results confirm that CA has an influence on the increase of the amount of portlandite. SEM sampling revealed differences in the microstructure of mortars both with and without crystalline additives after exposure to aggressive environments.

Acknowledgment

The paper was produced with the financial support of Czech Science Foundation, project No. 16-25472S "Secondary crystallization modified cement composites and their degradation dynamics".

References

[1] Bohus S and Drochytka R 2012 *Appl. Mech. Mater.* **166–169** 1773–1778
[2] Weng T L and Cheng A 2014 *Monatshefte fur Chemie* **145** 195–200
[3] Hewlett P C 2010 *Lea's Chemistry of Cement and Concrete* (Oxford: Elsevier Ltd)
[4] Page C L and Page M M 2007 *Durability of Concrete and Cement Composites* (Cambridge: Woodhead Publishing)
[5] Reiterman P and Pazderka J 2016 *Advances in Civil Engineering* **2016** 1-8
[6] Raki L et al 2010 *Materials* **3** 918–942
[7] Roig-Flores M Pirritano F Serna P and Ferrara L 2016 *Constr. Build. Mater.* **114** 447–457

Rheological characteristics of aerial lime-based pastes with addition of biopolymers

T Zizlavsky, M Vysvaril and P Rovnanikova

Faculty of Civil Engineering, Brno University of Technology, Veveri 331/95, Brno 602 00, Czech Republic

Abstract. Viscosity enhancing admixtures are widely used to improve the properties of concrete and ready-mix mortars. This study focuses on the influence of biopolymer (sodium salt of alginic acid, carrageenan, diutan gum, gellan gum, and xanthan gum) in doses of 1, 5, and 10 ‰ on the rheological properties of aerial lime-based pastes. The measurements were carried out on the Discover HR-1 hybrid rheometer with DIN concentric cylinders geometry. On the pastes the flow properties as yield stress, fluidity index and consistency coefficient as well as the viscoelastic properties (complex modulus, loss tangent, critical and flow strain) were studied. All pastes studied expressed shear-thinning behaviour and the flow curves indicated rheopectic character of samples. The significant differences in effectivity of admixtures used were observed. Xanthan and gellan gums were found less effective than the three others in influencing the rheology of aerial lime-based paste while diutan gum was so effective, that the future research with lower doses of the admixture are recommended to better explain its impact on the rheological properties.

1. Introduction

The viscosity enhancing admixtures (VEAs) are widely used in building industry to reduce bleeding and segregation in concrete [1] and to improve properties of ready-mix mortars. The most frequently used ones are the cellulose derivatives-based admixtures. Recently other biopolymer-based admixtures such as guar gum derivatives, diutan gum, and xanthan gum have occurred [2-4]. Most of these admixtures originate from the food industry, where they are used as stabilisers, thickeners and emulgators. The studies on their use, especially in concrete are lately being published [1-6].

The paper deals with the impact of addition of five less-known biopolymers on the rheological properties of aerial lime-based mortars. The sodium salt of alginic acid (ALGNA), a seaweed biopolymer that in cementitious materials surpassed commercial superabsorbent polymers (SAP) and met nominative strength requirements even though higher water demand [5, 6]. Carrageenan (CG) is the second seaweed biopolymer used in these experiments. CG used in fly ash geopolymer considerably increased strengths, and allowed the creation of a more condensed structure [7]. In cement-based materials the use of carrageenan was only reported as the foam-stabiliser for preparation of a high porosity cementitious foam, where the CG addition did not affect the hydratation of cement binder [8]. CG is reported to create more densely formed gels in $CaCl_2$ solution in comparison with gels prepared solely in water, thus we can assume that $Ca(OH)_2$ solution in the case of lime pastes could have similar effect [9]. The other three admixtures used are of microbial origin. Xanthan gum (XG) is used in SCC as VEA to reduce bleeding and segregation with slump-reduction side effect, thus the XG behaviour should be thinning-thickening [2-4, 10, 11]. The use of gellan gum (GeG) in building industry is mainly studied as a soil-stabilising agent. The application of GeG with cement or other building binders is reported only as a base for a hydrogel slurry used to create cementitious porous materials [12]. Diuthan gum (DG) has the same backbone as GeG but different substituents on the side chain, thus different properties. DG is used in SCC in similar applications as XG. DG is independent on temperature and concentration of ions, making it more suitable for use in building materials due to the higher concentration of Ca^{2+} cations. DG stabilises SCC by increasing the yield stress of the mixture [1, 2-4, 13].

This work studies the behaviour of the above-mentioned biopolymers in various doses in the lime-paste environment.

2. Materials and methods

2.1. Materials and mixing procedure
The dry mixture consisted of commercial dry hydrated lime (Carmeuse Czech Republic) of the CL 90 S class according to EN 459-1, and the admixture of dosage 1‰, 5‰, and 10‰ of binder weight. The admixtures used were: sodium salt of alginic acid (Sigma-Aldrich, Co.), diutan gum (Kelco-crete DG-F, BASF Construction Solutions GmbH), xanthan gum (Kelzan AP-AS), carrageenan (Genuvisco CG-131), and gellan gum (Kelcogel CG-LA) three last named biopolymers supplied by Biesterfeld Silcom s.r.o. but all with the exception of alginate being products of CP Kelco. The lime and admixture were mixed together with the specified amount of water (water/binder radio of 1:1) for one minute. Then the paste was introduced into the measuring cup.

2.2. Methods of rheological measurements
Rheological measurements were carried out on the hybrid rheometer Discovery HR-1 from TA instruments. The DIN concentric cylinders geometry has been chosen with the gap of 5.917 mm for all the measurements. The testing was executed at 20 °C. The results should not be reproduced quantitatively.

2.2.1. Flow properties. After 1 minute of mixing the paste was introduced into the measurement system. Five minutes after the start of mixing pre-shear was performed for 60 s at 100 s^{-1} to re-homogenize the sample. After pre-shear the sample rested for another 60 s, after that the following measuring procedure was carried out. Shear rate increase from 0.1 to 100 s^{-1} was applied through 30 steps consisting of 5 s of equilibration and 10 s of averaging. After reaching the highest shear rate similar procedure was performed with decreasing shear rate within the same values. The results have been expressed as flow curves (shear rate vs. shear stress) and the Herschel-Bulkley model was applied to downward curves to describe rheological behaviour:

$$\tau = \tau_0 + k\dot{\gamma}^n \tag{1}$$

where τ - shear stress, $\dot{\gamma}$ - shear rate, τ_0 - yield stress, k consistency coefficient and n - fluidity index which characterizes shear-thinning ($n < 1$) or shear-thickening ($n > 1$) behaviour of a material.

2.2.1. Viscoelastic properties. To determine viscoelastic properties of lime pastes, the small amplitude (0.005%) oscillation tests at 1 Hz frequency, measuring the storage modulus G' and G'' were carried out. Using this method the critical strain γ_C which determines the end of the linear viscoelastic region, where the decrease in G' is observed, and the flow point γ_F, where $G'=G''$ (from this point onward the sample shows the character of a liquid without a consistent chemical or physical network-of-forces [14]) were measured. The frequency sweep tests in linear conditions (strain value was set to be lower than the critical strain of any of tested samples) were conducted. The material behaviour was observed by increasing the frequency from 0.1 Hz to 10 Hz. The results were expressed as complex modulus G^* and loss tangent $tan\ (\delta)$ defined by following equations:

$$G^* = \sqrt{(G')^2 + (G'')^2} \tag{2}$$

IOP Conf. Series: Materials Science and Engineering **385** (2018) 012067 doi:10.1088/1757-899X/385/1/012067

$$\tan(\delta) = G''/G' \tag{3}$$

3. Results and discussion

3.1. Flow properties

The flow curves of lime paste with different amount of XG added were chosen for the illustrational purposes (figure 1). The shape of expressed curves indicates rheopectic (time-dependent thickening) character and shear-thinning (pseudoplastic) behaviour of aerial lime-based pastes, which was also observed on the lime-based mortar [15, 16]. Because of large amount of data, the simple comparison using bar charts (figures 2 to 4) was chosen. The DG 10 lime paste prepared with selected water-binder ratio was so dense that it was not possible to introduce it properly to the measuring instrument, thus none of the measurements was carried out on this mixture.

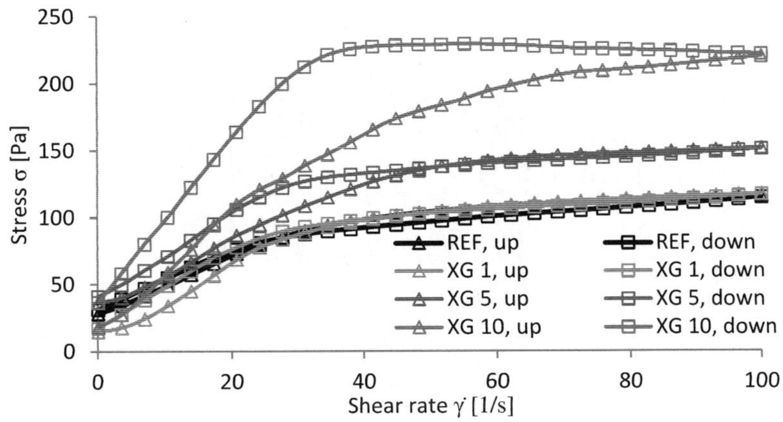

Figure 1. Flow curves of lime pastes with different XG addition.

The addition of biopolymers caused the decrease in yield stress in most cases, but with exception of GeG, the increasing dose of admixture caused the growth of yield stress. This, as reported by Cappellari et al. [17], is caused by the competition between two effects of biopolymers, the dispersing and lubricating effect, which is in this case dominant within low dose of admixture, and the associative property which prevails within the high doses of ALGNA, CG, and DG. The yield stress increase was reported also for carboxymethyl chitosan (CMCH) added to both lime and cement, and guar gum derivatives, hydroxypropyl guar (HPG) and carboxymethylhydroxypropyl guar (CMHPG) [15, 16,18]. Worse results of XG are probably caused by high concentration of Ca^{2+} ions which is reported to notably influence the XG performance [13]. The consistency coefficient (k) values are also shown in figure 2. The higher doses of ALGNA, CG, and DG reached really high values. Most of the pastes came up to higher values then the reference one. The increase in consistency coefficient is desirable, because it improves the adhesive effect of the paste. The fluidity index (n), which describes shear-thinning ($n < 1$) or shear-thickening ($n > 1$) behaviour of the pastes reached not higher than 0.57 in the case of ALGNA 10, thus we can say that all of the tested pastes expressed the shear-thinning behaviour, and the shape of all the curves is similar to the ones presented in figure 1.

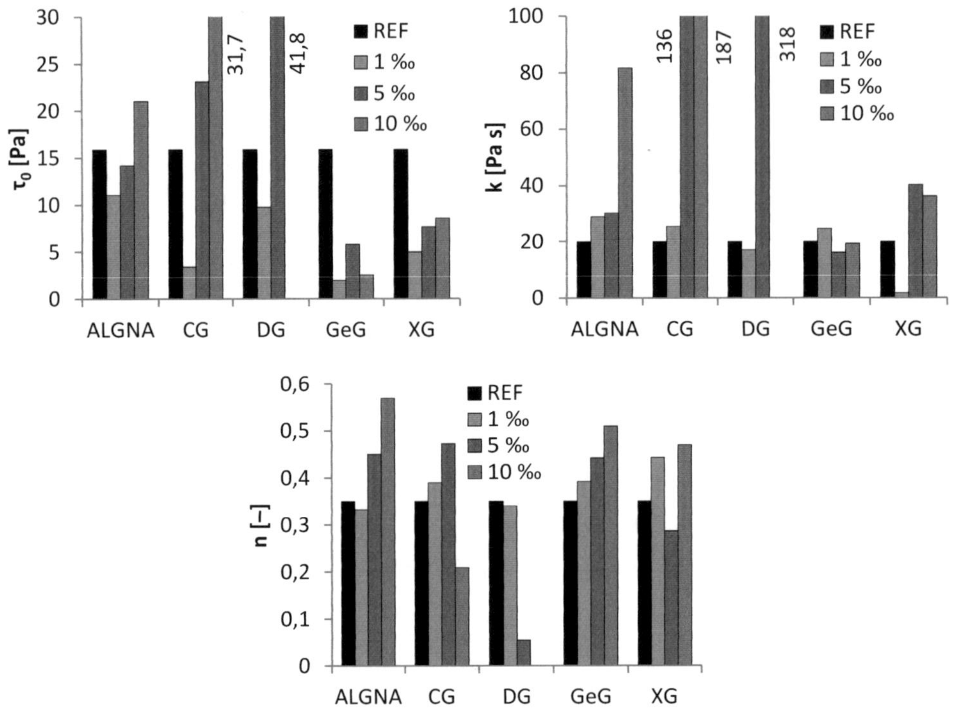

Figure 2. Comparison of yield stress (τ_0), consistency coefficient (k), and fluidity index (n) of lime pastes with biopolymer admixtures.

3.2. Viscoelastic properties.

3.2.1. Amplitude sweep test. Oscillation test was carried out to determine the end of the linear viscoelastic region (LVR), which is represented as critical strain (γ_c). From this point onwards the sample until now behaving as a solid starts to act as a gel. The tests carried out in the linear conditions are non-destructive to the microstructure of the material. The critical strain of the tested pastes does not overcome the value of reference paste (CG 5 and 10, and DG 5 reached similar value), so the biopolymer addition does not increase the stability of the lime paste (the LVR is the same or shorter as reference). Higher dosage of ALGNA or XG furtherly shortened the LVR whereas CG and DG with increasing dose improved stability in comparison with lower doses. All the samples tested had the storage modulus values (G') larger than the loss modulus (G''), which means that the material consists of a strongly multi-connected network of elements; this is quite typical for lime or cement based pastes but also for aqueous solutions of some of the biopolymers [14-16, 18, 19]. The flow strain γ_F tells us how hard it is to make the sample become liquid. With increasing dosage of CG and DG, the value of flow strain increases significantly; in the case of DG 5 the curves of the dynamic moduli did not intersect in the studied range of oscillation strain, thus being probably much higher than the value of CG 10. Uncertain dosage dependence was observed in the case of ALGNA and XG, where in the first case the 5‰ addition caused the lowest flow point of the doses studied, and in the case of CG the value was the highest. The addition of CG decreases the flow point value.

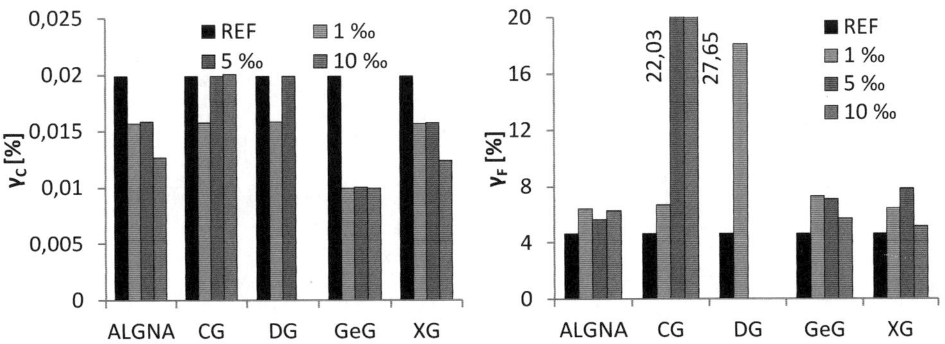

Figure 3. Comparison of critical strain (γ_C) and flow strain (γ_F) values of tested samples.

3.2.2. Frequency sweep test. Frequency sweep test at 0.005%. strain was performed to determine the viscoelastic properties of pastes. In figure 4, the charts show the values of complex modulus (G^*) and loss factor (*tan* (δ)) of the tested samples measured at 1 Hz frequency. The uprising trend of G^* in all tested pastes reports the increasing resistance to deformation of the tested materials. The growing G^* correponds to the decreasing loss tangent, where the diminishing loss tangent reports more elastic behaviour of the material, thus greater resistance to external interference. The dropping loss tangent value with the growing admixture dosage was also reported for CM derivatives of guar gum and chitosan added to lime mortar as well as starch ether added to the cementitious mortar [15, 16, 18-20]. On the other hand cellulose ethers, the most common VEAs, cause the increase in loss tangent value thus loss of elasticity of materials [17].

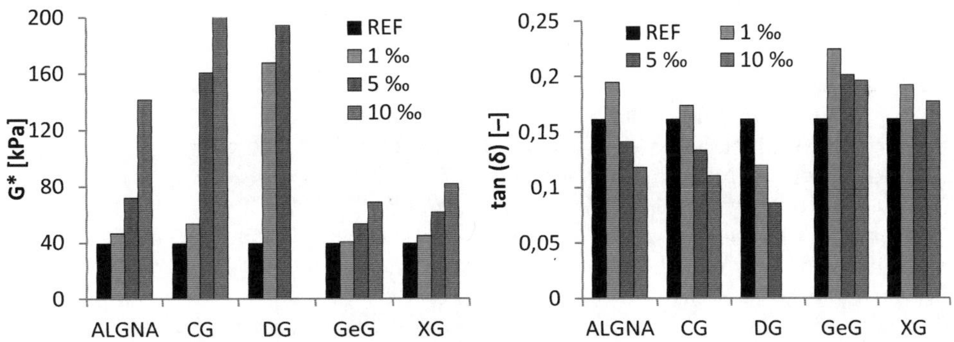

Figure 4. Complex modulus (G^*) and loss factor (*tan* (δ)) comparison of tested lime pastes.

4. Conclusions

The paper studies the influence of the addition of biopolymers, which are non-traditional in building industry in various doses on the rheological characteristics of aerial lime-based pastes. All the biopolymers tested fostered the rheopectic character and shear-thinning behaviour of aerial lime-based paste. The addition of xanthan gum and gellan gum was found to be far less effective than the addition of other three biopolymers. The yield stress decrease at low dosage was observed followed by improvement with increasing dosage. Only the most effective admixtures (CG and DG in higher doses) did not shorten the linear viscoelastic region, thus they did not reduce the stability of lime paste. The two aforementioned admixtures also significantly increased the value of flow strain making the specimen more difficult to become a liquid.

The increasing complex modulus with rising dose of any admixture reports the growing resistance to deformation, which is caused by developing elasticity of the materials represented by decreasing loss tangent. The xanthan and gellan gums were far less effective in the improvement of complex modulus mainly due to the increase in loss tangent.

The xanthan and gellan gums were found not very effective in influencing the properties of lime pastes in current doses. The sodium salt of alginic acid was found well useful if a small decrease in paste stability would be taken into account. Carrageenan and diutan gum were really effective in influencing lime pastes rheological characteristics but for better understanding, the future research on the addition of lower doses with finer differences between them should be conducted at least for these two admixtures.

Acknowledgements
This work has been financially supported by the BUT specific research project FAST-J-18-5274.

References

[1] Van Der Vurst F, Grünewald S, Feys D, Lesage K, Vandewalle L, Vantomme J and De Schutter G 2017 *Cem. Concr. Compos.* **82** 190-201
[2] Plank J 2004 *Appl. Microbiol. Biotechnol.* **66** 1-9
[3] Singh NK, Mishra PC, Singh VK and Narang KK 2003 *Cement Concrete Res.* **33** 1319-29
[4] Zhang Y, Zhao Q, Liu C, Zhou M 2016 *Constr. Build. Mater.* **102** 648-53
[5] Mignon A, Snoeck D, D'Halluin K, Balcaen L, Vanhaecke F, Dubruel P, Van Vlierberghe S and De Belie N 2016 *Costr. Build. Mater* **110** 169-74
[6] Bezerra U T 2016 *Biopolymers and Biotech Admixtures for Eco-Efficient Construction Materials* (Duxford: Woodhead Publishing)
[7] Li Z and Zhang L 2016 *Biopolymers and Biotech Admixtures for Eco-Efficient Construction Materials* (Duxford: Woodhead Publishing)
[8] Akthar F K and Evans J R G 2010 *Cem. Concr. Res.* **40** 352-8
[9] Tari Ö and Pekcan Ö 2008 *e-Polymers.* **8** 001
[10] Isik I E and Ozkul M H 2014 *Constr. Build. Mater.* **72** 239-47
[11] Vazquez A and Pique T M 2016 *Biopolymers and Biotech Admixtures for Eco-Efficient Construction Materials* (Duxford: Woodhead Publishing)
[12] Rutkevičius M, Munusami S K, Watson Z, Field A D, Salt M, Stoyanov S D, Petkov J, Mehl G H and Paunov V N 2012 *Mater. Res. Bull.* **47** 980-6
[13] Xu L, Gong H, Dong M and Li Y 2015 *Carbohydr. Polym.* **132** 620-9
[14] Mezger T G 2014 *The Rheology Handbook 4th ed.* (Hanover: Vincentz Network)
[15] Vyšvařil M and Žižlavský T 2017 *IOP Conf. Ser.: Mater. Sci. Eng.* **251** 012039
[16] Boháč M and Nečas R 2016 *Procedia Eng.* **151** 34-41
[17] Cappellari M, Daubresse A, Chaouche M 2013 *Constr. Build. Mater.* **38** 950-61
[18] Vyšvařil M, Hegrová M and Žižlavský T 2018 *Key Eng. Mater.* **760** 257-65
[19] Martín-Alfonso J E, Cuadri A A, Berta M, Stading M 2018 *Carbohydr. Polym.* **181** 63-70
[20] Izaguirre A, Lanas J, Álvarez J I 2010 *Carbohydr. Polym.* **80** 222-8

Repairing composite using hazardous waste containing heavy metals

T Zlebek, J Hodul and R Drochytka

AdMaS Centre, Faculty of Civil Engineering, Brno University of technology, Veveri 331/95, Brno 602 00, Czech Republic

Abstract. This research deals with the possibility of using hazardous waste in epoxy resin-based polymeric substances. More specifically, it is the use of neutralising galvanic sludge (NGS), which produces as a by-product in the surface treatment of metallic elements. The objective of the research was to incorporate as much as possible of this hazardous waste into a suitable polymeric substance because the further use of this waste is very problematic and is costly to treat and dump. Experiments carried out have shown that homogenisation of this hazardous waste with a suitable epoxy resin results in the coating of all filler particles in the form of waste by the binder and there is no release of heavy metals and other undesirable contaminants. From an ecological point of view, it is, therefore, possible to use the pre-treated neutralising sludge as suitable filler in the polymer-based repairing composite. On the basis of the assessment of the mechanical properties test evaluation, it can be stated that the most suitable seems the application of 40% quantity by weight of this hazardous waste type to be used in the repairing mortar. The perfect distribution of hazardous waste particles in the epoxy matrix was confirmed by the evaluation of scanning electron microscope (SEM) images. The distribution of the individual elements found in the filler was monitored by energy dispersive X-ray analysis (EDX).

1. Introduction

Repairing composite materials are materials used to repair various elements and structures from various materials (e.g. concrete, polymer, stone, etc.). Repairing materials may be silicate or polymer-based.

For the production of the highest quality composite materials, polymer-based binders are most commonly used and are characterised by very good adhesion to most materials [1]. For repairing composite materials, polymeric binders based on epoxy resins seem to be the most appropriate, as they exhibit excellent adhesion to most materials as well as excellent physical and mechanical properties, minimum shrinkage and good chemical resistance after curing [2]. For this reason, a polymeric substance based on epoxy resin was selected for this research. Hodul et al. [3] dealt with using epoxy resin as a binder for repairing material, where they watched the possibilities of using denitrified filter fly ash in the repairing materials.

In connection with the continuous production of hazardous wastes in the industry and the effort to avoid landfilling of these materials as much as possible, it is necessary to find appropriate use of such hazardous waste. Within this research, the use of hazardous waste as a filler in the polymeric repairing composite was experimentally checked. For this research, a neutralising galvanic sludge (NGS) was chosen as a by-product in the surface treatment of metallic elements. The purpose of the surface treatment of metals is to make metal coatings on mostly metallic-based materials, thereby creating a protective anticorrosive layer which protects metallic products from environmental influences during their use and thus prolongs their service life. Galvanic plating produces a variety of waste products, which can be sludges or filter cakes, from a neutralising station containing dangerous substances. In order to use the neutralising sludge as a filler in polymeric substances, it is necessary to adjust it accordingly. The most common type of hazardous waste treatment is solidification (stabilisation) [4]. It involves the mixing of hazardous waste with reagents that bind dangerous substances to a solid matrix to avoid the release of contaminated components into the environment. The most commonly used reagents are lime, cement or power fly ash. This type of adaptation was dealt with by Hodul et al. [5] in their experiment, which involved the solidification of neutralising sludges. The solidification of other hazardous wastes was investigated by Lopéz et al. [6] who tried to stabilise hazardous waste containing large amounts of mercury. The effort is to maximise the use of all waste in order to limit their very costly treatment and landfill, which contributes to the protection of the environment.

2. Identification of the materials

2.1. Neutralising galvanic sludge (NGS)

A neutralizing galvanic sludge (NGS) is produced as a by-product in the surface treatment of metallic elements, namely galvanic plating. For this reason, it contains a considerable amount of pollutants, predominantly heavy metals, which need to be successfully incorporated into a suitable matrix to avoid leaching them. According to the waste catalogue, this hazardous waste can be classified in group 19 02 05 Sludges from physical and chemical treatment containing dangerous substances. This neutralisation sludge exhibits several hazardous properties such as H5 (Harmfulness to health), H14 (Ecotoxicity), H15 (Ability to release dangerous substances to the environment during or after removal).

Since this sludge is produced with a high content of physically bound water (pasty consistency), it had to be dried (105 °C) before using as a filler into the polymer mass. Subsequently, the dried sludge (figure 2), milled to a desired fraction (figure 3). Figure 1 shows the distribution curve of the particle size distribution of the dried and grounded sludge. The density of this sludge is 2960 kg/m^3 and the specific surface is 880 m2/kg. The loss of drying at 105 °C was 34.7% and the pH of the sludge was 11.6. The high pH of the sludge is due to the neutralisation of the acid solutions used in the galvanic plating process when calcium hydroxide ($Ca(OH)_2$) is added to the solution. Table 1 presents the concentration of pollutants in the dry matter and table 2 presents the results of the leachability test carried out according to EN 12457-4 (Characterisation of waste. Leaching. Compliance test for the leasing of granular waste materials and sludges. One stage batch test at a liquid to solid ratio of 10 l/kg for materials with partied size below 10 mm (without or with size reduction). Determination of the pollutants concentration in dry matter and the leachability of waste results from Decree No. 294/2005 Coll. on the landfilling conditions of waste and its use on the terrain surface, and amending Decree No. 383/2001 Coll. on waste management details, which also includes limits for individual leachate classes. Based on the results of the leachability test, this waste can be classified in leachate class III because the limit for leachate class IIb for chlorides (2500 mg/l) and dissolved substances (6000 mg/l) has been exceeded.

Table 1. The concentration of specific heavy metals in the NGS sludge [mg/kg dry matter].

As	Cd	Cr	Ni	Pb	V
1.51	0.37	171	112	2110	8.36

Table 2. Results of the leaching test of the NGS sludge [mg/l].

DOC[1]	Chlorides	Fluorides	DS_{dried}[2]	SO_4	Ba	Cr	Cu	As	Mo	Pb
19.7	2950	0.2	6340	5.82	0.027	0.009	0.071	0.0014	0.019	< 0.001

Explanations: [1]Dissolved Organic Carbon, [2]Dissolved Substances.

IOP Conf. Series: Materials Science and Engineering **385** (2018) 012068 doi:10.1088/1757-899X/385/1/012068

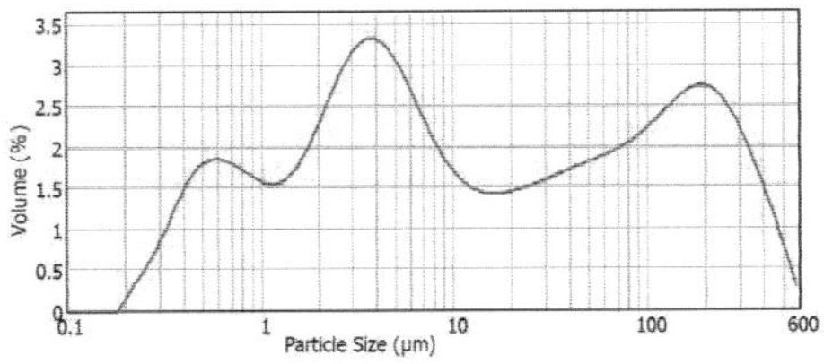

Figure 1. Particle size distribution of the pre-treated filler NGS.

Figure 2. Dried hazardous waste NGS (dryer: 105 °C).

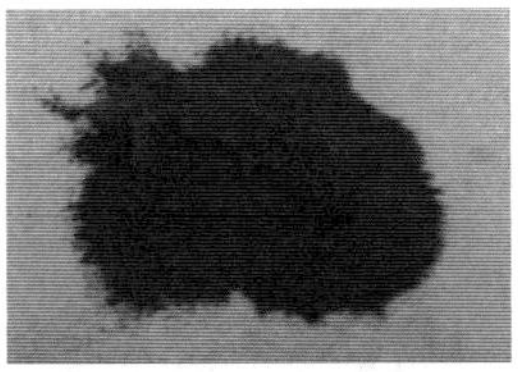

Figure 3. Prepared filler (milled NGS).

2.2. Reference filler – silica sand

It is perfectly pure silica sand with the ideal grain shape that makes it possible to create a perfectly dense structure in the polymer substance. It is commonly used as filler in various polymer composites, mortars, fillers, polymeric substances and polymer concrete. This sand contains 99.6% of SiO_2 and has a density of 2660 kg/m^3. As part of the research, this sand was used in reference samples to compare the parameters with the values of the developed repairing materials containing pre-treated hazardous waste.

2.3. Binder component – epoxy resin

As a binder the low viscosity solvent-free epoxy-based material with the polyamine (2,4,6-tris-(dimethylaminomethyl) phenol, and 4,4-methylenebis (cyclohexylamine)) based hardener was used. Epoxy resin is a very good binder to achieve high strength and minimal shrinkage [7]. A large amount of filler can be incorporated into this substance due to its very low viscosity.

3. Examinations carried out

3.1. Determination of compressive strength and three-point bending strength

These tests were carried out according to standard EN 13892-2 Methods of test for screed materials - Part 2: Determination of flexural and compressive strength. A filling of 40% and 45% of repairing substance by hazardous waste NGS was selected. Silica sand reference samples were created to compare the properties. The tests were carried out on 20 x 20 x 100 mm test joists (figure 4) made of epoxy-based polymeric substance and neutralisation sludge or reference sand with different percentages of filling [8].

3.2. Determination of hardness

The hardness of the repairing composite was measured using a Shore method, which is based on inserting the spike of a type A (use for softer materials) or type D (harder materials) durometer. In our case, a type D durometer was used (figure 5). The test was performed according to standard EN ISO 868 Plastics and ebonite - Determination of indentation hardness by means of a durometer (Shore hardness). For the test, a TQC durometer, model LD0551, was used.

Figure 4. Samples for determination of bending and compressive strength.

Figure 5. Determination of Shore D hardness.

IOP Conf. Series: Materials Science and Engineering **385** (2018) 012068 doi:10.1088/1757-899X/385/1/012068

3.3. Monitoring of microstructure using scanning electron microscope (SEM)

The distribution of the filler (neutralising sludge NGS) in the epoxy matrix was monitored using the Tescan Mira3 LM scanning electron microscope by supporting of energetic dispersion X-ray analysis (EDX), which is a function that allows the elemental composition of the material to be determined. This function has been used to determine the amount of representation of some elements (Fe, Ca) and their distribution in the substance. This microscope is equipped with a high brightness of Schottky emitter, which ensures high resolution and low noise imaging. The accelerating voltage of this SEM is from 200 V to 30 kV and provides a magnification from 3.5x to 1,000,000x.

4. Results and discussion

4.1. Compressive strength

Figure 6 below graphically illustrates the resulting compressive strength values of the verified repairing composite samples. As seen in figure 6, with the increasing percentage of the substance filling with the NGS neutralisation sludge, the compressive strength was reduced, unlike the reference substance where, with increasing percentage of filling, the tensile strengths increased slightly to over 120 MPa.

The developed repairing composite with hazardous waste exhibited the best compressive strength at 40% filling, evidenced by a compressive strength of almost 80 MPa, which is a relatively high strength. The reference substance has better properties mainly due to better physical-mechanical properties of the filler, namely silica sand, and also because of a substantially smaller specific surface of the sand, which results in a better coating of the sand grains of the binder component and thus the improvement of the resulting properties. Higher incorporation of the NGS filler into the epoxy matrix results in inhomogeneity of the cured repairing composite and in attenuation of the bonding of filler and binder. Above all, these facts result in a considerable deterioration of the physical-mechanical properties of the composite. However, 40% of filling by this hazardous waste is a significant ecological and economic benefit from the perspective of the repairing composite production.

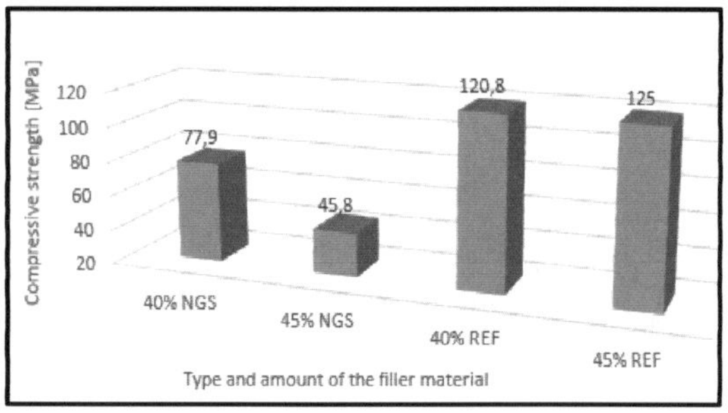

Figure 6. Compressive strength of the samples with NGS filler and reference sand.

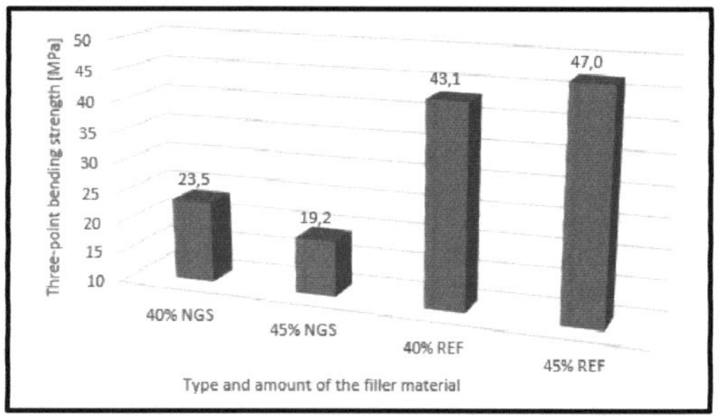

Figure 7. Three-point bending strength of the samples with NGS filler and reference sand.

4.2. Three-point bending strength

The resulting bending strength values of the test samples are graphically shown in figure 7. Here, it can be seen that the tendency of the resulting properties with the increasing percentage of filling is the same as that of the compressive strengths. Also, with the neutralising sludge NGS, the increasing filling percentage reduces workability and strength, with the reference substance increasing the three-point bending strength. The reference samples strengths ranged from 43 MPa to 48 MPa (figure 7). For the developed NGS repairing composite with hazardous waste, the flexural strength values ranged around 20 MPa. Better mechanical properties of the reference substance are also likely due to a more suitable particle size distribution and an optimal shape index of reference silica sand.

IOP Conf. Series: Materials Science and Engineering **385** (2018) 012068 doi:10.1088/1757-899X/385/1/012068

4.3. Shore hardness

The determination of the hardness of the developed composite with hazardous waste was done by measuring the shore-durometer at several locations on the test surface. It is generally known that the resulting hardness of epoxy resin-based materials depends largely on the curing time and the type of hardener. Composites with higher curing time in combination with polyamine-based hardeners typically have a high Shore D hardness. From the measured values, the average hardness of the repairing composite with 40% content of hazardous waste NGS was determined to be 90/100. It is, therefore, a material with a high surface hardness and is thus expected to be highly resistant to abrasion, impact and mechanical damage. For a reference sample containing 40% content of silica sand, a hardness value of 87/100 was recorded. Based on the evaluation of the hardness results, it can be said that the developed repairing composite will also be possible to use in places and operations where there is considerable mechanical stress in the form of inserting devices and sharp-edged objects.

4.4. Monitoring of microstructure using scanning electron microscope (SEM)

The scanning electron microscope (SEM) was used to determine the distribution of filler particles, namely hazardous waste NGS in the polymer matrix. With a magnification of 500x, it can be seen in figure 8 that some particles are substantially larger than the others, but no clumps are formed, and the particles are distributed evenly throughout the polymer matrix volume. Additionally, energy dispersive X-ray analysis (EDX), which is a supplement to SEM, has monitored the presence and amount of some elements contained in the sample with 40% NGS content. Figure 9 shows the distribution of iron (Fe) in individual NGS particles and figure 10 shows the distribution of calcium (Ca) in the neutralising sludge particles. The high calcium (Ca) presence in the NGS sludge is caused by the neutralisation of waste galvanic solutions using $Ca(OH)_2$. The presence of heavy metals could not be proven by EDX analysis because their amount was negligible compared to other elements, and a more sensitive EDX analysis would be needed to determine their distribution. However, it has been proved that the distribution of the individual elements in the epoxy matrix is uniform and the hazardous waste particles have been successfully incorporated into the closed solution structure while avoiding their leaching.

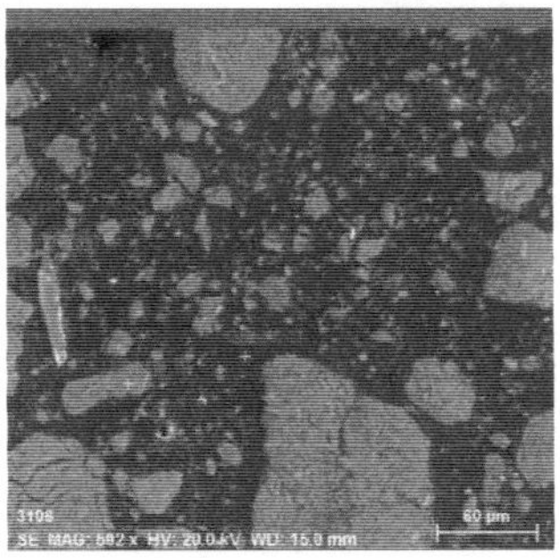

Figure 8. Distribution of the NGS particles in the epoxy matrix (40% NGS).

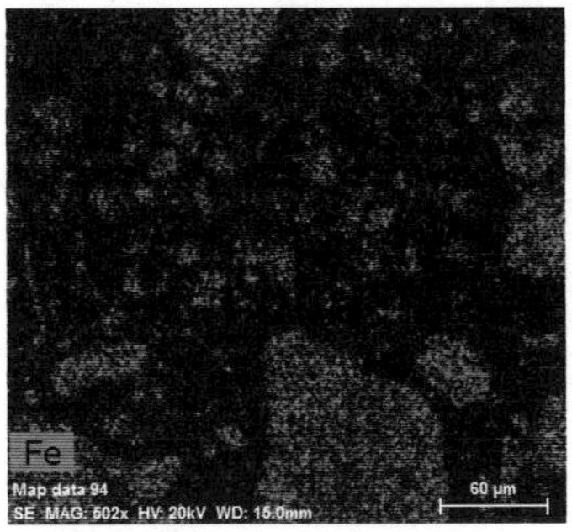

Figure 9. Distribution of iron (Fe) element in the composite with 40% NGS filler.

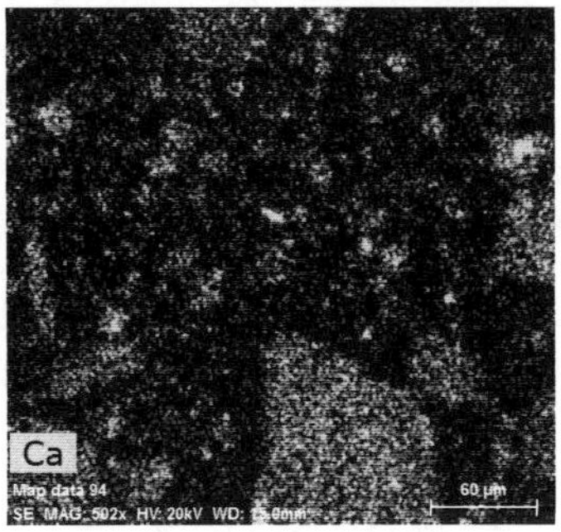

Figure 10. Distribution of calcium (Ca) element in the composite with 40% NGS filler.

5. Conclusion

Research on the possibilities and suitability of the use of hazardous waste, namely neutralisation galvanic sludge (NGS), has found that properly pre-treated hazardous waste can be used very efficiently and advantageously as a filler in the polymer repairing composite. The repairing composite with a relatively high content of pre-treated hazardous waste NGS (40%) exhibited compressive strength of about 80 MPa and three-point bending strength of about 25 MPa. These values are high enough for the use of this composite as a repairing substance for the rehabilitation of concrete structures or for the manufacture of prefabricated fittings. The repairing composite material has also been subjected to a Shore D surface hardness test, which has been found to be a material of high surface hardness and can therefore also be used in operations where a high resistance to surface deformation is required. From an ecological point of view, this repairing composite has a very positive environmental benefit because it uses a significant amount of hazardous waste that is perfectly integrated into the substance and there is no longer any release of hazardous substances that could otherwise get into the surrounding environment in the event of imperfect landfilling. By using this waste, there is an increase in environmental protection by limiting the very expensive and problematic landfilling of these hazardous wastes. In addition, using hazardous waste reduces the producer costs associated with its landfill and all in all the production of this repairing composite eliminates the use of fillers based on primary raw materials, so even the producers will reduce direct costs.

Acknowledgements

This work was financially supported by the junior specific research project No. FAST-J-18-5265 "Research of special grout joint material on polymeric basis using secondary raw materials and hazardous waste".

References

[1] Prolongo S G, Rosario G D and Urena A 2006 *Int. J. Adhes. Adhes.* **26** 125–132
[2] Cooke H G, MC Whorter W F and Strohscher A R 1964 *Ind. Eng. Chem.* **56** 38–41
[3] Hodul J, Drochytka R and Hodna J 2017 *Procedia Engineer.* **195** 134–141
[4] Bone B D, Bernard L H, Boardman D I, Carey P J, Hills C D, Jones H M, MacLeod C L and Tyrer M 2004 *Review of scientific literature on the use of stabilisation/solidification for the treatment of contamined soil, solid waste and sludges* (Bristol: Environment Agency)
[5] Hodul J, Dohnalková B and Drochytka R 2015 *Procedia Engineer.* **108** 639–646
[6] Lopéz F A, Alguacil F J, Rodríguez O, Sierra M J and Millán R 2015 *Waste Manage.* **35** 301–306
[7] Lee H and Neville K 1967 *Handbook of Epoxy Resins* (New York: McGraw-Hill)
[8] Sing L K, Azraai S N A, Yahaya N and Noor N M 2015 *Research Journal of Applied Sciences, Engineering and Technology* **11** 1430–1434

AUTHOR INDEX

AUTHOR INDEX

Institute of Physics
Dirac House, Temple Back
Bristol BS1 6BE UK

ISSN: 1757-8981
ISBN 978-1-5108-6754-3

9 781510 867543

CURRAN ASSOCIATES INC.
proceedings.com